T0317720

VCSEL Industry

IEEE Press
445 Hoes Lane
Piscataway, NJ 08854

VCSEL Industry

Communication and Sensing

Babu Dayal Padullaparthi
Photonic Components DFM Ltd., Hong Kong

Jim A. Tatum
Dallas Quantum Devices Inc, Texas, USA

Kenichi Iga
Tokyo Institute of Technology, Japan

The ComSoc Guides to Communications Technologies
Nim K. Cheung, *Series Editor*
Richard Lau, *Associate Series Editor*

Published by John Wiley & Sons, Inc., Hoboken, New Jersey.
Published simultaneously in Canada.

Library of Congress Cataloging-in-Publication Data is applied for

Hardback: 9781119782193

Cover Design: Wiley
Cover Image: © DrPixel/Getty Images

Set in 9.5/12.5pt STIXTwoText by Straive, Pondicherry, India

10 9 8 7 6 5 4 3 2 1

Dedication

With profound gratitude to all inspiring teachers, as well as scientists for the discoveries of light quanta, coherent photons, and engineered semiconductor laser devices, which are becoming blessing life-rays to humanity . . . to the Almighty (including late parents and current family) for providing an opportunity to see the brightness of invisible light in the darkest times of human tragedy . . . and to countless frontline medical staff, doctors, and law enforcement officials for selfless COVID-19 services to save this mighty planet.

Babu Dayal Padullaparthi

To my family, friends, and many co-workers that I have had the pleasure to work with and learn from over my lifetime. You have all been an inspiration to me. I especially thank my parents for the opportunities they afforded to me, and my daughter Destiny for being my best friend. Finally, to my mom, she is the strongest and most inspirational person I have ever known.

Jim A. Tatum

To my wife Tomoko for always supporting my research life and to my family, especially Soichiro, Hiroko, Saiko, and Yoichi for all the wonderful times we have spent together.

Kenichi Iga

Contents

About the Book and Authors Biographies

This book on Vertical-Cavity Surface-Emitting Laser or VCSEL is targeted for young entrepreneurs, managers, engineers, and researchers in a wide range of industries to understand how VCSELs are used in communication and sensing. With increasing Internet traffic, and large-scale manufacturing in Consumer, Automotive, and Industrial sectors, VCSELs are in high demand and the forecasted market size exceeds $10B with more than 5B chips to be shipped over the next 10 years. The authors present an unique style of illustrations and practical engineering tenets to be easily understood by non-engineering readers.

- **Introductory chapters** convey a unique history of VCSEL by the inventor, plus fundamental operating characteristics and industrial landscape including market size, chip demand and manufacturing challenges through the use of a few example products.
- **Multi-mode VCSEL chapters** discuss applications of high-speed VCSELs in modern data centers and high-performance computing; short range 3D sensing in consumer devices including Face ID and autofocus functions for augmented reality; flash- and scan-based LiDARs for long distance ranging in the automotive industry; and illuminators in night vision, security systems, and industrial heating modules.
- **Single-mode VCSEL chapters** present multi-wavelength and MEMS tunable VCSELs for long wavelength communication, and sensing applications in optical mice, atomic clocks, laser printers, and optical coherence tomography (OCT) and the book concludes with a summary of present trend and future insight for VCSELs.
- **Several appendices** are included that cover VCSEL design, epi-structure growth, wafer fab, testing, reliability, eye-safety, visible wavelength VCSEL applications, and photodetectors to serve as supplements to the main chapters.

Babu Dayal PADULLAPARTHI received his B. Sc./M. Sc., M. Phil., Ph. D. degrees from Andhra University (Govt. College at Rajahmundry), University of Hyderabad, IIT Delhi, all in Physics in 1994/1997, 1999, 2005, respectively. Dr. Padullaparthi worked in Tokyo Institute of Technology on VCSELs during 2005-09. Dr. Padullaparthi founded Photonic Components DFM Ltd., HK in 2017 to offer professional services in Datacom, 3D Sensing and LiDAR areas. He held many senior executive roles in HK and PRC companies in promoting multiple high volume VCSEL development & manufacturing projects. He received the Outstanding Poster Award from MRS in Fall 2004, JSPS Post-Doctoral Fellowship in 2005, and IEEE LEOS i-NOW / Japan Chapter Young Scientist Award in 2008. Dr. Padullaparthi has co-authored more than 50 peer reviewed technical papers on photonic materials and devices, is the co-inventor of 24 patents, and a member of IEEE (Photonics and Communications Societies), Optica (formerly OSA) & SPIE.

Jim TATUM received a BA from Austin College, and MS and PhD in electrical engineering from the University of Texas at Dallas. Dr. Tatum joined Honeywell in 1996 as a design engineer and was instrumental in the introduction of the world's first commercial VCSEL products. He later held roles in strategic marketing and finally as director of new product development at Finisar. Under his leadership, VCSELs operating at speeds up to 28Gbps were developed along with commercialization of single mode polarization locked VCSELs and high power VCSEL arrays for 3D sensing. More than 300M VCSELs based on his team's efforts are operating in the field today. In 2018 Dr. Tatum founded Dallas Quantum Devices and continues to develop even higher speed VCSELs along with a variety of VCSEL solutions in corollary sensor markets. Dr. Tatum has been granted more than 50 patents and authored many technical publications on VCSELs and related applications.

Kenichi IGA received his B. E., M. E., and Dr. Eng. Degrees from Tokyo Institute of Technology in 1963, 1965 and 1968, respectively. He joined Tokyo Institute of Technology in 1968, becoming Associate Professor in 1974, and Professor in 1984. In 2001 he was awarded Professor Emeritus. He served as the 27th President of Tokyo Institute of Technology for 2007–2012. From 1979 to 1980, he worked at Bell Labs., Holmdel. Professor Iga initiated pioneering research programs in surface emitting laser (VCSEL) and microoptics. He received 1992 IEEE/LEOS William Streifer Award, 1998 IEEE/OSA John Tyndall Award, 2003 IEEE Daniel E. Noble Award, 2002 Rank Prize, 2013 Franklin Medal with the Bower Award and Prize, and 2021 IEEE Edison Medal. He plays a contrabass as a hobby.

Foreword

"A conventional stripe semiconductor laser consists of two end mirrors perpendicular to the active layer so that the light output is parallel to the wafer surface. If a semiconductor laser with the light output perpendicular to the wafer surface could be obtained, it would be very easy to fabricate a 2-dimensional laser array. . ." This was proposed by Kenichi Iga in 1977, and the first paper on its realization, written by Haruhisa Soda, Kenichi Iga, Chiyuki Kitahara, and Yasuharu Suematsu, came out in 1979 [1]. This paper was a key milestone for the birth of the vertical-cavity surface-emitting laser, or VCSEL, as we now call it. As a longtime worker in fiber optics communications, I realize that the geometry of "the conventional laser" was deeply ingrained in the mindset of system engineers. The design, alignment, and packaging of an end-to-end fiber optics system were by and large catered to the concept that the laser light output is parallel to the wafer surface.

Following breakthroughs of the first room-temperature CW operation of VCSELs [2] and the first experimental demonstration of a 2D VCSEL array [3], worldwide VCSEL R&D was kick-started in the early 1990s. In 2000, DARPA managers Elias Towe, Robert F. Leheny, and Andrew Yang wrote the following about VCSEL in their review paper [4]: "Its size, manufacturability, and potential ease of heterogeneous integration of electronics promise a range of applications that have yet to be explored." This was the time DARPA invested considerable human and monetary resources in the R&D of VCSELs, in particular the massive integration of VCSELs, detectors, micro-optics, and driving electronics for optical backplanes, free-space optical interconnections, and all optical switching. While these applications did not emerge immediately, the DARPA funding impetus did drive widespread commercialization efforts.

The first commercial VCSELs, offered by Honeywell in 1996, were driven by the emerging demand for global bandwidth [5] that is now fueled by the need for low-cost data center communications with the worldwide construction of giant data centers by cloud computing companies such as Google, Amazon AWS, Facebook, and Alibaba. First generation VCSELs operated at 1 Gbps and have since evolved to datalink or active optical cables operating at 25 Gb/s NRZ modulation and 50 Gb/s PAM4 modulation for a single channel. Spectral and spatial multiplexing results in bandwidth expansions from 25 Gb/s to more than 400 Gb/s optical networks using multiple channels over a distance of several hundred meters. Another application in optical communications is in the widespread deployment of radio access network (RAN) in 4G and 5G LTE wireless fronthaul networks.

Yet another impetus for VCSEL applications is the progress in optical sensing and imaging in the fields of consumer electronics (mobile/smartphones) and automotive transport (autonomous vehicles through LiDARs). The property of optical signal processing perpendicular, rather than parallel, to the wafer surface makes VCSEL an ideal light source for many applications in computer vision and sensing. In 2017, Apple Inc. began to incorporate several kinds of VCSELs into its latest products starting with a structured light source for 3D depth sensing application (face ID) and now includes a LiDAR with 5 m range. We also witnessed the introduction of massively integrated VCSELs in other commercial LiDARs that has the potential to address the question of scalability in increasing the number of laser scanning lines beyond 16 and 32 to 64, 128, or even higher. The ability to integrate a large number of parallel beams perpendicular to the surface of the wafer also makes a VCSEL structure ideal for high-power (kW level) applications such as laser heating and laser cutting, among others. Other multi-mode and single-mode applications for VCSELs include atomic clocks, computer mice, laser printers, biometrics (Optical Coherence Tomography OCT), defense (drones), gas sensing, and polarization-controlled light sources for future quantum communication. Over the past 40 years, VCSELs have moved from a research curiosity to more than one billion devices in use today in a diverse range of applications. This commercial ramp-up has attracted a large commercial base that has heavily invested in high-volume manufacturing infrastructure [6].

With VCSELs now representing a multibillion-dollar global industry and worldwide interest in VCSEL's commercial usage, it is timely to write a book covering these wide range of applications and help to stimulate other novel applications. We are very fortunate and deeply honored to have Dr. Babu Dayal Padullaparthi, Dr. Jim Tatum, and Professor Kenichi Iga, a team of VCSEL experts who kindly agreed to write the book for Wiley-IEEE Press. Professor Iga is none other than the original VCSEL pioneer and author of the 1979 paper mentioned earlier. These three outstanding researchers have a collective amount of experience in academic and industrial research as well as commercial productization of VCSEL spanning the past four decades. They will provide the readers with a historical perspective and current state-of-the-art technology and product status of the field. Through the authors' extensive contacts with leading VCSEL manufacturers worldwide, the book will also present an up-to-date picture of R&D and product trends for both researchers and practicing engineers. I am confident that the readers will find this book an enjoyable tutorial and product reference covering the long and exciting journey of VCSELs.

Nim Cheung
Series Editor, Wiley-IEEE Press
November 23, 2020

References

1 H. Soda, K. Iga, C. Kitahara, and Y. Suematsu, 'GaInAsP/InP surface emitting injection lasers', *Jpn. J. App. Phys.*, Vol. **18**, No. 12, Pages 2329–2330 (1979)

2 F. Koyama, S. Kinoshita and K. Iga, 'Room temperature CW operation of GaAs vertical cavity surface emitting laser', *IEICE Trans.*, **E71**, 1089–1090 (1988)

3 Y. H. Lee, J. L. Jewell, A. Scherer et al., 'Room-Tempreature, Continuous-Wave Vertical-Cavity Single-Quantum-Well Microlaser Diodes', Electronics Letters **25**, No. 20 Pages 1377–1378 (1989)

4 E. Towe, R. F. Leheny, and A. Yang, 'A historical perspective of the development of the vertical-cavity surface-emitting laser', *IEEE J. Sel. Top Quantum Electron.*, Vol. **6**, No. 6, Pages 1458–1464 (2000)

5 J. A. Tatum, A. Clark, J. K. Guenter, R. A. Hawthorne, and R. H. Johnson, 'Commercialization of Honeywell's VCSEL technology,' *Proc. SPIE* **3946** 2–13 (2000)

6 B. D. Padullaparthi, R. Chen, A. Tan et al. 'High Volume Manufacturing of VCSELs for Datacom & Sensing', Industry Panel Discussions, Th4, pp: 51, International Nano-Optoelectronics Workshop (i-NOW) 2018, UC Berkeley, USA.

Preface

A vertical-cavity surface-emitting laser (VCSEL) is experiencing rapid growth followed by applications mostly in communication and sensing. This book describes the industrial technologies driven by the rapid expansion of the information age and the extension into rapidly emerging areas of AI (artificial intelligence) and IoT (Internet of Things).

The VCSEL was first conceived in 1977 by Kenichi Iga, one of the authors and while the device itself is smaller than a sesame seed, its impact on the world is profound. The unique advantages of small size, surface normal emission, and scalability into 1D and 2D arrays are hallmarks of the device. The name VCSEL was chosen because the light is emitted perpendicular to the wafer, and the pronunciation as "viksel" is reminiscent of a pixel in the scalability into large 2D structures.

Iga has continued basic research on VCSELs ever since its invention and witnessed commercialization in 1996 as the Internet began its rapid expansion. The first VCSEL products were in optical transceivers used in data centers, which support the modern information age. Along with the spread of the Internet, information-related companies now own huge information storage centers called the cloud (a mechanism for storing information in memory, such as hard disks, to read and use it). High-speed optical fiber networks exchange large volumes of information to and from data centers. Each of these data centers may contain many hundreds of thousands of VCSEL-based optical interconnects. In fact, VCSELs today provide more than 50 Mbps of connectivity for every person on earth.

In addition, many computer users have a mouse, which uses an optical source and a camera to track its position movement. The number of units produced so far is more than 1.1 billion. In 2017, the VCSEL was also adopted by the iPhone X for facial recognition. More recently, a laser RADAR (LiDAR) is being implemented in smartphones and iPads to enable AI and augmented reality. A new era of optical sensing was born taking advantage of the VCSEL's small volume, low power consumption, and scalability.

It can be said that most of us are using VCSELs even without recognizing them. This small device is becoming a powerful tool that supports the flow of information in everyday life.

The current technological progress has caused many paradigm-shifts in the information communication world, and we look at the new world of optoelectronics centering on VCSELs because they will be firmly rooted as the devices that enable the IoT and AI smarter. As for the descriptions of the theory and mechanism of the laser, we discuss the VCSEL from the industrial point of view in this book.

Applications of surface-emitting lasers are expanding, and the numbers of researchers, developers, and manufacturers who handle them are also increasing. The book is intended for a broader audience in industry and introduces the basic ideas and the scope of applications that have not necessarily been written in most of academic papers and textbooks.

We would be extremely pleased if this book could serve as an opening gate for emerging new VCSEL industries.

Kenichi Iga
Tokyo, Japan
July 11, 2021

Introduction

VCSEL, or vertical-cavity surface-emitting laser is now a buzzword in the science and engineering community. The excitement is driven by the promise of 3D applications everywhere, including smart phones, cars, gaming, and other forms of augmented or virtual reality. The new era for VCSELs was ushered in with the introduction of face recognition technology in the iPhone X. Since commercial introduction in 1996, VCSELs have long been the workhorse of data communication and data centers. As global Internet connectivity has grown, so has the VCSEL industry, expanded from producing a few tens of 75 mm wafers per year to thousands of 150 mm wafers today. Speeds have grown from 1 Gbps to over 56 Gbps in commercial data center products. VCSELs also found success in optical mice, autofocus assist, atomic clocks, and many other applications. But all of these taken together still did not drive interest of large-scale industrial production that consumer electronics could demand. In fact, despite the success of VCSELs, it was still considered a cottage industry until just a few years ago. One of the key advantages of VCSELs—small size—was limiting the wafer volume production to small boutique fabs. Early applications did not take advantage of another key attribute of VCSELs, scalability into 2D arrays of emitters. The typical VCSEL die used in a 3D sensing application is more than 20x the size of one used in data communications! This scaling of chip size and market opportunity has caused a huge surge in VCSEL wafer demand. Today high-volume applications in the consumer, automotive, and industrial sectors have driven VCSELs to become a multi-billion-dollar market, and that excludes traditional datacom and data center interconnects. With all of the activity around VCSELs, several engineering textbooks and references of VCSEL technology have been written, but none has focused on the basic operating principles, and none included aspects of manufacturing challenges and market dynamics.

This book is targeted to young entrepreneurs, managers, engineers, and researchers in a wide range of industries to understand how VCSELs are used in high-volume communication and sensing applications, to identify key manufacturing challenges, and future market prospects. In contrast to traditional academic textbooks, the technical content is focused on engineering design and the application of VCSELs with few mathematical expressions. The authors use a unique style of illustrations and practical engineering tenets to describe VCSEL operating principles and how they are used in a variety of applications. The book is a collection of experiences and the authors' views on topics that have and will drive the continued expansion of the VCSEL market. In other words, the book gives clear insight to understanding the overall landscape of the VCSEL industry and helps readers access the risks and rewards of the many segments. Readers are introduced to the basic operating principles of VCSELs that are relevant to application design, and a specific background in semiconductor lasers is not necessarily required. The book is focused on engineering understanding and industrial production of VCSELs and their applications. We are proud to include the historical perspectives and future insight in this book from Professor Kenichi Iga, the inventor of the VCSEL, and the 50th anniversary of the birth of VCSEL on March 22, 2027!

Chapter 1 introduces the history of semiconductor lasers, in particular VCSELs, and the potential of VCSELs as major components of choice in photonic applications. **Chapter 2** describes the basic structure of a VCSEL and its fundamental operating characteristics. **Chapter 3** describes the landscape of the VCSEL industry including leading participants, market size, chip demand, manufacturing challenges, and a few commercial products as examples. **Chapter 4** deals with the high-speed multi-mode VCSELs for datacom applications and its potential for 100G and 400G networks. **Chapter 5** focuses on multi-mode VCSEL arrays for short-range 3D sensing (up to 10 m) applications including proximity sensing in automobiles, autofocus functions for augmented reality, and facial recognition. **Chapter 6** considers multi-mode VCSEL array chips as light sources for flash- and scan-based LiDARs for long-distance ranging (up to 250 m) in the automotive industry. **Chapter 7** describes multi-mode VCSEL-based illuminators in night vision and security systems and continues the scaling of VCSELs to kW levels used in industrial heating modules. **Chapter 8** focuses on single-mode VCSELs for sensing applications, and **Chapter 9** deals with single-mode communications applications that will be the starting point for single- and entangled-photon sources for many quantum applications. **Chapter 10** summarizes the present trends and provides insight into the future directions for VCSELs and their impact on commercial products. The authors also provide several topics that are closely connected to the aforementioned chapters and cover some of the other considerations of VCSEL industrial production as **appendices**. These include design, epi-structure growth, wafer fab, testing, reliability, eye-safety, short (GaN and display), and visible-wavelength VCSEL applications, as well as photodetectors.

Babu Dayal Padullaparthi
Hong Kong, July 31, 2021

Jim Tatum
Celina, Texas/USA, July 30, 2021

Acknowledgments

The authors express their sincere gratitude to all who made this book possible in one of the most turbulent times in decades. Firstly, thanks to Dr. Nim K Cheung, IEEE Series Editor for inviting us to write a book on VCSELs in November 2019. Secondly, to the IEEE Press and Wiley Publishing staff, especially to Ms. Mary Hatcher and Ms. Teresa Netzler for their support in accepting our regular book proposal on July 2020 from multiple authors for a VCSEL industry reference that initially started as a short e-book! We thank all the reviewers of the proposal and manuscript production stages; without their timely support, the authors could not have completed this book in such a short period of time. The authors also sincerely thank all of the image contributors (companies and professional societies listed in page xxv, 311–312) in this book for their copyright permissions.

The authors would also like to thank the many co-workers, students, and colleagues that we have worked with over the years. We have learned a lot about VCSELs and especially have learned that these are one of the most complicated two-terminal devices ever manufactured. Writing this book has reinforced the breadth and depth of understanding needed in multi-disciplinary physics to design and manufacture VCSELs. They encompass aspects of mechanical, electrical, thermal, optical and solid-state physics are truly remarkable devices. Along with the technical challenges, the realization of market forces, the drive of standardization, and the necessity of meeting commercial objectives have been an amazing journey for all of us. The advances in so many interdisciplinary aspects of engineering have enabled more than 1 billion devices manufactured since its invention in 1977.

Babu Dayal is delighted and thankful for a unique and lifetime opportunity to work with the world's pioneering research group with a host of groundbreaking studies in optical communication and semiconductor lasers at Tokyo Institute of Technology (Tokyo Tech.), Japan, led by the renowned Professor Emeritus Yasuharu Suematsu and would like to express sincere thanks to the following:

- Professor Fumio Koyama from Tokyo Tech. for his gentle and generous support (then at the Microsystems Research Center in the Precision & Intelligence Laboratories) during 2005–2009 and advice during industry tenures (2011–2020). Professor Koyama kindly introduced me to the exciting world of VCSELs and its community that led to my continued professional journey in academics and industry.
- Esteemed co-authors of this book, Professor Emeritus Kenichi Iga (none other than the inventor of VCSEL in 1977 from Tokyo Tech.) and Dr. Jim A Tatum (industry veteran, formerly from Honeywell and Finisar from the mid-1990s) who truly brought academic and industry vigor to this book.
- Professor Kei Mei Lau (HKUST) for introducing me to Dr. Nim K Cheung, who brought up the idea of writing this book. Professor M. R. Shenoy (Department of Physics at IIT Delhi) for his first VCSEL assignment in the "Semiconductor Optoelectronics" OE703N course in 2000, Dr. L. Raghavendra Rao (former HoD Physics at Govt. Autonomous College at Rajahmundry, Andhra University, India), and Dr. Takemasa Tamanuki (formerly at Tokyo Tech. and Senior

Researcher at Yokohama National University, Japan) for their motivation with crystal clear concepts. Also, thanks to Prof. B. R. Mehta (IIT Delhi) and Dr. Takeshi Nakamura (formerly from Fuji Xerox and SAE [TDK]) for welcoming me to Nanomaterials R&D in 1999 and the VCSEL Industry in 2011, respectively.

- Dozens of researchers in the semiconductor lasers and VCSEL community formerly from AT&T Bell Labs, in particular the late Dr. Tingye Li, and Dr. T. P. Lee; Professor Yasuharu Suematsu and Professor Kohroh Kobayashi (Tokyo Tech., Japan), and Professor C S Sunandana (formerly at School of Physics, University of Hyderabad, India) for their inspiration and directly kindling the curiosity lamps with energetic sparks. Also special thanks to Professor Dieter Bimberg (TU Berlin, Germany), Professor Connie C Hasnain (UC Berkeley, USA), and Professor Anders Larsson (Chalmers University of Technology, Sweden) for their encouragement during industrial interactions. Thanks also to other co-workers from Tokyo Tech., SAE, and Sanan.
- Professor V. D. Vankar (IIT Delhi), late Dr. D. V. Sridhara Rao (Scientist G, DMRL, Hyderabad), Mr. G. Bhaskar Rao (Amazon, Vancouver, Canada), Mr. Kalyan G Pratap (TCS, Greater Hyderabad), and Mr. K. Pradeep (ProdyoVidhi Consulting, New Delhi) for their support that prompted to work in solid-state physics and semiconductor optoelectronics device R&D.

Jim Tatum would like to thank the following for their contributions to his professional and personal development:

- Dr. Luke Graham and Ms. Pritha Khurana for having the courage to start a new VCSEL company together. It is an honor to work with both of you on a daily basis, and you both teach me about the technical and business aspects of VCSELs every day.
- The VCSEL design group at Finisar, especially Jim Guenter, Ralph Johnson, Bob Biard, Gary Landry, Petter Westbergh, Deepa Gazula, Wenjuan Fan, Kent Wade, and Hao Chen, where we made some incredible devices over the years. There are too many others at Finisar to thank personally, but all of you helped!
- Dr. Duncan MacFarlane for taking a chance on a student and teaching me about lasers but more importantly, becoming a true friend, mentor, and role model.

Kenichi Iga would like to express thanks to the following people:

- Professor Yasuharu Suematsu, Honorary Professor at Tokyo Institute of Technology, for his continuous advice.
- Fumio Koyama, Professor of Institute of Innovative Research, Tokyo Institute of Technology, for VCSEL research together.
- Hiroyuki Uenohara, Susumu Kinoshita, Tomoyuki Miyamoto, Nobuhiko Nishiyama, Takahiro Sakaguchi, and Akihiro Matsutani, staff, graduates, and researchers of my laboratory who have been studying together.
- Kohroh Kobayashi, Professor Emeritus of Tokyo Institute of Technology, Akihiko Kasukawa of Furukawa Denko, and Yuichi Tohmori of former NTT and now Tsurugi Photonics Foundation for their encouragement of VCSEL research.
- For all members of the Microoptics Group, Japan Society of Applied Physics, who have been promoting the field of microoptics for 40 years. In particular, Dr. Genichi Hatakoshi, who is the co-author of *Treasure Microbox of Optoelectronics*, published from Adcom-Media Co. Ltd. and my playing buddy of Duo21.
- Dr. Connie Chang-Hasnain, Professor Emeritus of UC Berkeley, and Dr. Waguih Ishak of Corning R&D Organization for providing the data and having valuable discussions.

List of Image Contributions

Company	Contacts	Country
II-VI Inc.	Mark Lourie Gerald Dahlmann	USA Switzerland
Allos Semiconductors GmbH	Alex Loesing	Germany
ams AG	Jean Francois Seurin	USA/Austria
CISCO Systems	Ginger McCormick Achanta K Mohan	USA
Cosemi Technologies Inc.	Samir Desai Devang Parekh	USA
Dallas Quantum Devices	Jim A Tatum (author)	USA
LeddarTech Inc.	David Cheskis Stephane Duquet	Canada
Lumentum Operations LLC	AI Yuen Nataljia Filipovic	USA
Multilane Inc.	Kees Propstra	USA
Photonic Components DFM Ltd.	Babu D Padullaparthi (author)	Hong Kong
OSRAM Opto Semiconductors GmbH	Simon Thaler	Germany
Sense Photonics Inc.	Mark Sandoval Hod Finkelstrin	USA
Sino Semiconductor Photonics Integrated Circuit Co., Ltd.	Ling Yong Peng Yao Shun	China
SONY Corporation	Tatsushi Hamaguchi	Japan
TriLumina Inc.	Gianluca Bacchin	USA
Trumpf Photonic Components GmbH	Holger Moench	Germany
VI Systems GmbH	Nikolay Ledentsov Jr	Germany
Laser Focus World	John Lewis	USA
MSSCORPS CO., LTD	Winnie (汎銓科技)	Taiwan
Łukasiewicz Research Network-Institute of Microelectronics and Photonics, Warsaw	Paweł Piotr Michałowski	Poland
Everbright Photonics, Suzhou	Mike Wang	China
Person Publishers	Prince	USA

Professional Society/Individuals	Contacts	Country
Ethernet Alliance (EA)	David J Rodgers	USA
Institute of Electrical and Electronic Engineers (IEEE)	By default (Publisher)	USA
InfiniBand Trade Association (IBTA)	By web	USA
Fiber Channel Industry Association (FCIA)	By web	USA
Optica (Formerly Optical Society of America-OSA)	Hannah Greenwood	USA
The International Society for Optics and Photonics (SPIE)	Katie Sinclair	USA
Progress in Electromagnetics Research (PIER)	Editorial and Production Office	USA
Elsevier Copyright Clearance Center	Online	USA
University of Kiel	Helmut Foll (Retired)	Germany
International Monetary Fund	Patricia Loo	USA
Toshiba Corporation	Genichi Hatakoshi*	Japan
Tokyo Institute of Technology	Tomoyuki Miyamoto*	Japan
Georgia Institute of Technology	Stephen E Ralph*	USA
Society for Automotive Engineers (SAE)	By web	USA
Alibaba	By web	USA
FLIR	By web	USA
Novanta Inc.	By web	USA
VERTILAS GmbH	By web	Germany
Microchip Technology Inc.	By web	USA
UC Berkeley	Connie Chang Hasnain*	USA
Tokyo Institute of Technology	Nobuhiko Nishiyama*	Japan
Walter Schottky Institute and Technical University of Munich	Late Markus C Amann	Germany
Corning Inc.	Ishak Waguih*	USA
FUJIFILM Business Innivation (formerly Fuji Xerox)	Nobuaki Ueki*	Japan
RICOH Corporation	Shunichi Sato*	Japan
Santec Corporation	Chung Daiko*	USA/Japan
Tokyo Institute of Technology	Kenichi Okada*	Japan
Tokyo Institute of Technology	Kenichi Iga (author)	Japan
Novalux / Ushio Inc.	Hidekazu Hatanaka*	Japan
Tokyo Institute of Technology	Fumio Koyama*	Japan
Inneos	Brain Peters*	USA
Tokyo Institute of Technology	Yasuharu Suematsu*	Japan
National Institute of Information and Communications Technology (NICT)	Motoaki Hara*	Japan

*Courtesy of .....

1

Semiconductor Lasers and VCSEL History

Kenichi Iga

1.1 History and Basics of Semiconductor Lasers

The vertical-cavity surface-emitting laser (VCSEL, pronounced "vic-cell") is a special class of semiconductor laser that emits coherent, or laser, light orthogonally to the semiconductor substrate as shown in the schematic in Figure 1.1. The VCSEL device and its applications are the core subject of this book. This VCSEL structure generates a host of industrial applications that will be introduced comprehensively throughout this book. Before starting with a history of VCSELs, let us begin with some basic concepts of semiconductor lasers to aid in the understanding of the following chapters for general readers.

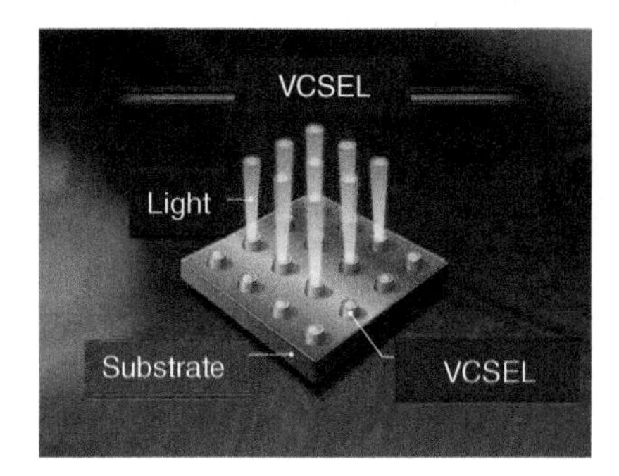

Figure 1.1 Schematic of a vertical-cavity surface-emitting laser (VCSEL). *Source:* Figure by K. Iga [copyright reserved by author].

1.1.1 Categorization of Semiconductor Lasers

Semiconductor lasers can be categorized into several types depending on:

i) Emission wavelength and materials;
ii) Resonant cavity configuration;
iii) Single mode or multi-mode;
iv) Direction of emitted light;
v) Direct modulation speed;
vi) Power output;
vii) Footprint of device;
viii) Beam form and connectivity to optics such as optical fibers;

VCSEL Industry: Communication and Sensing, First Edition. Babu Dayal Padullaparthi, Jim A. Tatum and Kenichi Iga.

ix) Price of device;
x) Manufacture volume;
xi) 1D or 2D array; and so on.

A simple category is outlined in Table 1.1. In this case, we categorized into two types on the basis of cavity configuration, i.e., edge- and surface-emitting cavities. This book is focused on the research, development, and industrial sectors to demonstrate the impact of **VCSEL**.

A schematic of an edge-emitting laser (EEL) is shown in Figure 1.2a. In edge-emitting lasers, the light emits through the edges of the wafer that is in-plane with the substrate and the optical cavity is horizontal for laser resonance. On the other hand, the structure of VCSEL is illustrated in Figure 1.2b. The resonance of light is vertical to the substrate and taken from the surface. The laser structure is formed on the substrate in just one manufacturing process.

Table 1.1 Categorization of semiconductor lasers.

Type of cavity	Edge-emitting laser	Surface-emitting laser
Single Mode		
Transverse	narrow stripe	narrow aperture
Wavelength	DFB, DBR	Fabry-Pérot
Multi-Mode		
Transverse	broad area	wide aperture
Wavelength	Fabry-Pérot	Fabry-Pérot
Array	one-dimensional	two-dimensional

Source: Table by K. Iga [copyright reserved by author].

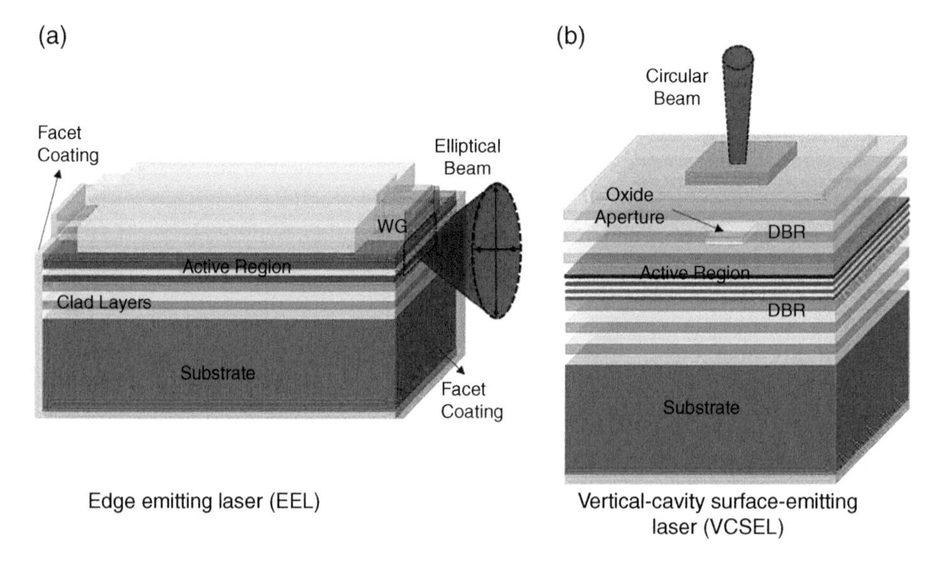

Figure 1.2 Schematic of semiconductor lasers (a) Edge-emitting laser (EEL). (b) Vertical-cavity surface-emitting laser (VCSEL). *Source:* Figure by B. D. Padullaparthi [copyright reserved by authors].

In both cases, the emission and amplification of light in semiconductor lasers is due to the recombination of electrons and holes that exist in the active region. This is what most of the standard textbooks on optical properties of semiconductors teach. Later, we will show the readers a different explanation.

1.1.2 Light Emission and Absorption in Semiconductors

The characteristics of the semiconductor diode current (*I*) as a function of the applied voltage (*V*) are used to create several different types of optical devices. The voltage–current (*V-I*) characteristic of a diode is shown in Figure 1.3. The first quadrant Q1 is known as forward bias, and the current increases exponentially with the applied voltage. When an electron and a hole recombine near the p-n junction, the energy is released as a photon. This regime is where light-emitting diodes (LEDs) and lasers operate.

The third quadrant Q3 is the reverse bias region, which is used as photodetectors. When a photon is absorbed near the p-n junction, the light energy creates an electron and hole pair. The electrons drift to the positive electrode and the holes move to the negative electrode under reverse bias.

The fourth quadrant Q4 is where photoconduction occurs and is the operating regime for solar cells. When a photon is absorbed near the p-n region, an electron hole pair is created. The resultant current times voltage (power) generates an electrical energy in the solar cell. The second quadrant is not used for practical optical devices.

The first and third quadrants play critical roles as key optoelectronics components for optical communication and sensing applications. To understand how light (photons) interacts with semiconductors, a deeper understanding of light emissions and absorption is needed.

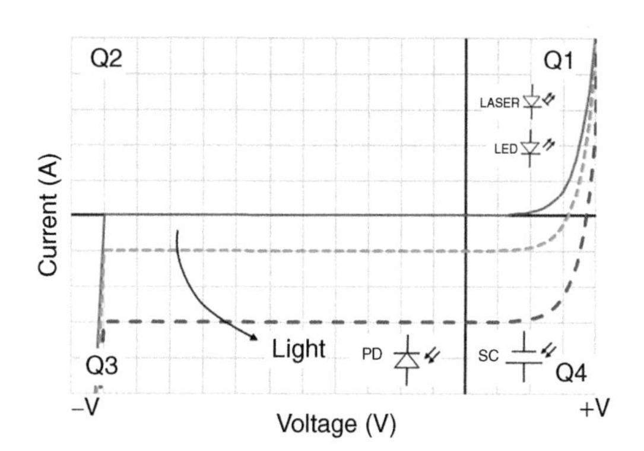

Figure 1.3 Voltage–current (V–I) characteristic of a p-n junction with no incident light (solid curve) and with incident light (dashed curves). *Source:* Figure by K. Iga and J. A. Tatum [copyright reserved by authors].

1.1.3 Birth of Semiconductor Lasers

1.1.3.1 Homostructure and Double Heterostructure Lasers

Based on the principle of light amplification in semiconductors, the first semiconductor laser was realized by four groups almost simultaneously in 1962. This was two years later after Therdore Maiman demonstrated the first laser [1]. The optical gain layer was located near the p-n homojunction parallel to the substrate [2–5]. The light resonance occurs between the mirrors formed at the edge of the substrate by cleaving or polishing the semiconductor crystal. These lasers are known as edge-emitting lasers (EELs).

In 1970, eight years later, two groups reported a double heterostructure (DH)-based laser that enabled room-temperature continuous operation [6, 7]. The device using an AlGaAs-GaAs DH

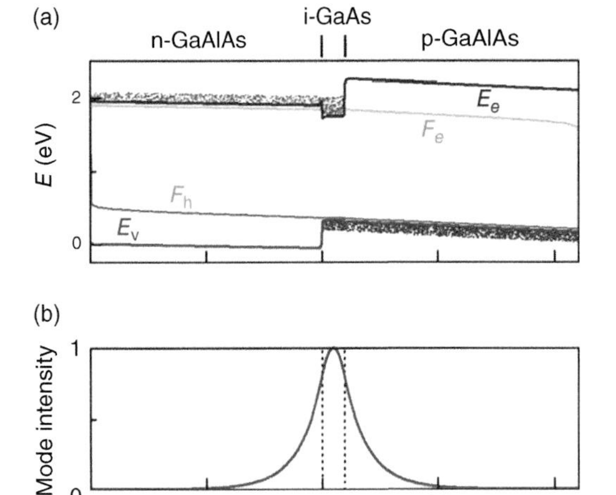

Figure 1.4 Double heterostructure laser. (a) The carrier densities of electron and holes in p, i, and n regions. (b) The optical field intensity near the double heterostructure. *Source:* [8]. [Image courtesy of Genichi Hatakoshi.]

structure, as shown in Figure 1.4, introduced by Hayashi and Panish reduced the threshold current density to about 1 kA/cm^2 at 300 K, a major breakthrough over the initial homo-junction devices.

Back in 1963, a double heterostructure laser had been proposed by Herbert Kromer by using semiconductor layers with varying bandgap and refractive index [9]. Herbert Kroemer and Zhores Alferov [7] both received the 2000 Nobel Prize in Physics for the idea.

This DH structure is used only for confining light field as shown in Figure 1.4b. More descriptions can be found in related textbooks [10–14].

1.1.3.2 Quantum Well Lasers

Present semiconductor lasers utilize quantum wells (QWs) as shown in Figure 1.5 for the light emitting and amplification layer for almost all semiconductor lasers and LEDs. The QW [15] has a thickness of several nanometers (1 nm = 10^{-9} m). Grain-shaped quantum dots [16, 17] that have a several-nanometer box size can be used to further reduce the threshold current in semiconductor lasers. The optical gain necessary for lasing using multiple quantum wells (MQWs) is further explained in Chapter 2.

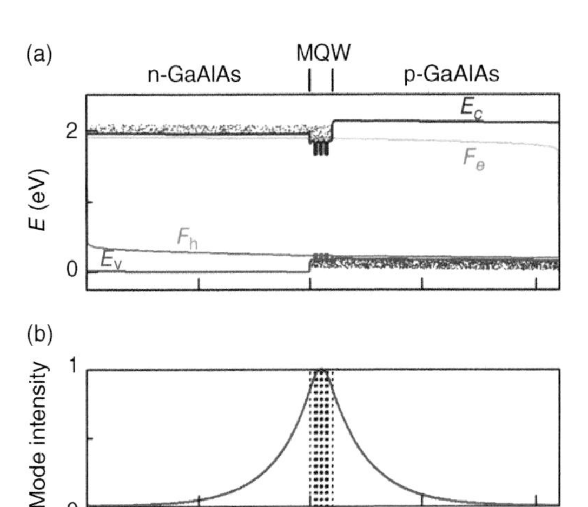

Figure 1.5 Quantum well structure with separate optical confinement. (a) The carrier densities of electron and holes in a quantum well structure. (b) The quantum well laser where the optical field is confined by the double heterostructure. *Source:* [8]. [Image Courtesy of Genichi Hatakoshi.]

1.1.4 Amplification of Light in Semiconductors

The relationship between the density of states and energy in semiconductors is shown in Figure 1.6. This is determined by the product of the parabolic density of states in the conduction band and the valence band and the Fermi-Dirac distribution according to Pauli exclusion principle. In the case of bulk semiconductors, it has the shape shown in the Figure 1.6. E_c and E_v are Fermi levels in thermal equilibrium in the conduction band and valence band, respectively, and E_g is the bandgap energy. If the electrons and holes are excessive due to photoexcitation or current injection, the distribution changes. The electron and hole levels in each band are called quasi- or degenerate-Fermi level and are indicated by E_{fc} and E_{fv}. The condition for the population inversion from the thermal equilibrium state (also called quasi equilibrium) is expressed by:

$E_{fc} - E_{fv} > E_c - E_v$: negative temperature or amplifying (condition for gain).

When the light with angular frequency ω comes in the semiconductor, the following characteristics appear, where $\hbar$ is the reduced Planck's constant: $\hbar = h/2\pi$ (h: Planck's constant). Here, $\hbar\omega$ is a quantized photon energy, which appears in Chapter 8.

$\hbar\omega < E_c - E_v$: transparent
$\hbar\omega > E_c - E_v$: absorptive

As shown in Figure 1.6, in the excited state, the carrier distribution representing the gain is distributed in a mountain shape with respect to energy. The result of the calculation is also shown in Figure 1.7.

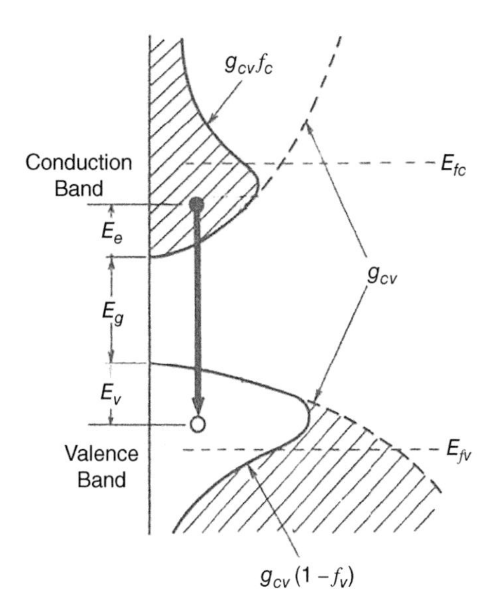

Figure 1.6 Population distribution of electrons and holes in semiconductor well vs. energy. *Source:* [18]. (After Masahiro Asada and Yasuharu Suematsu)

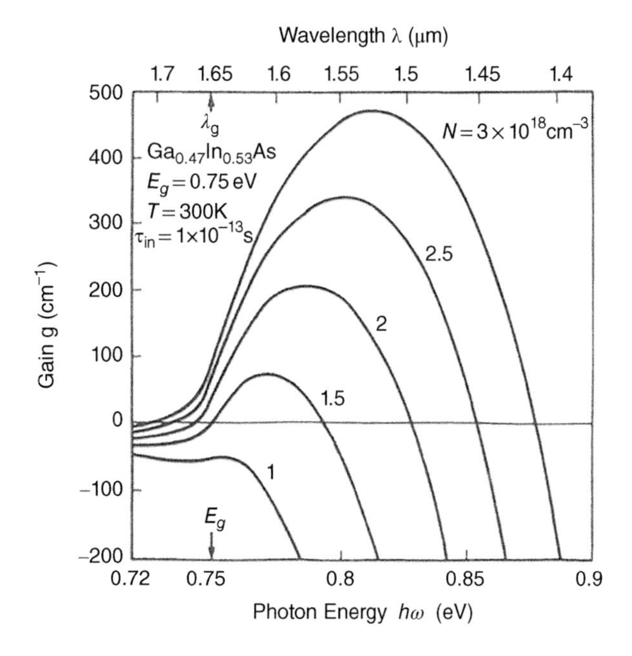

Figure 1.7 Optical gain of semiconductor vs. energy. *Source:* [19]. (After Masahiro Asada and Yasuharu Suematsu)

1.1.5 Oscillation Conditions in Semiconductor Lasers

1.1.5.1 Laser Resonators

The schematics of the edge- and surface-emitting laser (VCSEL) are shown in Figure 1.8. The edge-emitting (EE) laser, shown in Figure 1.8(a), includes Fabry-Pérot (FP) laser, distributed feedback (DFB) laser, and distributed Bragg reflector (DBR) laser. These laser types will be compared with VCSEL in Figure 1.8b in subsequent chapters.

Let us consider the oscillation conditions of a semiconductor laser. As shown in Figure 1.8(a) and (b), the target semiconductor laser is composed of a Fabry-Pérot (FP) cavity. Both resonators are generalized in Figure 1.9(a). By using the field reflectance including phase shift, which will be given later, this model can be applied to edge-emitting Fabry-Pérot lasers, DFB lasers, DBR lasers, and VCSELs as well.

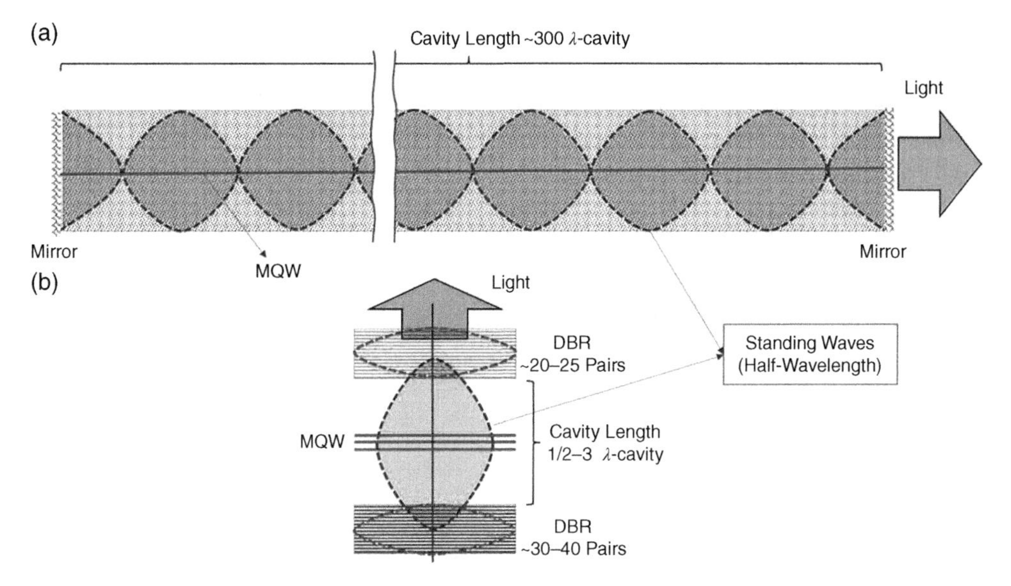

Figure 1.8 Schematic of edge-emitting Fabry-Pérot lasers and VCSELs. The coupling of light through standing waves in laser resonators (a) Fabry-Pérot (edge-emitting laser/EEL), and (b) surface emitting laser (VCSEL). *Source:* Figure by K. Iga and B. D. Padullaparthi [copyright reserved by authors].

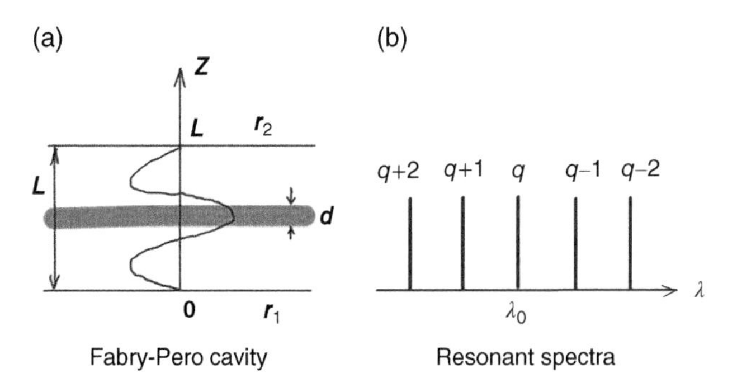

Figure 1.9 Fabry-Pérot cavity and resonant spectra. (a) Fabry-Pérot cavity. (b) Resonant spectra.

<Parameters>

L: cavity length
d: thickness of active layer
ϕ_1, ϕ_2: phase shift of each reflection
r_1, r_2: electric field reflectance coefficients of the mirrors at both ends
R_1, R_2,: power reflection coefficients of the mirrors at both ends

$$r_1 = \sqrt{R_1}\mathrm{e}^{-j\phi_1}$$

$$r_2 = \sqrt{R_2}\mathrm{e}^{-j\phi_2}$$

g: gain coefficient
α: loss coefficient
β: propagation constant ($= w/c = 2\pi f/c$)
ω: angular frequency

Consider that the light wave in the resonator travels in the *z* direction from $z = 0$ and is reflected by the reflector r_2 at $z = L$; then it goes backward by the length of *L* and returns to the starting point $z = 0$. If the electric field is sustainable, we should have:

$$\sqrt{R_1 R_2} e^{-j(2\beta L + \phi_1 + \phi_2)} e^{(gd - \alpha L)} = 1 \tag{1.1}$$

By comparing the imaginary and real parts, we have:

$$2\beta L + \phi_1 + \phi_2 = 2\pi q \left(q = \text{integer}\right)\langle \text{resonance conditions} \rangle \tag{1.2a}$$

$$\sqrt{R_1 R_2} \exp\left[\left(gd - \alpha L\right)\right] = 1 \langle \text{gain condition} \rangle. \tag{1.2b}$$

For the threshold gain g_{th} required for oscillation, use *ln*, the natural logarithm; from Eq. (1.2b),

$$g_{th} = \alpha \frac{L}{d} + \frac{1}{2d} \ln\left(\frac{1}{R_1 R_2}\right). \tag{1.3}$$

The first term is absorption by the medium, and in GaAs, absorption by the free carrier has a magnitude of about 10 cm^{-1}. In the second term, the reflectance of a reflector made by cleaving the surface of a semiconductor is

$$R_1 = R_2 = R = \left\{\left(n-1\right)/\left(n+1\right)\right\}^2 \quad \left(n: \text{Equivalent refractive index of semiconductor}\right) \tag{1.4}$$

Therefore, in the case of a GaAs edge-emitting laser ($n = 3.5$) with $L = d = 300$ μm, it is about 39 cm^{-1}. To oscillate, a threshold gain of $10 + 39 = 49$ cm^{-1} or more is required.

The electric field E_{out} of the output light, with E_0: field at the end of cavity, is given by:

$$E_{out} = \sqrt{1 - R_2}\, E_0. \tag{1.5}$$

1.1.5.2 Resonant Wavelength

Now, if λ denotes the wavelength and *n* the equivalent refractive index, from Eq. (1.2a) we have:

$$2\left(2\pi n/\lambda\right)L + \phi_1\left(\lambda\right) + \phi_2\left(\lambda\right) = 2\pi \mathrm{q}\left(\mathrm{q}: \text{integer}\right). \tag{1.6}$$

When the phase shift is 0 and the reflector is at the fixed end, for example, the standing wave is as shown in Figure 1.8b. The total length L is an integer q times the half wavelength $\lambda/(2n)$ in the medium:

$$(\lambda/2n)q = L. \tag{1.7}$$

Now, in a laser cavity with a resonator length L longer than the wavelength, waves of many wavelengths with slightly different lengths can resonate. These modes are called longitudinal modes. On the other hand, the modes in the perpendicular direction are called the transverse modes.

Considering a normal semiconductor laser, if λ is 1.3 μm, $n = 3.5$, and $L = 3$ μm, then $q = 16$.

Therefore, even if q differs by 1, the resonance wavelength changes only slightly as $\Delta\lambda$. With $|\Delta\lambda| \ll \lambda$ in mind, if $\lambda \to \lambda_0 + \Delta\lambda$, $q \to q + 1$, then we obtain,

$$\frac{\Delta\lambda}{\lambda_0} = -2\frac{\lambda_0}{n_{eff} L}. \tag{1.8}$$

This $|\Delta\lambda|$ is called free spectral range (FSR) and is inversely proportional to cavity length, L.

Here, n_{eff} is the effective index considering the dispersion of the medium and is given by the following expression:

$$n_{eff} = n\left\{1 - (\lambda_0/n)(\partial n/\partial\lambda)\Big|_{\lambda=\lambda_0}\right\}. \tag{1.9}$$

Since $\partial n/\partial\lambda < 0$ in ordinary semiconductors, n_{eff} is usually larger than n. In the above example, $n_{\text{eff}} = 4.0$ and $|\Delta\lambda| = 70$ nm.

1.1.5.3 Cavity Formation

To achieve laser oscillation, a resonator that provides optical feedback to the gain medium is required. The laser resonator is formed by a pair of mirrors; a so-called Fabry-Pérot (FP) resonator is shown in Figure 1.8(a). In an edge-emitting laser, the gain width is w, the cavity length is equal to L, and the mirror is usually made by simply cleaving the semiconductor crystal. In this case the refractive index of about 3.5 is higher than outside air, and the resonator edges look like open termination. (Here $\phi = 2\pi$. ϕ is defined in Eq. (1.1)).

In Table 1.1 we have touched on DFB and DBR structures for single-mode operation of edge-emitting lasers [20–23]. In both cases, we utilize a pair of Bragg mirrors having an electric field reflectivity expressed by $r = \sqrt{R}\exp(-j\phi)$ that sandwich some space or active region. These wavelength-selective cavities can provide single longitudinal-mode operation. For using those lasers in optical pulse code modulation (PCM) for optical fiber communications, they should maintain single mode under high-speed modulation (~100 Gb/s). Moreover, in the case of coherent digital communications, the laser should operate with narrow spectrum (~kHz). This kind of lasers is called a dynamic single-mode laser [24].

In the case of VCSELs the mirrors are formed by semiconductor Bragg reflectors or dielectric mirrors, and therefore, we can design the resonator as open or short terminations. We can use its large free spectral range (FSR) for pure single longitudinal-mode operation and wide-range wavelength tuning. The details will be described in Chapters 2 and 8.

1.2 Semiconductor Lasers and Manufacturing

1.2.1 Manufacturing Process of Edge-Emitting Lasers

A schematic of EEL manufacturing and testing processes is shown in Figure 1.10. The edge-emitting lasers (such as FP and DFB lasers) require facet coatings after cleaving and often need regrowth of specific steps.

The FP-EELs with cleaved laser mirrors have only 33% reflectivity. The reflectivity of the cleaved surface can be modified to be either higher or lower reflectivity by coating multi-layer films that can be used to optimize the performance characteristics and to protect the surfaces of the EEL. The facet coatings are applied after the lasers are cleaved from the substrate and require extensive handling, which makes them more difficult to manufacture.

Another approach is to form the distributed reflectors through multiple epitaxial steps with intermediate fabrication steps. The regrowth and the intermediate processing result in a nonmonolithic epitaxial growth making this process very complex and not manufacturing-friendly compared to a fully monolithic VCSEL fabrication.

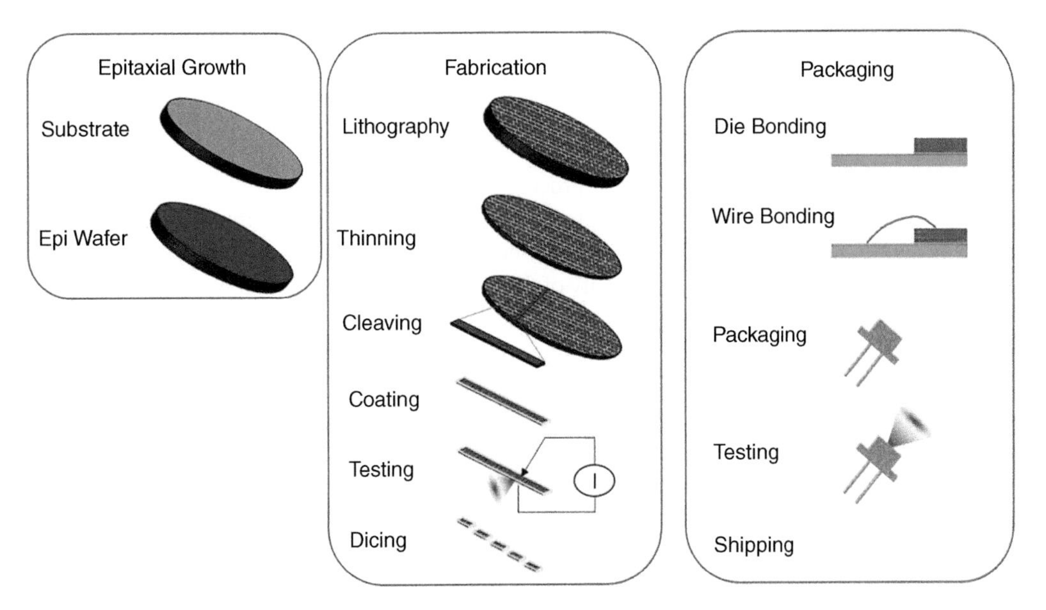

Figure 1.10 The manufacturing processes of edge-emitting lasers. *Source:* Figure by K. Iga and J. A. Tatum [copyright reserved by authors].

1.2.2 Vertical-Cavity Surface-Emitting Laser

In vertical-cavity surface-emitting lasers, the optical cavity is designed to be normal to the wafer surface, and the light emits vertically from the surface, as shown in Figure 1.2(b). High reflectivity mirrors (>99%) can be obtained with the growth of multiple epitaxial layers just above and below the active cavity without regrowth, resulting in an optical resonator cavity on the order of the emission wavelength, which is often referred to as a microcavity resonator. The lateral optical and electrical confinement is achieved by an oxidation process and may have sizes from one to several

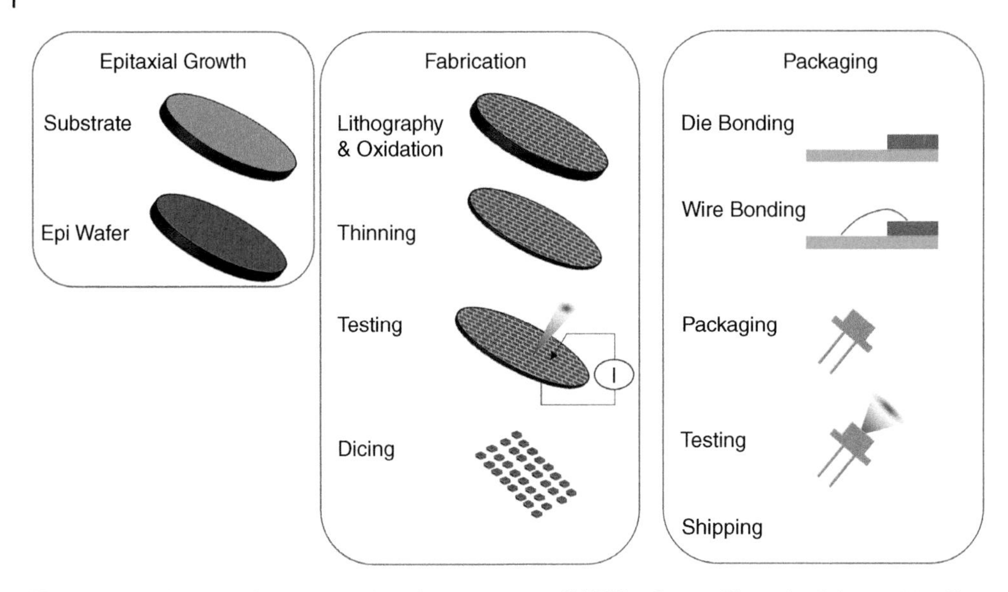

Figure 1.11 The manufacturing and testing processes of VCSELs. *Source:* Figure by K. Iga and J. A. Tatum [copyright reserved by authors].

tens of microns. This lateral size controls the mode profile of the laser and will be further described in Chapter 2.

As shown in Figure 1.11, the epi-growth process is fully monolithic and device fabrication is manufacturing-friendly and scalable. Mass production of VCSELs thus appears more like modern LED and IC manufacturing. In contrast to FP-edge-emitting lasers, the handling of full wafers only (no bar handling or facet coatings), the ability of fully testing on wafer (see Appendix D), and the knowledge of yield at that point result in a lower cost for VCSELs.

Since 2020, VCSELs are manufactured on wafer diameters of 150 mm diameters and as previously described are compatible with high-volume III–V semiconductor manufacturing processes. Many tens to hundreds of thousands of VCSELs can be fabricated on a single 150 mm wafer. With dramatic improvements in the epitaxial and fabrication processes, the device yields are routinely in excess of 90%. Details of wafer size and die counts will be given in Chapter 3. An example of fully processed VCSEL epitaxial wafer (so-called deliverable-wafer) is shown in Figure 1.12.

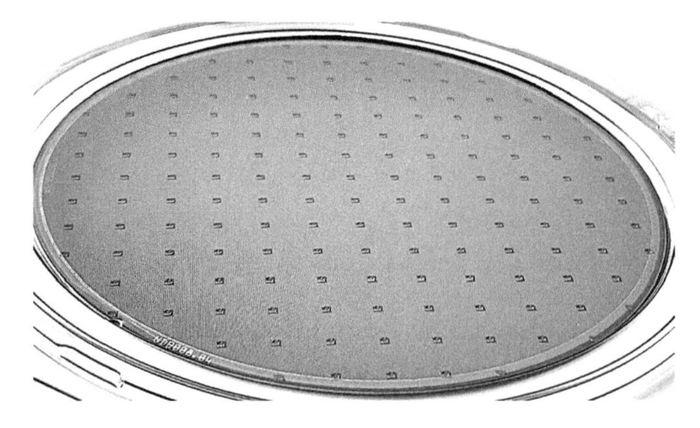

Figure 1.12 A fully processed VCSEL layer structure on 6″ (150 mm) GaAs substrate. *Source:* Wafer photo by Jim A. Tatum, Dallas, Texas, USA. [copyright reserved].

1.3 VCSEL History and Development

During the four decades since the initial VCSEL conception in 1977, many hundreds of millions of VCSELs have been shipped from a large number of VCSEL manufacturers. Some of the more ubiquitous examples are laser printers, bar code scanners, computer mice, high-speed data communications over fiber optic networks, 3D sensing in consumer and automotive electronics, night vision equipment, and many more industrial and consumer devices. It is not surprising to say that VCSELs have affected the life of nearly every person and household. With this background, we present a brief history of VCSELs from its birth to today's use in many commercial products. We divide the time in five different periods to describe the generations of VCSEL development as shown in Figure 1.13 and detail the stages in the following paragraphs.

1.3.1 Stage I: Initial Concept and Invention

1.3.1.1 Stage Ia: Invention and Initial Demonstration

Now, what is this new surface-emitting laser (SEL) or the vertical-cavity surface-emitting laser (VCSEL)? The structure is substantially different from conventional edge-emitting lasers (EELs), i.e., the vertical cavity is formed by the surfaces of epitaxial layers, and light output is from one of the mirror surfaces orthogonal to the substrate as has been shown in Figure 1.13. It is recognized that one of the authors (Iga, from Tokyo Institute of Technology) invented VCSEL in 1977 [25–28] as shown in the inset of Figure 1.13. This new invention was coined VCSEL (vertical-cavity surface-emitting laser), following the naming of a "pixel," which means any of the small discrete elements that together constitute an image (as on a television or digital

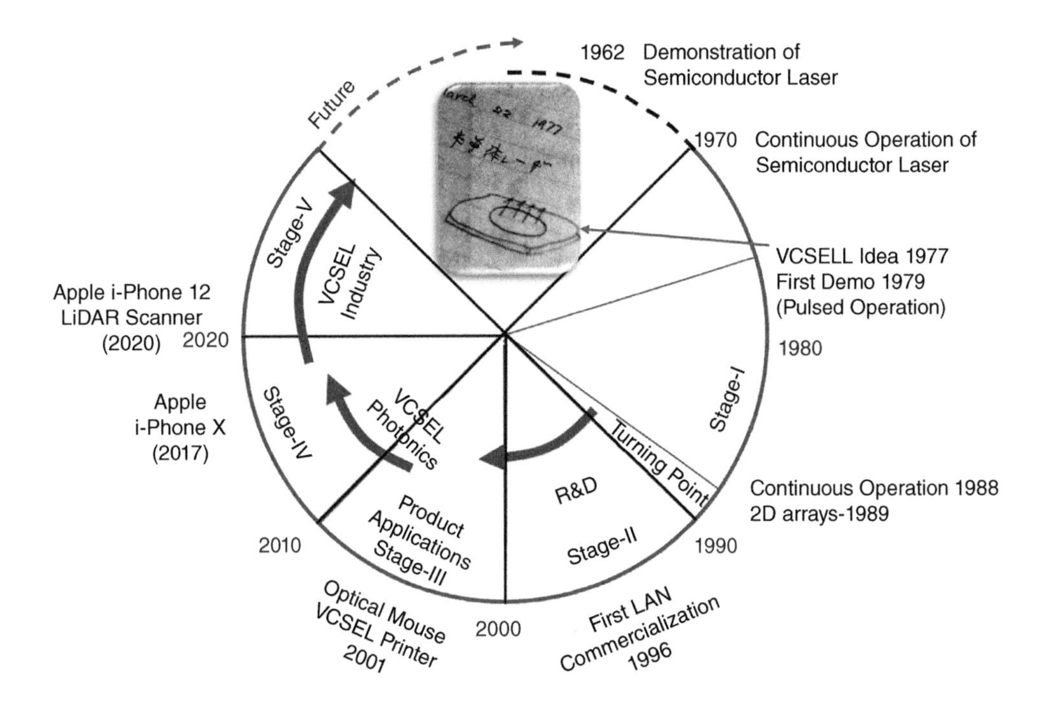

Figure 1.13 Stages of VCSEL development. The inset figure shows the sketch of VCSEL drawn by Kenichi Iga on March 22, 1977. *Source:* Figure by K. Iga and B. D. Padullaparthi [copyright reserved by authors].

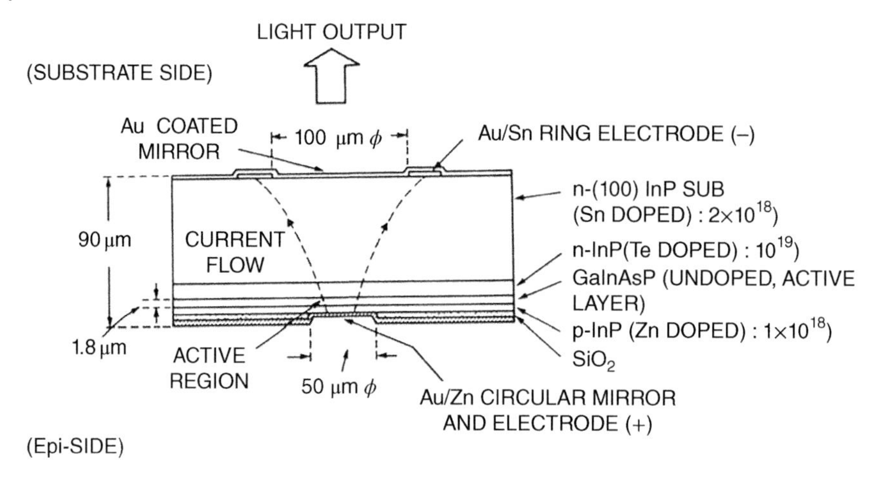

Figure 1.14 The first demonstration of a surface-emitting laser. *Source:* Figure by K. Iga [29] [copyright reserved by author].

screen). In the first stage, Ia, there were many technical challenges to overcome to realize this new device. The main challenges were the relatively low optical gain, overall mirror quality, and efficient current injection.

The first device (prototype) was realized in 1979 using a GalnAsP-InP material for the active region. The VCSEL operated at a 1300 nm wavelength [29]; a schematic cross section of the device is shown in Figure 1.14. This VCSEL used a double heterostructure with GalnAsP as an active layer, which was grown on an InP substrate. Light is emitted by injecting current from circular electrodes, and metal reflectors are formed above and below the substrate to form a resonator. This laser was driven by a pulsed current and was cooled to 77 K using liquid nitrogen. At 800 mA the device lased. When we looked at the light coming out of the device, it flashed rapidly at a certain current. It was possible to finally measure the spectrum, and it was much narrower than LEDs, which indicated laser oscillation. As mentioned above, the device was named surface-emitting laser. The threshold was very high, more than 20 times that of a normal laser, and as such, the device was out of order immediately!

1.3.1.2 Stage Ib: First Room-Temperature Continuous-Wave Operation

In 1982, Iga and co-workers made a VCSEL with 10 μm length cavity and confirmed the clear VCSEL oscillation [30]. In 1982, Iga's group made a buried confinement VCSEL with a 6 mA threshold GaAs device using liquid phase epitaxy (LPE) [31]. A major breakthrough was the achievement of continuous-wave (CW) operation at room temperature (RT) at 820 nm wavelength on GaAs substrate by Iga and Koyama (also from Tokyo Institute of Technology) in 1988 [32, 33]. The device structure is shown in Figure 1.15(a). The device was grown by metal organic chemical vapor deposition (MOCVD). With this achievement, global R&D of VCSELs has outperformed ordinary semiconductor lasers in the area of expertise. The concept of semiconductor DBR demonstration in 1988 [34] and the introduction of multi-quantum wells into VCSEL [35] contributed to the improvement of VCSEL development in later years.

After this breakthrough from Tokyo Institute of Technology, continuous room-temperature operation of the VCSEL, as shown in Figure 1.15(b), was also achieved by Jack Jewell and co-workers at Bell Laboratories in 1989 [36, 37]. The concept of periodic gain or matched gain in quantum wells contributed to reduce the threshold by Larry Coldren and co-workers [38, 39].

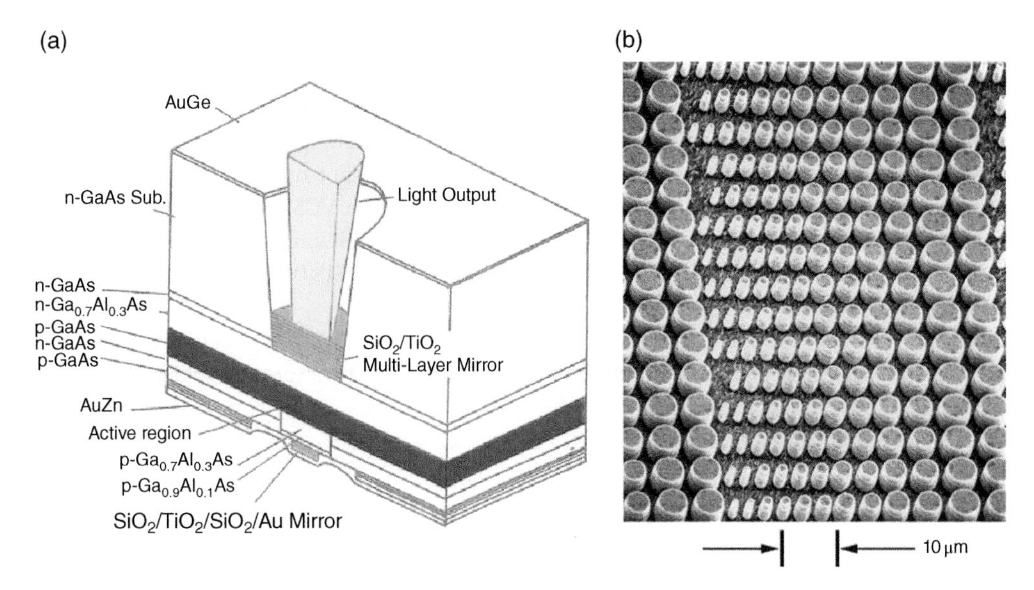

Figure 1.15 Initial VCSELs achieving room-temperature continuous operation. (a) The VCSEL device that exhibited the first room-temperature continuous-wave operation by Koyama and Iga in 1988 [32]. *Source:* Copyright reserved by Fumio Koyama and Kenichi Iga (b) A 2D micro-post array by Jewell and Lee in 1989 [36, 37]. *Source:* Adapted from IEEE.

1.3.2 Stage-II: Spread of Worldwide R&D

The second stage (1991–2000) covers the expansion of VCSEL research, the advancement in growth technology, and the emerging application needs in data communications. The first DARPA funding was driven by the Joint Strike Fighter (JSF) program. Three centers for optoelectronics were started in universities. Honeywell, Motorola, and HP were the primary companies working on the programs. More details can be found in the wiki page:

https://en.wikipedia.org/wiki/Vertical-cavity_surface-emitting_laser.

Areas of emphasis in Stage II include mass production technology [40], threshold current reduction[39–42], transverse mode control, oxidation [43, 44], polarization control, initial tunable VCSELs [45], MEMS elements [46], 2D arrays [47], high-speed and high-power VCSELs, InP-based device with continuous operation [48] and quantum wells VCSELs, and so on. This was a golden period in the VCSEL journey to mass production, and many technical and manufacturing advances contributed to the foundation of VCSEL technology.

1.3.3 Stage III: Extension of Applications and Initial Commercialization

Stage-III of VCSEL development started in 1999, shown in Figure 1.12, as we entered a new information and technology era in 2000. The third stage (1999–2010) brought on new development of wavelengths, single mode, VCSEL arrays, volume manufacturing driven by Internet traffic demand, autofocus, and so forth, and the focus has shifted to commercial efforts. Why did 1310–1550 nm VCSELs not become widely adopted? That was primarily due to the technical difficulty of making mirrors and overcoming optical loss in materials.

In 2000 one of the authors (Iga) wrote a VCSEL review paper [49] and in the same special issue, DARPA managers Elias Towe, Robert F. Leheny, and Andrew Yang wrote the following about VCSEL in their review paper [50]: "Its size, manufacturability, and potential ease of heterogeneous integration of electronics promise a range of applications that have yet to be explored." This was the time when DARPA invested considerable human and monetary resources in the R&D of VCSEL, in particular the massive integration of VCSELs, detectors, micro-optics and driving electronics for free-space optical interconnects and all optical switching. However, practical and commercial free-space interconnect and all optical switching did not really pick up during the subsequent years. This investment nonetheless continued to drive VCSEL innovations such as high-power VCSEL arrays, high-contrast gratings, athermal VCSELs, coupled cavity VCSELs, VCSELs-based slow light waveguide devices, multi-wavelength VCSELs/WDM [51], quantum dot VCSELs, high-bandwidth VCSELs (>20 GHz), and so forth.

VCSELs are currently applied in various optical systems, such as optical networks, parallel optical interconnects, laser printers, computer mice, and so on. The three critical application areas that provided the commercial impetus for continued VCSEL expansion were high-speed data connectivity, computer mice, and laser printing.

1.3.3.1 LAN for Internet

A first large market for VCSELs with large-scale production had begun in 1995 [40]. Around 1999, the Internet spread rapidly worldwide. The dramatic growth of data centers created communication networks that support the Internet, including long-distance optical fiber networks and local area networks (LANs). Metaphorically, the artery of the blood vessel is the long-distance line, and the LAN is a capillary. VCSEL was adopted as a light source for LANs operating at 1 Gbit/s and running Fiber Channel and Ethernet protocols.

The protocols were further standardized (10G; IEEE802.3ae in 2002, 100G; IEEE802.3ba in 2011) for the optical fiber communication that constitutes the LAN of the Internet. In 2020, high speed VCSELs and the pulse amplitude modulation (PAM) scheme have been developed for 400 Gbit/s high-speed Ethernet. Information flows through the capillaries of companies and universities. Details on VCSELs in data communications will be covered in Chapter 4.

In addition to applications in data centers owned and operated by IT companies, national organizations and universities began to include optical interconnects in computing architectures. For the supercomputer TSUBAME 3.0 by Tokyo Institute of Technology more than 16 000 VCSELs were included in the system. More than 300 000 VCSELs are used in IBM's top supercomputer. In the world's fastest (2019 and 2020) supercomputer Fugaku of RIKEN of Japan, it was reported that the numbers of VCSEL chips used was 640 000.

https://www.r-ccs.riken.jp/en/fugaku/project/outline.
https://www.fujitsu.com/downloads/SUPER/primehpc-fx1000-hard-en.pdf.

1.3.3.2 Computer Mouse

A computer is one way to access the Internet, and a computer mouse is useful for operating a computer. Since around the year 2000, VCSELs have also been applied to computer mice by Hewlett-Packard; this represented the first high volume use of VCSELs in a consumer market. The use of VCSELs in consumer electronics may be comparable to the development of electronic devices such as LSIs and semiconductor memories. VCSELs used in computer mice will be further discussed in Chapter 8.

1.3.3.3 Laser Printers

VCSEL arrays were introduced into laser printers in 2001 by companies such as Fuji Xerox, Ricoh, and Canon/Sony. Since then, VCSELs have largely replaced edge-emitting lasers and LEDs in printers. The aforementioned companies have about 80% market share of integrated laser printers in the world. VCSELs in laser printing will be further discussed in Chapter 8.

1.3.4 Stage IV: Spread of VCSEL Photonics

In the fourth stage of VCSEL history, from 2010 to 2020, the true scaling of VCSEL production has been realized. In data communication, highly reliable 850 nm VCSELs are made using InGaAs quantum wells with 3 dB bandwidths exceeding 25 GHz and operated above 70 Gb/s NRZ. New modulation (PAM-4) standards are made for higher speeds to meet the continued network demand. This is also supported by development of short wavelength WDM VCSELs (in the 850–980 nm band).

In the optical sensing area, high-power 940 nm VCSELs arrays have been made with optimized designs with power conversion efficiency > 50%, slope efficiency = 1.0 W/A, and with new trends of using multi-tunnel junctions that offer a power conversion efficiency > 60%, slope efficiency = 3.0 W/A, with power densities of 1 kW/mm^2. This facilitated the use of VCSEL arrays in consumer devices (mobile/smart phones/smart homes), infrastructure and transport applications incorporating LiDAR, surveillance and night vision products, robots, drones, IoT, and so on. Further applications include multi-mode VCSEL arrays in LiDARs at 905 nm, 850 nm, and 1060 nm; large-scale 940 nm arrays in industrial heating systems. On the other hand, single-mode VCSELs are being applied to optical coherence tomography (1060 nm) and atomic clocks. Details and references can be found in related chapters.

The multi-function ability of VCSELs further expanded manufacturing bases across the world with investments prompting high market demands never seen before. This also triggered high-volume manufacturing from 4″ (100 mm) to 6″ (150 mm) for optical sensor products. All these items will be discussed in Chapters 3–9.

1.3.5 Stage V: VCSEL Industry

In the fifth stage (2020 onward) VCSELs will continue to expand in global volume production and find new application areas. The ever-increasing demand for data communication and emerging technologies in machine vision, artificial intelligence (AI), augmented reality (AR), mixed reality (MR), and the Internet of Things (IoT) will drive global demand for VCSELs [52]. In this stage, VCSELs will affect nearly all aspects of human life. Details on computer vision (AR, MR, VR) will be introduced in Chapter 5. The application to automotive LiDARs and autonomous shuttles will be covered in Chapter 6.

1.4 Timeline and Milestones

1.4.1 Milestones of VCSEL Research and Development

In Table 1.2 we show a list of key benchmark events based on 44 years of basic research and commercial product developments on semiconductor lasers and VCSELs.

1.4.2 Single-Mode and Multi-Mode Behavior

Semiconductor lasers oscillate in different modes (power radiation patterns) that depend on the dimensions of the optical resonator. Especially for VCSELs, the mode structure or pattern depends on the size and shape of the oxide aperture (mode) diameter used for current and optical confinement [53, 54]. Several kinds of modes appear in the emission spectra, namely longitudinal, transverse, single, and multi-modes. In Figure 1.16 we show how the behavior of VCSEL in single longitudinal and multi-transverse (and longitudinal) operation. More details on the VCSEL mode structure will be discussed in Chapter 2.

Table 1.2 Milestones of surface emitting laser research and development.

1977	Surface-emitting laser concept proposed
1979	First device demonstrated (77 K, pulsed)
1988	First RT CW operation
1988	Semiconductor DBR
1989	QW VCSEL
1989	Micro-post QW laser RT CW (Bell Labs)
1989	Periodic gain proposed (UCSB)
1990	AlGaAs hydro-oxidization (UIUC)
1992	VCSEL mechanical tuning demonstrated
1995-	Low threshold device competition $I_{th} < 0.1$ mA
1995	MEMS tunable VCSEL (USB)
1996	VCSEL commercialization (Honeywell)
1999	VCSEL LAN
2000	Oxide aperture device reliability
2001	VCSEL printer (Fuji, Xerox),
2001	Computer mouse (HP)
2002	10G Ethernet standard IEEE802.3ae
2003	4Gb/s VCSELs (Honeywell)
2006	High-contrast grating for VCSEL (UC Berkeley)
2010	100G Ethernet standard IEEE802.3ba
2011-	VCSEL photonics
2016	VCSEL arrays for sensing
2016	VCSEL SWDM in datacom
2017	VCSEL face ID 3D sensor (iPhone X from Apple)
2018	400G AOC (Finisar/II–VI)
2020	LiDAR scanner (iPhone 12 Pro, iPad Pro 11″/12″ from Apple)
2021-	VCSEL Industry

Italic: From K. Iga's group.
RT CW: Room-temperature continuous wave
QW: Quantum well.

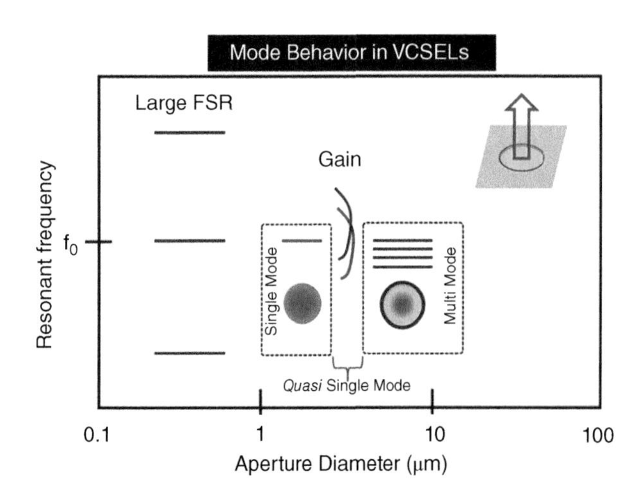

Figure 1.16 Origin of single-mode and multi-mode behavior in VCSELs. *Source:* [55, 8]. Figure by K. Iga [copyright reserved by author].

The origin of multi-mode operation in a VCSEL is not from longitudinal mode behavior but multiple transverse mode oscillation. When the lateral extent of the optical resonator diameter is extended to larger than several microns, multi-transverse with multispectral-mode operation is achieved. On the other hand, small diameters lead to single-transverse and single spectral-mode operation.

1.4.3 Major Features of VCSELs

VCSELs have become the light source of choice for many applications and are rapidly replacing edge-emitting lasers (EELs) and LEDs in many more every day. The inherent advantages in manufacturing and the ability to tailor the VCSEL properties to different applications has been key to their success. The ability to scale the optical power and emission pattern with 2D arrays of emitters has enabled the widespread adoption of 3D sensors. Table 1.3 summarizes the typical operating characteristics of VCSELs and EELs. Even with the long list of advantages, there are still many applications where EELs offer a better solution. Some areas where EELs are still dominant are single-mode fiber optic communications and extremely high-power applications.

The authors rearranged Table 1.3 to provide a simple picture of advantageous and key characteristics along with performance attributes into Table 1.4.

1.4.4 VCSELs as Major Optical Components

The single-mode and multi-mode VCSELs categorized in Section 1.2.1 show decisive advantages as optical components in communication and sensing as shown in Table 1.5.

i) Single or 1D VCSEL emitters are used as light source in active optical cable (AOC) or optical interconnects for data communication.
ii) 2D VCSEL arrays found high volume applications as printer and time of flight (ToF) or structured light sources in 3D sensing. Proximity illumination in Face Recognition (FR), Gesture Recognition (GR). Time-of-Flight (ToF), Phase shift, FMCW light sources in 3D ranging. They are also used for scanning mode or flash mode LiDARs for automotive, surveillance/night vision ranging, robotics, and drones.
iii) Large scale VCSEL arrays in industrial heating etc.

All these items will be discussed in detail in Chapters 4–9 as classified in Table 1.5.

1.4.5 VCSELs in Optical Communication and Sensing

1.4.5.1 The Concept of VCSEL Communication and Sensing

As depicted in Figure 1.16, in most of the applications, VCSELs are used either as light transmitters for communication or light sources for sensors. Data transmission in optical communication systems requires a light transmitter and a receiver together with a transmission media such as optical fibers or air. Along with these core components, driver/receiver ICs with connecting optics are incorporated.

Similarly, in an optical sensing system, a VCSEL light source illuminates 2D/3D objects, and a receiver is used to capture the reflected rays from the objects. Figure 1.17 shows the concepts of optical communication and sensing that are required to understand Chapters 4–9.

1.4.5.2 VCSELs in Optical Communications

The two-way communication concept described in Figure 1.16(a) is used in optical transceivers made from VCSELs, such as pluggable transceiver, active optical cable (AOC), HDMI-AOC, active direct attach copper (DAC), and USB-3 or USB-C. These are high-volume applications especially in 100 and 400 Gb/s networks in data centers with ranges from <3 m to >100 m. Normally,

Table 1.3 Differences between VCSEL and edge-emitting lasers (EEL).

			VCSEL			Edge-Emitting Lasers	
	Structure/Parameter	Units	Single Mode	Multi-Mode	Multi-Mode Array	DFB/DBR	Fabry-Pérot
Electro-Optical	Operating current	mA	6 mA		depends on the numbers of emitters	30 mA	
	Threshold current	mA	<1 mA			25 mA	
	Series resistance	ohm	50 Ω			3 Ω	
	PCE/WPE	%	35–40%		>40%	>50%	>55%
	Slope efficiency	Watt/Amp	0.4–0.7 W/A		>0.45 W/A	0.3 W/A	
	Output power	mW	1–10 mW		>1000 mW	<120 mW	
	Rise and fall time	nano sec	<1 ns			5–10 ns	
	Modulation speed	Gbit/s	>40 Gb/s	>200 Gb/s	not reported	>200 Gb/s	>25 Gb/s
	3 dB down S21 bandwidth	GHz	>20 GHz	>40 GHz	not reported	> ~ 100 GHz	>30 GHz
Spectrum	Linewidth	nm	<0.1 nm	0.2–0.6 nm	1–3 nm	<1 nm	1–2 nm
	Beam divergence (angle)/ quality	degree	symmetric (2–20°)/no astigmatism			elliptical (15/40°)/ astigmatism	
	Speckle	looking	high	low	moderate	high	
	Single/multi-mode behavior	looking	pure single		multi-transverse	quasi-single/ multi-longitudinal	
Thermal	Wavelength stability (shift)	nm/Kelvin	0.06 nm/K			0.3 nm/K	0.3
	Reliability (lifetime)	hours	high			high	
Manufacture	Array scaling	dimension	2D			1D	
	Wafer diameter	inch(mm)	4″ (100 mm) & 6″ (150 mm) ready			4″ (100 mm) to be ready	
	Assembly and packaging	complexity	simple and easy			**complex**	
	Growth and processing	complexity	monolithic/standard CMOS			regrowth/facet coating needed	
	Cost	amount	low			high	

Source: [Table by B. D. Padullaparthi and K. Iga] [copyright reserved by authors].

Table 1.4 Attributes of VCSEL in datacom, sensing, and manufacturing.

Datacom	Sensing	Manufacturing	Others (Performance)
high fiber coupling efficiency	high peak pulse powers	vertical integration	low threshold currents
high bandwidths	high PCE and SE	array scalability and small footprint	circular beam (Low divergence)
high modulation speed	high rise and fall time (integration time)	easy alignment and packaging	narrow linewidth
low power consumption (energy efficient in data centers)	low power consumption (long battery operating times consumer)	monolithic process handling (epi-growth and wafer fabrication)	operation in single-mode and multi-mode
wavelength tunability (WDM)	high beam quality	low cost and high yield	high reliability (auto grade)
low thermal impedance (efficient heat dissipation)		on-wafer testing	wavelength stability

Source: [Table by B. D. Padullaparthi and K. Iga] [copyright reserved by authors].

Table 1.5 Mode dependent VCSEL applications.

	Optical Communications	Optical Sensing	Others
Single mode	mid distance transceiver	optical mouse	printers
		gas sensing	displays
		OCT	atomic clock
		bio sensing	
		motion sensing	
Multi-Mode	LAN		
	short-reach transceiver	face recognition	manufacturing
	interconnects	illumination	heating
		LiDAR	
		robotics	

Source: [Arranged by K. Iga and B. D. Padullaparthi] [copyright reserved by authors].

multi-mode fibers are used for short-reach (<100 m) applications, and single-mode fibers are employed for long-reach (2–10 km) applications.

Using high performance VCSELs with sufficient output power, low-loss optical fibers, and efficient high-speed detectors, data can be transmitted at speeds of 25–100 Gb/s over 300 m at temperatures well over 85°C [56]. The details of multi-mode VCSELs for data communications/data-center applications are discussed in Chapter 4.

1.4.5.3 VCSELs in Optical Sensing

Referring to the optical sensing system in Figure 1.16(b), short-range (<10 m) image sensing in consumer electronics use 2D VCSEL arrays as light source with two different schemes. In the first scheme, intense uniform light from 2D VCSEL arrays illuminates on an 2D/3D object, is reflected

(a)

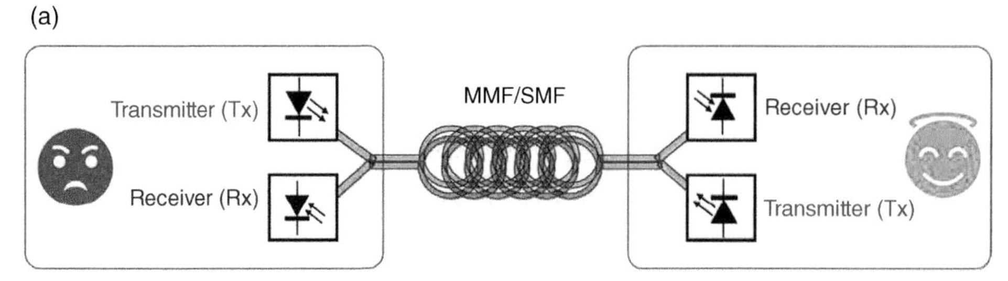

VCSEL based optical communication system (Transceiver)

(b)

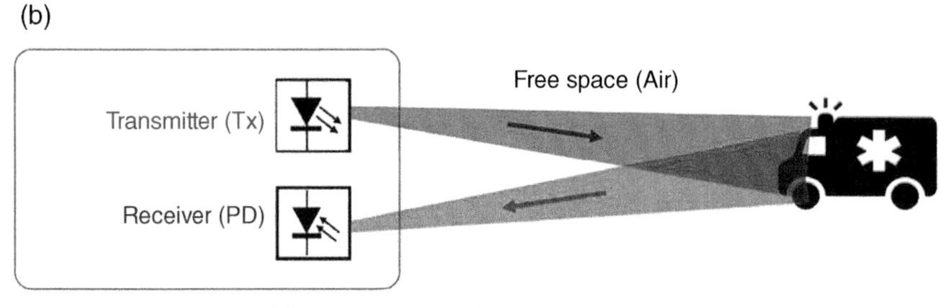

VCSEL based optical Sensing system (Optical Sensor)

Figure 1.17 VCSEL-based optical communication and sensing systems. *Source:* Figure by K. Iga [copyright reserved by author].

back to image sensors, and the distance of the object is measured by a method called time of flight (ToF).

Besides, a simple 3–6-emitter VCSEL array is also used for proximity sensing in smart phones using the ToF technique.

In the second scheme, intense distributed light from a 2D VCSEL array illuminated on a 3D object is reflected back to the image sensors, and the depth of the object is measured by a method called structured light. This is the mechanism of face unlocking in smart phones. There are a host of emerging applications from VCSEL arrays as sensors in AR, robotics, smart-home appliances, and so on. Details are discussed in Chapter 5.

Long-distance ranging or object detection (~250 m or longer) can also be done using the sensing concept shown in Figure 1.16 for automobile by using LiDARs. This is sometimes known as vehicle-to-everything (V2X). LiDARs use individually or row/column addressable arrays of VCSELs or edge-emitting laser arrays to illuminate the scene either through a single flash, sequential flashes by selectively addressing the emitters, or scan functions, and the object image is created through powerful signal processing and artificial intelligence (AI).

Besides ToF, more precise object measurement techniques such as optical phased arrays (OPA) or frequency modulations (FMCW) are also used for advanced driver-assistance systems (ADAS). Details of other applications are discussed in Chapters 6–9.

To facilitate comprehension, supplementary information is given as appendices on generic VCSEL design (Appendix A), epitaxial growth (Appendix B), wafer processing (Appendix C), testing (Appendix D), reliability and qualification (Appendix E), and eye-safety issues (Appendix F). Special notes on display (Appendix G), red VCSELs (Appendix H), photodetectors (Appendix J), and GaN VCSELs (Appendix I) and are also provided.

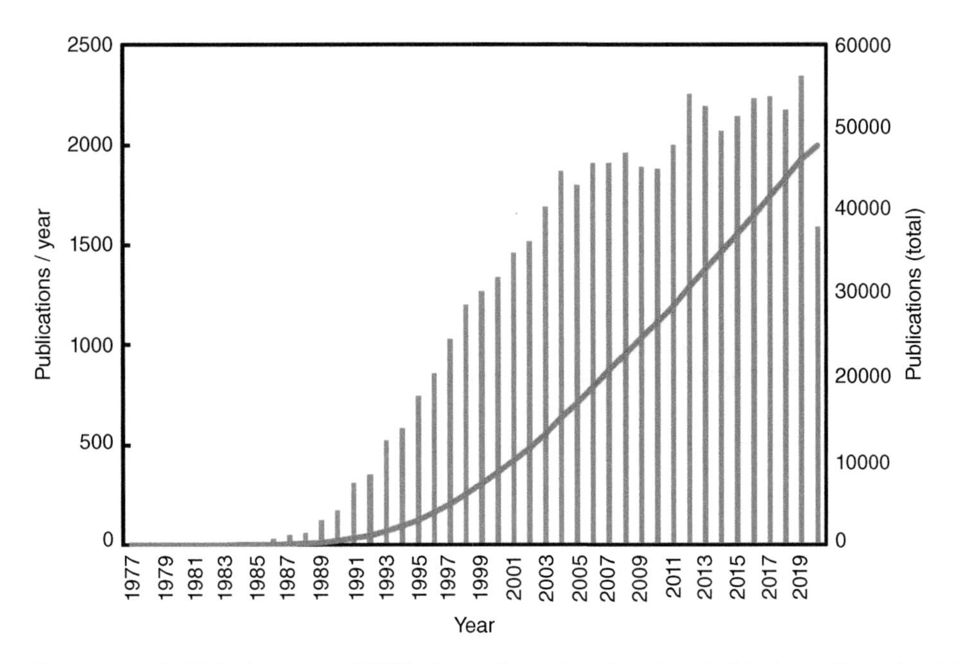

Figure 1.18 Published papers on VCSEL. *Source:* Data taken from Google Scholar on November 2, 2020. (Searching key words: "VCSEL" OR "vertical-cavity surface-emitting laser" OR "surface emitting laser" in the text or title.) [Image courtesy of Tomoyuki Miyamoto, Tokyo Institute of Technology.]

1.5 State of VCSEL Development

1.5.1 Published Papers

Figure 1.18 shows the number of published papers per year and accumulated statistics. Note that the total number of related papers has reached about 50,000 as of 2020. The yearly publications have skyrocketed over the last decade, indicating that commercialization is taking place.

1.5.2 Toward VCSEL Photonics

Ever since Honeywell started VCSEL commercialization and introduced the first reliable product in around 1996 [57], VCSEL technology has made a huge impact on several key industries with multiple growth windows. Thanks to several commercial epi-houses, III-V opto-foundries, and other equipment vendors, researchers and engineers have overcome great challenges to make VCSEL-based commercial products a practical reality since the beginning of 2021.

After 44 years of VCSEL invention [25] and the marathon industry efforts to realize volume manufacturing, it is not surprising that most people carry a VCSEL device along with them, if not a few! This means VCSELs have rapidly grown up, fully matured, and penetrated into commercial products that affect the daily lives of humans.

Four major industries dominate most of the VCSEL applications space (called core products), while few other neighboring fields are also emerging in the entire VCSEL application seabed. On core applications, the first is high-speed VCSELs for data communications and data center applications. There is a steady demand for 100 to 300 m active optical cables in 100–400 Gb/s data center needs, HDMI AOC, USB-3 (c-type), and so on. The world's fastest (as of 2019 and 2020) high-performance computer, Fugaku, uses more than 6 400 000 chips of VCSELs for AOCs.

The second and biggest opportunity is high-power VCSEL arrays for consumer electronics (3D sensing and imaging up to 10 m), including illumination, face recognition (FR), and augmented reality (AR) needs. This extends to object detection for autonomous vehicles and intelligent transport through powerful long-range LiDARs up to 250 m or beyond. Further high-power VCSEL arrays (where 100 000 to a few million units of emitters are needed) find applications in industrial heating.

Furthermore, we find VCSEL applications in neighboring areas such as coherent communication, laser printers, additive manufacturing, gas sensing and spectroscopy, biometrics such as optical coherence tomography (OCT) and iris scans, gaming (VR and MR), robotics, and drones attracting considerable investments, particularly on AI programming, smart home and IoT, and automotive Ethernet and even to quantum computing.

1.5.3 Toward VCSEL High-Volume Manufacturing

The proliferation of VCSELs into several industries and their use in countless commercial products make the VCSEL industry bright and vibrant. In Figure 1.19 we show the market of VCSELs estimated in 2021. It is quite amazing to realize that hundreds of millions of VCSELs have been deployed. The overall market is forecast to continue to grow in current core applications, and it is expected that even more areas will be uncovered to edge areas in the future. The large shipping volume has created an ecosystem of vertically integrated manufacturers as well as robust foundry vendors. With millions of epi-wafers processed, multibillion-dollar market forecasts, and nearly a trillion of VSCEL units shipped, undoubtedly VCSELs are the future optical components of choice and are cementing their presence in the photonics industry!

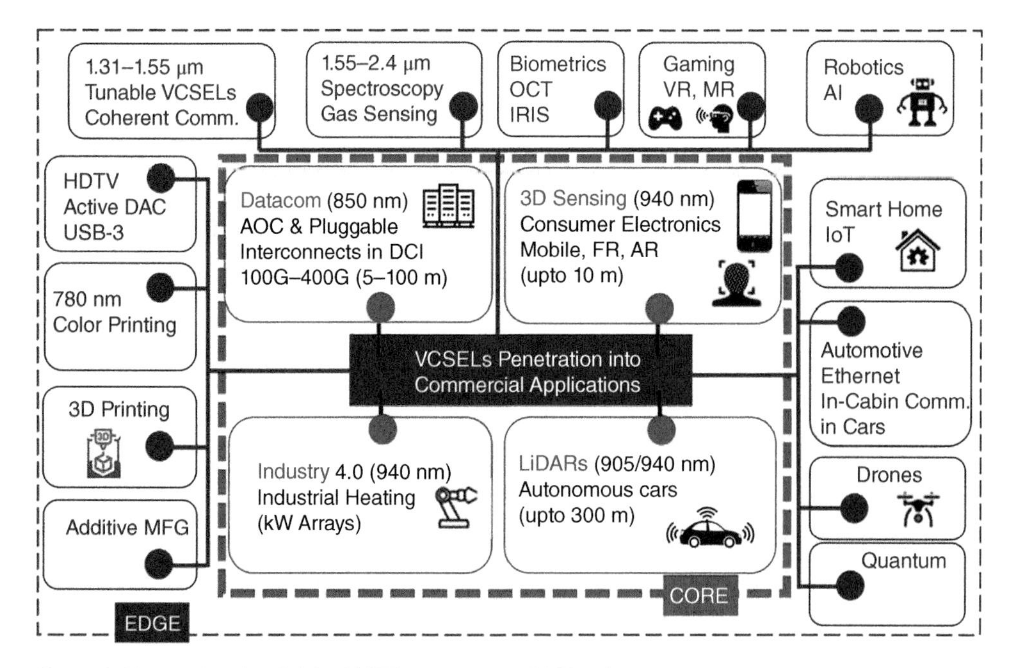

Figure 1.19 Application fields of VCSEL market as of 2021; data taken from various sources. *Source:* Figure by B. D. Padullaparthi and K. Iga [copyright reserved by authors].

In Figure 1.20, we show a modified hype cycle of VCSEL industrialization. This shows the timeline of VCSEL industrialization based on Iga's personal limited knowledge. In Chapter 10, we will discuss this matter after reviewing all the items in technical chapters by all the authors and editor by including other applications.

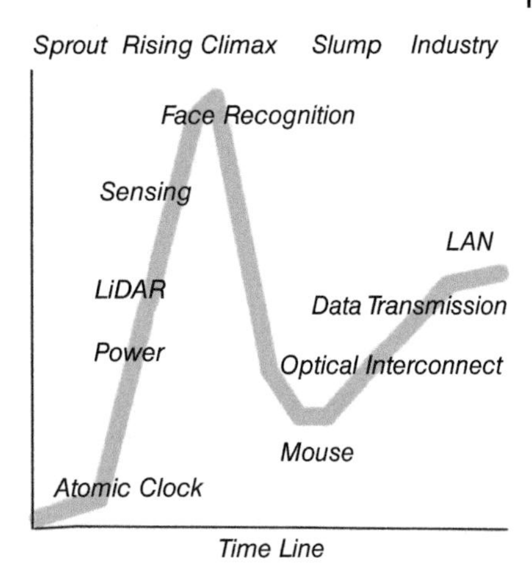

Figure 1.20 The modified hype cycle of VCSEL industrialization. *Source:* K. Iga's observation in the middle of 2020 [copyright reserved by author].

1.5.4 Prospects of VCSEL Market

At the beginning of 2021, VCSEL technology has emerged as a low-cost, high-volume business opportunity for large-scale manufacturers, comparable to other semiconductor technologies such as GaN, SiC RF, and power devices, LEDs, displays, photovoltaic solar cells, Si-photonics, and InP-based semiconductor lasers. A schematic of market sizes for different VCSEL technologies are shown in Figure 1.21. The total photonics market is expected to reach about $80 billion by 2025 from $24 billion in 2020. The authors consider two types of opportunities for VCSEL-based commercial products, namely **core** and **edge** markets.

The **core** market includes four major areas: datacom, 3D sensing (mobile), 3D imaging LiDAR (automotive), and industrial heating. In these areas, VCSEL is a proven technology addressing strong societal needs and appears to be gaining a major market share with readily available commercial products.

The **edge** markets include defense and aerospace, medical (OCT and iris scans), gas sensing, gaming, global AR/VR/MR, surveillance (IP and CCTV cameras), 3D printing, laser printing, and many other applications. Other **edge** markets such as robotics, AI, and smart-home appliances

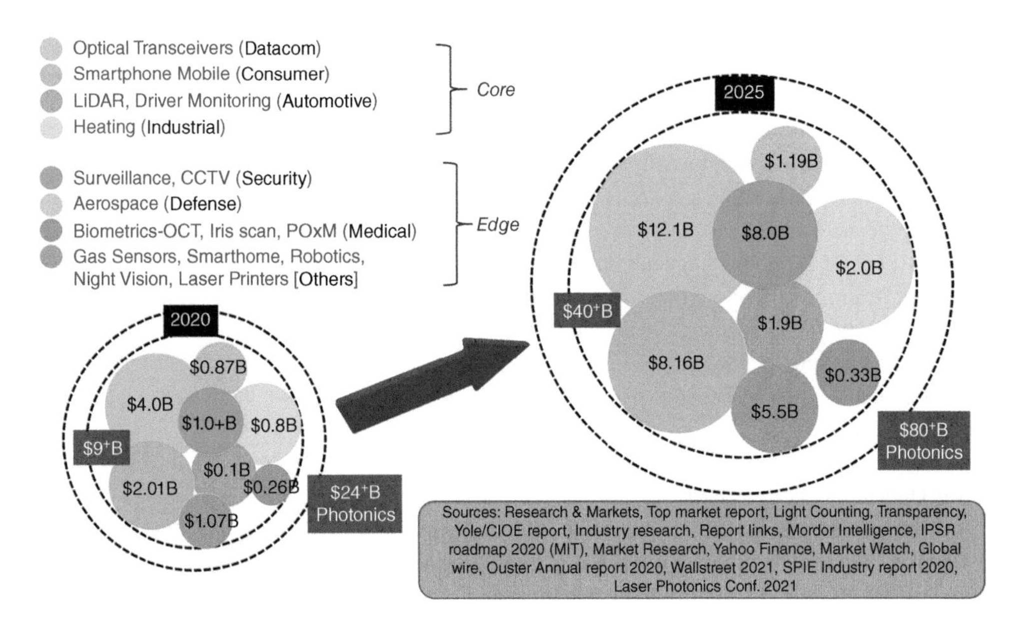

Figure 1.21 Total addressable market of VCSELs at module levels till 2025. *Source:* B. D. Padullaparthi [copyright reserved by author].

either have small market share or are still developing prototypes for final stages of release for commercial products [58].

The total addressable market projected for VCSEL's **core** and **edge** areas at module level is expected to be around $40 billion by 2025, where the **core** part alone is forecast to be about a $24 billion by 2025 market, as shown in Figure 1.20.

The chip level projection for datacom, telecom, mobile consumer, automotive, medical, industrial, and defense fields is estimated to be about $4.8 billion by 2025, that is 24% of the corresponding module level projections.

With an increasing number of autonomous cars with LiDARs by 2030, it is anticipated that the market size of the automotive industry will exceed that of consumer electronics, prompting a large number of VCSELs to be used for long-distance ranging as flash or scan LiDARs. Further, several **edge** and other markets projected a total reaching $80 billion. When a fraction of 15% (about $8.4 billion) is assumed for add value to **core** fields, the total addressable market size at module level will be at least about $32.4 B, as shown in Figure 1.20. Some chip level details are given in Chapter 3.

In summary, it is concluded that VCSEL is finding a vibrant commercial prospect for high-volume manufacturing and product demands that are further expanding.

References

1 T. H. Maiman, "Stimulated optical radiation in ruby," *Nature*, **187** 4736, pp. 493–494 (1960).

2 R. N. Hall, G. E. Fenner, J. D. Kingsley, T. J. Soltys and R. O. Carlson, "Coherent light emission from GaAs junctions," *Phys. Rev. Lett.*, Vol. **9**, No. 9, pp. 366–368 (1962).

3 T. M. Quist, R. H. Rediker, R. J. Keyes, W. E. Krag, B. Lax, A. L. McWhorter and H. J. Zeigler, "Semiconductor maser of GaAs," *Appl. Phys. Lett.*, Vol. **1**, No. 4, pp. 91–92 (1962).

4 M. I. Nathan, W. P. Dumke, G. Burns, F. H. Dill, Jr. and G. Lasher, "Stimulated emission of radiation from GaAs p-n junctions," *Appl. Phys. Lett.* Vol. **1**, No. 3, pp. 62–64 (1962).

5 N. Holonyak, Jr. and S. F. Bevacqua, "Coherent (visible) light emission from $Ga(As_{1-x}P_x)$ junctions," *Appl. Phys. Lett.*, Vol. **1**, No. 4, pp. 82–83 (1962).

6 I. Hayashi, P.B. Panish, P.W. Foy and S. Sumski, "Junction lasers which operate continuously at room temperature," *Appl. Phys. Lett.*, Vol. **17**, pp. 109–111 (1970).

7 Zh. I. Alferov, V. M. Andreev, E. L. Portnoi and M. K. Trukan, "AlAs-GaAs heterojunction injection lasers with a low room-temperature threshold," *Fiz. Tekh. Poluprovodn.*, Vol. **3**, pp. 1328–1332 (1969); Sov. Phys. Semicond., Vol. 3, pp. 1107–1110 (1970).

8 K. Iga and G. Hatakoshi, "Treasure Microbox of Optoelectronics," Adcom-Media Co. Ltd. Tokyo, April 25, 2020. (PDF Japanese language version).

9 H. Kroemer, "A proposed class of heterojunction injection lasers," *Proc. IEEE*, Vol. **51**, No. 12, pp. 1782–1783 (1963) H. Kroemer: "Solid state radiation emitters," US Patent 3309553 (Application date: Aug. 16, 1963).

10 Y. Suematsu and K. Iga, "*Introduction of Optical Fiber Communication*," John Wiley and sons, 1976.

11 H. C. Casey and M. B. Panish, *Heterostructure Lasers*, Academic Press, New York (1978).

12 A. Yariv: Optical Electronics, 1991.

13 L. A. Coldren, S. Corzine, and M. L. Masanovic, "*Diode Lasers and Integrated Optics*," Wiley, 1994.

14 S. L. Chuang, "*Physics of Optoelectronic Devices*," John Wiley & Sons, New York, 1995.

15 J.P. van der Ziel, R. Dingle, R.C. Miller, W. Wiegman and W.A. Nordland, Jr., "Laser oscillation from quantum states, in very thin $GaAs\text{-}Al_{0.2}Ga_{0.8}As$ multilayer structures," *Appl. Phys. Lett.*, Vol. **26**, No. 8, pp. 463–465, 1975.

16 Y. Arakawa and A. Yariv, "Theory of gain, modulation response and spectral linewidth in AlGaAs quantum well lasers," *IEEE J. Quantum Electron.*, Vol. **QE-21**, No. 10, pp. 1666–1674, Oct. 1985.

17 M. Asada, Y. Miyamoto and Y. Suematsu, "Gain and the threshold of three-dimensional quantum-box lasers," *IEEE J. Quantum Electron.*, Vol. **QE-22**, pp. 1915–1921, 1986.

18 K. Iga, "*Fundamentals of Laser Optics*," Plenum, p. 173, (1994).

19 M. Asada and Y. Suematsu, IEEE J. QE, vol. QE-21, p. 434(1985).

20 H. Kogelnik and C.V. Shank, "Coupled wave theory of distributed feedback lasers," *J. Appl. Phys.*, Vol. **43**, No. 5, pp. 2327–2335, May 1972.

21 Y. Suematsu and K. Hayashi, "General analysis of distributed Bragg reflector and laser resonator using it," *National Conv. Inst. Electron. Comm. Eng*, **1200**, p. 1203, (July 25–27, 1974).

22 W. Tsang and S. Wang, "GaAs-$Ga_{1-x}Al_xAs$ double-heterostructure injection lasers with distributed Bragg reflectors," 9th IQEC, p. 38 June 1976.

23 K. Utaka, Y. Suematsu, K. Kobayashi and H. Kawanishi, "GaInAsP/InP integrated twin-guide lasers with first-order distributed Bragg reflectors at 1.3μm wavelength," *Jpn. J. Appl. Phys.*, Vol. **19**, No. 2, pp. L137–L140, Feb. 1980.

24 Y. Suematsu, "Dynamic single mode lasers," *J. Lightwave Technol.*, vol. **32**, no. 6, pp. 1144–1158, March 2014.

25 K. Iga, Research Notebook, March 22, 1977.

26 K. Iga, T. Kambayashi, C. Kitahara, "Surface-emitting GaInAsP / InP laser (I)," 25th Joint Conference on Applied Physics (at Musashi Institute of Technology), 27-p-11, p. 63, March 27, 1978.

27 K. Iga, Y. Suematsu, K. Kishino and H. Soda, "Surface emitting semiconductor laser," Japan. Patent, Hei 1-56547, Jan. (1980).

28 K. Iga and G. Hatakoshi, "The Principle and Application Systems of Vertical Cavity Surface Emitting Laser," Adcom-Media Co. Ltd. Tokyo, Sept. 25, 2020. (PDF Japanese language version)

29 H. Soda, K. Iga, C. Kitahara, and Y. Suematsu, "GaInAsP/InP surface emitting injection lasers," *Jpn. J. Appl. Phys.*, vol. **18**, no. 12, pp. 2329–2330 (1979).

30 Y. Motegi, H. Soda, and K. Iga, "Surface emitting GaInAsP/InP injection laser with short cavity length," *Electron. Lett.*, Vol. **18**, No. 11, pp. 461–463 (1982).

31 K. Iga, S. Kinoshita, and F. Koyama, "Microcavity GaAlAs/GaAs surface-emitting laser with I_{th}=6 mA," *Electron. Lett.*, vol. **23**, no. 3, pp. 134–136, Jan. (1987).

32 F. Koyama, S. Kinoshita, and K. Iga, "Room temperature cw operation of GaAs vertical cavity surface emitting laser," *Trans. IEICE*, vol. **E71**, No. 11, pp. 1089–1090 (1988).

33 F. Koyama, S. Kinoshita, and K. Iga: "Room-temperature continuous wave lasing characteristics of GaAs vertical cavity surface-emitting laser," *Appl. Phys. Lett.* vol. **55**, no. 3, pp. 221–222 (1989).

34 T. Sakaguchi, F. Koyama, and K. Iga, "Vertical cavity surface-emitting laser with an AlGaAs/AlAs Bragg reflector," *Electron. Lett.*, vol. **24**, no. 15, pp. 928–929, July (1988).

35 H. Uenohara, F. Koyama, and K. Iga: "Application of the multiquantum well (MQW) to a surface emitting laser," *Jpn. J. Appl. Phys.*, vol. **28**, no. 4, pp. 740–741, April (1989).

36 J. L. Jewell, S. L. McCall, A. Scherer, H. H. Houh, N. A. Whitaker, A. C. Gossard, and J. H. English, "Transverse modes, waveguide dispersion and 30 ps recovery in submicron GaAs/AlAs micro-resonators," *Appl. Phys. Lett.* vol. **55**, no. 1, pp. 22–24 (1989).

37 Y. H. Lee, J. L. Jewell, A. Scherer, S. L. McCall, J. P. Harbison, and L. T. Florez: "Room- temperature continuous-wave vertical-cavity single-quantum-well micro-laser diodes," *Electron. Lett.*, vol. **25**, no. 20, pp. 1377–1378 (1989).

38 S. W. Corzine, R. S. Geels, R. H. Yan, J. W. Scott, and L. A. Coldren, "Efficient, narrow-linewidth distributed-Bragg reflector surface emitting laser with periodic gain," *IEEE Photon. Technol. Lett.*, vol. **1**, no. 3, pp. 52–54 (1989).

39 R. S. Geels, and L. A. Coldren: "Sub-milliamp threshold vertical-cavity laser diodes," *Appl. Phys. Lett.*, Vol. **57**, pp. 1605–1607, (1991)

40 R. A. Morgan, "High-performance, producible vertical-cavity lasers for optical interconnect," in *Current Trends in Vertical Cavity Surface Emitting Lasers*, T. P. Lee Ed., World Scientific, pp. 65–95 (1995).

41 T. Wipiejewski, K. Panzlaf, E. Zeeb, and K. J. Ebeling, "Sub-milliamp vertical cavity laser diode structure with 2.2 nm continuous tuning," 18th European Conf. Opt. Comm. '1992, PD II-4, Sept. 1992.

42 Y. Hayashi, T. Mukaihara, N. Hatori, Ohnoki, A. Matsutani, F. Koyama, and K. Iga, "Record low-threshold index-guided InGaAs/GaAlAs vertical-cavity surface-emitting laser with a native oxide confinement structure," *Electron. Lett.*, vol. **31**, no. 7, pp. 560–561, Mar. (1995).

43 J. M. Dallesasse, N. Holonyak Jr., A. R. Sugg, T. A. Richard, and N. El-Zein: "Hydrolyzation-oxidation of $Al_xGa_{1-x}As$-AlAs-GaAs quantum well heterostructures and superlattices," *Appl. Phys. Lett.*, vol. **57**, no. 26, pp. 2844–2846 (1990).

44 M. H. Crawford, K. D. Choquette, R. J. Hickman, and K. M. Geib, "Performances of selective oxidized AlGaInP-based visible VCSELs," in *OSA Trends in Optics and Photonics (Advances in Vertical Cavity Surface Emitting Laser)*, Ed. C. Chang-Hasnain, vol. **TOPS15**, pp. 112–117 (1997).

45 N. Yokouchi, T. Miyamoto, T Uchida, Y Inaba, F. Koyama, and K. Iga, "40 Å continuous tuning of a GaInAsP/InP vertical-cavity surface-emitting laser using an external mirror," *IEEE Photon. Technol. Lett.*, vol. **4**, no. 7, pp. 701–703, July (1992).

46 M. S. Wu, E. C. Vail, G. S. Li, W. Yuen, and C. J. Chang-Hasnain, "Tunable micromachined vertical cavity surface emitting laser," *Electron. Lett.*, vol. **31**, no. 19, pp. 1671–1672 (1995).

47 E. Ho, F. Koyama, and K. Iga: "Effective reflectivity from self-imaging in a Talbot cavity and its effect on the threshold of a finite 2-D surface emitting laser array," *Appl. Opt.*, vol. **29**, no. 34, pp. 5080–5085 (1990).

48 T. Baba, Y. Yogo, K. Suzuki, F. Koyama, and K. Iga, "Near room temperature continuous wave lasing characteristics of GalnAsP/lnP surface emitting laser," *Electron. Lett.*, vol. **29**, no. 10, pp. 913–914 (1993).

49 K. Iga, "Surface emitting laser-its birth and generation of new optoelectronic fields," *IEEE J. Sel. Top. Quantum Electron.*, Invited paper, vol. **6**, No. 6, pp. 1201–1215, Nov./Dec., 2000.

50 E. Towe, R. F. Leheny, and A. Yang, (December 2000). "A historical perspective of the development of the vertical-cavity surface-emitting laser". *IEEE J. Sel. Top. Quantum Electron.*, vol. **6**, no. 6, pp. 1458–1464, Nov./Dec., 2000.

51 C. J. Chang-Hasnain, J. P. Harbison, C. E. Zah, M. W. Maeda, L. T. Florez, N. G. Stoffel, and T. P. Lee: "Multiple wavelength tunable surface-emitting laser arrays,," *IEEE J. Quantum Electron.*, vol. **27**, no. 6, pp. 1368–1376 (1991).

52 B. D. Padullaparthi, R. Chen and A. Tan et al., "High Volume Manufacturing of VCSELs for Datacom & Sensing," Industry Panel Discussions, Th4, pp: 51, International Nano-Optoelectronics Workshop (i-NOW) 2018, UC Berkeley, USA.

53 K. Iga, F. Koyama, and S. Kinoshita, "Surface emitting semiconductor laser," *IEEE J. Quantum. Electron.*, vol. **QE-24**, no. 9, pp. 1845–1855, Sept. (1988).

54 C. Jung, R. Jäger, M. Grabherr, P. Schnitzer, R. Michalzik, B. Weigl, S. Muller, and K. J. Ebeling, "4.8 mW single mode oxide confined top-surface emitting vertical-cavity laser diodes," *Electron. Lett.*, vol. **33**, no. 21, pp. 1790–1791 (1997).

55 K. Iga, "Forty years of VCSEL: Invention and innovation," *Jpn. J. Appl. Phys.* Vol. **57**, No. 8S2, pp. 1–7, Aug. (2018).

56 B. D. Padullaparthi, "Impact of Δn_{eff} of 850nm VCSEL cavity on low noise for 100G eSR4 transmission and its potential for ≥400G datacenter optical interconnects," *Proc. SPIE* **11704** 11704–24 (2021) doi:10.1117/12.475724.

57 J. A. Tatum and J. K. Guenter, "The VCSELS are coming," Proc. SPIE 4994, Vertical-Cavity Surface-Emitting Lasers VII, 17 June 2003;

58 K. Iga, "VCSEL Odyssey," SPIE No. PM318, September 1, 2020.

2

VCSEL Fundamentals

Jim Tatum

2.1 Introduction to Lasers

All lasers operate on the principle of positive feedback to an amplifier. One familiar example of positive feedback is when an audio signal from a speaker is picked up by a microphone and amplified and then returned to the speaker. When this happens continuously, a very loud signal is quickly generated. For a laser, the audio feedback is equivalent to a mirror reflecting light, and the microphone and electrical amplifier are equivalent to the optical gain material. This chapter describes the basic building blocks of a VCSEL, the distributed Bragg reflector (DBR) mirror, which provides the optical feedback, and quantum well (QW) gain region. In the following sections, typical operating characteristics are described such as the light output as a function of current (L-I), forward voltage as a function of current (V-I), spectral and spatial characteristics of the optical emission, frequency modulation characteristics, and how these properties vary over temperature. Definition of industry terms and performance specifications are described that will be relevant to applications described in the remaining chapters of this book. This chapter will also review the materials used to make various wavelength VCSELs and describe several alternate VCSEL structures.

2.2 Basic VCSEL Structure

A typical schematic cross section of an oxide confined typical VCSEL is presented in Figure 2.1. The oxide-confined VCSEL is the most common structure in production today. While the main focus of this chapter will be on this structure, there are other methods to fabricate VCSELs that will be described in Section 2.11. The basic components of an oxide-confined VCSEL are (i) the N-electrical contact, (ii) the GaAs substrate, (iii) the n-type DBR, (iv) the QW active region, (v) the oxidation layer, (vi) the p-type DBR, and (vii) the P-electrical contact. Also shown in Figure 2.1 is the current flow from the P-electrical contact through the active region, the direction of heat flow from the structure through the substrate, and the optical output normal to the surface of the structure.

VCSEL design presents a multi-physics design problem, and engineering trade-offs in the optical, electrical, and thermal characteristics have to be carefully balanced with complexity and limitations of the epitaxial and fabrication processes.

VCSEL Industry: Communication and Sensing, First Edition. Babu Dayal Padullaparthi, Jim A. Tatum and Kenichi Iga.

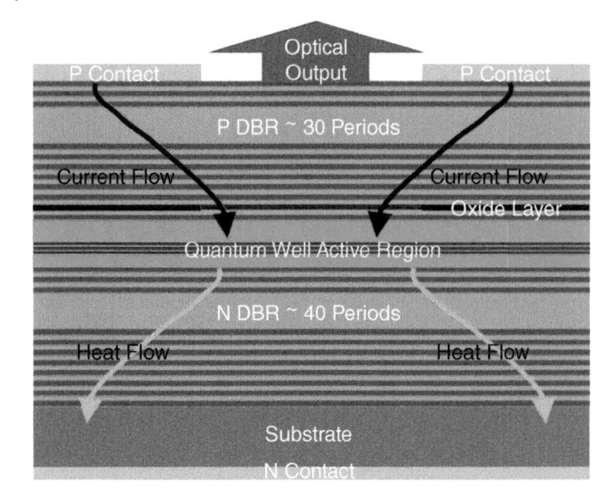

Figure 2.1 Cross section of an oxide-confined VCSEL showing the basic components of the device.

2.3 Quantum Well Gain Region (Active Region)

The first semiconductor lasers utilized a so-called bulk active region where the thickness was on the order of 1 μm. The relatively thick active regions required high operating currents and were inefficient. Nearly 20 years after the invention of the VCSEL, the QW active region emerged; this region enhances the optical gain by more tightly confining the available electron and hole energy states. By confining the available electrons and holes in one dimension, the QW active region reduces the available density of states from a continuous parabolic-type function to a single energy level. Detailed description of QW laser can be found in many textbooks [1]. Figure 2.2 shows an energy band diagram of three 60 Å GaAs QWs surrounded by $Al_{0.2}Ga_{0.8}As$ barriers. The conduction band level E_C = 1.446 eV and the valence band level E_V = −0.017 eV, which is a bandgap energy of E_{BG} 1.463 eV, or 848 nm. This is a basic design for an 850 nm VCSEL active region.

Note that both E_C and E_V are less than the bandgap of the $Al_{0.2}Ga_{0.8}As$ ($E_{BG} = 1.442 + 1.2475x$, $x \leq 0.45$, for $Al_{0.2}Ga_{0.8}As$ E_{BG} = 1.61 eV), and thus injected carriers are confined within the QWs. The electron and hole wavefunctions are also shown in Figure 2.2 demonstrating the confinement of the carriers in the QWs. The optical gain of the QW can be calculated as described in [1] and is shown in Figure 2.3a as a function of wavelength for an injected current density of $N = 1.6 \times 10^{19}/cm^3$ and $N = 1.7 \times 10^{19}/cm^3$ at 300°C. Figure 2.3b shows the calculated gain for $N = 1.6 \times 10^{19}/cm^3$ at several temperatures. Both the bandgap and peak gain are strong functions of temperature and amount of carrier injection.

2.4 Distributed Bragg Reflector Mirrors

The first semiconductor lasers used mirrors that were formed by cleaving the edge of the semiconductor and are known as the laser facets. These facets provide about 30% reflection of the optical energy and hence require much more optical gain (carrier injection) to achieve lasing. The high current requirement causes heating. To manage the heat dissipation, many of the early

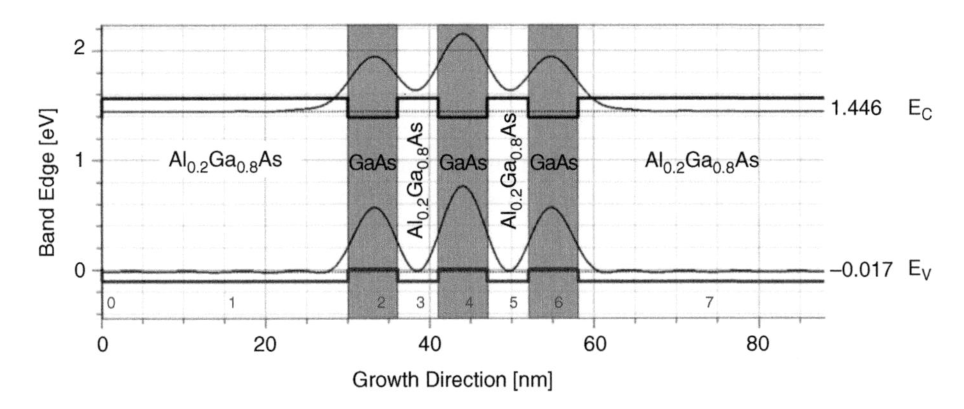

Figure 2.2 Energy band diagram of a 60 Å GaAs QW showing the quantum wells (shaded region), the conduction band (black) and valence band (blue), and the electron and hole wavefunctions. The band gap (E_C − E_V) of the quantum well structure is 1.463 eV (~850 nm).

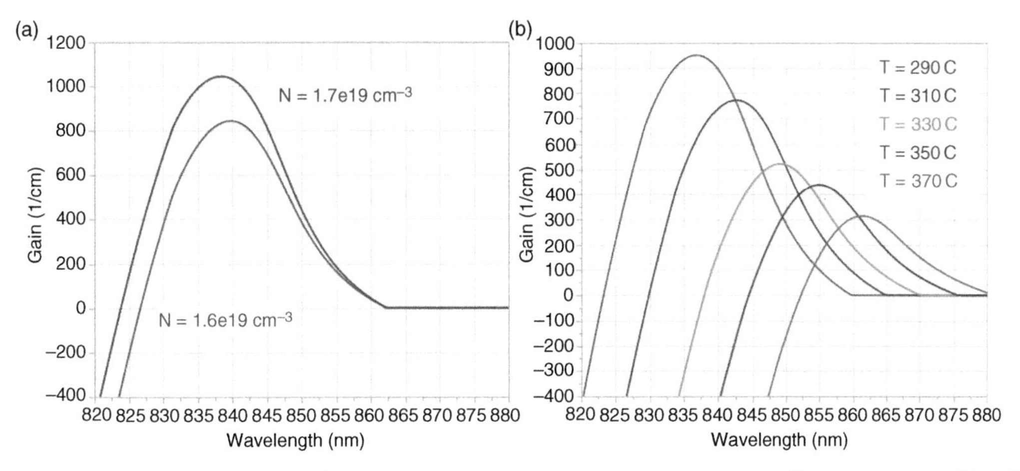

Figure 2.3 (a) Optical gain (cm^{-1}) for a 60 Å GaAs quantum well with N = 1.6 × 10^{19} and N = 1.7 × $10^{19}/cm^3$ carrier injection as a function of wavelength. (b) Optical gain over temperature for N = 1.6 × $10^{19}/cm^3$.

lasers could only be operated pulsed. The reflectivity can be enhanced by adding optical coatings to the facets but at the cost of manufacturing complexity. Other advances in the mirrors for edge-emitting lasers (EELs) include the use of DBR gratings and distributed feedback (DFB) gratings. The challenges presented to make better mirrors in an EEL lie primarily in the device fabrication and not in the epitaxial growth. VCSELs utilize highly reflective DBR mirrors to provide optical feedback to the QW active region. One of the key differences between an EEL and a VCSEL is that the mirrors can be grown directly on the material during the epitaxial process. DBR mirrors consist of alternating layers of high- and low-index materials and in a VCSEL are most commonly formed by varying the mole fraction x in $Al_xG_{1-x}As$. The index of refraction, n, varies with x as $n(x) = 3.3 - 0.53x + 0.09x^2$ for wavelengths above the bandgap. A simplified schematic of a DBR structure is shown in Figure 2.4. In this example, the optical energy starts in the active substrate with index of refraction n_{SUB} and is propagating to the right. At each material interface, there is a reflection of a portion of the incident energy. To calculate the total reflection of the structure, the reflection magnitude and phase must be summed. A complete mathematical treatment of a DBR can be found in many textbooks [2, 3]. The DBR is designed to operate at the Bragg wavelength, $\lambda_{BRAGG} = 2n_1d_1 + 2n_2d_2$, where n and d are the respective refractive index and thickness of layer i (i = 1, 2).

One period of the DBR is defined as one pair of material 1 and 2. Neglecting any optical loss, the total power reflection at λ_{BRAGG} for m periods with $n_1 < n_2$ can be expressed as

$$R = \left[\frac{1 - \left(n_1 / n_2 \right)^{2m}}{1 + \left(n_1 / n_2 \right)^{2m}} \right]^2 \tag{2.1}$$

A VCSEL requires highly reflective mirrors to operate, typically > 99% reflectivity for both sides of the structure. Figure 2.5a is a plot of the power reflectivity of a typical DBR mirror for an 850 nm VCSEL composed of $Al_{0.15}Ga_{0.85}As/Al_{0.85}Ga_{0.15}As$ mirrors. To achieve the required reflectivity, between 30 and 40 mirror periods are needed. Figure 2.5b shows the maximum power reflectivity at the Bragg wavelength as a function of the number of mirror pairs.

The requirement to make highly reflective mirrors is one of the primary challenges in manufacturing VCSELs. The mirror must be carefully designed for high reflectivity but also generally needs to conduct current to the active region and carry heat away from the active region. The balance between reflection, electrical conductivity, thermal conductivity, and optical absorption on one hand and epitaxial growth complexity on the other remains one of the key design and

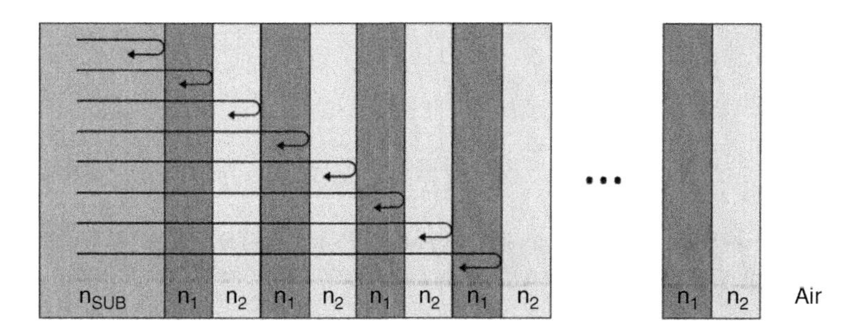

Figure 2.4 Simplified schematic of a DBR mirror showing the reflections from the material interfaces.

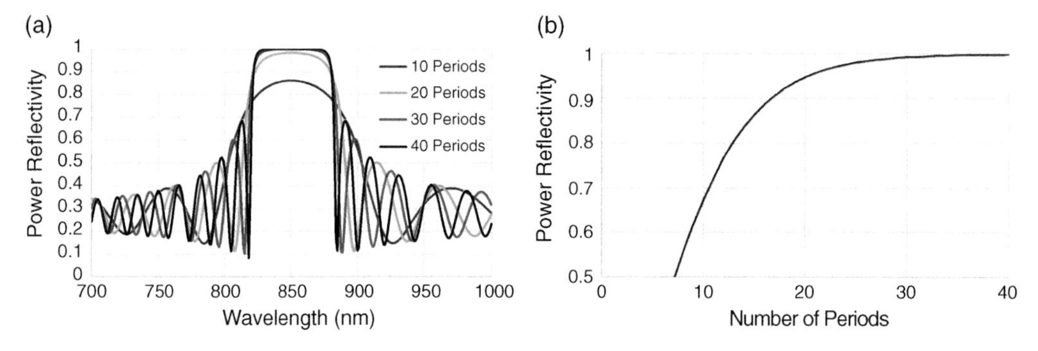

Figure 2.5 (a) Power reflectivity for 10, 20, 30, and 40 periods of a DBR mirror with $Al_{0.15}Ga_{0.85}As$ (n = 3.222) and $Al_{0.85}Ga_{0.15}As$ (n = 2.914). (b) Power reflectivity at the Bragg wavelength as a function of the number of periods.

manufacturing challenges in VCSELs. Advances in epitaxial growth technology since the first VCSELs have enabled the high-performance VCSELs available today.

2.5 Light Output Characteristic

The light output from a semiconductor laser is a function of the injected current and operating temperature. Figure 2.6 shows a typical light output characteristic as a function of the injection current. There are several salient features of this chart to expand upon.

As the current is increased, there is a modest increase in optical power until the threshold current (I_{TH}) is achieved. At this point, the gain in the optical cavity equals the sum of all of the electrical and optical losses. Prior to this current, light from the VCSEL is incoherent, spontaneous emission. Above I_{TH} stimulated emission occurs, and the emission from the VCSEL becomes coherent. As the current increases, the optical power increases linearly, and the change in optical power with current is called the slope efficiency (η) and measured in watts per ampere. As the current is further increased, the optical power saturates and begins to decrease as the current is further increased. This is known as rollover, and the current and power this occurs at are denoted as I_{MAX} and P_{MAX}, respectively. Power rollover happens in a VCSEL when the temperature of the active region pushes the gain characteristic (Section 2.2) past the maximum value at the lasing wavelength. To maximize the reliability (Appendix E), VCSELs are generally operated in the linear regime of the power characteristic, in this example less than 8 mA. The values shown in the graph are typical of a VCSEL used in data communications applications as further described in Chapter 3.

2.6 Forward Voltage Characteristic

VCSELs are PN junction diodes that are operated under forward bias. Figure 2.7 shows a typical forward voltage as a function of current in a VCSEL. The series resistance (R_S) is defined as the change in voltage with change in current. One major difference in a VCSEL versus other laser devices is that current flows through the DBR mirrors. The doping in the mirrors is kept relatively

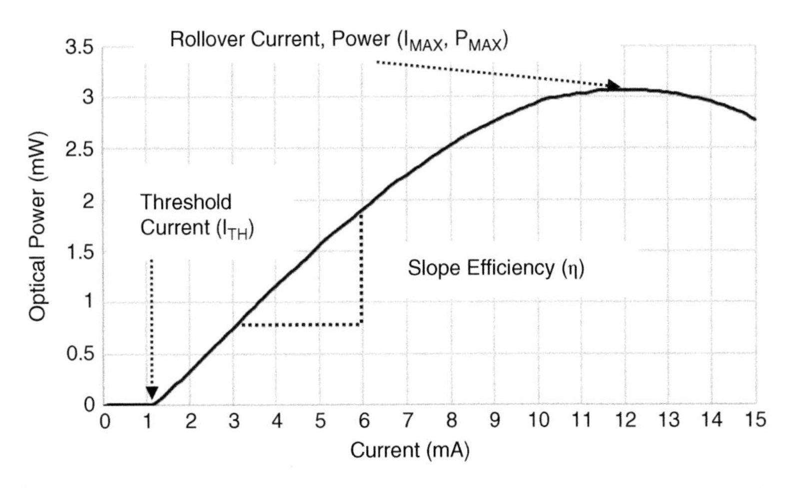

Figure 2.6 Light output as a function of current with the threshold current, slope efficiency, and rollover current and power defined.

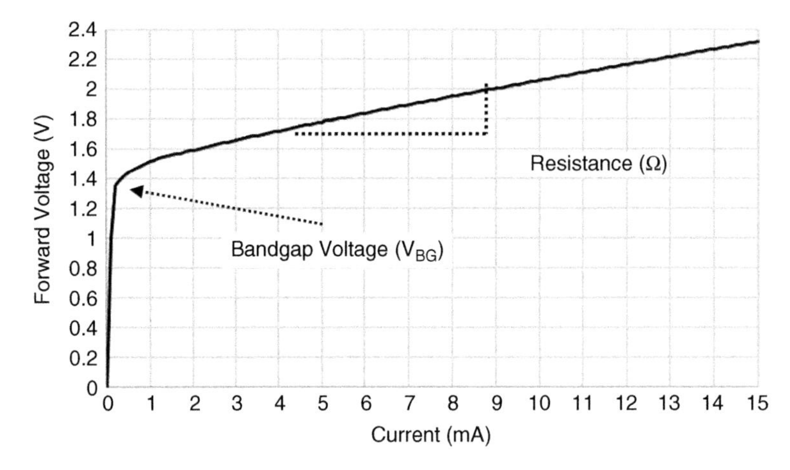

Figure 2.7 Forward voltage as a function of current with the series resistance and bandgap voltage defined.

low to reduce the optical absorption, and each of the heterojunction interfaces that define the DBR add to the total series resistance. To conduct current through the active region, the bias voltage must exceed the bandgap of the PN junction, and extrapolating the resistance to 0 current, the bandgap voltage (V_{BG}) can be determined. The VCSEL operating wavelength (λ) is approximated by $\lambda = 1.242/V_{BG}$.

2.7 Optical Modes

The optical modes of a VCSEL are defined by both the vertical cavity and the transverse extent of the oxide aperture. The mode spacing, $\Delta\lambda$, of an optical cavity is given by $\Delta\lambda = \lambda^2/(2n_{AVG}L_C)$ where n_{AVG} is the average index of the cavity, and L_C is the cavity length. Since the roundtrip phase of the electric field must be maintained, meaning that $n_{AVG}L_C$ must be an integer, the cavity longitudinal mode spacing reduces to $\Delta\lambda = \lambda/(2M)$ where M is an integer greater than 1. Figure 2.8 is a plot of the vertical electric field in an 850 nm VCSEL structure.

To determine the cavity length, it is necessary to include the penetration depth of the electric field into the n and p DBR structures in addition to the active region thickness. Analog expressions

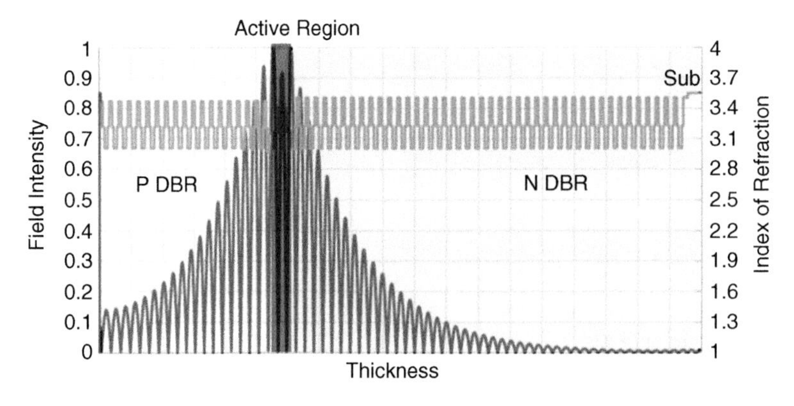

Figure 2.8 Plot of the electric field and index of refraction of a typical VCSEL.

for the penetration depth were derived [4] but can be empirically estimated form the graph. The penetration depth is defined as the point where the electric field has decayed to $1/e^2$ (13.5%) of its peak value. We know that each period of the DBR is $\lambda/2$, so we can estimate the penetration in both the P and N side to be approximately 10 and 8 periods (5λ and 4λ), respectively. The active region is usually 1λ, so the total cavity length is approximately 10λ. For an 850 nm VCSEL, the longitudinal mode spacing is 85 nm. This wide separation is well outside the DBR reflectance band (Figure 2.5a) and the optical gain needed to establish lasing. In contrast, a typical Fabry-Pérot (FP) edge-emitting laser may have a cavity length of hundreds of microns with $\Delta\lambda_{FP} \sim 1$ nm leading to multiple longitudinal mode operation. Figure 2.9 shows the optical emission spectrum of a single-mode VCSEL. Note that there is a second peak in the optical spectrum that is 40 dB below the primary peak. The difference in peak height is referred to as the side-mode suppression ratio (SMSR) and is a common figure of merit for single-mode VCSELs.

VCSELs also support higher-order transverse modes in the optical cavity when the lateral extent of the opening in the oxidation region is greater than approximately 3 μm. The second (lower) peak in Figure 2.9 is from a second transverse mode in the VCSEL. In a round VCSEL aperture, these mode solutions are described as Laguerre-Gaussian and in rectangular VCSEL apertures as Hermite-Gaussians. Solutions to these modes can be found in numerous textbooks [5, 6], and the reader is referred to these more complete treatments. Each of the transverse modes will operate at a slightly different wavelength and can be observed in both the optical spectrum and the near-field image of the VCSEL aperture. The previous discussion has focused on a single transverse electric (TE) mode of the VCSEL cavity. Transverse magnetic (TM) modes can also be supported in the cavity, and because of the difference in IOR of the polarization modes, there can be polarization degenerate modes in the cavity. Figure 2.10a shows the measured optical spectrum of a 25 Gbps VCSEL on a log scale to demonstrate the presence of the several optical mode types. In an oxide-confined VCSEL the longest wavelength mode is the fundamental, or lowest-order, transverse mode as labeled in Figure 2.10a. The higher-order transverse modes are present at shorter wavelengths, and the mode spacing is related to the size and placement of the oxide aperture. Closer examination of the peaks shows a slight separation within the major peaks that are indicative of polarization splitting (TE/TM) of the transverse mode and is also indicated in Figure 2.10a. The emission spectra is most often characterized by fitting a Gaussian function to the measured data. The fit excludes any emission that is 20 dB below the maximum power level. Figure 2.10b shows the same optical spectrum as 2.10a on a linear scale and includes the Gaussian fit to the spectrum. A common figure of merit is the standard deviation, σ, of the fit and is referred to as the RMS

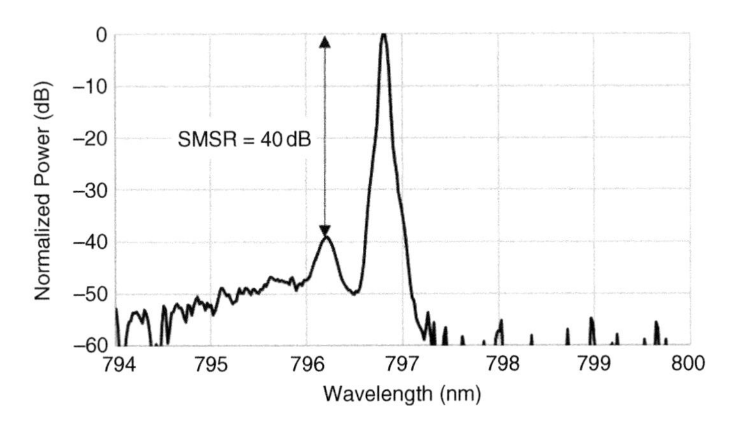

Figure 2.9 Optical emission spectra from a single-mode VCSEL showing SMSR = 40 dB.

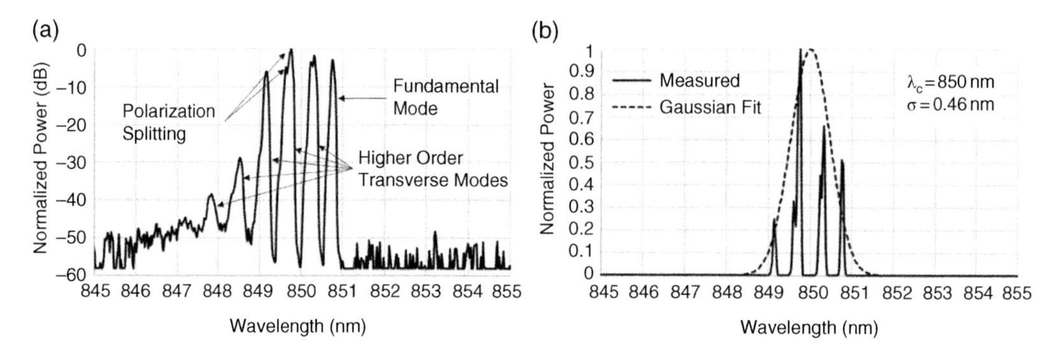

Figure 2.10 (a) Optical emission spectrum of a multimode oxide isolated VCSEL showing the several types of optical modes. (b) Emission spectrum on a linear scale with the Gaussian fit indicating the center wavelength and spectral width.

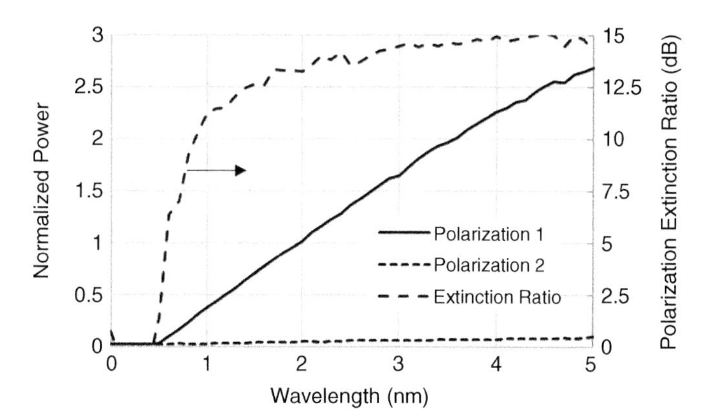

Figure 2.11 Measured L-I characteristic of a polarization-controlled VCSEL with PER = 15 dB. Note that the PER is limited by the polarizer used in the measurement.

spectral bandwidth. The mean value of the fit is often referred to as the center wavelength λ_C. The spectral content of a VCSEL can be a source of noise in optical communications systems, and its effects will be discussed in more detail in Chapter 3.

To control the polarization modes, some form of polarization-dependent loss or gain must be included in the VCSEL design. There are many methods to introduce polarization-dependent loss or gain, which have been reviewed in detail [7]. Control of the VCSEL polarization is a requirement in many applications as further described in Chapter 8. Figure 2.11 shows the L-I characteristic of a polarization-controlled single-mode VCSEL. The ratio of power in the orthogonal polarization direction is referred to as the polarization extinction ratio (PER), PER $= 10 \times \log(P_1/P_2)$, where P_1 and P_2 are the powers in the two polarization directions.

2.8 Beam Divergence

One of the drawbacks of EELs is the laser output beam is elliptical, highly divergent, and astigmatic, which requires more complex optical components to collimate or focus. VCSELs far-field profiles are generally circularly symmetric, have relative low divergence, and do not have astigmatism. Single transverse mode VCSELs have near Gaussian beam profiles and generally lower

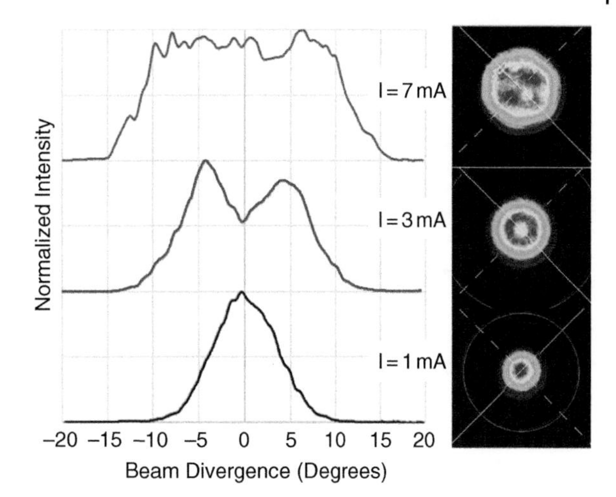

Figure 2.12 Measured far-field beam profile of a 25 Gbps VCSEL at I = 1, 3, and 7 mA. The profiles shown are a cross section view of the full two-dimensional data shown on the right side of the figure.

divergence when compared to multi-mode VCSELs. The near- and far-field profiles of a multi-mode VCSEL are shown in Figure 2.12. Single-mode VCSELs generally have emission shapes as shown in the 1 mA case in Figure 2.12.

2.9 Modulation Characteristics

The first application that pushed VCSELs into high-volume production was data communications, and their application in this area will be discussed in more detail in Chapter 4. A rate equation model for a laser can be solved in steady state and perturbation analysis applied to determine the small signal bandwidth. This analysis is useful to compare and characterize VCSEL designs and can be related to the large signal response required in a communication system. A more detailed analysis can be found in several textbooks [3, 8]. The small signal response of a VCSEL can be written as

$$f_r^2 = \frac{v_g\left(dg/dN\right)}{V_m} \cdot \frac{\eta_i}{q}\left(I - I_{th}\right) = \left[\frac{v_g}{qL_{cav}}\right] \cdot \eta_i \cdot \left(dg/dN\right) \cdot \left(J - J_{th}\right), \tag{2.2}$$

$$f_d = \Gamma f_r^2 + \gamma_0, \tag{2.3}$$

$$\Gamma = 2\pi\left[\tau_p + \varepsilon\chi / \left(v_g\left(dg/dN\right)\right)\right], \tag{2.4}$$

where f_r is the resonance frequency, ν_g is the group velocity (c/n_{AVG}), q is electron charge, η_i is the internal carrier injection efficiency, dg/dN is the change in gain with current at a fixed energy, J is the current density, J_{TH} is the threshold current density, γ_0 is the damping constant, f_d is the damping frequency, Γ is the damping coefficient, t_p is the photon lifetime, ε is the gain saturation parameter, and χ is the carrier transport factor. The VCSEL modulation transfer function is modeled as a two-pole filter and can be written as

$$H(f) = \left| A\left(\frac{f_r^2}{f_r^2 - f^2 + j\left(\frac{f}{2\pi}\right)\Gamma}\right)\left(\frac{1}{1 + j\left(\frac{f}{f_p}\right)}\right)\right|^2, \tag{2.5}$$

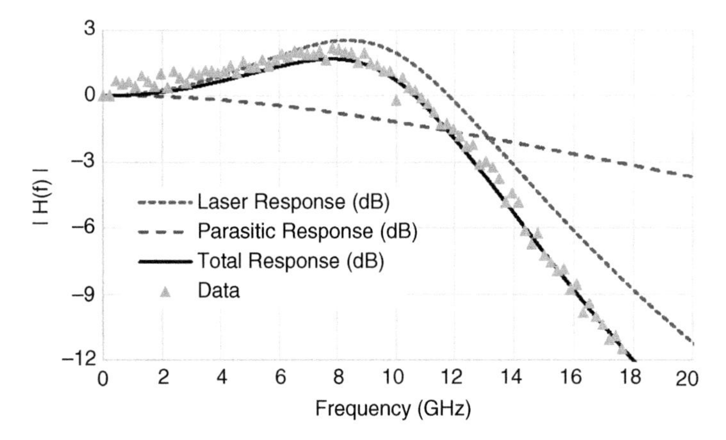

Figure 2.13 Measured modulation transfer function (triangles), the intrinsic laser response (dotted line), the parasitic electrical response (dashed line), and the total response (solid line). The 3 dB bandwidth is 13 GHz.

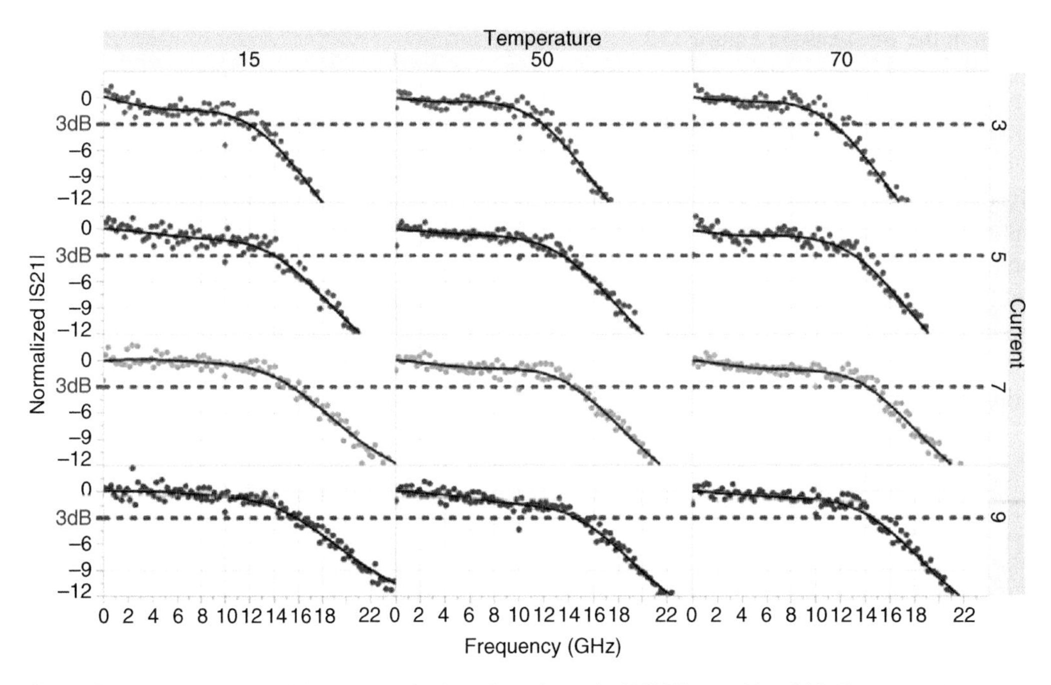

Figure 2.14 Measured modulation transfer function of a typical VCSEL capable of 25 Gbps operation.

where the electrical parasitic frequency $f_p = 1/(R_S C_T)$, and C_T is the total capacitance of the VCSEL. H(f) is known as the modulation frequency and can be measured directly as S21 when using a network analyzer. An example of measured H(f) for a VCSEL capable of > 10 Gbps modulation is shown in Figure 2.13 along with the derived fit with the intrinsic laser response and the parasitic electrical response separated. In this example, $f_r = 10.5$ GHz, $\Gamma = 57$ GHz, and $f_p = 19$ GHz.

Figure 2.14 shows the dependence of the modulation bandwidth on current and temperature for a typical VCSEL used in 25 Gbps operation. Note the VCSEL bandwidth is a function of the bias current and the operating temperature, and it is important to adjust the operating point to maintain consistent operation.

These equations are the basis for designing and optimizing VCSELs for high-speed operation. The parameters in these equations are primarily through the QW and DBR design and are balanced with the other desired performance parameters and operating lifetime. In data communications networks, lifetimes in excess of 10 years at maximum rated current and temperature are required. More details on reliability modeling can be found in Appendix E.

2.10 Temperature Characteristics

All of the previously described operating characteristics will vary as a function of temperature and need to be included in the application design. Many VCSELs are co-packaged with a power-monitoring photodiode that can be used to provide feedback to control the power over temperature by adjusting the drive current. Figure 2.15 shows a typical set of power and voltage characteristics as a function of the drive current for temperatures ranging from 15°C to 125°C.

Operation at temperatures below −20°C are often required, and parametric extraction can be used to predict the performance at lower temperatures. Figure 2.16 shows the extracted parametrics from the power and voltage characteristics along with simple extrapolation fits to operation at −20C.

The temperature dependence of a VCSEL can be modeled with a few simple equations and can even be put into a SPICE model for application development. The power from a VCSEL as a function of temperature (above threshold current and before rollover) is given by $P_{OUT}(T) = \eta(T)(I - I_{TH}(T))$, where the temperature dependence of the slope efficiency relative to 25°C is $\eta(T) = \eta_{25}(1 - d\eta dT(T - 25))$ and the temperature dependence of the threshold current relative to 25°C is given by $I_{TH}(T) = I_0(1 - A(T - T_0)^2)$. The new parameters I_0 and T_0 are the threshold current and temperature at which the threshold current is at the minimum of the parabola, and A is a fitting parameter. It is typically for commercial VCSEL data sheets to specify the values of dηdT and T_0 with typical values being −0.4%/°C and 40°C, respectively. These parameters are specific to the VCSEL design. The forward voltage can be expressed as $V(T) = V_{BG}(T) + IR_S(T)$. Here the bandgap voltage is a material parameter and changes at approximately 2.5 mV/°C, and the series resistance normalized to the 25°C value is $R_S(T) = R_{25}(1 - dRdT(T - 25))$ with typical values for dRdT being −0.3%/°C. Finally, the optical wavelength can also be modeled as a linear variation with temperature as $\lambda(T) = \lambda_{25} + d\lambda dT(T - 25)$ with a typical value of 0.06 nm/°C. In recent years, much more detailed SPICE and ADS models have been created from the measured values of a VCSEL, but these simple equations are often sufficient.

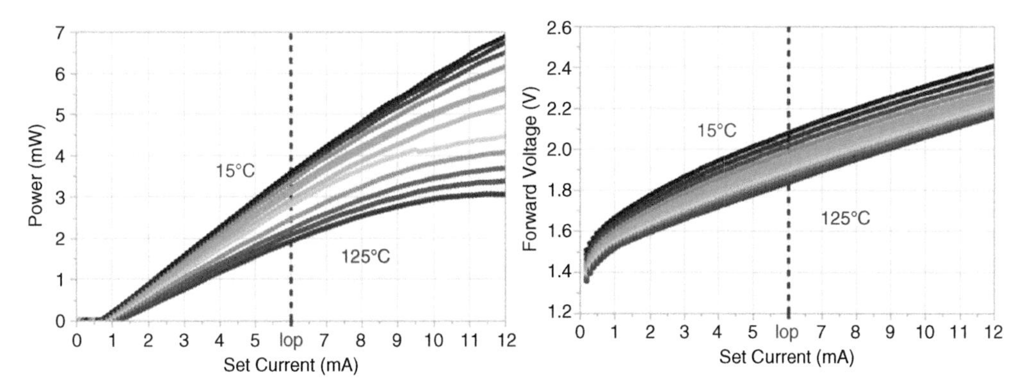

Figure 2.15 Light output and forward voltage as a function of current at temperatures from 15°C to 125°C.

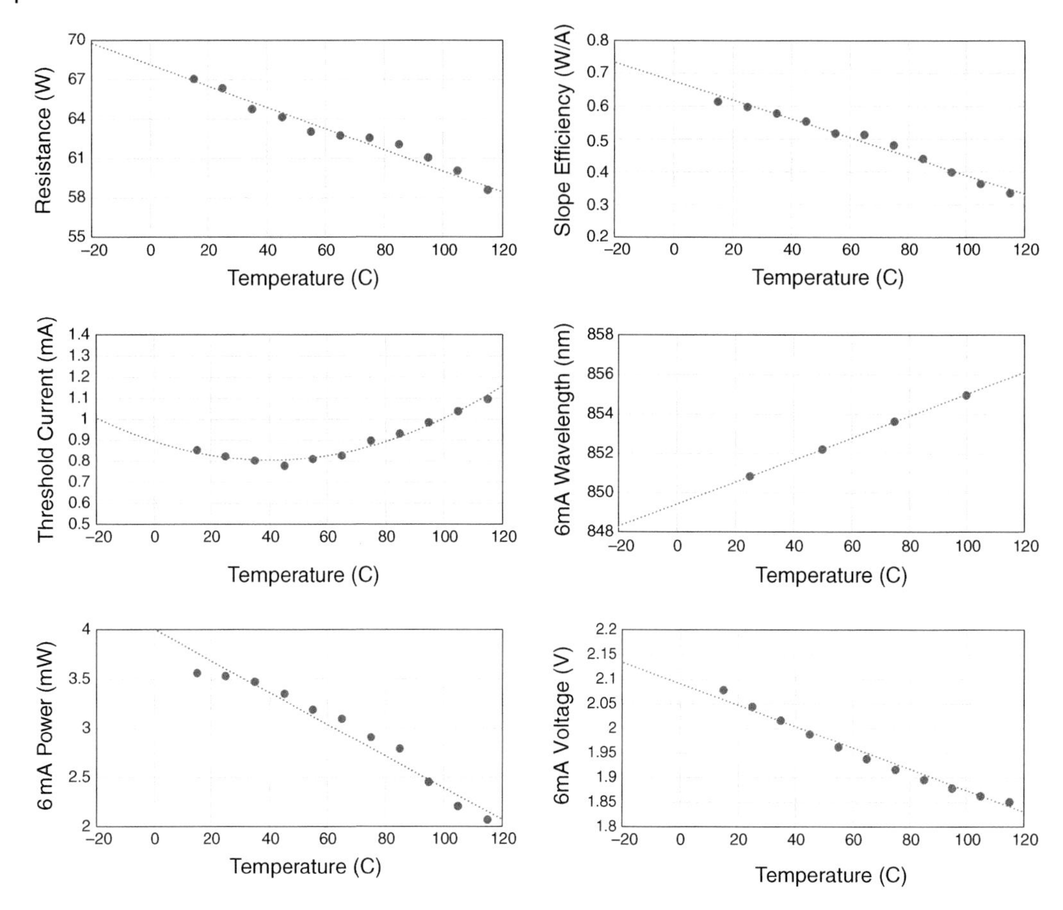

Figure 2.16 Extracted performance metrics from Figure 2.15 as a function of temperature.

2.11 Thermal Transient Behavior and Short-Pulse Operation

An operating regime of particular interest in 3D sensing is the response of the VCSEL to short current pulses at low-duty cycles. As described in Section 2.5, the maximum optical power is limited by the thermal response of the VCSEL. One way to increase the peak optical power is to drive the VCSEL with optical pulses that are shorter than the thermal time constant of the VCSEL and using a low-duty cycle, thereby reducing the overall temperature rise of the VCSEL. The thermal rise of a VCSEL active region during a pulse can be written as

$$\Theta_{\mathrm{MAX}} = \Theta_{\infty} \exp\left[\frac{-\left(T - \tau_{PULSE}\right)}{\tau_{THERMAL}}\right], \tag{2.6}$$

where Θ_{MAX} is the maximum temperature during the pulse, Θ_{∞} is the temperature of the active region under continuous wave (CW) operation, T is the pulse period, τ_{PULSE} is the electrical pulsewidth, and τ_{THERMAL} is the thermal time constant of the VCSEL active region. The current I_{MAX} that will result in the junction heating to Θ_{MAX} can be written as

$$I_{\mathrm{MAX}} = 2A_{j}\left(\frac{\Theta_{\mathrm{MAX}}\sqrt{\pi\kappa\gamma c}}{\rho\sqrt{\tau_{\mathrm{PULSE}}}}\right)^{r}, \tag{2.7}$$

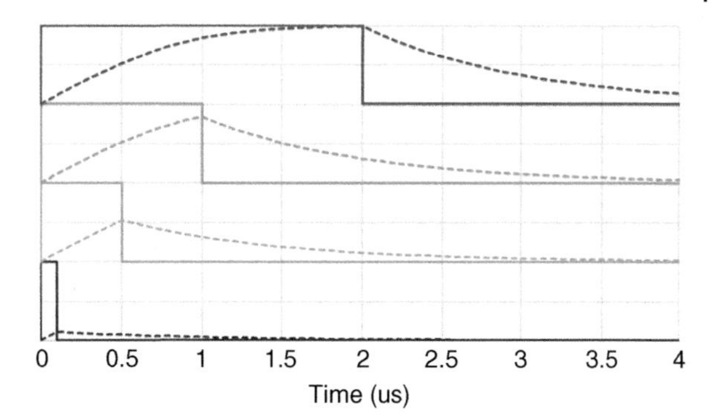

Figure 2.17 Thermal response of a VCSEL active region (dashed lines) to electrical current pulses (solid lines) of equal magnitude. The thermal time constant was set to 1 μs for this calculation.

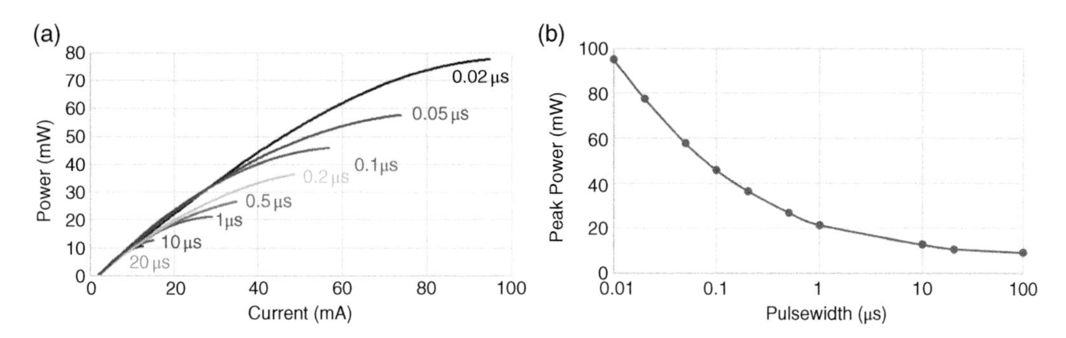

Figure 2.18 (a) Light output as a function of current for pulsewidths ranging from 0.02 to 20 μs. The duty cycle was held at a constant 1% for this measurement. (b) Peak optical power as a function of the pulsewidth.

where A_j is the area of the VCSEL junction, κ is the thermal conductivity, ρ is the thermal resistivity, γ is the specific mass, c is the specific heat, and r is an empirically derived parameter for GaAs. Figure 2.17 shows the thermal response of a VCSEL active region (dashed line) to current pulses ranging from 100 ns to 2 μs (solid lines) for equal maximum current pulse magnitude. The shorter pulses show less temperature rise in the VCSEL active region.

The reduction in temperature rise of the VCSEL can be used to increase the peak power emitted from the VCSEL. Figure 2.18a shows the L-I characteristic for a VCSEL driven with several different pulsewidths at a constant pulse duty cycle of 1%. The peak power obtained can be significantly higher than P_{MAX} measured under CW drive conditions, as indicated in Figure 2.18b.

The ability to achieve a huge increase in the peak power by managing the thermal response of the VCSEL active region is critical to 3D sensor operation and in particular for time-of-flight methods, as will be described in Chapter 5.

2.12 Other VCSEL Structures

The basic VCSEL described in Figure 2.1 has been the primary structure used in commercial devices. There are many variations on the design that have been produced, but all have the general elements of an active region and mirrors. When beginning the design process of a VCSEL, there are several basic choices to be made. The first requirement is often the wavelength, and this will dictate the choice of material systems as described further in Section 2.13. The design choices to be made next include the direction of laser emission, the type of optical gain material, the type of

Table 2.1 Design choices for VCSEL sections.

Emission Direction	Mirror Type	Active Region	Electrical Guiding	Optical Guiding
epi side	semiconductor	quantum well	proton implant	none
substrate side	dielectric	quantum dot	oxidation	oxide
	high-contrast grating	type II QW	bandgap (regrowth)	bandgap (regrowth)
	grating	quantum cascade		photonic crystal
	movable mirror	tunnel junction		
	external cavity			

mirrors, the method of confining electrical carriers, and finally the method of confining the optical radiation. Table 2.1 is a summary of the many possible design choices available for these several options. Some choices may be mutually exclusive from each other, but many can be readily mixed. In practice, the most common variations are in the mirrors, and Figure 2.19 shows some of the possible combinations.

The first design choice for the VCSEL structure is the direction for the light output. If the substrate is not transparent to the operating wavelength, light will almost always be taken from the

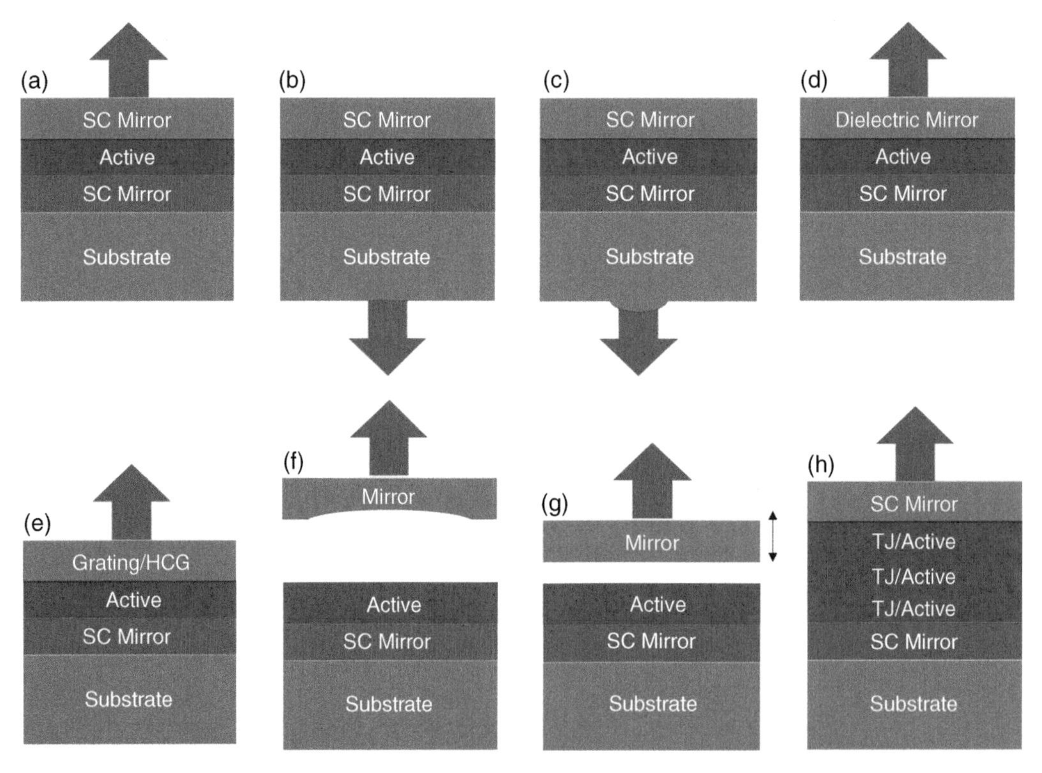

Figure 2.19 (a) Common top-emitting VCSEL structure. (b) Bottom-emitting VCSEL structure. (c) Bottom-emitting VCSEL structure with lens. (d) Top-emitting VCSEL structure with dielectric mirrors (common for longer wavelength VCSELs). (e) VCSEL with top mirror formed with a grating or high-contrast grating structure. (f) A vertical extended cavity surface emitting laser (VECSEL) with an external mirror typically used for high power. (g) VCSEL with an external mirror that can be adjusted to tune the VCSEL wavelength. (h) A VCSEL with multiple active regions connected by tunnel junctions.

epitaxial side (often referred to as the top side). Back side emission with absorbing substrates is possible, but substrate removal is complex and results in extremely thin material that can be delicate to handle. Designing the VCSEL to emit from the substrate enables flip chip bonding of the VCSEL to an electrical connection or allows the addition of optical components such as lenses, gratings, or diffusers to the substrate [9]. (Note top-emitting VCSELs may be flip-chip mounted to transparent substrates such as glass.)

The second major design decision involves the choice of mirror technology. The most common construction for VCSELs in the 650–1200 nm range is to use epitaxially grown semiconductor mirrors. VCSELs in the 1200–1600 nm do not have a good material that is lattice matched to InP to make the mirrors. The mirrors suffer from low index contrast and absorption. One method to address this issue is to utilize a DBR mirror made from dielectric materials such as silicon nitride, silicon dioxide, or titanium oxide. The index contrast can be quite high and may require only a relatively few periods to achieve high reflectivity. The primary drawback is the potential mechanical stress of the dielectric on the semiconductor material may impose some operational limitations. If polarization control of the optical emission is a requirement, a grating may be used in conjunction with a partial semiconductor mirror [10]. Gratings can be made in either dielectric or semiconductor and typically have a period on the order of the wavelength. They provide polarization-dependent reflection and can replace multiple periods of semiconductor DBR. Another type of grating known as a high-contrast grating (HCG) has dimensions much less than the optical wavelength and is typically made in metal. HCGs can be very broadband reflectors with polarization selectivity but are difficult to fabricate [11]. When very high power and single-mode operation are required, the mirror can be formed with an external optical component. This class of lasers is often referred to as vertical-external-cavity surface-emitting lasers (VCSELs) [12]. Finally, if broad or rapid tunability of the laser wavelength is a requirement, then an external mirror that is moveable with electrostatic control has been realized [13]. These lasers are useful in optical coherence tomography (OCT, Chapter 8) and in wavelength division multiplexing (WDM, Chapter 4) systems.

The third major choice for the VCSEL designer is the type of active region needed to provide the required optical gain. QW active regions are by far the most prevalent, but alternatives such as quantum dots, type II QWs, and quantum cascade devices have been proposed and, in some cases, realized [14, 15]. For longer wavelength VCSELs, the active region may be grown on InP substrates and then wafer bonded to AlGaAs-based mirrors [16]. The primary objective to these alternate optical gain regions is to extend the operating wavelength, or to extend the laser tunability. For example, type II QWs have been utilized to make VCSELs with wavelengths > 3 μm and extended even further to > 5 μm using quantum cascade structures [15]. The main disadvantage to these types of active regions is the lower optical gain compared to QWs. In high-power VCSELs, where a very high slope efficiency may be desired, tunnel junctions have been used to stack several active regions back-to-back to recycle the carriers [17]. The resulting device has a much higher operating voltage but has a higher overall efficiency because the resistance of the DBR mirrors is only encountered once by the carriers.

The fourth major design choice is how the electrical carriers will be injected and confined to the active region. Some of the earliest VCSELs used a simple etch to define the active region, but this has been abandoned due to the impact on reliability and relatively high surface recombination current and scattering losses of the etch [18]. The first commercial VCSELs used proton implantation to guide current by reducing the conductivity of semiconductor mirrors [19]. Oxidation of AlAs was first introduced to VCSELs in the mid-1990s [18] and has since become the primary method of directing carriers to the active region. A final choice for electrical guiding is bandgap isolation [20], a technique often used in EELs. The primary disadvantage of this method is the requirement for a partial VCSEL growth, processing steps, and then regrowth of the remaining VCSEL structure.

Table 2.2 Semiconductor material choices for VCSEL active regions and mirrors.

Wavelength range	λ < 500 nm	600 nm < λ < 1300 nm	1300 nm < λ < 1800 nm	λ > 1800 nm
Substrate	GaN	GaAs	InP	GaSb
Active Region	$In_xGa_{1-x}N$ $Al_xGa_{1-x}N$	$Al_xGa_{1-x}As$ $In_xGa_{1-x}As$ $In_xGa_yAl_{1-x-y}As$ $In_xGa_{1-x}P$ $GaAs_xP_{1-x}$ $In_{1-x}(Al_yGa_{1-y})_xP$	$Al_xGa_yIn_{1-x-y}As$ $In_xGa_{1-x}As_yP_{1-y}$	$Al_xGa_{1-x}Sb$ $In_xGa_{1-x}As_ySb_{1-y}$
Mirrors	$Al_xGa_{1-x}N$	$Al_xGa_{1-x}As$	$Al_xGa_{1-x}As$	$Al_xGa_{1-x}As_ySb_{1-y}$

The final major design selection, optical confinement, is coupled with the choice of electrical confinement. The most common method in production today is the oxide aperture, and the design thickness, placement, composition, and growth profile can all have profound effects on the VCSEL performance. Ion implantation has very little optical guiding, and as mentioned previously, the air post VCSELs have generally high optical losses. The use of semiconductor regrowth isolation for the electrical confinement can also act as optical confinement [20]. Photonic crystal structures have also been used to separate the optical and electrical confinement and have been used to make large-aperture single-mode devices [21].

The design choices and processes presented in this section represent the majority of possible opportunities but are certainly not an exhaustive list, and more options are being developed to meet specific engineering requirements for the VCSEL.

2.13 VCSEL Materials

Commercial VCSELs are generally made with direct bandgap group III and group V semiconductor materials and emit in the 650–1550 nm range as summarized in table 2.2. Research and development continue on both longer and shorter wavelengths and in some cases involve the use of group II and group VI elements. The table 2.2 summarizes the most often used materials in making VCSEL active regions and semiconductor mirrors. (Note, as discussed in Sections 2.11 and 2.12, there are many other potential options for the mirrors.)

2.14 Summary

This chapter has provided a review of the basic components of a VCSEL along with the fundamental VCSEL characterization parameters. Insight into the design selection process for the VCSEL materials and the various options for the optical gain, the mirrors, output direction, and the choice of optical and electrical confinement has been reviewed. The design and manufacture of VCSELs is a complex multi-physics problem that requires engineering trade-offs in the complexity and operating characteristics of the VCSEL. Many of the applications that will be further described in the following chapters take advantage of the VCSEL operating characteristics, but still others are limited by them.

References

1 P. S. Zory, Jr. *Quantum Well Lasers*. Boston, Mass: Academic Press, 1993.
2 G. R. Fowles, "*Introduction to Modern Optics*," 2, Dover Publications, Inc., New York, 1989.
3 L. A. Coldren, S. W. Corzine, and Milan Mashanovitch. *Diode Lasers and Photonic Integrated Circuits*. 2. Hoboken, N.J.: Wiley, 2012.
4 D. I. Babic and S. W. Corzine, "Analytic expressions for the reflection delay, penetration depth, and absorptance of quarter-wave dielectric mirrors," in *IEEE J. Quantum Electron.*, vol. **28**, no. 2, pp. 514–524 (1992)
5 A. Siegman, *Lasers*, University Science Books (1986)
6 E. A. Saleh, Bahaa, and Malvin Carl Teich. *Fundamentals of Photonics*. 2. Hoboken, N.J.: Wiley-Interscience, 2007.
7 B. D. Padullaparthi and F. Koyama, "A review on polarization control of vertical-cavity surface-emitting semiconductor lasers," Recent Patents on Electrical Engineering 4 (2) pp. 81–97 (2011).
8 S. F. Yu, "Analysis and Design of Vertical Cavity Surface Emitting Lasers: YU/Surface Emitting Lasers." (2005).
9 R. Carson, M. Warren, P. Dacha, T. Wilcox, J. Maynard, D. Abell, K. Otis, and J. Lott "Progress in high-power high-speed VCSEL arrays," *Proc. SPIE* **9766**, (2016)
10 R. Michalzik, "VCSELs: Fundamentals, Technology and Applications of Vertical-Cavity Surface-Emitting Lasers." (2012).
11 C. Chang-Hasnain, Y. Zhou, M. Huang and C. Chase, "High-contrast grating VCSELs," in *IEEE J. Sel. Top. Quantum Electron.*, vol. **15**, no. 3, pp. 869–878 (2009).
12 M. Herper, S. Gronenborn, X. Gu, J. Kolb, M. Miller, and H. Moench "VECSEL for 3D LiDAR applications," *Proc. SPIE* **10901** (2019).
13 V. Jayaraman, D. John, C. Burgner, M. Robertson, B. Potsaid, J. Jiang, T. Tsai, W. Choi, C. Lu, P. S. Heim, J. Fujimoto, and A. Cable "Recent advances in MEMS-VCSELs for high performance structural and functional SS-OCT imaging," *Proc. SPIE* **8934**, (2014).
14 G. Veerabathran, S. Sprengel, A. Andrejew, and M. Amann "Electrically pumped VCSELs using type-II quantum wells for the mid-infrared," *Proc. SPIE* **10536** (2018).
15 S. Grzempa, W. Nakwaski, and T. Czyszanowski "Designing principles of quantum-cascade vertical-cavity surface-emitting lasers," *Proc. SPIE* **10938** (2019).
16 A. Karim, P. Abraham, D. Lofgreen, Y. Chiu, J. Piprek, and J. Bowers. "Wafer bonded 1.55 μm vertical-cavity lasers with continuous-wave operation up to 105°C," *Appl. Phys. Lett.* **78**, 2632 (2001).
17 M. Dummer, B. Olson, K. Tatah, K. Johnson, M. Hibbs-Brenner "High efficiency multijunction VCSEL arrays for 3D sensing," *Proc. SPIE* **11300** (2020)
18 K. Choquette and H. Hou, "Vertical-cavity surface emitting lasers: moving from research to manufacturing," in *Proc. IEEE*, vol. **85**, no. 11 pp. 1730–1739 (1997).
19 R. Morgan, M. Hibbs-Brenner, "Vertical-cavity surface-emitting laser arrays," Proc. SPIE 2398 (1995).
20 L. Chirovsky, W. Hobson, R. Leibenguth, S. Hui, J. Lopata, G. Zydzik, G. Giaretta, K. Goossen, J. Wynn, A. Krishnmaoorthy, B. Tseng, J. Vandenberg and L. D'Asaro, "Implant-apertured and index-guided vertical-cavity surface-emitting lasers (I2-VCSELs)," in *IEEE Photon. Technol. Lett.*, vol. **11**, no. 5, pp. 500–502 (1999).
21 D. Siriani, M. Tan, A. Kasten, A. Lehman, P. Leisher, J. Raftery, Jr., A. Danner, and K. Choquette, "Mode control in photonic crystal vertical cavity surface emitting lasers and coherent arrays," *IEEE J. Sel. Top. Quantum Electron.* **15**. pp. 909–917 (2009).

3

VCSEL Industry: Prospects and Products

Babu Dayal Padullaparthi

3.1 Industry Background

The breakthrough demonstrations of (i) continuous wave (CW) operation of VCSEL at RT [1], (ii) monolithic fabrication of millions of etched microlasers on a tiny (~0.5 cm^2) chip [2], and (iii) current confinement of injection carriers through gain-guided structures using proton (H^+ ions) implantation [3], all set the foundation for large-scale VCSEL development. Furthermore, the invention of native oxidation of $Al_xGa_{1-x}As$ [4] and highly reflecting semiconductor layer stacks as DBRs [5] have further accelerated VCSEL development. All these research efforts have sparked worldwide industrial development and commercialization, as listed in the timeline in Table 1.2. Commercial VCSELs based on GaAs QWs with H^+ implantation as gain-guided structures were first introduced by Honeywell in 1996 for datacom applications [6]. This opened the gates for large-scale VCSEL manufacturing with dozens of companies following suit. The market segment for VCSELs was first presented in the early 2000s for commercial application in the gigabit Ethernet (GBE) standard [7, 8]. Due to the speed limitation of implanted VCSELs, native oxidation VCSELs with index guiding was quickly implemented to keep up with the speed demands of fiber channel and Ethernet (see Chapter 4). After over four decades of academic and industrial R&D, VCSELs are now the laser of choice as light transmitters for many applications, as shown in Table 1.2. Continuous improvements include higher bandwidth [9], narrow line width with low noise [10, 11], scalability to 2D arrays with high peak powers in pulsed operations [12, 13], epi-growth on large-sized wafers [14], better process uniformities [15], high reliability [16, 17, 18], and high single-mode powers [19]; these improvements have significantly changed the face of VCSEL communication and sensing applications. VCSELs are now present in many areas of science and engineering; these include areas such as medical, defense, aerospace environment, biometrics, photonics, electronics, consumer, and automotive industries.

As described in Figure 1.21, the strong demand for network bandwidth (capacity) in the exploding Internet traffic [20], 3D image detection and sensing in consumer electronics and automotive industries [21, 18], and future needs of high volumes of VCSELs for industrial applications [22] are the main factors that have propelled VCSELs to emerge as the primary optical component in the semiconductor laser industry. For example, Internet traffic is driven by development of large-scale data centers by companies such as Google, Facebook, Apple, Microsoft, IBM, Amazon, Intel, and Cisco, which have fueled the demand for high-speed VCSELs (Chapter 4). Moreover, the

VCSEL Industry: Communication and Sensing, First Edition. Babu Dayal Padullaparthi, Jim A. Tatum and Kenichi Iga.

adoption of VCSELs in iOS and android based gadgets/devices for proximity, illumination, and face-sensing functions (Chapter 5) has triggered a massive window (second wave after datacom bust) of opportunity in consumer electronics. This growth has created new demand for high-power VCSELs and OEMs (epi-houses, foundries) and development of internal capability by several LED makers around the world. Further, in self-driving vehicle sectors, innovations facilitated the use of VCSELs for imaging (scan and flash) LiDARs in the automotive industry (Chapter 6). While details about these topics will be covered in subsequent chapters, this chapter focuses on VCSEL (chip and module level) market size, its landscape, and a few examples of commercial products to aid the VCs and investors.

3.1.1 VCSEL Market

Driven by the total addressable market (TAM) projected for VCSELs in Section 1.5.4, this chapter discusses the core market prospects shared by datacom, consumer mobile, automotive LiDARs, and industrial heating that are established for mass manufacturing of VCSELs. The market for datacom transceivers is projected to grow to \$12.1 billion in 2025 from \$4.0 billion as of 2020, with a CAGR of 25%, and the CAGR for the combined TAM in the same period is 15%. The total market includes transceivers operating on both single- and multi-mode optical fiber and covers link distances from a few meters to many kilometers. The ratio of SMF to MMF transceivers is approximately 55 to 45. The market for mobile consumer (3D sensing) at module level is projected to grow to \$8.16 billion in 2025 from \$2.01 billion in 2019, with a CAGR of 26%. The use of LiDAR in the automotive segment (3D LiDARs) at module level is projected to grow to \$1.9 billion in 2025 from \$0.1 billion in 2020 with a CAGR of 80%. Finally, in the last major area of industrial heating, the market is projected to grow to \$2.0 billion in 2025 from \$0.8 billion in 2020, with a CAGR of 20%. The projected market growth for these segments is shown in Figure 3.1. Market share was a major driving force for semiconductor manufacturers to balance investment (risks) and business fortunes (returns).

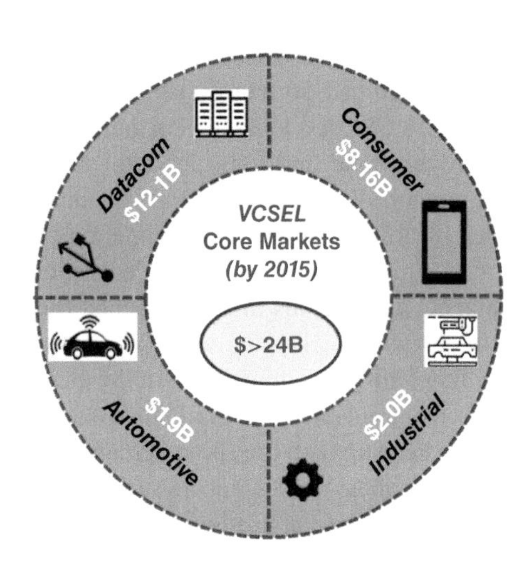

Figure 3.1 Core fields module-level VCSELs market projection for the next five years. *Source:* © Photonic Components DFM Ltd. *Sources:* Yole forecasts (2019–2020), Ouster Annual report-2020.

3.1.2 VCSEL Chip Demands

The forecasted growth of the several market segments is driving demand for VCSEL chips around the world for the next five years and even beyond. Each field and each product has its custom performance and physical dimension requirements that are dictated by either speed (bandwidth) or power (illumination distance); it is important to know the origins and drivers of requirements for at least the core fields.

The chip demands for datacom, consumer, automotive, and industrial fields, shown in Figure 3.2, have totally different origins. In datacom the key driver is the ever-increasing demand for high-bandwidth connectivity; this drives the number of data centers (DCs) that are built for the future.

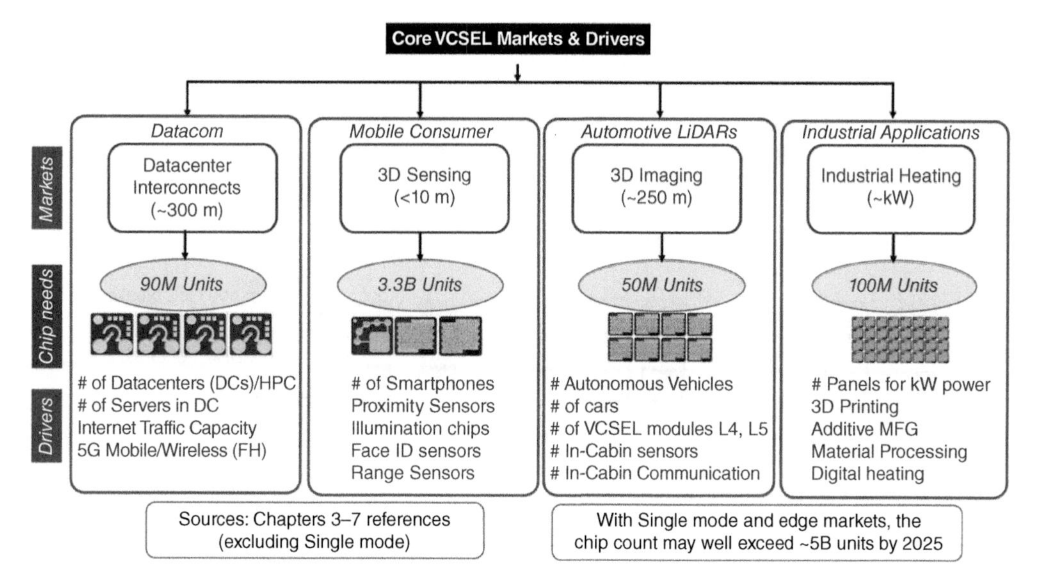

Figure 3.2 Chip demands of VCSELs for the next five years. *Source:* © Photonic Components DFM Ltd.

Most DC architectures have either three-layer or spine-leaf network topologies for their edge and core computing needs with racks, servers, switches, ports, and so forth. For example, in a three-layer topology, a DC has 1000 racks and 10K servers, use ~8K optical transceivers (OT) units for 40 Gb/s and ~800 OT units for 100 Gb/s. Similarly for a spine-leaf topology, the DC has the same number of racks and servers but uses ~40K OT units for 25 Gb/s and ~4K OT units for 100 Gb/s [23]. An example of spine-leaf architecture in DC considered by Cisco Systems is given in Figure 3.3. There are 1000s of operating data centers in the world now, and the number is expected to increase due to demand for data storage, cloud services, and overall Internet traffic, as described in section 4.4.3 The total number of OTs forecasted by Yole are 211M in 2025 including both datacom and telecom [24]. Datacom alone will need at least 94M OTs per annum in the next five years. This exact number of OTs are estimated by multiplication of the number of data centers, the number of OTs per DC, and the SM/MM share. The associated form factors, mode type, and distances are explained in Chapter 4.

In 3D sensing applications in consumer electronics, there are multiple types of VCSEL devices used in smartphones. These include proximity sensors, illuminator sensors, face recognition sensors, and short-distance LiDAR scanners (<10 m). The sensors can have front- and/or back-facing (i.e., looking at the world) configurations, as shown in Figure 3.4a and b. In short, every smartphone will use at least four VCSEL chips, each with different number of emitters and different physical dimensions.

As the number of smartphones projected by 2024 reaches 1.5 billion units [25], a simple estimate is there will be at least 6.0 billion VCSEL chips needed over the next four years. This whopping numbers leads to demand for over 600M 4″/6″ wafers in the next four years. This dramatic capacity expansion can't be handled by existing producers with 100s of MOCVD systems and process capability of a few 1000 wafers/week! This massive task has created new business opportunities for OEMS (device MFG and III-V foundries), equipment vendors, epi-houses, and so forth. Several mergers and acquisitions (M&As) took place to quickly capture the market share. In fact this is the precise reason for Apple's commitment to invest $380 million in Finisar in 2018 [26],

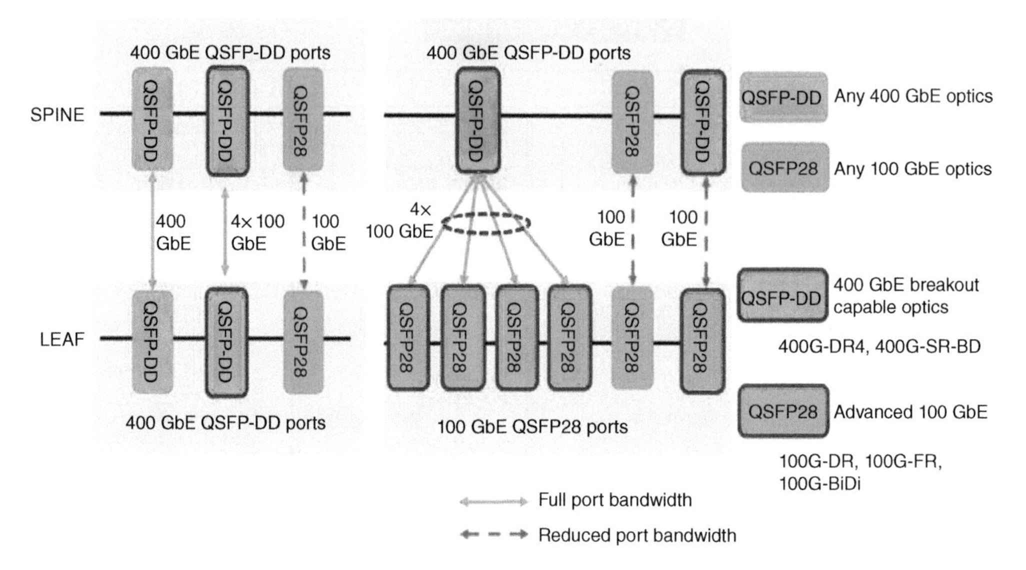

Figure 3.3 Spine-leaf DC architecture considerations in Cisco Systems. *Source:* Reprinted with permission from Cisco Systems, USA.

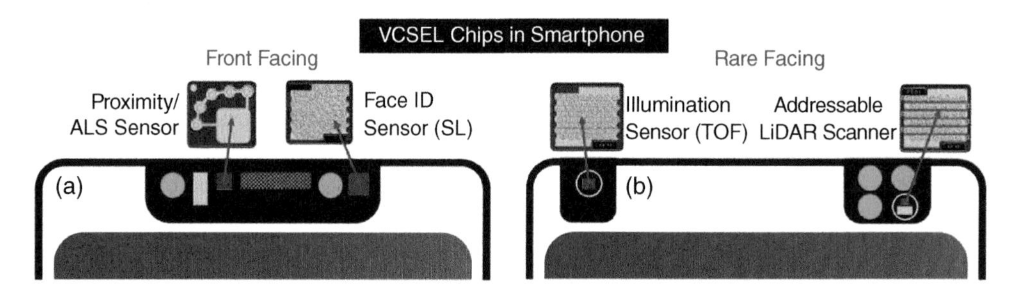

Figure 3.4 Typical (a) front- and (b) back-facing smartphone layouts. *Source:* © Photonic Components DFM Ltd.

AMS expanding their Singapore plant with a $200 million investment [27], IQE investing in a mega foundry at its Newport facility with advanced MOCVD machines and providing 6″ VCSEL wafer growth [28], and LED makers such as OSRAM and several Asian companies producers of LED and EEL establishing fabrication facilities, investing heavily in VCSEL mass manufacturing. More details of VCSEL optical sensors for distance (<10 m) are explained in Chapter 5.

Another area of VCSEL application is automotive LiDARs in autonomous vehicles (AVs). Unlike fixed wavelengths in datacom (850 nm) and 3D consumer mobile sensing applications (940 nm), the light sources for LiDARs are split among different wavelengths (NIR 808, 850, 905, 940 nm, and short-wavelength infrared [SWIR] 1550 nm) with three different light sources (VCSELs, EELs, and fiber lasers). Automotive LiDARs are aimed at long-range (~200 m or more) object detection while both light sources and objects are in motion. Precise reflection mappings (point clouds) from objects and background solar irradiance are crucial aspects to process the image; both are strongly dependent on weather condition and wavelength. For example, different wavelengths have different water absorption, and the reflectivity of light from objects varies with wavelength. Furthermore,

CMOS detector sensitivity, process compatibility, and eye safety issues make some preferred choices of LiDAR wavelengths 905 and 940 nm. Further, InP-based 1550 nm VCSELs provide eye safety, and their maximum permissible exposure (MPE) is several orders of magnitude higher than NIR counterparts (905 or 940 nm), but they are not in practical use with very expensive InGaAsP SWIR PDs. The exception is that 1550 nm based telecom-grade EELs (tunable DFBs, DBR lasers, or EDFA) with high-power densities are becoming popular for FWCM LiDARs and widely accepted in the automotive industry.

The number of LiDAR units projected by 2025 is 3.8 million and by 2030 is 19.3 million [29], derived from the number of vehicles (2.5M by 2030) on the road. The major attraction comes from NIR wavelengths 905 and 940 nm VCSELs that are directly competing with high power density F-P EELs. However, due to matrix- or individual zones–addressable nature of 2D arrays with moderate peak power densities (kW/mm^2), VCSELs are becoming attractive as popular light sources for long-distance ranging using flash/sequential flashes (electronic scanning) [30, 31], [32], and [33]. As each LiDAR unit will have at least 3–4 VCSEL modules at L4 and L5 levels of advanced driver assistance systems (ADAS), and each module will have at least 8–10 VCSEL dies, this will drive at least 50 million VCSEL chips by 2025 (under the assumption of 50% share of VCSELs). More details of VCSELs for automotive LiDARs are explained in Chapter 6.

Lastly in the industrial heating sector, 2D VCSEL arrays are used in large-scale heating modules or panels to produce IR radiation anywhere from a few W to few 100s of kW. Here, individually addressable VCSEL arrays use flood illumination (intense and narrow light often from VCSELs integrated with micro-lenses) and offer flexible, large-scale (digital) heating systems with power densities of more than 1 kW/cm^2 without the use of external optics. The number of VCSEL arrays depends on the total power of the module/panel for a given application. Normally, a few 1000s VCSEL emitters are used in one array chip and at least few 10s (if not 100s) of VCSEL array chips are needed for making a heating system. In addition to heating systems, 2D arrays of VCSELs are also used in night vision systems for security and surveillance applications. Thus, there is a massive need of 2D MM VCSEL arrays in future. Both of these sectors are explained in Chapter 7.

There was a forecast from Yole Développement predicting a requirement of a whopping 3.3 billion VCSEL chips for 2017–2023 [34]. This forecast is mainly focused on core markets that have major market shares and likely do not include neighboring-edge and single-mode market sectors, described in Figure 1.21. Overall, with core and edge markets, the authors predict that there will be much higher chip demands in the future due to exploding VCSEL applications and total chip count may reach 5 billion by 2025 and it may well exceed 10 billion units by 2030. These market dynamics are encouraging strong investments in VCSEL mass manufacturing and attracting new players that may dramatically change the existing VCSEL industry landscape.

3.1.3 VCSEL Attractiveness

During the mid-1990s to early 2000s, there were more than 30 companies for VCSEL commercial development and their decades-long investments couldn't generate the revenue from commercial products except optical mice. When it comes to the 2012–2016 period, only a handful of companies existed and continued their R&D investments in then-promising 16 Gb/s and 28 Gb/s high-speed VCSELs in datacom AOCs for 100 GBE networks. Later, high-bandwidth VCSEL pushed datacom to 200 G and 400 G networks including 5G fronthaul. But the situation dramatically changed when Apple announced iPhone X in September 2017 with 3 VCSEL products inside its smartphone, which opened a large window of opportunity. Traditional, older companies looked for M&As to capture the market share of 3D sensing (consumer mobile) and automotive LiDARs. Interestingly,

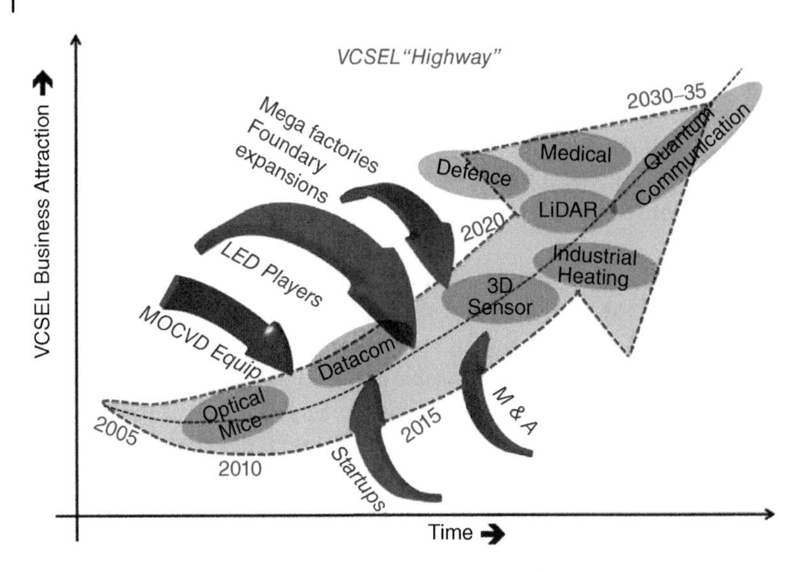

Figure 3.5 Sources of VCSEL business attractiveness ("highway"). *Source:* © Photonic Components DFM Ltd.

some players disappeared due to lack of manufacturing infrastructure for 4″ or 6″ VCSEL development, but on the other hand, it attracted several players into the *VCSEL highway*. Together, these factors have made the VCSEL business attractive, as shown in Figure 3.5.

This also created an opportunity for epi-growers and III-V opto foundries to expand their MOCVD facilities and fabrication infrastructure, respectively, for future VCSEL capacity needs in the industry. Besides, there are a few dozens start-ups as design houses, R&D centers, consulting firms, fabrication facilities, and testing labs that popped up to cater to the needs of VCSEL chip and module demands. Therefore, it is evident from the above discussion that there is a fierce competition among VCSEL players to catch the momentum of VCSEL attraction (~$12.8 billion for the next 10 years from Figure 10.1) as described in Figure 3.5. For the future chip demand, the VCSEL highway appears to be a fertile ground for business, even under COVID-19 devastated market and broken supply chain situations. At the end of 2020 Lumentum acquired Trilumina [35] and NVIDIA acquired Optigot [36], both true indicators of an ever-expanding landscape for a brighter VCSEL industry. Overall, by end of 2020, the VCSEL industry appeared to be a mix of old folks evolved as big giants after M&As and with new entrants as well. This is also evident from a number of companies that have provided multiple images in this book to showcase their manufacturing strengths and commercial products.

3.1.4 VCSEL Die Cost and Foundry Economics

As described above, with increased demand in VCSEL quantities for the next 5–10 years, it is worthwhile to understand the VCSEL die cost structure and foundry (economic) models. Many current VCSEL chip manufacturers are vertically integrated with millions (if not billions) in chip capacity, and both epi-wafer and process foundries are working on low-cost production models to meet future capacity needs. Practically speaking, the VCSEL die cost is directly related to the foundry model used by the manufacturers including the size of the wafer and its associated production (bill of materials, or BOM) costs and business model (fabless/IDM/in-house development). A 6″ GaAs wafer (150 mm) processing was first demonstrated by Anadigics Inc. in 1999 [37] and has now become the high-volume standard format, with 4″ wafer (100 mm) generally used for low-volume application cases. An example of a fully processed 6″ GaAs VCSEL wafer for 3D sensing applications is given in Figure 3.6.

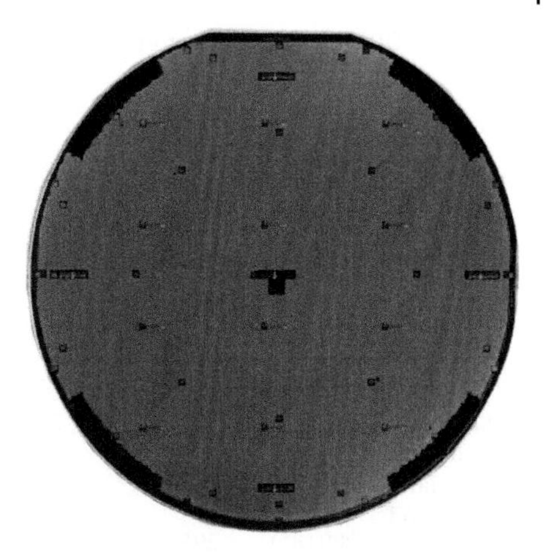

Figure 3.6 Fully processed 6″ (150 mm) GaAs VCSEL wafer. *Source:* Reprinted with permission from II-VI Incorporated.

Since the announcement of VCSELs for the iPhone X by Apple in 2017 and given that several new applications are emerging, the volume of VCSELs chips justifies increased operation and production costs. Thus, in high-volume applications, the ability to add scalability to the production capacity is the key for mass manufacturing of products. This implies that the production costs are able to meet the same requirements of low-cost LEDs vertical (planar) manufacturing without compromising the product performance and quality of devices. The following sections will explore the details of both foundry and business models.

In a foundry model, the key aspects are the wafer size and process costs associated with BOM. For VCSELs, over the past few years, 4″ wafer (100 mm) was an established model, and now due to the increased demand for VCSEL chips, it has switched to larger a wafer size of 6″ wafer (150 mm). Both are now used in III-V foundries for specific product depending on volume requirements. But with exploding market demand discussed previously, future production demand is unlikely to be met with existing 6″ wafer fabs, and development of VCSELs on 8″ wafer (200 m) looks imminent. This may open a new window for the use already available resources from vertically integrated Si- and LED (200 or 300 mm) foundry tools for future 8″ (200 mm) or even 12″ (300 mm) wafer processing. A schematic of future VCSEL demand and its expected cost trend as a function of wafer size is shown in Figure 3.7.

The die quantity from individual wafers depends on die size and the edge exclusion zone, typically 5 mm from the outer edge of the wafer. Using some existing die and wafer sizes, a few die count scenarios are presented in Table 3.1. Due to the lower volume and smaller size of datacom chips when compared to consumer mobile (smartphones), 4″ wafer foundry mostly meet the needs

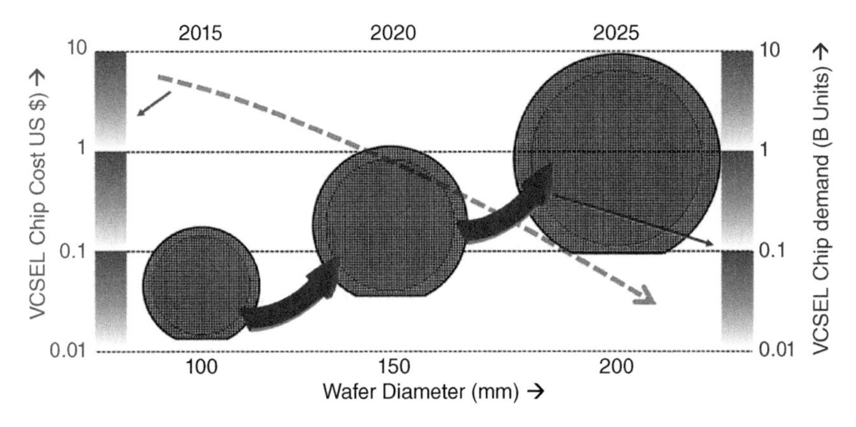

Figure 3.7 VCSEL demand and cost trends as a function of wafer size. *Source:* © Photonic Components DFM Ltd.

Table 3.1 Die quantity and cost estimates for different product (die) sizes.

Field	Wafer diameter (mm)	Wafer area w/o edge exclusion (mm^2)	Chip count (# of units) for given die size					
			0.25 × 0.25	0.25 × 1.0	0.5 × 0.5	1.0 × 1.0	2.0 × 2.0	4.0 × 4.0
Datacom	100 (4″)	7084	113 354	28 335				
	Chip cost (US$)	10G (SFP)	~1.0	~5.0				
		25G (QSFP)	~3.0	~15.0				
Consumer, Automotive, Industrial heating and others	150 (6″)	16 504	264 074		66 018	16 504	4126	1031
	~Chip cost (US$)		~0.1		~0.5–1.0	~1.0–2.0	~3.0–5.0	~6.0–8.0
	200 (8″)	29 849	477 594		119 398	29 849	7462	1845

Source: © Photonic Components DFM Ltd.

of the datacom industry. However, for consumer mobile (smartphones), the 6″ wafer foundry is warranted and even 8″ wafer (200 mm) foundry may also be needed in the future. The idea of moving from small to larger wafer diameter is to ensure optimized epi-growth and process parameters (BOM cost, fab yield, etc.) to achieve the best overall cost-benefit ratio with the aim of achieving cost structures comparable to those of successful LED models.

The purpose of Table 3.1 is to give a general idea of chip costs to potential investors, not to represent an exact picture of current industry. These are based on announcements from professional forecast agencies and on first-hand authors' knowledge and experience in the industry. From Table 3.1, it is apparent that there will be a >400% wafer area use from 4″ to 8″ wafer, and gross dies per wafer will also increase in the same proportion. In datacom, the standard size of a VCSEL die is generally 0.25 mm × 0.25 mm for singlet applications and 0.25 mm × 1.0 mm or 0.25 mm × 3 mm for 4 and 12 channel applications, respectively. Currently the market price of 10G and 25G chips for are approximately $1 and $5 for single channel, and $3 and $15 for 4 channels, respectively. Speeds higher than for 25 or 28 Gb/s applications (say, 40 Gb/s or 56 Gb/s NRZ/PAM-4) will have higher chip value.

In contrast, the consumer electronics (3D sensing-mobile), automotive LiDARs, and industrial heating applications use custom chip dimensions that are generally much larger. One aspect of the die cost is the number of photomask layers used for chip processing. For example, a 28 Gb/s datacom VCSEL may have 10–11 mask layers while a 10 W high-power VCSEL may have only 6–7 layers. The custom size in many non-datacom applications is dictated by the number of VCSEL emitters needed to provide specific requirements of the product (power, illumination distance, far-field angle, thermal management, etc.). Here, the pitch is mostly constant for an illuminator (say, 20–40 μm) using TOF and may be random for face ID using structured light schemes. A TOF scheme can also be used for face sensing; the technical details of these topics are explained in Chapter 5. Currently, the market price of high-power VCSEL chips closest to the dimensions of 0.25 × 0.25 mm^2, 0.5 × 0.5 mm^2, 1.0 × 1.0 mm^2, 2.0 × 2.0 mm^2, 4.0 × 4.0 mm^2 are approximately $0.1, $0.5–1.0, $1.0–2.0, $3.0–5.0, $6.0–8.0, respectively. Further, most of the manufacturers for sensor and LiDAR VCSELs use 6″ wafer foundry, and it is expected that there will be an additional cost reduction if 8″ wafer foundries become a commercial reality.

3.2 VCSEL Industry Landscape

When it comes to VCSELs as products in large-scale manufacturing, several aspects play key roles, and it becomes strategically important to run the production and business smoothly. Factors such as (i) abilities of the VCSEL devices, (ii) manufacturing (design, growth, and fab) specifications and tolerances, (iii) the challenges involved in meeting those specifications and tolerances, (iv) the business models followed, (v) working with collaborations and crossing barriers against competitors, (vi) the readiness of the global supply chain, and (vii) trade control practices all effect the manufacturing and operations of product delivery. Each of these aspects are explored to more clearly understand the VCSEL industry landscape in the following sections.

3.2.1 The Key "Abilities" of VCSELs

As discussed earlier, in Chapter 1, VCSELs have several unique abilities that have surpassed the commercial product developments of their counterpart LEDs and EELs. All these "abilities" are outlined in Schematic 3.1.

3.2.2 High-Volume Manufacturing Challenges

There are some critical challenges that every manufacturer will encounter during growth and processing of VCSEL wafers. Alike any semiconductor device development, meeting specifications with precise tolerances in the layer structure by epitaxial growth and chip processing is the fundamental requirement in VCSELs. For VCSELs, some of the critical manufacturing steps include (i) epi-wafer Fabry-Pérot (F-P) and photoluminescence (PL) wavelength uniformities across the entire wafer diameter, (ii) inductively coupled plasma (ICP)/reactive ion etching (RIE) dry etching of top-mesa to maintain etch depth uniformity across the wafer, (iii) very unique feature of oxidation aperture shape, its size tolerance, and uniformity, and (iv) chip qualification and reliability tests as well as meeting performance specifications.

Schematic 3.1 Multiple technological abilities of VCSELs.

Growability	: Full VCSEL epi-structure grown monolithic on single substrate without re-growths
Scalability	: Scalable in terms of emitting wavelength and output power with 1D or 2D arrays
Processability	: III-V opto foundry extendible from 3″ (75 mm) to 6″ (150 mm) wafers
Integrability	: Compatible with CMOS, Si photonics for electronic components integration
Testability	: Allows on-wafer testing, burn-in, and inspection
Packageability	: Diced chips allows packaging on to TO headers, PCB/driver ICs, TOSA, OTs, and modules
Bondability	: Can be easily wired and die-bonded during assembly stage with established processes
Reliability	: Highly reliable with long wear-out lifetimes and high activation energy for failure
Manufacturability	: Vertical structure allows planar layer-by-layer processing with traditional semiconductor manufacturing equipment (metallization, dry-etching, oxidation, passivation, etc.)
Wavelength Stability	: Wavelength changes with temperature are lower and more stable than EEs and LEDs

3.2.2.1 Epi-Wafer Growth and F-P and PL Uniformities

For the first part of epitaxial growth, VCSEL layer structures (with typically few 100s of semiconductor layers) are grown by MOCVD systems with complex control of hazardous gases and materials in vapor forms so precise that groups of atoms and molecules of precursor materials settle as thin layers (typically few nm) onto lattice matched substrates. The background of MOCVD and its growth kinetics can be studied from the reference [38]. In order to qualify the VCSEL epi-wafer, it must meet some stringent specifications and tolerance tests before passing it to the chip processing, called growth acceptance criteria (GAC). Some of the key specifications and tolerances are given in Table 3.2.

Table 3.2 shows two scenarios for epi-growth for 4″ (100 mm) for datacom and 6″ (150 mm) for sensing wafer diameters that are currently used in the industry. The focus point is the importance of controlling the key VCSEL epitaxial wafer growth conditions to meet given specifications. The success of qualifying an epitaxially grown VCSEL wafer for device processing lies in tight control of MOCVD system parameters such as growth temperature, materials, and gases mass flow rates, material composition chosen for laser structure, and so forth, as explained in Appendix B.

It is well known that the device yield of a VCSEL wafer mainly depends on the FP and PL uniformities. An example of a GaN QW grown on a Si substrate for 8″ (200 mm) wafer, best in the industry, shows PL uniformity >99% with (with std. dev. = 0.56 nm) across the wafer is shown in Figure 3.8a [39]. It is difficult to maintain such excellent uniformity even for 4″ (100 mm) and 6″ (150 mm) GaAs wafers. With top and bottom DBRs and multiple QWs (MQW), all create stresses and often thickness nonuniformity exists, a serious drawback for VCSEL epi-wafer. On the other hand, an FP (etalon) dip map of full VCSEL wafer, shown in Figure 3.8b, exhibits a large wavelength shift from the center wavelength of 850 nm, implying post processing device yield may be affected. As the substrate wafer size becomes larger, maintaining PL and FP epi-growth uniformities under specifications is a challenge indeed.

Table 3.2 Key specifications for VCSEL epi-wafer growth.

Emission/Center wavelength	NIR (800–1100 nm)	NIR (800–1100 nm)
Wafer (GaAs substrate) diameter (mm)	100	150
Al mole fraction (x)	≤0.02	≤0.02
PL wavelength (nm)_	±1	±1
CV-level (%)	±20 or less	±20 or less
Thickness uniformity (%)	≤5	≤5
Doping level uniformity (%)	±20 or less	±20 or less
Al mole fraction uniformity (x)	±0.02 or less	±0.02 or less
Target PL wavelength (nm) from ($In_xGa_{1-x}As$)	Depends on emission wavelength (x=0-1)	
PL wavelength uniformity (nm)	≤3	≤3
PL linewidth-FWHM (meV)	<30	<30
Defect density (cm^{-2})	<25	<25
Minimum defect size (μm^2)	<3	<3
FP etalon tolerance (nm)	≤5 (from emission wavelength)	
% area to lie within etalon spec. (%)	>60	>80
center wavelength offset (nm)	<2	<2

Source: © Photonic Components DFM Ltd.

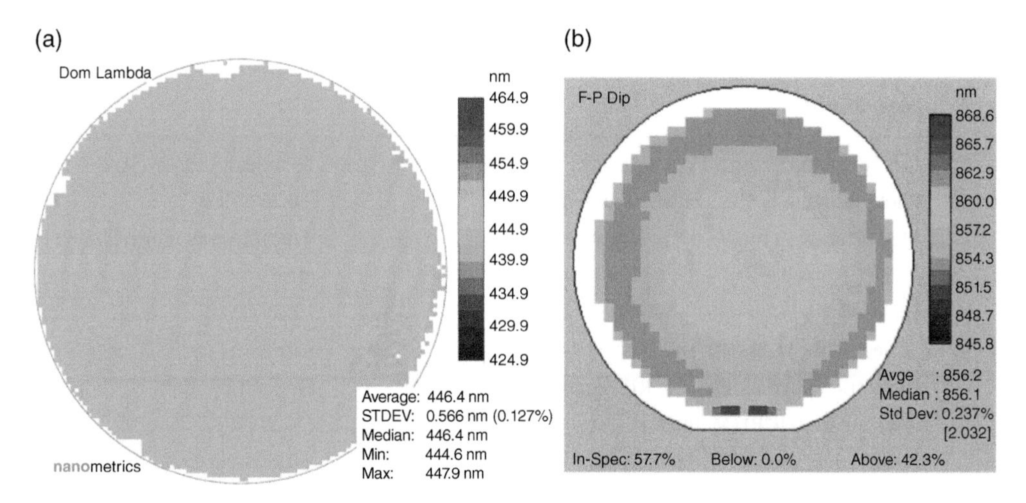

Figure 3.8 (a) PL map of GaN QWs grown on Si 8″ (200 mm) substrate. (Reprinted with permission from Allos Semiconductors. [39]); (b) FP etalon map of 850 nm full VCSEL grown on unknown size of GaAs wafer (Adapted from IEEE [40]).

3.2.2.2 Wafer-Fab (Processing) Specifications

Table 3.3 shows specifications for front- (FE) and backend (BE) processing for a 6″ (150 mm) wafer commonly used in the industry. Tight control of key VCSEL process specifications are essential to high yield, quality, and performance of VCSELs. For high-volume manufacturing, fully automated cassette-to-cassette fabrication processes with a capacity of few 100s (if not 1000s) wafers per week are needed. The key FE processing parameters in VCSEL manufacturing include (i) ICP dry etch (mesa or multi-layer stack of semiconductors) capabilities up to 4–5 μm depth with etch tolerance of ±0.2 μm, (ii) plasma-enhanced chemical vapor deposition (PECVD)/sputter dielectric deposition thickness tolerances (±10% or less) for mesa passivation (PASS) and protection (PROT) of emitting surface, (iii) metallization tools (such as e-beam, sputter, and electro-plating) for ohmic contacts, thick metals for heat dissipations, and proper adhesion for wire-bond strengths, (iv) tight control of oxide aperture diameter (±0.5 nm or less) with oxidation uniformities <5%,) cured thickness control of polymer coatings for reducing electrical parasitics and improving device speeds, and so forth, as shown in Table 3.3. The BE processing steps include substrate thinning to typically 150 μm or less, final optical inspection and wafer level probe tests, dicing, assembly, and packaging. Items such as dicing, expansion, sorting, and binning, for assembly pick-up are critical stages to catch electrostatic discharge (ESD) that may damage VCSEL chips, as discussed in Appendix C. Strict ESD protocols (electrically grounded lines in fab, floor, humans and equipment required specific voltage levels) are to be implemented for VCSEL manufacturing lines where individual VCSELs devices are handled.

3.2.2.3 Dry Etch Depth Uniformity

Dry etching of semiconductors layers with etch depth of a few microns by ICP, RIE (using reactive ions BCl3, Cl2, SF6), or its combination ICP-RIE is well known in the CMOS/Si/LED/PV industry and can achieve excellent etch depth uniformities on 12″ (300 mm) wafers. However, for multi-layer semiconductor stacks with 100s of layers with varying refractive indices, deeper etch requirements, and maintaining an etch tolerance <5% with smooth sidewalls to enable passivation coverage on 150 mm wafers is not a trivial task. In VCSELs, often this is called mesa etch process and contains multiple etch steps to translate the shape of the photomask pattern onto the

Table 3.3 Key specifications for wafer-fab (processing).

	Wafer diameter (mm)	150
FAB	Fab ESD requirements (V)	50 or less
	Photolithography CD/stepper alignment (μm)	±0.25 or less
FE	Metallization / ohmic Contacts (Methods)	E-beam, sputter and Plating
	P-Ti/Pt/Au and n-Au-AuGe (resistivity) $\Omega\cdot cm^2$	<1e5
	Dielectric deposition ($SiN/SiO_xN/SiO2$)	PECVD
	Dielectric thickness variation (%)	≤10
	Dielectric Film etching (ICP/RIE/chemical)	Dry and Wet
	Mesa etching (ICP) (Methods)	Dry (ICP, RIE or both)
	Mesa Height (depth) Variation (μm)	±0.2 (Angled and vertical)
	Wet thermal oxidation tolerance (μm)	±0.5 or less
	Wet thermal oxidation uniformity (%)	<5
	Polymer coating (PI/BCB/PBO etc.) cured thickness (μm)	4 or higher
	Au-plating thickness (μm)	2.0–4.0
BE	Wafer to be thinned (μm)	150 ± 10
	Final optical inspection (%)	100
	On-wafer testing: LIV and FFP, NFP, WL (%)	(100 and <1)
	Wire-bond test requirements	Telcordia/MIL

Source: © Photonic Components DFM Ltd.

semiconductor multilayer stack. These steps include (i) photoresist (PR) etch (soft mask etch), (ii) a thin (100–200 nm) or thick (> 1.0 μm) dielectric etch (hard mask), and then (iii) VCSEL DBR (semiconductor multilayer stack) etch in the same order [41]. The specific shape (vertical or tapered) of mesa can affect device performance and reliability [42] too. Maintaining smooth sidewalls and etch depth tolerance is mandatory for VCSEL processing as going above the upper specs limit (USL) or below the lower specs limit (LSL) values seriously affects the device performance. Mesa etch process, explained in this section, is for a generic case, and some tier-1 players use holes/trenches using dry-etching in compliance with planar CMOS processes without fully exposing mesa sidewalls. Both are shown in Figure 3.9b.

A schematic cross section of generic (fully exposed) VCSEL mesa with its sidewalls is shown in Figure 3.9a It also shows (i) typical dimensions (total thickness, etch depths, mesa diameter) for a NIR VCSEL, (ii) a thick hard (SiN/SiOxN) mask, (iii) fully exposed oxidation layer [with Al(x) between 0.98 and 1.0)] for current confinement, and (iv) high Al(x) layers in bottom DBR for low thermal impedance. Normally a multipoint sampling procedure is followed to check the etch depth tolerance by focused ion beam (FIB) etching to see the x-section of respective mesas. Due to thickness nonuniformities, stresses in multilayer stacks, ICP, and RIE plasma etch conditions (gas flow rates and temperatures), very often the etch depth tolerance is not uniform across the wafer area: It has under etch or over etch or both, and this seriously affects the performance of VCSELs. For the under-etch scenario, the QWs below the oxide layers will be un-etched across the wafer, and crystal defects such as dislocations and dark line defects continue to move into the

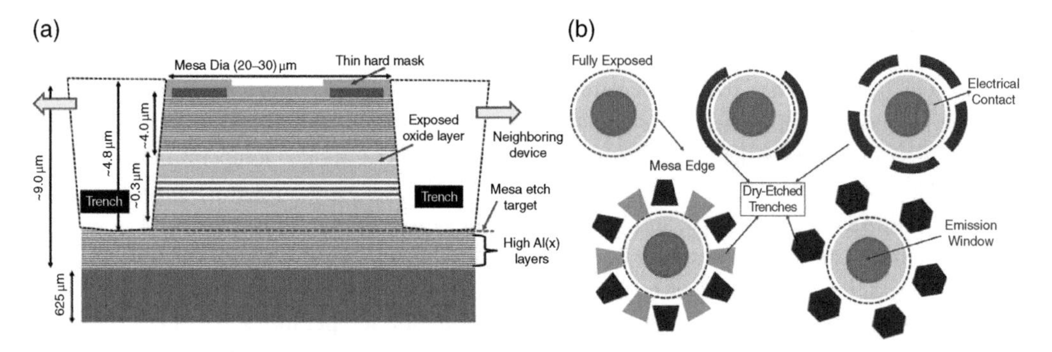

Figure 3.9 A schematic (a) cross section of generic, and (b) top-views of some VCSEL mesas surrounded by trenches. *Source:* © Photonic Components DFM Ltd.

rest of the VCSEL material layers. For the over-etch scenario, the high Al(x) DBR layers below QWs layers will be exposed and get oxidized during the oxidation process. Here selectivity (difference of wet/dry etch rates among DBR mirror, SiN/SiOxN, and photoresist [polymer]) plays a crucial role and needs to be optimized for efficient and repeatable process control. So there exists an optimum etch depth (dotted line in Figure 3.9a as mesa etch target) with a tolerance of ±0.2 μm or less.

An example of such multipoint sampling is shown in the left part of Figure 3.10. For each point, the mesa x-section is checked by cutting the mesa to determine the etch depth and mesa sidewall profile. The etch depth variation for a nine-point wafer is given in the right part of Figure 3.10. As the FIB system is destructive and expensive, not all manufacturers have it in the VCSEL fab line for checking etch depth and profiles. A much simpler nondestructive and in-situ method that uses optical reflection specifically to monitor the refractive index contrasts is endpoint detection/monitors (EPM). The key aspect when using an FIB system or EPM is to strictly comply with the etch depth tolerance. Also, EPMs are mainly single-point detection systems, and it is difficult to get wafer level sampling averages. A schematic of VCSEL cross section with its FIB etched mesa is shown in Figure C.1.1.

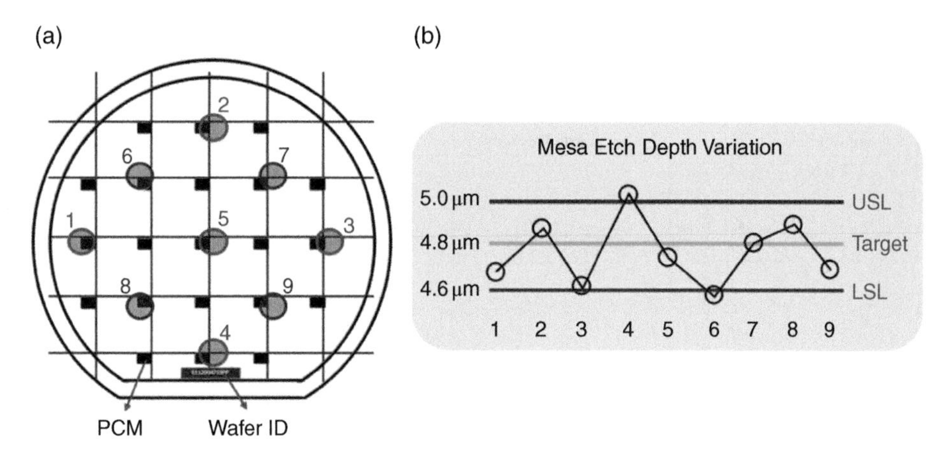

Figure 3.10 (a) Etch depth sampling points and (b) its depth variation across the wafer. *Source:* © Photonic Components DFM Ltd.

3.2.2.4 Wet Thermal Oxidation, Aperture Control and Uniformity

After mesa etch, the fully exposed $Al_xGa_{1-x}As$ oxide layer with Al(x) [x = 0.98 to 1.0] (shown in Figure 3.9) will be subjected to wet thermal ambient to convert this semiconductor layer into an insulating layer for current confinement. The oxidation process in $Al_xGa_{1-x}As$ alloys is well controlled [43] and oxidation depth strongly depends on $Al_xGa_{1-x}As$ layer thickness, applied temperature, and mole fraction of Al(x), as shown in Figure 3.11 [44]. This process is also extended to other III-V alloy systems AlInSb [45], AlInAS [46], InAlP [47], and InAlGaP [48] for various device applications.

As of January 2020, the wet thermal oxidation of $Al_xGa_{1-x}As$ alloys is a well-established and widely used process in fabrication of microelectronic and optoelectronic devices and has become essential in VCSEL development and manufacturing. Oxidation temperatures are in the range of 360–450°C. The wet oxidation process changes the crystal structure of (crystalline) $Al_xGa_{1-x}As$ (x = 0.98 to 1.0) with refractive index ~3.0 to noncrystalline (amorphous) aluminum oxide layers with refractive index ~1.6. As it can be seen from Figure 3.11c, there will be more than an order of magnitude of oxidation rate difference between x = 0.98 to 1.0 in $Al_xGa_{1-x}As$ alloys, highly dependent on temperature and thickness, as shown in Figure 3.11a and b, respectively.

Figure 3.12a and b show a schematic of oxidation rate as a function of Al(x) and experimentally observed oxidation aperture shapes from CNRS group, respectively. In Figure 3.12b one can clearly note that the oxidation speeds in [010] and [110] are different; these different speeds are the prime source of the noncircular aperture shape. One critical challenge in VCSEL development is to control the circular shape of oxide aperture on the entire wafer area. Poorly shaped apertures create modal disturbances (noise) and nonuniform intensities (speckle) and degrade device performance [49]. There are multiple commercial vendors supplying vertical- (chamber) and horizontal- (tubular) type oxidation systems (furnaces) with advanced functions such as in-situ monitoring of the oxide, precise shape and size control software models, batch processing for 4″ (100 mm) to 8″ (200 mm) wafers. The vendors include Koyo Thermo Systems co. Ltd. (Japan), AET (France), and CSL (USA) [50].

3.2.2.5 Chip Qualification and Reliability Tests

After the wafer fabrication, the chips at die or module level are subjected to standard qualification tests. Devices must show a low failure rate and long life and wear-out times to qualify into final products. The metric used is called failures-in-time (FIT), defined as the number of failures that can be expected in one billion device-hours of operation (i.e., FIT = probability of failure as ppm/1000 hours of operation, or number of failures per 1000 devices operating 1 million hours; for

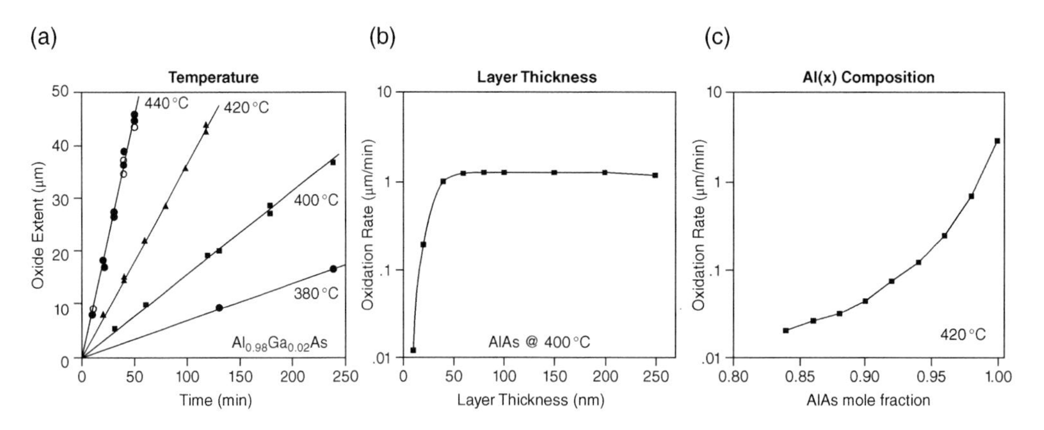

Figure 3.11 Oxidation rate dependence of (a) time, (b) thickness, and (c) Al composition. *Source:* Adapted from IEEE (44).

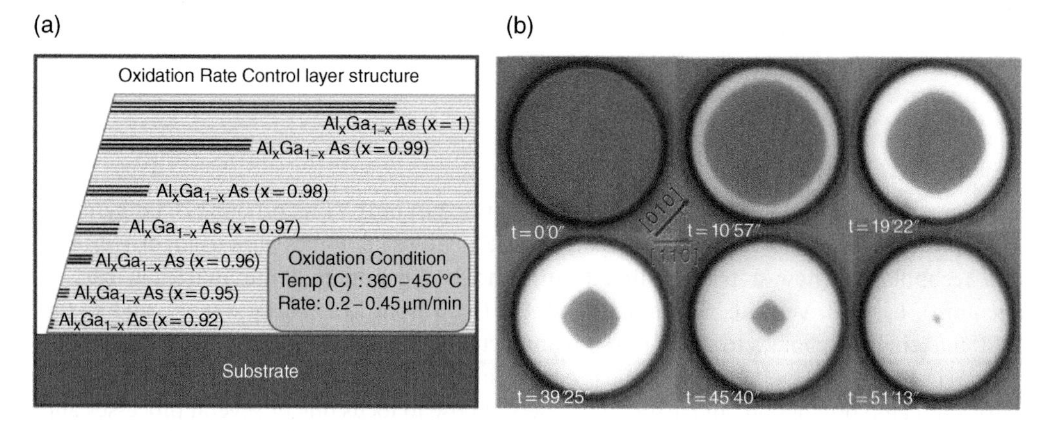

Figure 3.12 Schematic of (a) oxidation depth as a function of Al(x) mole fraction from 0.95 to 1.0. *Source:* © Photonic Components DFM Ltd., and (b) creation of oxide aperture as a function of time with different oxidation speeds in two crystal directions. *Source:* Reprinted with permission from Optica (formerly OSA) [51].

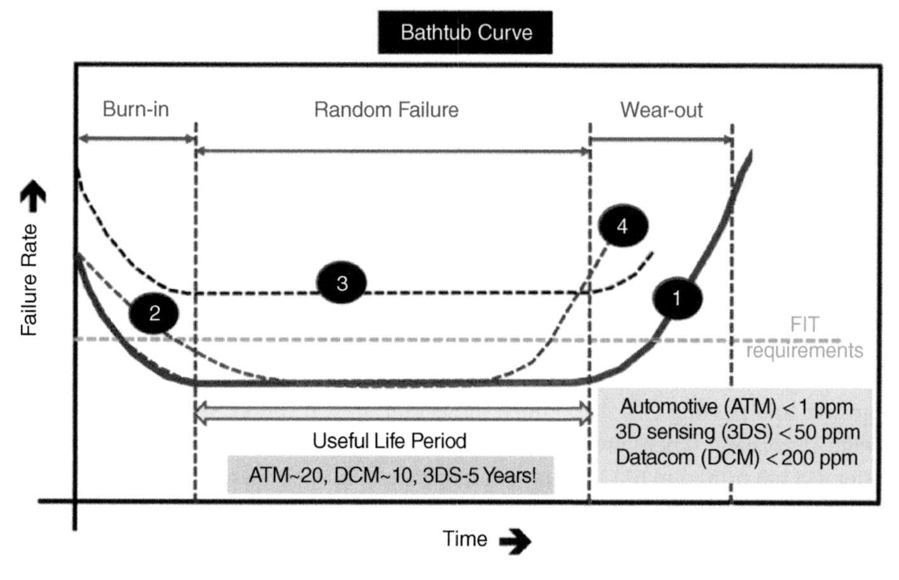

Figure 3.13 Bathtub curve failure distribution trends of an optical component (VCSEL). *Source:* © Photonic Components DFM Ltd.

a single device, the probability of failure in 1 hour of operation is FIT × 10^{-9}). The bathtub curve in Figure 3.13 shows an example of maximum accepted failure distribution label-1 of an optical component (VCSEL). If this follows the trend of labels 2, 3, and 4, they are called early life failure, excess random failures, and early wear out, respectively.

The FIT requirement and long useful lifetimes are very stringent for VCSELs, the specific requirements varying depending on the application. Figure 3.13 shows that for automotive (ATM), 3D sensing (3DS), and datacom (DCM) VCSELs, typical industry acceptable (i) FIT numbers are less than 1, 50, and 200 ppm, and (ii) lifetimes are 20, 5, and 10 years, respectively. For VCSEL manufacturers, it's quite hard to reach these numbers with short cycle times and come up with the highest quality for their products, especially for automotive and datacom VCSELs. For the consumer segment, a bit lower lifetimes are required, as the average use of a smartphone varies from two to four years [52, 53]. Details of qualification and reliability tests for datacom, 3D sensing, and automotive VCSELs are described in Appendix E.

3.2.3 Industry Players

The major commercial VCSEL players are Finisar (now a part of II-VI Incorporated), Lumentum, Princeton Optronics and Vixar (both now a part of AMS), Philips Photonics (now a part of Trumpf), Avago technologies (now a part of Broadcom), and late commers Trilumina (now part of Lumentum), Lasertel, Vertilite, Sanan, and OSRAM (both from LED backgrounds), and so on. VCSEL makers have choices to make in the manufacturing process. Some choose to completely insource (in-house development) all of the processes including epitaxial growth, fabrication, testing, backend chip production, and assembly. Others choose to focus one or several processes and outsource the rest to external parties. With the growth in overall VCSEL demand, viable outsourcing options exist for all levels of the supply chain. The two key processes in VCSEL supply chain are epitaxial growth and FE fabrication, and this section focuses on these two items.

3.2.3.1 Epi-Houses

The epi-growth service starts from VCSEL layer structure, and it is grown monolithically on a single crystal substrate. Depending on the wavelength, this might be InP, GaN, GaSb, or GaAs substrate. Today, GaAs substrates dominate VCSEL production, so the following discussion will be limited to this material. There are several well-known GaAs single crystal substrate vendors, namely Freiberger, AXT, and Sumitomo, among others. Most of the VCSEL growers use high-quality (low EPD 4″ [100 mm] to 6″ [150 mm]) GaAs substrates with various configurations (doping, thickness, major/minor flats, polishing, crystal orientations, etc.) explained in section B.3.1.3, Table B.2. GaAs VCSELs are grown mostly by MOCVD and occasionally with MBE, as described in section B.2.1, and there are multiple vendors including, IQE, VPEC, Intelliepi, (II-VI) Epi-works, and Landmark, shown in Table 3.4.

3.2.3.2 Process Foundries

After epi-layer growth and passing GAC, explained in Section 3.2.2.1, the VCSEL wafers will be transferred to wafer processing. There are several commercial VCSEL foundry service providers that currently offer 4″ (100 mm) and 6″ (150 mm) wafer processing. Commercial GaAs foundries include Win-Semi, GCS, (part of Unikorn), Sanan, and SinoSemic. Foundries can offer varying levels of service, including epitaxial growth, fabrication, testing, and BE processing, as further explained in the ODM/IDM discussion in Section 3.2.5. Many of the foundries originated from LED, Si-technology (RF), and PV manufacturing backgrounds and have grown to include laser diodes and photodetectors. For VCSEL specific manufacturing, these foundries added wet thermal oxidation systems and appropriate metrology tools including high-resolution microscopes/nasoscopes, field emission scanning electron microscope (FE-SEM), FIB, and transmission electron microscope (TEM) for inline process monitoring and material verification. Usually VCSEL companies are the design and sales channels from the fabs to end users. The process foundries provide customized VCSEL chips to integrators in data centers, 3D sensing (smartphone mobile), automotive imaging sector, and so forth. A schematic of VCSEL players is provided in Table 3.4.

3.2.4 Business Models

A business model is a plan by which a company operates and monetizes value to the stakeholders, which involves a set of choices and consequences [54]. Industry product lifecycles follow the four stages of introduction, growth, maturity, and decline [55]. Based on this model, the VCSEL

Table 3.4 Appearance of VCSEL players post Apple's i-phone X announcement* [© Photonic Components DFM Ltd.]

Layer Designers	Bare substrate	Epi-wafer Growth	Process Foundry	Fabless/IDM	Software	Internal Use
DQD	Freiberger	IQE	Win-Semi	II-VI	Crosslight	Fuji Xerox
PCDL	AXT	VPEC	AWSC	ams	Synopsys	Ricoh
	Sumitomo	Intilli-Epi	GCS	Broadcom	Photon Design	SONY
	Chemicals	Epi-works	Sanan	Lumentum	COMSOL	NEC
	Umicore	Landmark	SinoSemic	Trumpf	Silvaco	
		Masimo	Unikorn	LaserTel		
		Win-Semi		OSRAM		
		Sanan		VI Systems		
		SinoSemic		Sanan		
		VIGO		ROHM		
		Unikorn		Vertilite		
				Vertilas		
				FLIR		
				Bandwidth10		
				Astrum		
				Modulight		

*Non-Exhaustive list

industry entered the maturity stage in 2017 with rapid growth. With the many changes in the VCSEL market over the last several decades, it is worth looking at some of the business models adopted by the VCSEL manufacturing industry.

There are several types of firms that act as investors, namely innovators, followers, and attackers [56, 57]. Depending on the market demands and investment returns, firms make some choices to adopt strategies as business models. Investors follow one of the three core and well-known business models of IDM, fabless, and foundry (including internal sourcing) along with a supporting model of outsourced semiconductor assembly and test (OSAT) and equipment. As there exist several combinations among business models, as shown in Schematic 3.1, companies involve strategic thinkers (investors, technologists, and market analysists) to define a successful business model.

As fabless and foundry models have grown above the industry average more than IDM models, they are much more likely to be adopted by VCSEL makers [58]. Most outsource to avoid capitalization costs and/or improve time to market. If we apply the above business model to the firms

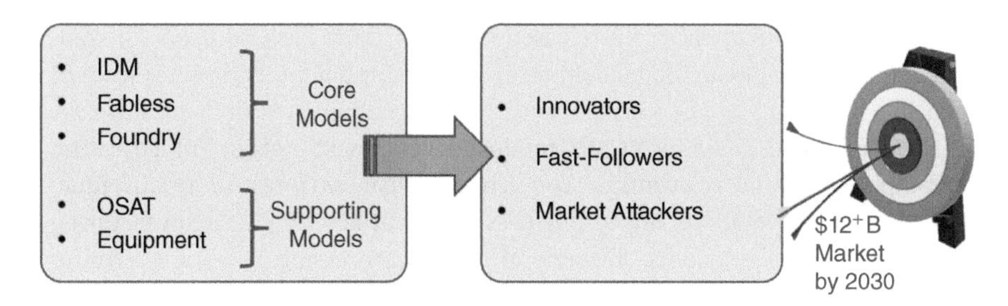

Schematic 3.1 Business models for III-V opto foundries. *Source:* © Photonic Components DFM Ltd.

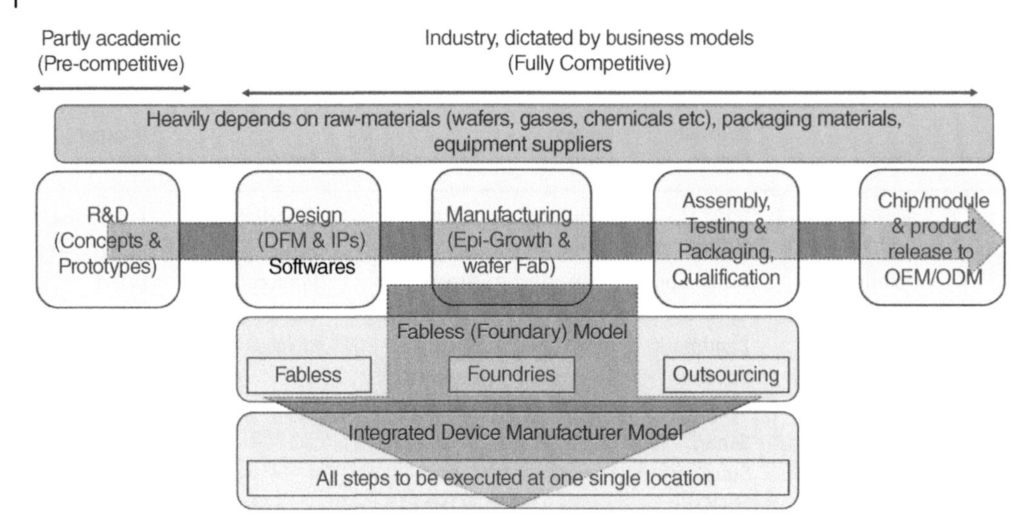

Figure 3.14 VCSEL supply chain (with business models overlapped). *Source:* © Photonic Components DFM Ltd.

mentioned in Table 3.4, it is evident that with the exception of Lumentum (fabless), most of the VCSEL makers source internally, and a few follow the IDM model. Commercial players such as Sanan lower the barriers to entry for small companies and provide scalable manufacturing to big (sub-system level) companies. This can also be understood from Figure 3.14 below on supply chain flow. A classic example shows the importance of business model and value-for-money product offering. Within 18 months after Apple's iPhone X announcement in 2017, many android players widely deployed autofocus functions, developing multiple versions of smartphones with face ID and illumination and attacked the market. All these firms are fast followers to wait and watch until the product and technology through a given application reaches global volume and then strike the market with maximum innovation and low cost. Here intellectual property (IP) also plays a big role in the decision and technology to be implemented.

3.2.5 Supply Chain

The process of making a VCSEL commercial product is represented by a supply chain, as shown in Figure 3.14. In the R&D stage, concepts are made, and prototypes are created to determine the functionalities of proposed devices; this stage is somewhat academic and noncompetitive in nature. Often this step involves the device modeling with advanced software. Once the device function through a prototype is established, firms focus on more detailed designs and secure intellectual property (IP). Then the designs (layer structures) are released to epi-growth, and VCSEL epi-wafers are passed to chip fabrication (processing) followed by chip testing and qualification/reliability tests. Finally, fully qualified chips are released to OEMs/ODMs as end users for commercial product realization. All stages are dependent on each other, and production line (section) managers are responsible for readiness of their part of actions for passing on to the next stages. A schematic with the stages of supply chain for VCSEL product development is given in Figure 3.14. In reality when the complex supply chain events are coupled with business (profit/nonprofit) functions, fabless and IDM models operate as mixed business model at key VCSEL manufacturing stages and lose their individual significance. Generic MFG and processing flows are given in Figure 3.15 and Appendix C.

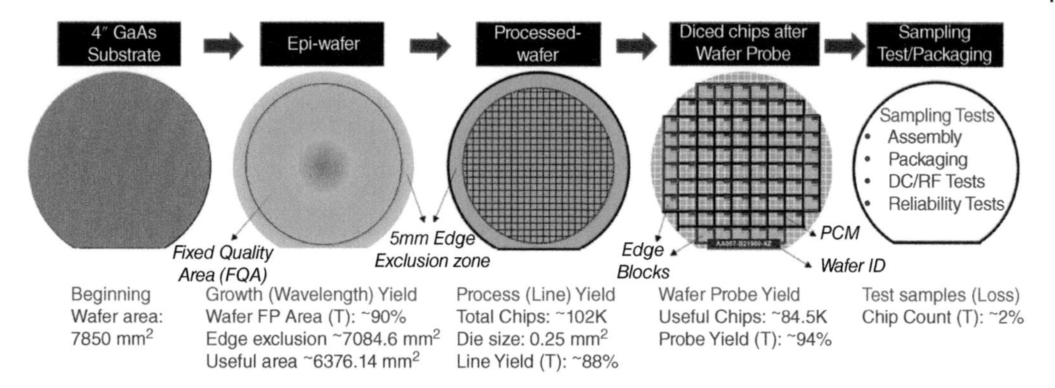

Figure 3.15 Schematic of VCSEL product development or manufacturing stages for yield estimation from wafer area and number of useful chips (T = Tentative 4″ = 100 mm). *Source:* © Photonic Components DFM Ltd.

3.2.6 Yield Improvements

Yield optimization is a critical factor for quantitative estimates on return on investment (ROI) for a profitable business. Die yield is defined as the ratio of the number of good devices (working units at the end) to total number of available dies on a wafer. Here good dies refers to dies that pass all performance specifications and visual inspection criteria. Yield is a measure of effectiveness of design, manufacturing processes, and process controls. Yield improvement is a critical task of all semiconductor device manufacturing as it directly affects profitability. A small yield improvement can significantly reduce the manufacturing cost per wafer, and these are seriously monitored by production line managers. Advanced SPC tools are used to verify possible defects and improve yields at every stage using many different metrology tools. Most of the big players in datacom, 3D sensing, automotive LiDAR, and industrial heating sectors focus on high-volume manufacturing through wafer fab optimization for the overall cost reduction, trying to improve yield through excursions. Often, yield improvements require costly production resources [59].

Defect density is one of the major parameters that defines net yield in semiconductor manufacturing. In VCSEL manufacturing, defects may come from several stages, including (i) epi-layer growth (wavelength yield), (ii) wafer fab (line yield), (iii) post-fab wafer level tests (probe yield), (iv) assembly (packaging yield), (v) DC, RF, and module tests (test yield), (vi) reliability or qualification (burn-in yield), and (vii) final visual inspection (physical defects). Normally, the stages (i, ii, iii, vii) are considered to define the die yield of the wafer, which involves 100% of the wafer area, and all available dies are included in the calculations; the remaining stages (iv, v, vi) are on sampling basis, roughly representing a few percent of the chips (throughput yield) across the wafer area. As VCSEL manufacturing consists of the series of stages mentioned above, the net yield is defined as the product of yields at individual stages. Unlike Si processing, most of the defects in GaAs technology arise from epi-layer growth, specifically during the wafer fabrication stage. Here the authors consider primary defects to illustrate an example and assume sampling chips to be a fixed sum for throughput yields. Figure 3.15 shows a schematic of how each stage contributes to the overall yield estimation in VCSEL manufacturing.

As shown in Table 3.5, a 4″ (100 mm) GaAs wafer with edge exclusion of 5 mm has an available product area of 7084.6 mm^2. Wavelength uniformity is one yield loss at the epitaxial level, and while a minimum of 60% FP center wavelength within 5 nm tolerance may be allowed for wafer fabrication, generally the FP useful area should be well above 90%. This implies that ~10% of useful

Table 3.5 Typical scenario of net yield estimation.

	Area (mm^2)	Chip count	Min	Med	Max	Yield	Loss
Substrate (4″, or 100 mm, diameter)	7850.0						
Epi-wafer (5 mm edge exclusion)	7084.625						9% Area
Reticle design (die size 0.2 mm^2)	[a]	113.35K (total)	101.9K (~102K)[b] (exclude edge blocks and PCM)			100%	
Fully processed wafer (inspection)		(100%) 101.9K	(80%)	(88%)	(96%)	88% 89.67K	12% Yield loss (12.22K)
Wafer level probe (100% LIV + <2% WL)		89.67K	90%	94%	98%	94% 84.32K	6% Yield loss 5.35K
Test and reliability (sampling chips)		84.32K	1%	2%	3%	2% 82.63K	2% (sampling) 1.69K
Typical case	net yield (0.88 × 0.94 =) 0.827 ≈ 83% (~ 84, 577 chips)						

Source: © Photonic Components DFM Ltd.
[a] Dicing street 50 μm and device pitch 0.25 mm.
[b] Included edge blocks and PCM.

area is already lost even before wafer fabrication. In some cases VCSEL epi-wafers may be scrapped if FP area is below a certain percentage based on business circumstances. Useful wafer area is also lost in the fabrication process to make room for PCM for every shot, alignment marks, wafer process ID, and mask labels prints. For a 4″ VCSEL epi-wafer with a 5 mm edge exclusion zone, PCM, alignment marks, wafer ID, dicing street, and so forth, occupy significant wafer area. For example, a 100 mm wafer can produce a chip count of ~102K dies; these can be made with each die measuring 0.25 mm × 0.25 mm with a dicing street 50 μm and pitch 0.25 mm among the dies. Immediately after full wafer fab (including FEP and BEP), the wafer will be subjected to (i) 100% inspection (from contaminations, particles, other defects chipping from dicing, etc.) and (ii) 100% LIV probe and <2% of selective wavelength tests simultaneously.

Manufacturing excellency can be witnessed by minimum number of defects during inspection (2–3%) and high level of LIV performance spec (wafer level probe) yield (>94%). In short, the process (line) yield is typically well above 88% for standard III-V opto foundries with controlled process capabilities. However, this is not the true yield of dies to be considered for final use, and there are several tests to be done through <2% dies as sampling for assembly and packaging, chip level LIV and RF tests, and final life (reliability) tests. The net yield can be approximated (the product of wavelength yield and line yield and assembly yield, viz. 0.88 × 0.94) to 0.83 (83%). This excludes the number of samples tested that pass all performance tests (chip level tests and reliability qualifications), typically ~1–3% of total chip count. In other words, from ~102K chips, the useful (sellable) chip count is roughly 84.6K.

For consumer sensing (<10 m) and autonomous LiDARs (~250 m) as 2D arrays of VCSEL emitters are used, it is more critical that all emitters should light up with uniform intensity, and even a single dead emitter can disqualify that particular (full) array/chip as defective. This shows how critical are the two aspects of (i) tight control of process parameters with minimum number of defects and (ii) yield improvements from lot-to-lot that paves the way for use of good dies in module/systems. By using advanced Industry 4.0 automated tools as end-to-end yield improvement solutions, more accurate analysis for the statistical root cause of defects can be achieved.

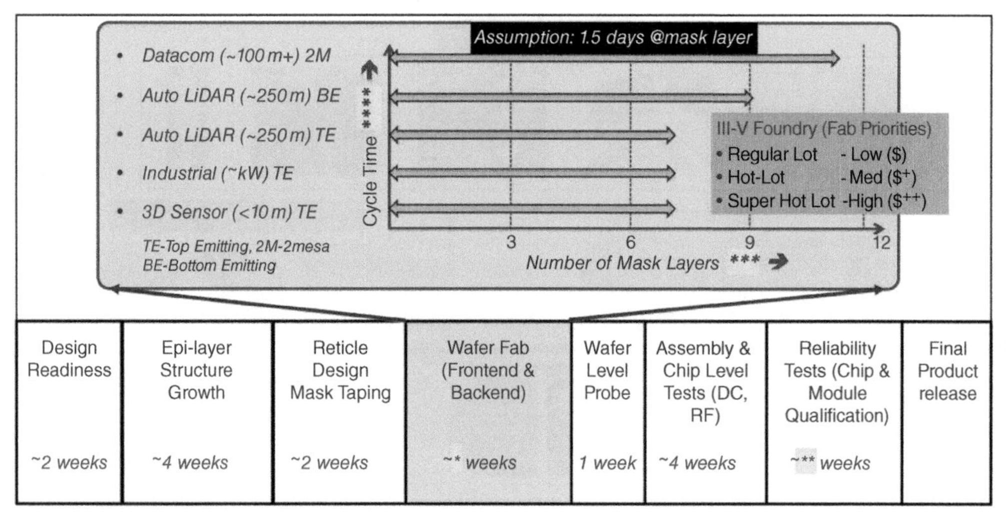

Figure 3.16 Essential stages of cycle time during VCSEL product development. *Source:* © Photonic Components DFM Ltd.

3.2.7 Cycle Times

Besides yield improvements, manufacturing cycle time also affects the manufacturing cost, and it should be carefully planned. Cycle time includes machine and queue time in the factory. It is normally defined as number of days per mask layer, and it generally varies between 1.0 and 1.5 days/mask layer. Excellent fabs can make it to ~0.8! In VCSELs the number of mask layers are different for different products; this can be understood from an example given in Figure 3.16. The total VCSEL product manufacturing lead time consists of individual cycle times of (i) VCSEL layer growth, (ii) photomask design and taping, (iii) FE and BE processes, (iv) wafer level probe, (v) assembly (sorting, expansion, wire/die bonding) and packaging, (vi) chip level tests (DC and RF), and (viii) reliability qualification tests. Typical VCSEL manufacturing cycle takes at least 18 weeks (excluding reliability tests), and for a 7-mask layer process, cycle time takes about 2 weeks (with no re-works) with smooth fab operations. For product manufacturing, the cycle, or lead time, also depends on fab priorities and increasing costs customer dependent from regular to hot-lots. So a careful production planning is needed for efficient use of a fab's human, machine, and economic resources to realize products in the shortest cycle times and with the lowest cost.

3.2.8 COVID-19 Effects

Recent trade sanctions and the COVID-19 pandemic have seriously disrupted the global supply and value chains in communication, electronics, semiconductors, optical components, software, IC chips, and so forth [60]. Some industries are more exposed to shocks due to the geographically dependent manufacturing bases. As value chains comprise a large number of companies with different sizes, models, and network relationships, multiple resilience strategies are needed to reduce vulnerable shocks.

VCSEL manufacturing depends on high quality substrates, and wafer-fab is also affected by cross-country networks. New operational choices such as multi-supplier sourcing, additional inventory stockpiling, reducing product complexity, and flexible production protocols are emerging. New business strategies such as avoiding non-delivery suppliers, monitoring multiple layers

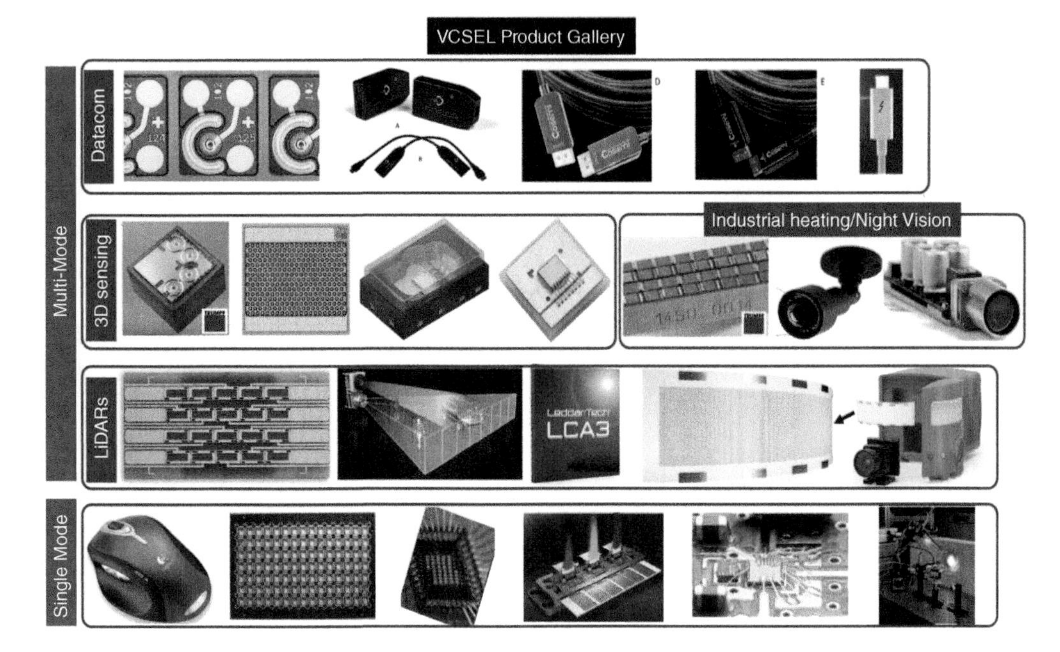

Figure 3.17 Examples of few MM and SM products. *Source:* Note: Necessary copyright permissions for images are obtained for all images shown above.

of supplier networks, and accelerating response times have been effectively implemented. Further, new trends of digitalization (non-production staff work from home) and smart automation (remote access to machines and people) are critical factors to optimize financial results. In nutshell, all these operating and business models need to be designed, calibrated, and tested against all possible force majeure situations for resilience and transparency for business efficiency and cost control [61].

3.3 VCSEL Commercial Products

In this Figure 3.17, the authors show some commercial examples of MM VCSEL-based products in datacom, 3D sensing, automotive LiDARs, and industrial heating and other SM products in atomic clocks, laser printing, optical mice, and so forth. Details of these products are discussed in respective chapters and appendices.

3.4 Summary

This chapter has provided VCSEL industry background and the essential ingredients of VCSEL manufacturing such as landscape, abilities, epi-growth, wafer-fab control, high-volume manufacturing challenges, industry players, business models, related supply chain, yield improvements, cycle times, etc. Insights into market prospects, chip demands, and approximate die costs have also been explained. Examples of VCSEL chips and single- and multi-mode VCSEL-based commercial products in datacom, 3D-sensing, automotive LiDAR, industrial heating, laser printers, atomic clocks, gas sensors, etc are presented.

References

1 F. Koyama, S. Kinoshita and K. Iga, 'Room temperature CW operation of GaAs vertical cavity surface emitting laser', *Trans. IEICE,* **E71**, 1089–1090 (1988).
2 J. L. Jewell, J. P. Haribson, A. Scherer et al., 'Vertical-cavity surface-emitting lasers: design, growth, fabrication, characterization,' *IEEE J. Quantum Electron.* **27** 1332–1346 (1991).
3 R. A. Morgan and M. K. Hibbs-Brenner, "Vertical-cavity surface-emitting laser arrays," *Proc. SPIE* **2398**, (1995); doi:10.1117/12.206347.
4 J. M. Dallesasse, N. Holonyak Jr., A. R. Sugg et al., 'Hydrolyzation oxidation of AlxGa1-xAs-AlAs-GaAs quantum well heterostructures and superlattices,' *Appl. Phys. Lett.*, **57**, 2844–2846 (1990).
5 R. S. Geels, S. W. Corzine, and L. A. Coldren., 'InGaAs vertical-cavity surface-emitting lasers' *IEEE J. Quantum Electron.* **21**, 1359 (1991).
6 J. A. Tatum, A. Clark, J. K. Guenter et al., "Commercialization of Honeywell's VCSEL technology," *Proc. SPIE* **3946** 2–13 (2000).
7 J. A. Tatum, J. K. Guenter, 'The VCSELs are coming' *Proc. SPIE* **4994** 1–11 (2003).
8 J. A. Tatum, 'VCSEL proliferation' *Proc. SPIE* **6484** 12–18 (2007).
9 D. M. Kuchta, A. V. Rylyakov, F. E. Doany et al., 'A 71-Gb/s NRZ modulated 850-nm VCSEL-based optical link' *IEEE Photon. Technol. Lett.,* **27**, 577 (2015).
10 B. D. Padullaparthi, R. Chen, J. R. Fei et al., 'Directly Modulated, 40Gb/s@25C &28Gb/s@85C 850 nm Multimode VCSELs for 100GBE, DCI &Green Photonics', **TuE3–5,** *OECC-2019*, Fukuoka, Japan.
11 M. V. R. Murty, D. Cunningham, L. Giovane et al., 'Mode partition noise characterization of 25Gb/s VCSELs' *Proc. SPIE* **9381** 938104 (2015).
12 https://www.laserfocusworld.com/lasers-sources/article/16548239/verticalcavity-surfaceemitting-lasers-vcsel-arrays-provide-leadingedge-illumination-for-3d-sensing.
13 https://www.lumentum.com/en/products/multi-junction-vcsel-array.
14 https://optics.org/news/11/11/11.
15 https://xueqiu.com/9231373161/135486811.
16 J. Guenter, L. Graham, B. Hawkins et al., "The range of VCSEL wear out reliability acceleration behavior and its effects on applications," *Proc. SPIE* **8639**, 86390I (2013) doi:10.1117/12.2002840.
17 https://www.laserfocusworld.com/lasers-sources/article/16564586/ultrahighreliability-highpower-vcsel-array-for-3d-sensing-and-gesture-recognition-unveiled-by-princeton-optronics.
18 T. R. Fanning, J. Maynard, C. J. Helms et al., "Performance, manufacturability and qualification advances of high power VCSEL arrays at TriLumina corporation," *Proc. SPIE* **11300**, 1130002 (2020).
19 A. Haglund, J. S. Gutsavsson, J. Vukusic et al., *'Single fundamental mode output power exceeding 6 mW from VCSELs with a shallow surface relief,' IEEE Photon. Technol. Lett.* **16**, 368–370 (2004).
20 Cisco Public, white paper 'Cisco Virtual Networking Index: Global Mobile Data Traffic Forecast Update 2017–2022', Feb. 2019 https://www.cisco.com/c/en/us/solutions/collateral/service-provider/visual-networking-index-vni/white-paper-c11-741490.html.
21 H. Moench, M. Carpaij, P. Gerlach et al., "VCSEL based sensors for distance and velocity," *Proc. SPIE* **9766**, 97660A (2016).
22 H. Moench, R. Conrads, C. Deppe et al., "High power VCSEL systems and applications," *Proc. SPIE* **9348**, 93480W (2015)
23 R. Wu, Technology: Optical Transceivers: How it differs in 5G and cloud, Jefferies Franchise Note (2017) https://www.jefferies.com/CMSFiles/Jefferies.com/files/Insights/Technology.pdf.

24 M. Vallo, P. Mukish, and E. Mounier, Yole Development Market Forecast on Optical Transceivers for Datacom & Telecom (2020) https://s3.i-micronews.com/uploads/2020/06/YDR20120-Optical-Transceivers-for-Datacom-Telecom-2020-Sample.pdf.

25 J. Mongardini and A. Radzikowski, 'Global Smartphone Sales May Have Peaked: What Next?' IMF Working paper WP/20/70 (2020) www.imf.org › Files › English › wpiea2020070-print-pdf.

26 Apple investment of 380M$ in Finisar in 2018 https://www.apple.com/newsroom/2017/12/apple-awards-finisar-390-million-from-its-advanced-manufacturing-fund.

27 AMS expanding their Singapore plant with 200M$ https://www.eenewseurope.com/news/ams-opens-singapore-site-under-200-million-plan-0.

28 IQE expands its capacity at Newport Mega Foundry 2019 https://www.iqep.com/media/2019/2019/05/iqe-announces-full-product-qualification-and-first-mass-production-order-for-newport-mega-epi-foundry-(1)/.

29 Yole Development & Woodside Capticals, The Automotive LiDAR Market (2018) http://www.woodsidecap.com/wp-content/uploads/2018/04/Yole_WCP-LiDAR-Report_April-2018-FINAL.pdf.

30 https://www.trilumina.com/latest-news/press-releases/trilumina-to-demonstrate-3d-solid-state-lidar-using-its-3d-sensing-vcsel-illumination-solutions-at-ces-2018.

31 https://www.st.com/en/imaging-and-photonics-solutions/vl53l0x.html.

32 https://www.lumentum.com/en/products/multi-junction-vcsel-array.

33 https://www.lumentum.com/en/media-room/videos/autosens-2020-high-power-vcsels-automotive-applications.

34 P. Boulay and P. Mukish., Yole Development, VCSELs: technology and market trends (2018). https://www.slideshare.net/Yole_Developpement/vcsels-market-and-technology-trends-2019.

35 https://www.lumentum.com/en/media-room/news-releases/lumentum-acquires-trilumina-assets.

36 https://globaluniversityventuring.com/nvidia-sweeps-up-optigot.

37 D. Cheskis, '6-inch VCSEL Wafer Fabrication Foundry Economics,' CS MANTECH Conference, May 18th - 21st, 37–39 (2015).

38 C. Wilmsen, H. Temkin and L. A. Coldren, "*Vertical-Cavity Surface-Emitting Lasers-Design, Fabrication, Characterization and Applications*, Cambridge University Press, 109–192 (1999) ISBN 0 521 00629 5.

39 https://www.allos-semiconductors.com.

40 K. L. Chi, J. L. Yen, J. M. Wun et al., 'Strong wavelength detuning of 850 nm vertical-cavity surface-emitting lasers for high-speed (>40 Gbit/s) and low-energy consumption operation,' *IEEE J. Sel. Top. Quantum Electron.* **21**, 1701510 (2015).

41 Y. S. Lee, M. W. DeVre, R. Westerman et al., 'Characterization of GaAs/AlGaAs non-selective ICP etch process for VCSELs applications,' http://plasma-therm.com/pdfs/papers/2002_CSMAX_VCSEL_Etch.pdf.

42 D. Thomas, "Plasma Processes for High Volume Manufacturing of VCSELs" https://www.novuslight.com/plasma-processes-for-high-volume-manufacturing-of-vcsels_N9982.html.

43 J. M. Dallesasse et al., 'Hydrolyzation oxidation of AlxGa1-xAs-AlAs-GaAs quantum well heterostructures and superlattices,' *Appl. Phys. Lett.*, **57** pp. 2844–2846 (1990).

44 K. D. Choquette, K. M. Gieb, C. I. H. Ashby et al., 'Advances in selective wet oxidation of AlGaAs alloys,' *IEEE J. Sel. Top. Quantum Electron.* **3**, 916 (1997).

45 K. Meneou, H. C. Lin, and K. Y. Cheng , 'Wet thermal oxidation of AlAsSb alloys lattice matched to GaSb, *J. Appl. Phys.* **95**, 5131 (2004).

46 R. Zhang, J. J. G. M. Van der Tol, H. Ambrosius et al., 'Oxidation of AlInAs for Current Blocking in a Photonic Crystal Laser' 978–1–4244-7333–5/10 @ IEEE 95–96 (2010).

47 Y. Cao, 'Investigation of InAlP native oxides for GaAs metal-oxide semiconductor device applications,' Ph. D. Thesis, University of Notre Dame, July (2006).

48 S. A. Maranowski, F. A. Kish, S. J. Caracci et al., 'Native oxide defined In0.5(AlxGa1?x)0.5P quantum well heterostructure window lasers (660 nm),' *Appl. Phys. Lett.* **61**, 1688 (1992).
49 B. D. Padullaparthi, P. Chen, Q. L. Qin, J. R. Fei; 'Low speckle laser array and image display thereof', US 10910791, February (2021).
50 A. Pizzagalli and G. Giusti, https://s3.i-micronews.com/uploads/2020/06/Thinning-Equipment-Technology-and-Market-Trends-for-Semiconductor-Devices_sample-2.pdf.
51 S. Calvez, G. Lafleur, A. Arnoult et al., 'Modelling anisotropic lateral oxidation from circular mesas' *Opt. Mater. Express* **8**, 1762 (2018).
52 https://resource.lumentum.com/s3fs-public/technical-library-items/diodelaser3d-wp-cl-ae.pdf.
53 R. P. Aranda, 'VCSEL reliability analysis for technical feasibility assessment' IEEE 802.3 OMEGA Study Group - November (2019) http://www.ieee802.org/3/OMEGA/public/nov_2019/perezaranda_OMEGA_05a_1119_VCSEL_Reliability.pdf.
54 S. J. Oh, 'A Study of the Foundry Industry Dynamics' Master's Thesis Submitted to the MIT Sloan School of Management May 7 (2010) https://dspace.mit.edu/bitstream/handle/1721.1/59145/659531602-MIT.pdf;sequence=2.
55 T. Levitt, 'Exploit the product life cycle' *Harv. Bus. Rev.* (1965) https://hbr.org/1965/11/exploit-the-product-life-cycle.
56 R. C. Masanell, and J. E. Ricart, 'From Strategy to Business Models and to Tactics *Harvard Business School*', Working Paper 10–036 (2009).
57 C. Gilligan and R. M. S. Wilson, Strategic Marketing Planning 'The Formulation of Strategy 3: Strategies for Leaders, Followers, Challengers and Nichers' **Chapter-12** 455–518 (2009).
58 U. Naeher, S. Suzuki and B. Wisemen, 'The evolution of business models in a disrupted value chain, 2011 https://www.mckinsey.com/~/media/mckinsey/dotcom/client_service/semiconductors/pdfs/mosc_1_business_models.ashx.
59 K. D. Bacaer, M. Mancini, R. J. Huan et al., 'Taking next leap forward in semiconductor yield improvement' *Mckinsey & Company*, April (2018) https://www.mckinsey.com/industries/semiconductors/our-insights/taking-the-next-leap-forward-in-semiconductor-yield-improvement.
60 https://www.mckinsey.com/business-functions/operations/our-insights/risk-resilience-and-rebalancing-in-global-value-chains.
61 https://www.mckinsey.com/business-functions/operations/our-insights/risk-resilience-and-rebalancing-in-global-value-chains.

Bibliography

[1989JEW] J. L. Jewell et al., ' ...... ' *Proc. Integrated Optics and Optical Fiber Communication Conf. IOOC* '89, Postdeadline paper 18B2–6 (1989).
[2015GUN] https://spectrum.ieee.org/computing/hardware/intel-talks-thunderbolt-3.
[2019SPI] T. Spillett, 'Is it better to be a first mover or a fast follower?' (2019) www.bdo.co.uk/en-gb/plugdin/insights/is-it-better-to-be-a-first-mover-or-a-fast-follower.
[2020II-VIN] https://optics.org/news/11/11/11.
[2020IQEN-2] https://www.londonstockexchange.com/news-article/IQE/iqe-plc-development-of-iqgevcsel-150-tm-technology/14739534.
[2020MSS(PRC)] https://en.msscorps.com/archive/MSS%20newsletter%2020f_e.pdf.

4

Data Communications Applications

Jim Tatum

4.1 Introduction

In the early 1990s, fiber optic networks were defined by the Fiber Distributed Data Interface (FDDI) operating at 100 Mbps and utilized light-emitting diodes (LEDs) and multi-mode optical fiber (MMF). As the Internet began to grow and the world was becoming more connected, there was a clear need to increase the bandwidth of data interconnections. Both Ethernet and Fiber Channel standards organizations began to develop protocols for 1 Gbps connections. The logical light choice was edge-emitting lasers (EELs), which were widely available in other commercial applications. However, these lasers were used primarily in consumer devices and lacked the reliability needed in data communications systems. (Aspects of reliability are further discussed in Appendix E.) Standardization using MMF was highly desirable because it offered a lower total cost of ownership in comparison to single-mode fiber (SMF) solutions at that time. Some economic aspects of optical interconnects are described in Section 4.4.5. The need for a new laser type for MMF was clear, and vertical-cavity surface-emitting lasers (VCSELs) emerged to cover the burgeoning demand. As described in Chapter 1, initial government funding in the United States for VCSELs was driven by military demand for high-speed interconnects. This funding, coupled with the growing Internet demand, led to the first commercial application for VCSELs in short-distance optical communications on MMF running Ethernet and Fiber Channel protocols. The choice of 850 nm devices was driven by the embedded optical fiber networks that had been manufactured and characterized at 850 and 1310 nm. There was natural skepticism in the industry about the use of VCSELs until the first published results on VCSEL reliability in [1]. These first VCSELs utilized proton implantation to provide the electrical current confinement and were somewhat difficult to drive at higher speeds. As the demand for worldwide data continued to expand, the need for even higher speed VCSELs was clear. A new technology using an intracavity oxide layer to improve the carrier injection, optical confinement, and absorption loss from the implant layer was introduced [2]. The first reliability studies quickly followed [3, 4]. Early oxide isolated VCSELs were capable of operating at up to 4 Gbps. Subsequent improvements in design and manufacturing have produced widespread availability of VCSELs capable of 28 Gbps and emerging capability for 28 GBd (56Gbps) using pulse-amplitude modulation (PAM), and now serial data rates of 56 Gbps are on the near horizon. Through 2020, nearly 1B VCSELs are in operation, carrying Internet and other data traffic worldwide. As the global demand for bandwidth and higher-speed computing continues to grow, so will the overall data communications market, and VCSELs will continue to

VCSEL Industry: Communication and Sensing, First Edition. Babu Dayal Padullaparthi, Jim A. Tatum and Kenichi Iga.

be dominant in short-distance interconnects. This chapter will cover the operation of VCSELs in data communication networks and introduce the various standards and modulation types, operating speeds, transceiver types, form factors, and data center architecture and discuss the current trends in the market.

4.2 Growing Data

The demand for data, and the delivery of that data instantaneously, continues to drive the build-out of massive data centers and delivery infrastructure. To understand how that demand is evolving, consider a simple example of a single live event being streamed just in the United States. For simplicity, we assume that there are 100M households and that the HD video stream requires approximately 5 Mbps. This single event requires 500 000 Gbps in Internet capacity, or the entire Internet in 1997! This does not include the overhead for switching, storage, and any other data traffic demand. It is easy to understand how the demand for data has grown and will continue to grow over the coming years. The demand for even more video on demand (VOD), the emergence of 4K television broadcasts, the Internet of Things (IoT), augmented reality (AR) gaming and conferencing, and autonomous driving are all new drivers of global bandwidth. Figure 4.1 shows the exponential growth of Internet traffic over the last 30 years in both fixed line and mobile connections.

From Figure 4.1, the Internet traffic has grown by a factor of 10^7 from 1995 to 2020 [5]. In the same time period, the predominant single-line rate of data interconnects has grown from 1 to 25 Gbps. Figure 4.2 combines the total (fixed + mobile) Internet traffic growth from Figure 4.1 with the increase in the single-channel line rate over the same time frame. The figure also includes lines for Moore's law [6]. We expand Moore's observation of the doubling of transistors every 18 months to determine the time for doubling in both Internet traffic and line rate. Through 2005, the total Internet traffic was doubling every year, but has since slowed to a doubling every two years. In aggregate, the blue line in Figure 4.2 shows Internet traffic doubling every 1.5 years. When the single-channel line rates are plotted over the same time period, the doubling period is

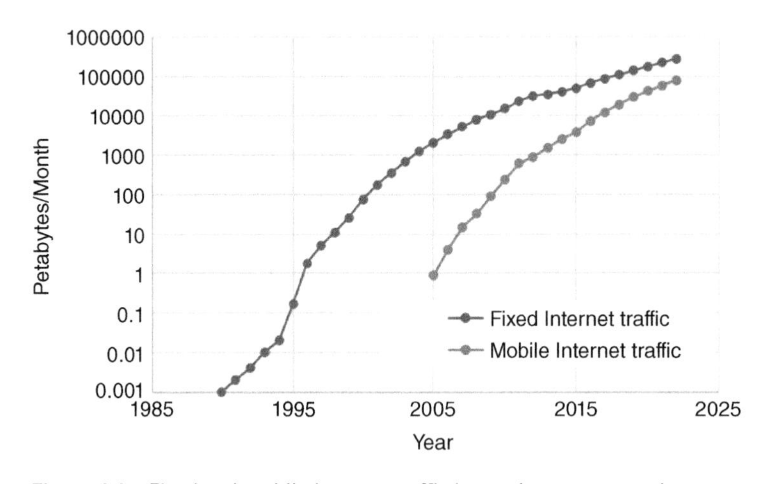

Figure 4.1 Fixed and mobile Internet traffic in petabytes per month.

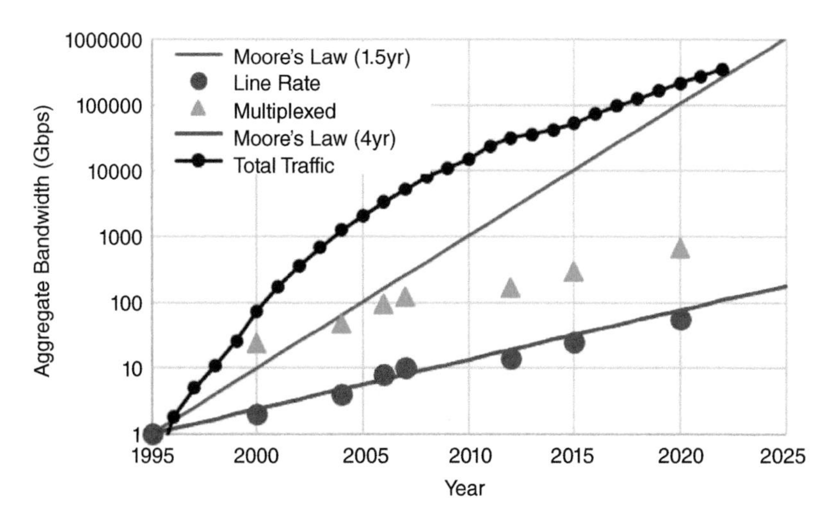

Figure 4.2 Total Internet traffic (black line), Moore's law at 1.5 years (blue line), single-channel data rate (red circles), Moore's law at 4 years (red line), and 12x multiplexed data rate (green triangles) as a function of time.

every four years. This divergence in demand and supply has driven the adoption of various multiplexed solutions and other innovations. Still the fundamental line rate has not kept up with overall demand growth, which has in turn driven demand for a larger number of interconnects to fill the demand gap.

4.3 Data Centers and High-Performance Computing

There has been a tremendous build-out of infrastructure to deliver the bandwidth described in Section 4.2. The visible connectivity growth for most people has been at the home or business side (referred to as the client side). Connectivity has come in many forms, including wireless (5G), satellite, and in many cases fiber to the home (FTTH) or business. Networks within a home or a business are referred to as a local area network (LAN). MMF and VCSELs are common within a LAN. The FTTH installation is typically SMF connections up to 10 km and not a normal application for multi-mode VCSELs. All of those connections are aggregated in data centers filled with racks of switches and servers. These local connections are then aggregated onto a metropolitan area network (MAN), which can span hundreds of kilometers and several cities. The MAN is interconnected to create a wide area network (WAN), which can span entire countries or continents. Data centers have been constructed globally to handle the hosting and storage requirements driven by the bandwidth expansion, and this is where the vast majority of VCSEL-based data links are deployed. The second major communications sector is in high-performance computing (HPC) and has been driven by the race to exaflop (10^{18} floating-point operations per second) computing. HPC centers can have many hundreds of thousands of VCSEL-based interconnects. It is estimated that nearly 1B VCSELs are currently in use in data centers and HPC centers around the world [7].

4.3.1 Data Centers

To meet the bandwidth demand driven by applications such as e-commerce, VOD, videoconferencing, data storage, and IoT, massive data centers have been constructed round the world. Data centers can range in size from a small closet in an office building to many thousands or even millions of square feet. The largest one in 2020, The Citadel, located near Reno, Nevada, is over 7 200 000 ft^2 (670 000 m^2), and the top ten data centers total more than 25M ft^2 (2.3M m^2). Datacenters are classified in several different ways, including size and operational reliability, as defined in Tables 4.1 and 4.2 [8].

With the continued expansion of data center size and required uptime, two of the principal concerns are the availability of low-cost power and efficient cooling resources. Cooling has become such a power concern that several data center providers have begun building completely submergible pods to house the equipment underwater. Section 4.8 describes some of the recent advances in energy-efficient VCSELs to help reduce the cooling requirements in the data center.

4.3.2 High-Performance Computing

The growth in HPC capacity has averaged a 10x improvement every four years since 1990, as shown in Figure 4.3 [9], and the race to the first exaflop computer is on and expected to conclude within the next year or two. Optical interconnects were first widely adopted in the IBM

Table 4.1 Classification of data centers by size.

Data center type	Typical size (ft^2)	Typical equipment
Server closet	100	1–2 servers No external storage
Server room	<500	<12 servers Limited external storage
Localized data center	<1000	<100 servers Moderate external storage
Mid-tier	<20 000	<500 servers Extensive external storage
Enterprise	<50 000	Several thousand servers Extensive external storage
Hyperscale	>50 000	Many thousands of servers Extensive external storage

Table 4.2 Classification of datacenters by reliability.

Metric	Tier 1	Tier 2	Tier 3	Tier 4
Single point failure	possible	rare	unlikely	none
Annual uptime (%)	99.671	99.741	99.982	99.995
Redundancy	N	N + 1	2N	2N + 1
Downtime per year	28.8 h	22 h	1.6 h	26.3 min
Power fault protection	none	unspecified	72 h	96 h

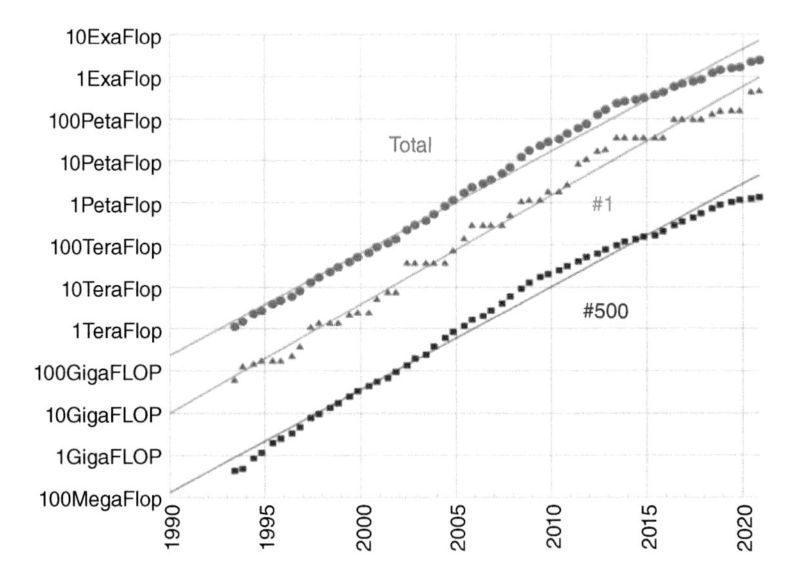

Figure 4.3 Total computing capacity (circles), the fastest (triangles), and the 500th fastest of the top 500 supercomputers over the last three decades.

Road Runner in 2008 with 40 000 links, and in 2011 IBM Sequoia employed 620 000 fibers with MMF and VCSELs being dominant. The ability to compactly integrate transceivers onto the processor board and the mechanical flexibility of the optical fibers were major reasons for adoption. A general rule in HPC centers is that the connectivity bandwidth scales directly with total processing capability (1B/FLOP). With this projection, a single exaflop computer will need 1M 100 Gbps links! Over the course of about six to eight years, the top HPC slides to ~500th fastest in the world. Total computing speed can be limited by latency (time delay) in transmitting data between processors and memory, so there is a strong preference for simple on-off keying (OOK) to minimize latency and power consumption, as further described in Section 4.5.

4.3.3 Structure of Data Centers and HPC Centers

Optimization of network traffic in a data center and HPC environment requires different considerations. Data centers are generally optimized for the flow of data into and out of the data center, so called north-south traffic. The preferred data center network topology is often referred to as a fat tree or leaf and spine, as depicted in Figure 4.4a. Data coming into the data center has a direct path to its destination, and the amount of switching is minimized in order to maximize the throughput. This is advantageous for accessing and serving information to a user but not optimized for computation. In an HPC environment, the overriding concern is communication between processor and memory or other processors, or so-called east-west traffic. The choice of architecture drives the number and length of data links. In an HPC center, for example, there are few links over 30 m and are dominated by multi-mode optics and VCSELs [7]. As data centers have grown, the number of links over 100 m has also grown significantly. Figure 4.5 shows the leaf and spine structure used in a Facebook data center [10]. The bandwidth demand in the core of the leaf and spine networks has continued to expand and is now deploying 400 Gbps optical links.

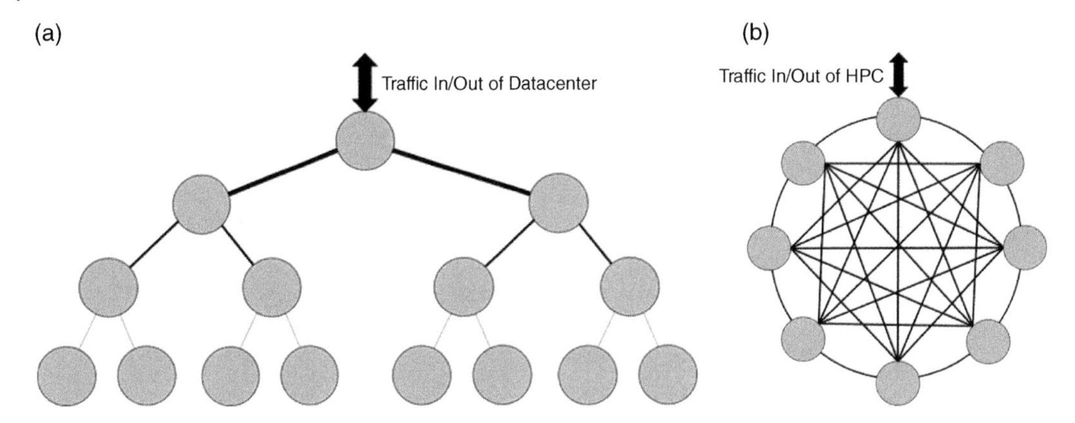

Figure 4.4 (a) Fat tree or spine and leaf architecture preferred by data centers. (b) Dragonfly architecture often used in modern HPCs.

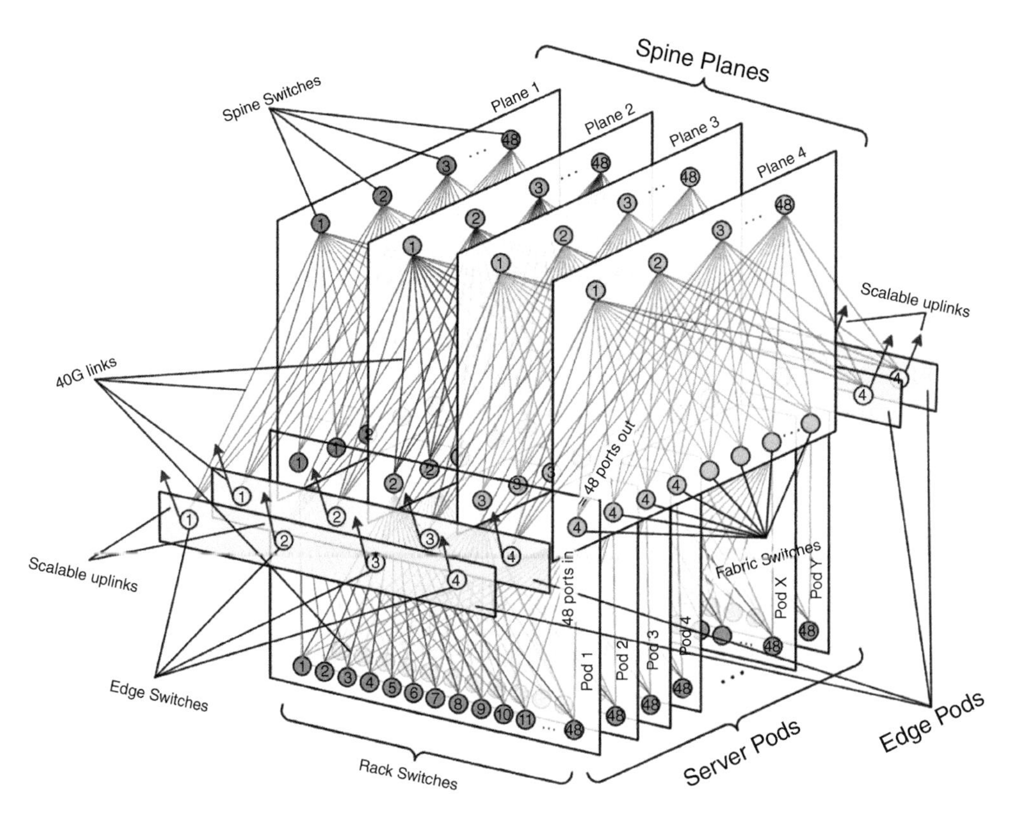

Figure 4.5 Architecture of a modern Facebook data center [10].

4.4 Optical Interconnects

4.4.1 Introduction

Traditionally, the cost of optical interconnects dictated that they were only used when required for distance. As the data rates have increased, optical interconnects have been more widely deployed and the cost difference with electrical interconnects has eroded. Today, some of the biggest

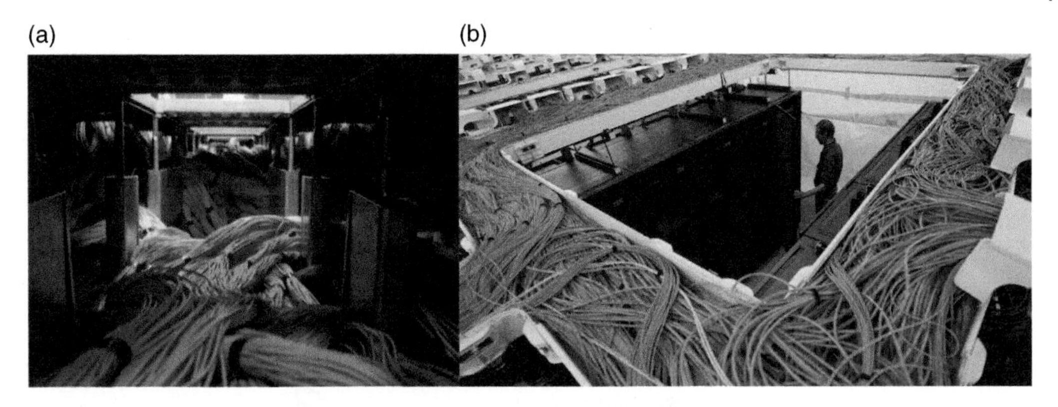

Figure 4.6 Typical cable carrier racks in a large data center. (a) This particular installation has more than 650 000 multimode fiber connections [7]. (b) The SuperMUC at the Leibniz Supercomputing Center. *Source:* Image taken from www.redditt.com.

Table 4.3 Classification of data centers by reliability.

Connection type	Wire diameter	Bend radius (mm)	Weight (g/m)	Length (m)
DAC	30 AWG	38	>200	<3
Active DAC	24 AWG	60	>200	<10
Optical fiber	2 mm	20	<5	<100

advantages to fiber optic cabling is reduction in size and weight, flexibility of the cabling, and reduction of electromagnetic interference. These concerns are driving the adoption of even more optical cabling in both data centers and high-performance computers. For example, Figure 4.6 shows the very congested cabling that can occur in a large data center, where all the cables shown are multi-mode fiber cables. Table 4.3 shows the comparative mechanical characteristics of Direct Attach Copper (DAC) and fiber cabling operating at 25 Gbps.

4.4.2 Networking Communications Standards

To preserve the integrity of data connections and system interoperability, communications standards have long been used to define the electrical and optical requirements of a port. Other industry standards and manufacturing sourcing agreements have been developed to address the mechanical and thermal requirements. Ethernet is focused on LAN and WAN Internet traffic, ANSI X3.T11 (Fibre Channel) focused on storage area networks (SANs), and InfiniBand focused on chip-level connectivity; all of them have come to the be the most important standards, though several others such as Optical Interconnect Forum (OIF) and PCI Express are also commonly used protocols. Figure 4.7 shows the evolution of Ethernet, Fibre Channel, and InfiniBand speeds over time [11, 12, 13].

4.4.3 Optical Transceiver Types

Optical transceivers (TRX) can be classified by how they connect to the equipment. The most ubiquitous TRX are at the edge of the equipment. Early TRXs were soldered directly to the network boards and have now evolved to be hot pluggable. These TRX typically have an electrical connector

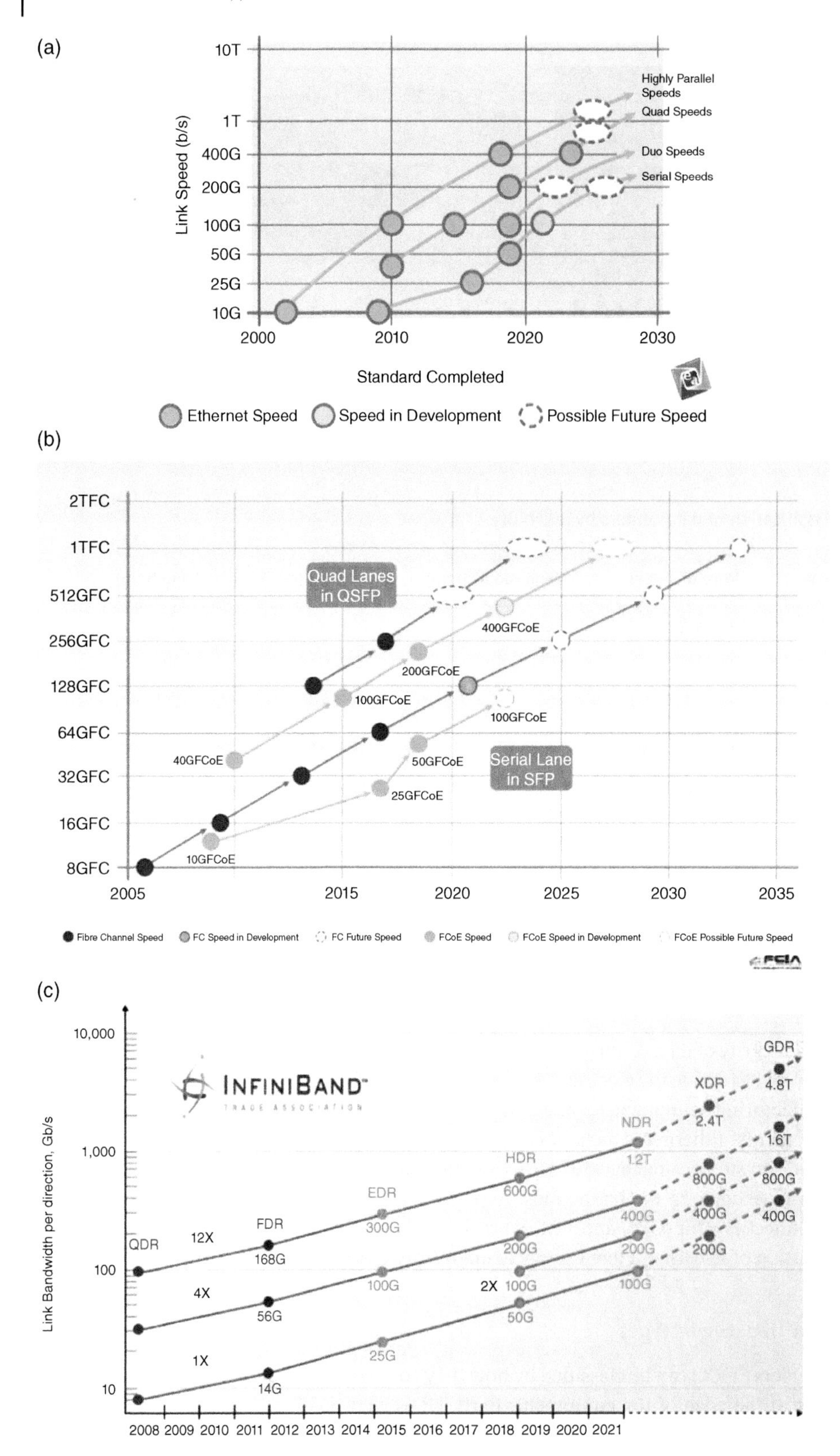

Figure 4.7 Evolution of communications standards speeds (a) Ethernet speeds [11], (b) Fibre Channel [12], and (c) InfiniBand [13]

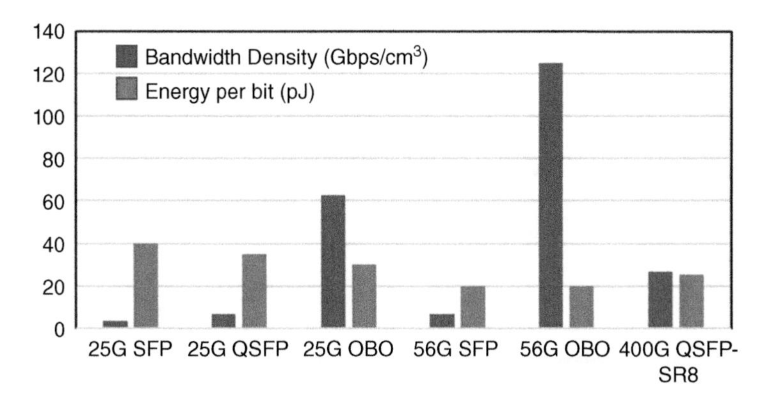

Figure 4.8 Comparison of bandwidth density and energy per bit for some common VCSEL transceiver types.

and a fiber optic connector. More recently, TRXs with permanently attached fiber cables have emerged and are known as active optical cables (AOCs) and are in direct competition with passive and active DACs. The primary advantage of AOCs is that they are tested as a link, electrical input to electrical output. This eliminates the requirement to independently verify the optical output of the transmitter (TX) and the receiver (RX) functionality as defined in the many standards. The reduced testing translates directly to reduced cost of an AOC link when compared to two TRX and associated fiber cabling. The electrical form factor of an AOC is identical to a TRX, and they are electrically interchangeable. TRX and AOCs are at the edge of the equipment. TRXs and AOCs can be single- or multi-channel and employ different signaling types, as described in Section 4.5. One metric of AOCs and TRX is the maximum faceplate density, defined as the total bandwidth connectivity that can be placed in a common 19 in. equipment rack. For example, one of the early transceivers was the gigabit interface converter (GBIC), and 12 could fit on a 1U rack, creating a 12 Gbps faceplate density. The GBIC was replaced by the Small Form Factor Pluggable (SFF/SFP) that allowed up to 36 transceivers and with speeds now evolved to 50 Gbps, thus creating a faceplate density of 1.8 Tbps in a single 1U rack. Packaging transceivers together in parallel introduced the Quad Small Form Factor (QSFP) pluggable with four channels per transceiver; with 24 TRX in a 1U rack operating at 50 Gbps per channel it yields 4.8 Tbps in faceplate density. While these are impressive numbers, the desire to further increase the faceplate density and simplify on-board electrical routing has led to the development of on-board optical (OBO) TRXs that further increase the faceplate density to more than 10 Tbps. Figure 4.8 compares some common footprint transceivers for bandwidth density measured in Gbps/cm^3, and the total energy per bit required for transmission. Figures 4.9 and 4.10 show how some of the available form factors of optical TRXs, AOCs, and OBOs are utilized in a network interface.

4.4.4 Consumer Connectivity

Optical interconnects are not only used in large data centers and supercomputers, they are also finding application in some consumer electronic connections such as High-Definition Media Interface (HDMI) and long-distance Universal Serial Bus (USB) connections. The total serial bandwidth needs of a 4K television with 3840 × 2160 resolution, 8-bit color, and without chroma subsampling with 60 Hz refresh rate is 17.92 Gbps. The speed doubles at 8K resolution, and doubles yet again for 16K. HDMI cables are generally short, 1 or 2 m, and utilize passive copper cabling. To transmit longer distance, active HDMI cables can be used, and recently AOCs have been

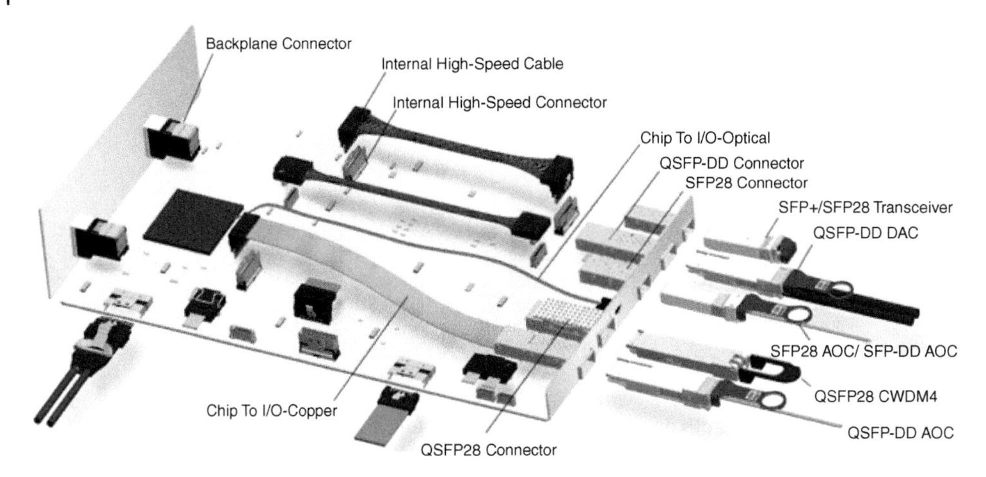

Figure 4.9 Schematic of common optical interfaces used in networking equipment. *Source:* Image courtesy of LR-Link [43].

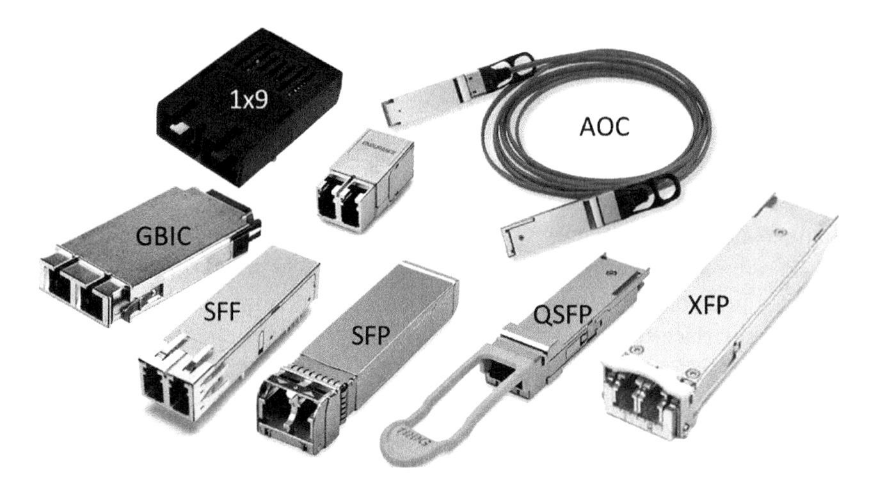

Figure 4.10 Some common VCSEL-based optical transceivers, active optical cables, and on-board optical components [43].

introduced to transmit even longer distance and to minimize the cable presence (size) in home media or theater environments. USB 3.0 operates at speeds up to 5 Gbps and the maximum recommended cable length is 3 m for passive copper. Some applications may require longer distances, for example in an auditorium, and AOCs with direct USB connections are now available. Other cables such as FireWire are also available as AOCs. All of these utilize VCSELs and MMF. Figure 4.11 shows some examples of HDMI and USB AOCs and adaptors that are currently available.

4.4.5 Techno-Economic Comparison of Transceiver Technology

Data centers continue to get larger and more of them are using SMF links, and that has driven the cost down and improved availability of the transceivers. However, VCSEL-based transceivers continue to be the lowest-cost optical interface, and single-mode transceivers are more than 2x the price of multimode transceivers [14]. The choice of technology comes down to distances needed and system complexity. In data centers, more than 90% of the links made with multi-mode fiber are under 100 m, and the average length is under 50 m [14]. In supercomputer applications, more than

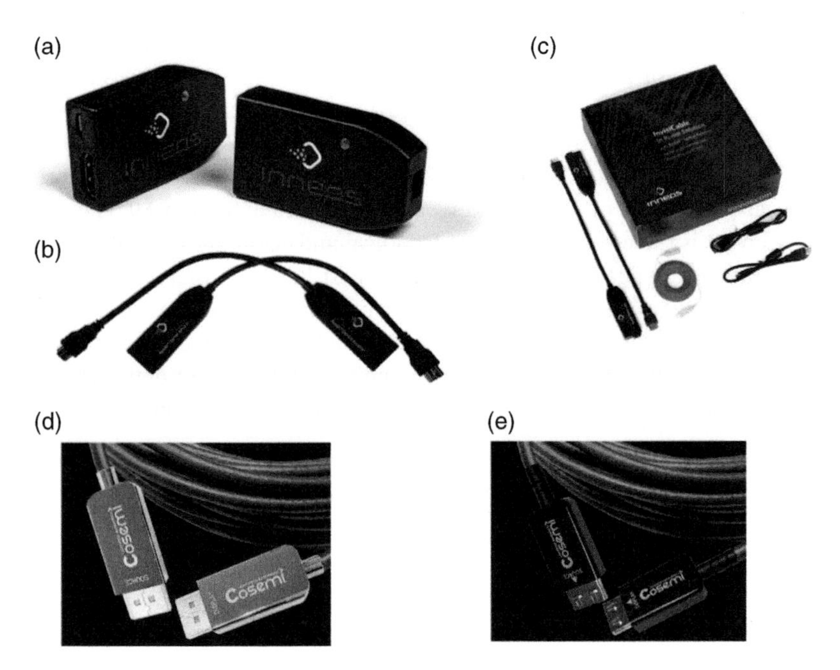

Figure 4.11 (a, b) Converter for HDMI electrical to optical link extension capable of supporting the full 18 Gbps bandwidth of 4K resolution. (c) Complete HDMI link extension with InvisiCable™ to reduce cable visibility in installation. (d) USB AOC. (e) HDMI AOC. *Source:* Photos courtesy of Inneos Corp. and Cosemi Technologies.

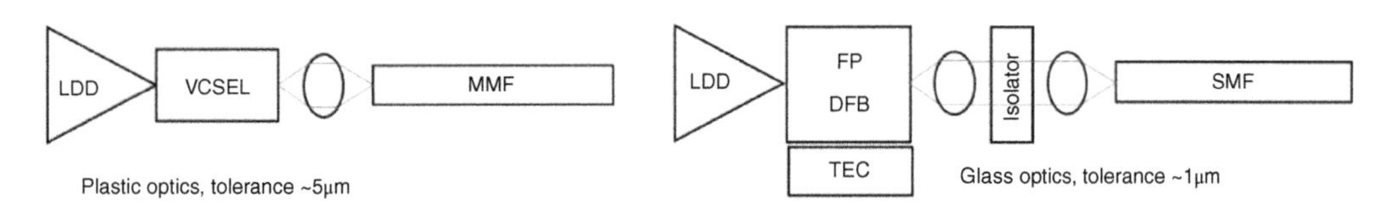

Figure 4.12 Comparison of components for a VCSEL transmitter and an EEL-based transmitter.

90% of the fiber links are less than 30 m [7]. So why are multi-mode transceivers so much less expensive? The answer to this question is driven by both technology and manufacturing issues. VCSEL-based TRXs tend to be the simplest to manufacture and have the lowest bill of materials, and that comes down to the Tx and Rx assemblies, as indicated in Figure 4.12. A VCSEL-based transmitter consists of the laser, a power monitor photodiode, and a plastic optical component to align the laser to the fiber. The alignment tolerances allow for the use of low-cost epoxy assembly. An SMF transceiver includes a laser, a monitor photodiode, an optical isolator, and metal assemblies that are welded together with sub-micron tolerances. Additionally, distributed feedback (DFB) lasers may require a thermo-electric temperature controller to operate over a wide temperature range. In many cases, the VCSEL die is self-hermetic, whereas indium phosphide (InP) –based EELs may require a hermetic housing for reliable operation. On the receiver side, the MMF parts can again use plastic components and epoxy assembly processes whereas the SMF again requires more precision and cost. For practical purposes, the associated electronics of a laser driver, transimpedance amplifier, limiting amplifier, and printed circuit board are essentially the same. Similarly, the mechanical components such as the housing and connectors are essentially the same.

The manufacturing cost of the VCSEL compared to an FP or DFB laser is significantly lower due to material cost, ease of testing, and back-end chip handling, as described in Chapter 3. The advent of silicon photonics was set to change this economic paradigm but has not yet been realized [14].

With AOCs becoming more prominent, the cost curve of VCSEL-based TRX continues to be ahead of SMF TRXs. The practical choice of much lower power dissipation for VCSEL TRXs is also a compelling reason for their continued use in data centers and supercomputers. In addition, the ability to readily make parallel fiber interconnects, break out cables, and flexible routing are other key technical advantages. Historically the economics have also favored VCSEL transceivers because of the higher total volume, about 10x that of SMF transceivers, as described in Section 4.9. The real question for VCSEL transceivers is the crossover point in technical and economic considerations when compared to Direct Attach Copper. A general rule in the past has been that when optics competes with copper cables, copper wins. This is still largely true today, but the crossover point has moved from 10 m in 2000 to just 1 or 2 m today. The practical implementation of copper cable management has also become an issue, and the size, weight, and flexibility of fiber optic cabling coupled with low-power dissipation and bandwidth density has driven the adoption of VCSEL-based MMF interconnects in many cases.

4.5 Data Encoding and Multiplexing

4.5.1 Introduction

The simplest optical communications system to consider is the presence (digital “1”) or absence (digital “0”) of light, and in its most primitive form was used by early man with candles and lamps to signal ships passing in the night. The bandwidth of those networks was at most a few hertz (Hz). The operating principle is the same for higher-speed networks, and this simple protocol is referred to as OOK or more generally today as two-level pulse-amplitude modulation (PAM-2). A digital byte is transmitted as a series of PAM-2-encoded bits of light. The line rate of the data is equal to half of the data rate. For example, a 1 Gbps communications channel operates at a fundamental frequency of 500 MHz. A general rule for PAM-2 coding is that the individual components need a fundamental frequency response of 80% of the data rate. In this example, the minimum bandwidth of a laser transmitter and corresponding receiver would need to be 0.8 GHz. At low data rates, it is relatively easy to achieve the required bandwidth, but as PAM-2 data rates are now approaching 56 Gbps, the required bandwidth is more than 40 GHz, a much more challenging task for the optical components especially when operation over time and environmental conditions are considered.

4.5.2 Spatial and Wavelength Multiplexing

One approach to scaling aggregate date rates is to multiplex the optical signal on to multiple channels. With VCSELs this can be done with either spatial division multiplexing (SDM) using parallel ribbon fiber cables or using wavelength division multiplexing (WDM) by incorporating multiple wavelengths on a single optical fiber. Figure 4.13 shows examples of single channel along with both spatial and spectrally multiplexed signals.

In a single-channel connection, a single data line connects directly to the receiver. In a spatially multiplexed system, a single data channel is split into multiple lower data-rate channels, and each lower-rate channel is transmitted on an independent optical fiber. It is not a requirement that the channels operate in concert in either the WDM or SDM example, and in fact many system designers operate the channels independently and take advantage of the increase in bandwidth density offered in these transceivers, as described in Section 4.4.3. The network architectures described in Section 4.3.3 lend themselves well to SDM and WDM transceivers. In this case, each of the channels can operate independently and be routed to different locations with a specialized breakout cable.

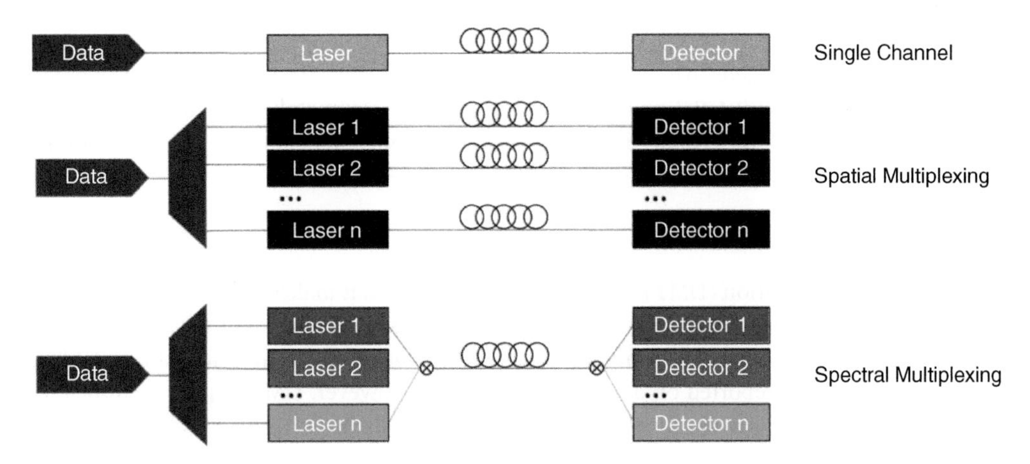

Figure 4.13 Schematic of single, SDM, and WDM optical channels.

4.5.3 Pulse-Amplitude Modulation (PAM-n)

For serial data rates up to 25 Gbps, pulse-amplitude modulation with two levels (PAM-2, or often referred to as non-return to zero, NRZ) is the preferred method of data encoding. This has the distinct advantages of providing the lowest data latency and generally lowest cost and system complexity. The disadvantage to PAM-2 is it requires the highest bandwidth of the laser and detector per bit transmitted. However, with tailoring of the electrical drive signal, lower bandwidth components can be used. Efficiently driving VCSELs at high speed will be further discussed in Section 4.6. The limited laser and detector bandwidth can also be overcome by using more advanced data encoding methods to more efficiently utilize the available bandwidth. For example, to extend an Ethernet system to 56 Gbps the IEEE has adopted four-level pulse amplitude modulation (PAM-4) to encode multiple bits into one symbol. Figure 4.14 shows an example of the PAM-2 and PAM-4 level encoding of multiple signal levels in a single symbol and the corresponding eye diagrams.

PAM-4 encoding has the distinct advantage of transmitting twice the number of bits in the same channel bandwidth when compared to PAM-2. This is especially useful as the bandwidth of the VCSELs and detectors are becoming a limitation in overall system performance. However, the trade-off in bandwidth places other requirements on the lasers. As can be seen in the optical eye diagrams, the multi-level signaling in PAM-4 decreases the amplitude difference between the relative signal levels. This, in turn, places more stringent requirements on laser relative intensity noise (RIN) and linearity of the modulation signal. As an example, consider a signal level of 1 mW

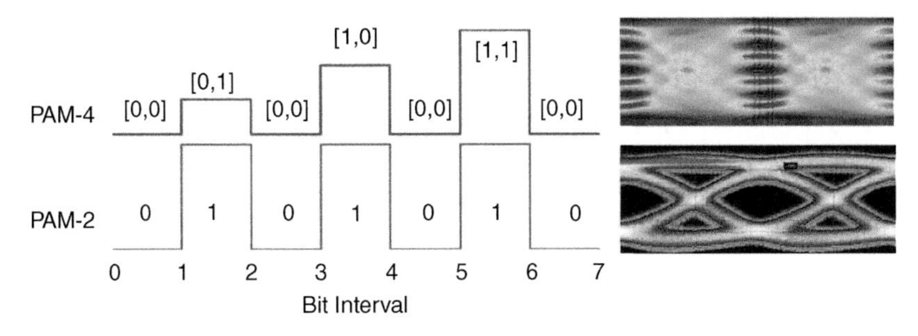

Figure 4.14 Schematic of PAM-2 and PAM-4 signaling with representative optical eye diagrams at 25 Gbps line rate.

as the optical 1 and 0.1 mW as the optical zero. To operate at a bit error rate (BER) of 1e−12, the signal-to-noise ratio (SNR) needs to be >7.04 on both logical levels 1 and 0. PAM-4 encoding would have logical levels at 0.1 mW, 0.4, 0.7, and 1 mW. This leads to approximately a 3.5 dB noise penalty in the link. Details of link budget analysis will be further discussed in Section 4.7. PAM encoding with VCSELs has been extended to even higher levels such as PAM-8 [15].

4.5.4 Discrete Multi-Tone Modulation (DMT)

Discrete multi-tone modulation (DMT) is a modulation method that makes the best use of available analog bandwidth of the components in the system by breaking the signal into frequency sub-bands less than the bit rate. Some of the highest speed VCSEL links have been achieved with DMT, with more than 160 Gbps reported on a single channel [16]. However, DMT requires significant signal processing to encode and decode the system, and the added power has essentially prevented adoption in modern interconnects.

4.5.5 Other Modulation Formats

As speeds continue to increase, modulation formats often move from amplitude modulation to schemes that incorporate the phase of the carrier signal. Phase-shift keying (PSK) and its associated forms of binary PSK (BPSK), quadrature PSK (QPSK), and others can add to the carrying capacity of a modulation format by adding phase states in a similar way that PAM added levels in amplitude. Coherent modulation formats are emerging commercially as a way to further increase the bandwidth. In this case orthogonal polarization signals are modulated to encode the data. Because PSK and coherent solutions require the preservation of phase, they are best used with single-mode optical fiber and are not commonly used with VCSELs and MMF.

4.5.6 Analog and Radio Access Modulation

The vast majority of data communications links with VCSELs use digital modulation techniques. As the 5G network is rolled out, there is a growing demand for analog fiber optic links that connect the data center or digital world to the analog transmission and vice versa [17]. This is the so-called fronthaul links and is a bottleneck in the 5G rollout. The distance between towers in a 5G network (~100 m for 28 GHz broadcast) is significantly less than in a 4G and LTE (~10 km) environment and requires more back-end connectivity to carry the traffic, as depicted in Figure 4.15. This results in a dramatic increase in the amount of fiber in the access network. With decreasing distances between radio points, multi-mode fiber is becoming a more viable option for connecting access points. Some implementations may use analog radio signals over fiber to connect to the cell towers. However, this may not be the best solution, and digitizing signals before transmission to the central tower may become a more attractive and robust configuration. Analog lasers and transceivers are more difficult to mass-produce and as the network is pushed closer to the user, simpler solutions will likely prevail. Over the next decade, many tens of millions of picocells that have coverage of 100 m or less will be deployed in densely populated areas. Femtocells, with coverage of a few tens of meters will also be installed in homes and office environments. These small cells will drive high-speed fiber optic connectivity to aggregation points in network closets and other stand-alone local facilities that connect to broader fiber optic networks.

4.5.7 Modulation Format Conclusion

Modulation formats have evolved from simple PAM-2 to PAM-4 as the data rate increased from 28 Gbps to 28 GBd (56 Gbps). Additionally, to close the link budgets at longer distances, error correction algorithms have been implemented. Forward error correction (FEC) has been included

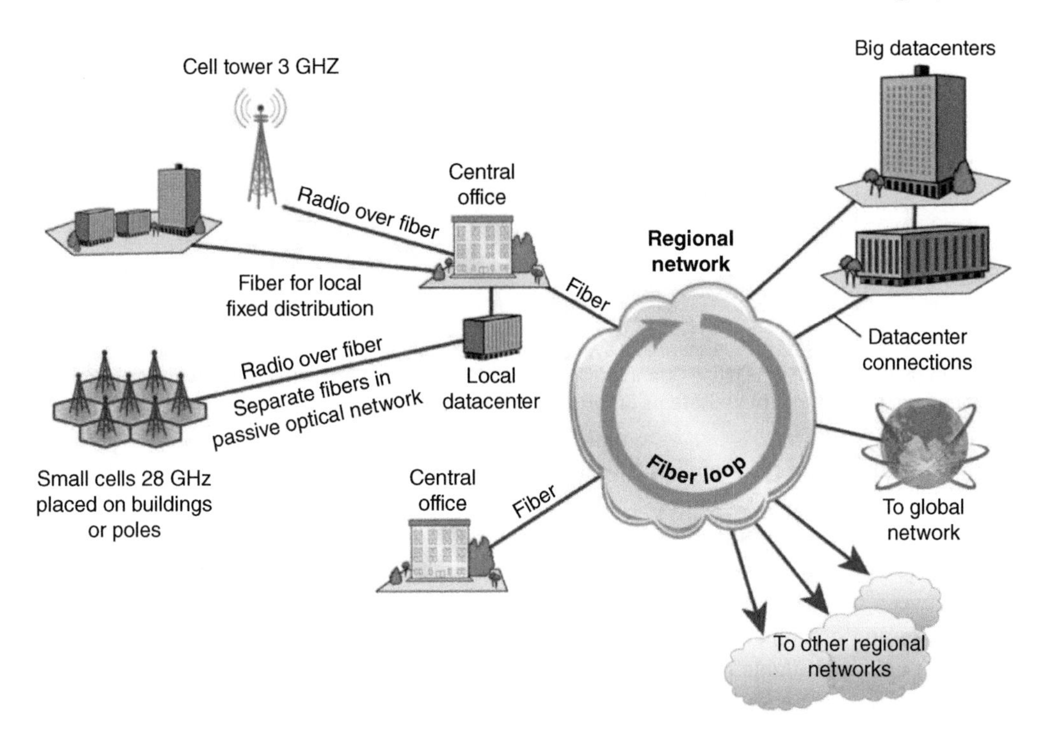

Figure 4.15 Schematic of a 5G network [17].

in the Ethernet and Fiber Channel standards, and together with PAM-4 signaling, has increased the complexity of laser and transceiver design. At the system level, this has added power dissipation and latency. While this is necessary to use higher-level modulation formats now, fundamentally higher-speed VCSELs are needed.

4.6 High-Speed VCSELs

As previously described, the need for VCSELs with fundamentally higher speed is clear. The VCSEL base bandwidth has grown substantially slower than the data rates and has driven the need for electronic equalization, pulse shaping, and error correction routines to be included in industry standards. In this section, current industry capability is reviewed and methods to increase the fundamental VCSEL bandwidth are explored along with a discussion on the various compensation techniques. The methods described are applicable to multiplexed configurations, but for simplicity the focus will be on single transmitter links.

4.6.1 Current Industry Capability

Today, 850 nm VCSELs and TRXs operating at 25 Gbps (PAM-2) and 25 GBd (PAM-4) are widely available. Figure 4.16 shows examples of VCSELs available from several vendors. For high-speed operation, the VCSELs have both anode and cathode contacts on the top. This configuration reduces electrical crosstalk between elements in an array and reduces the emitted radio frequency (RF) radiation from the circuit. Most VCSELs are manufactured on semiconducting substrates because of the availability of low defect density substrates. To achieve the required bandwidth, the quantum well active regions use $In_xGa_{1-x}As$ with $Al_xGa_{1-x}As$ quantum wells, as described in Section 4.6.2 [18, 19]. To reduce the capacitance, bond pads are placed on insulating material with

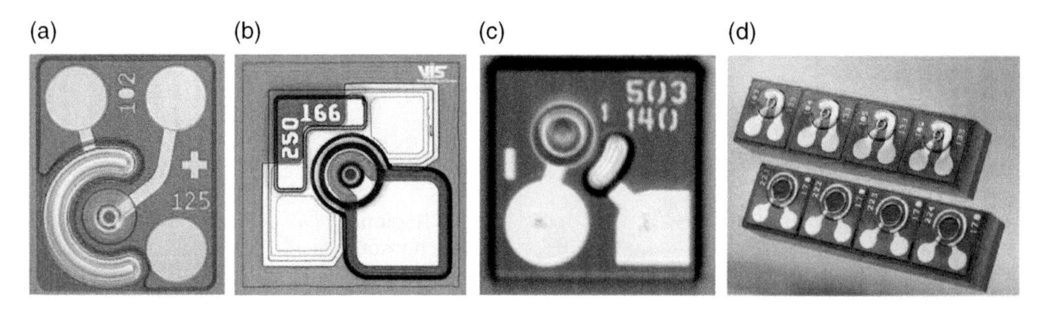

Figure 4.16 Examples of some commercially available VCSELs capable of 25 Gbps (PAM-2) and 25 GBd (PAM-4) from (a) II-VI, (b) VIS, (c) Dallas Quantum Devices, and (d) Trumpf. *Source:* Reprinted with respective permissions from (a) II-VI Inc, (b) VI Systems GmbH, (d) Trumpf Photonotic Components GmbH.

dielectric coverings or on polyimide planarization. Other design features of commercial VCSELs include the use of 0.5λ cavities and place the oxide aperture in very close proximity to the active region. This reduces the overall device capacitance and enhances the standing wave in the quantum wells. The emission diameters range from approximately 5 to 8 μm and are chosen as a compromise between modulation bandwidth, RIN, spectral bandwidth, and reliability. The specific design features and manufacturing details of these VCSELs are closely guarded with only limited information publicly available.

The predominant high-speed data rates today are 25 Gbps and 25 GBd. To address higher speed link requirements, standards for four-channel WDM using VCSELs (SWDM, for short-wavelength division multiplexing) and parallel fiber connections have emerged to cover 100, 200, and 400 Gbps requirements, and 800 Gbps standards are in development to support the roadmaps presented in Figure 4.7. The higher speed facilitates advanced data center architectures, as described in Figure 4.5.

4.6.2 VCSEL Bandwidth Improvement

The equations governing the VCSEL small signal bandwidth were previously described in Section 2.10. The keys to increasing the fundamental laser bandwidth are to increase the differential gain, dg/dN, reduce the photon lifetime, τ_p, and reduce the active area volume. The VCSEL quantum well active region was made using GaAs for speeds up to 10 Gbps. As the speed increased, the quantum wells transitioned to $In_xGa_{1-x}As$ to increase dg/dN. The strain induced by adding In increases dg/dN, and Figure 4.17 shows the calculated value as a function of the In concentration for a quantum well

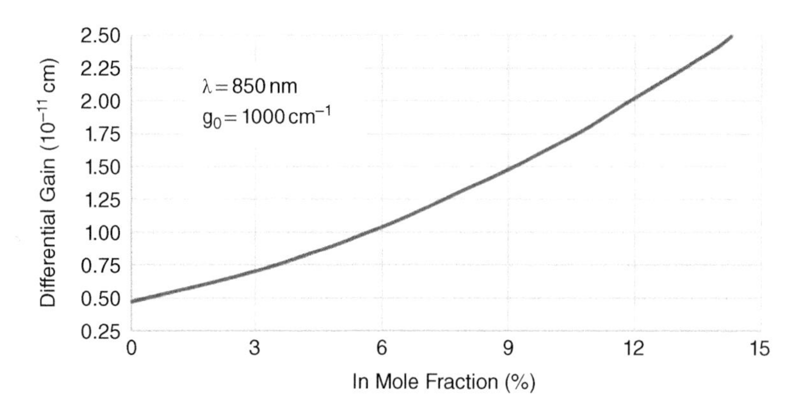

Figure 4.17 Differential gain as a function of In content in an 850 nm quantum well.

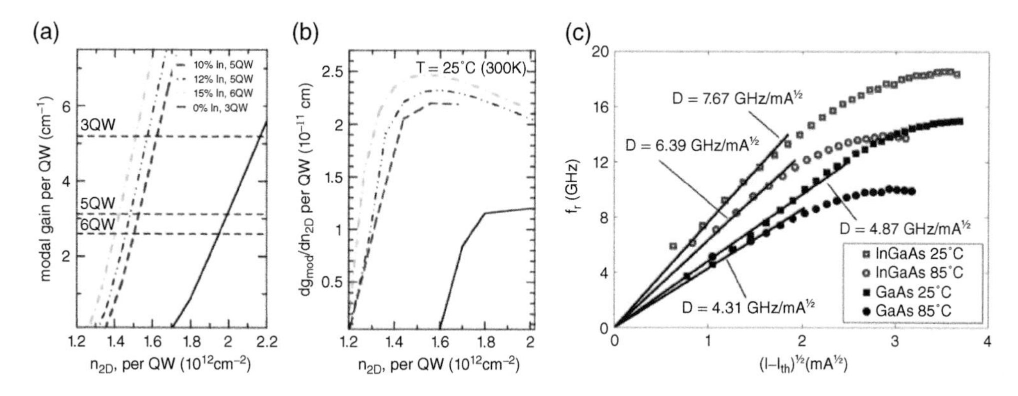

Figure 4.18 (a) Modal gain and (b) differential gain as a function of carrier injection for 0, 10, 12, and 15% In mole fraction. (c) Measured relaxation oscillation frequency as a function of relative threshold current for GaAs and $In_{0.10}Ga_{0.9}As$ quantum wells [20].

emitting at 850 nm. A nearly 5x increase in dg/dN can be obtained by adding In to the quantum wells, and lasers operating at 25 Gbps and higher now utilize $In_xGa_{1-x}As$ quantum wells [19].

The addition of In not only increases the differential gain, it also decreases the laser threshold, as shown in Figure 4.18a. The differential gain is a function of the carrier injection, as shown in Figure 4.18b, and the increases in dg/dN for In concentration decreases as the carrier density increases. Putting all of this together results in a significantly higher modulation efficiency known as the D factor, as shown in Figure 4.18c [20]. There are practical limits to increasing the In composition of the quantum wells. As the In mole fraction increases, the quantum well thickness has to be reduced to maintain emission at 850 nm, and this makes them difficult to grow and control in metal organic chemical vapor deposition (MOCVD) systems. Also, as the In mole fraction increases, the reliability of the VCSEL decreases, and careful control of growth conditions and design parameters are needed to maintain acceptable operation.

Another method to increase the VCSEL bandwidth is to reduce the photon lifetime. The direct effect of photon lifetime on bandwidth was measured by selectively etching a portion of the distributed Bragg reflector (DBR) mirror to increase the optical loss of the cavity [21]. The results are shown in Figure 4.19a, where the photon lifetime is calculated as a function of the amount of material removed from the DBR, and the resulting small signal modulation response (|S21|) is shown in Figure 4.19b. Reducing the photon lifetime effectively reduces the damping of the small signal and increases the amount of peaking in the modulation response. As will be discussed in Section 4.6.4 the peaked modulation response may not be the optimal for large signal modulation.

A third approach to increasing the fundamental VCSEL bandwidth is to decrease the active volume of the cavity. Here, the diameter of the cavity is reduced to single-, or nearly single mode diameter. The reduction in volume more tightly confines the injected carriers and the optical mode and reduces the effect of carrier diffusion on the modulation response. Single-mode VCSELs have been demonstrated with bandwidth of more than 30 GHz at 850 nm [22] and more than 35 GHz at 980 nm [23]. Single-mode VCSELs have been limited in deployment in transceivers due to operation at high temperature and their lower reliability when compared with multi-mode VCSELs.

As the modulation bandwidth increases in the laser, the effect of the parasitic resistance, capacitance, and inductance of the VCSEL become increasingly important, as described in Section 2.10. VCSEL parasitics can be modeled using the lumped circuit elements shown in Figure 4.20. The inductance is a result of the bond wire connection from the laser driver to the VCSEL die and is approximately 1 nH/mm of wire. It is best practice to minimize the bond wire

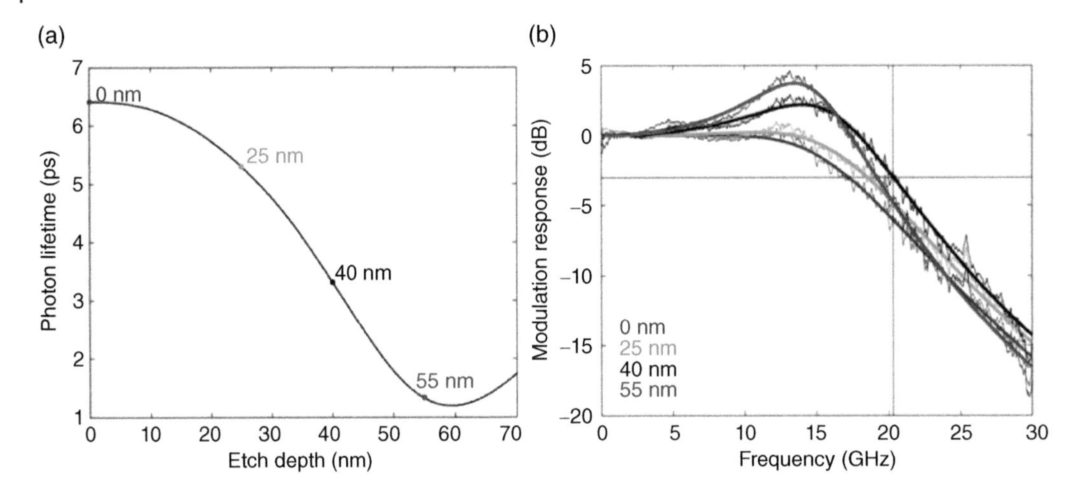

Figure 4.19 (a) Photon lifetime as a function of etch depth into the DBR. (b) Measured and fit |S21| characteristic for the several etch depths showing the dependence of bandwidth on photon lifetime [21].

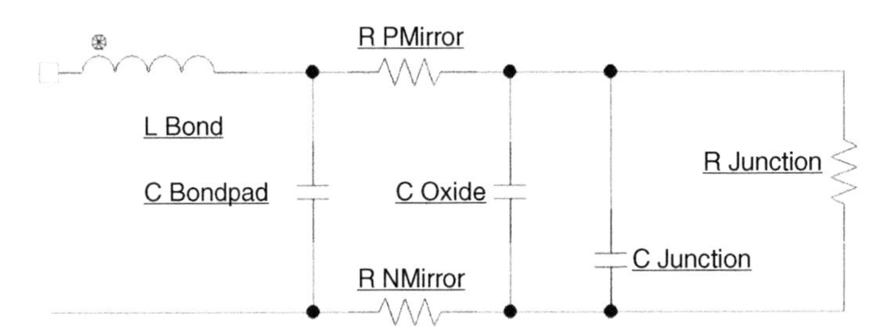

Figure 4.20 Schematic of the parasitic electrical components of a VCSEL.

length, and some transceivers have moved to flip chip bonding to eliminate the inductance. The bond pad capacitance is dependent on the size of the pad and the underlying material. Pads on polyimides such as benzocyclobutene (BCB) can reduce the capacitance. The P mirror resistance is a result of the specific VCSEL design and is a trade-off between optical absorption loss and resistance. The oxide capacitance depends on the thickness of the oxidation layer and the area under the oxide. Minimizing the area under the oxide is important to reducing the capacitance. The junction resistance and capacitance are functions of the diameter. Generally, the junction resistance and capacitance trade off almost evenly with area. The N mirror resistance is a function of the design parameters and is again a trade-off in resistance and absorption loss. The N mirror absorption is typically less than that of the P mirror, and higher doping can be used on the N side. With good VCSEL and package design, parasitic bandwidth should exceed 40 GHz.

4.6.3 Photonic Resonance VCSELs

Section 4.6.2 focused on the extension of VCSEL bandwidth by direct modulation of a single VCSEL cavity. Research has also progressed on other methods of achieving higher bandwidth operation. One method that has emerged is the use of a coupled-cavity VCSEL structure to create a photonic resonance to increase the laser bandwidth [24, 25, 26]. In this approach, two VCSEL cavities are placed in close proximity to each other, and the light form one cavity is leaked to the other and an additional resonance is created either as a result of a supermode of the two cavities or

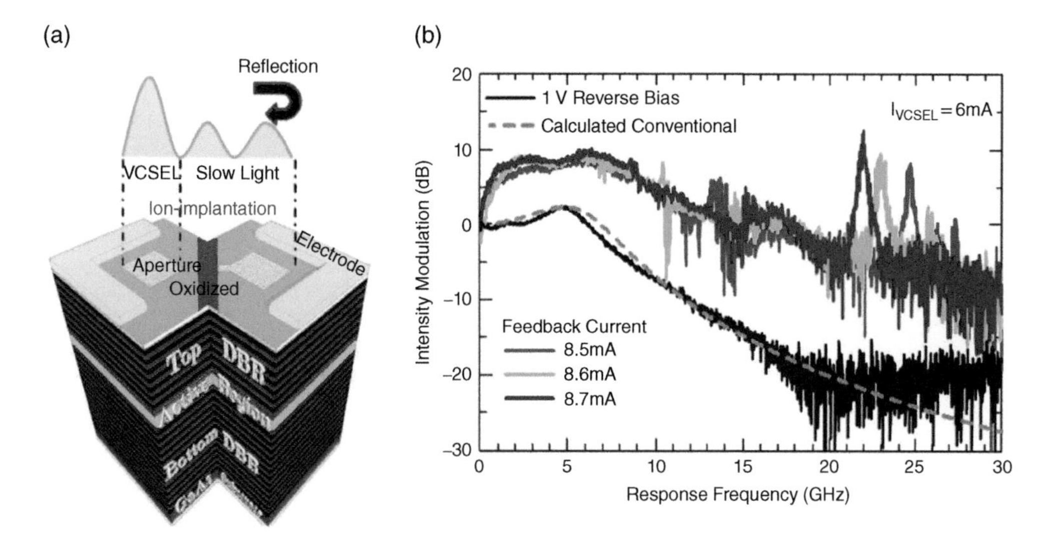

Figure 4.21 (a) Schematic of a photonic resonance oxide-confined VCSEL. (b) Modulation response from cavity A with 6 mA bias current and with negative bias (black line) and forward bias on the cavity B VCSEL [24].

as a result of delayed optical feedback. Figure 4.21 shows the results from [24]. The VCSEL cavities are shown in Figure 4.21a and are formed laterally with oxidation and separated electrically by an isolation implant so that the two cavities can be independently controlled. Light from cavity A leaks into cavity B and is eventually coherently fed back into cavity A creating a resonance when the feedback is in phase. Figure 4.21b shows the results of small signal modulation. In all cases the cavity A VCSEL is held at 6 mA bias current and modulation is applied. In the case of reverse bias applied to cavity B, all of the light form cavity A is absorbed, and the modulation response appears normal, as shown by the black line. When bias is added to cavity B, the modulation response is significantly enhanced, and clear second resonance peaks are apparent in the response in the 25 GHz region. The 3 dB bandwidth is increased from less than 10 GHz to nearly 20 GHz.

Another approach to making a photonic resonance VCSEL is through the use of a photonic crystal lattice [25]. A photonic crystal is used to create an index guide that forces the VCSEL to operate in certain spatial modes. The schematic cross section of the photonic crystal VCSEL with two cavities is shown in Figure 4.22a. The photonic crystal forces the lasers to operate in a single longitudinal mode, and the lateral extent is electrically controlled with an isolation implant. The cavities are electrically isolated from each other with the isolation implant and can be addressed independently. The results of small signal modulation experiments are shown in Figure 4.22b where the response for one cavity is measured with and without bias in the second cavity. The 3 dB frequency bandwidth increases from less than 10 GHz to more than 35 GHz. The behavior of the photonic crystal coupled-cavity VCSEL is different than the oxide structure defined previously. When the two laser cavities are operating at different wavelengths, as shown in Figure 4.22(A2), the far-field profile in Figure 4.22(A3) is Gaussian-like, and the resulting eye diagram in Figure 4.22(A1) is bandwidth limited. As the VCSELs are tuned into wavelength match, as shown in Figure 4.22(B2, C2), the far-field profile of the structure in Figure 4.22 (B3, C3) changes markedly and appears to couple the cavities in either in-phase or out of phase. This is evidence that a supermode of the structure is formed and that the lasers are now operating in concert and coherently coupled [26]. The large signal optical eye diagrams of the coherently coupled VCSELs shown in Figure 4.22(B2, C2) are now wide open. The ability to actively adjust the VCSEL frequency response with the addition of a photonics resonance may further enable high-speed VCSELs.

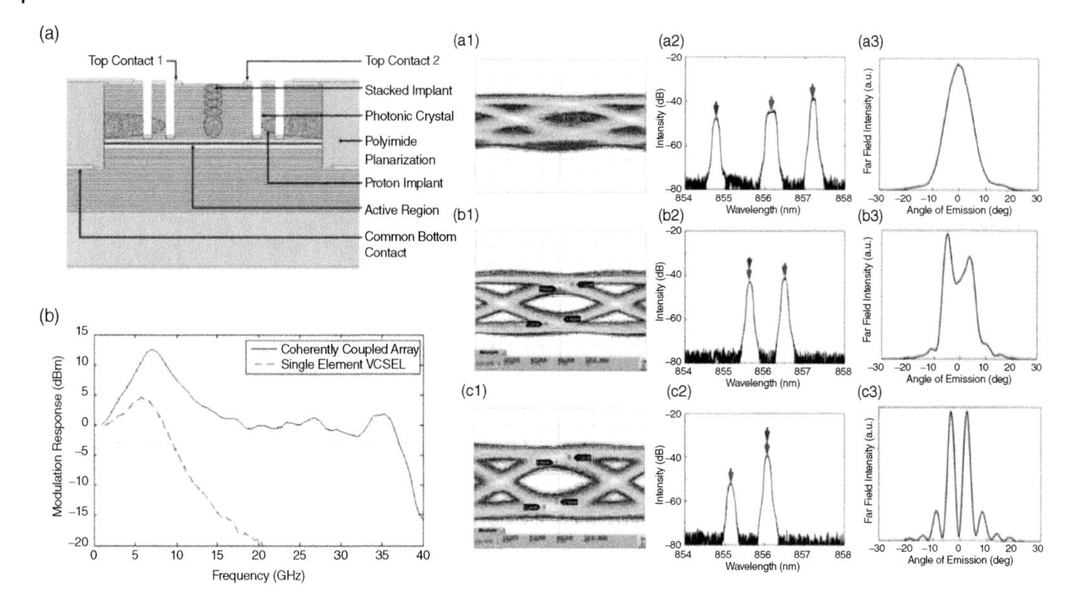

Figure 4.22 (a) Schematic of a photonic resonance VCSEL defined by a photonic crystal structure. (b) Modulation response from the photonic crystal VCSEL (a1, b1, c1) show the optical eye diagrams at 35 Gbps for different alignments of the VCSEL modes as shown in (a2, b2, c2). (a3, b3, c3) Show the measured far-field pattern and are indicative of different phase relationships of the supermode formed by the VCSELs [25, 26].

4.6.4 Laser Driver Compensation

As of 2020, optical transceivers operating at 25 Gbps (PAM-2) and 28 GBd (PAM-4) have become widely available. The VCSEL bandwidth needed to support this speed is approximately 16 GHz, and commercial VCSELs are available from multiple sources. The design of the VCSEL frequency response is critical to successful transceiver manufacturing. Consider the simulated VCSEL S21 response (see Chapter 2 for reference) and resultant eye diagrams shown in Figure 4.23 [27]. Both VCSELs have the same relaxation oscillation frequency (f_r) of 15 GHz but have different damping frequency (f_d), as shown in 4.23a. As a figure of merit, the 3 dB bandwidth of the underdamped VCSEL has significantly higher bandwidth ($f_{3\ dB} > 20$ GHz) than the more damped VCSEL ($f_{3\ dB} \approx 18$ GHz) but the resultant eye diagrams show more eye closure for the higher speed VCSEL in 4.23b than in the "slower" VCSEL in 4.23c. The overshoot in the optical eye diagram is a direct result of the underdamped resonance response. Modern laser driver circuits have incorporated electrical waveform shaping to compensate the VCSEL response. It is easier to design a compensation network for an overdamped VCSEL than one with underdamped frequency response. This frequency compensation is critical to achieving robust link performance with bandwidth challenged devices.

The primary method to compensate the limited bandwidth of the VCSEL is to adjust the frequency response of the laser driver. Consider the laser frequency response shown in Figure 4.24. The driver frequency response is adjusted to increase the amount of power at the higher frequencies to compensate for the laser response, and the result is a much flatter total response from the laser, as shown in Figure 4.24a. The impact of equalization of the eye shown in Figure 4.24b is demonstrated in the optical eye diagram shown in Figure 4.24c. In this example, equalization was done to 30 GHz, and a maximum of 12 dB of equalization was applied to the driver. There is a clear improvement in the vertical eye closure in the equalized eye diagram. It is difficult to equalize the

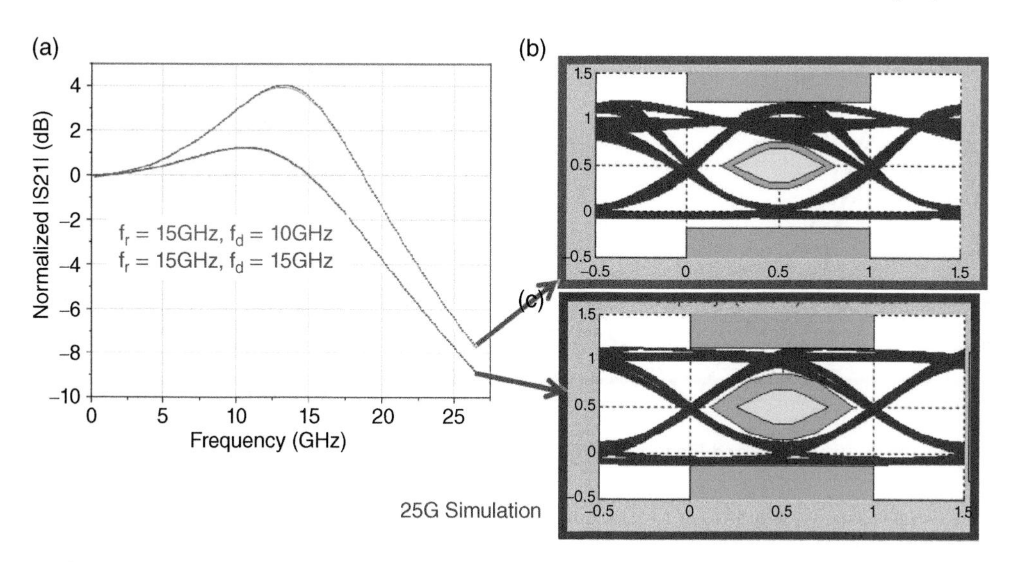

Figure 4.23 Simulated S21 response and resulting eye diagrams for two different VCSEL characteristics [27].

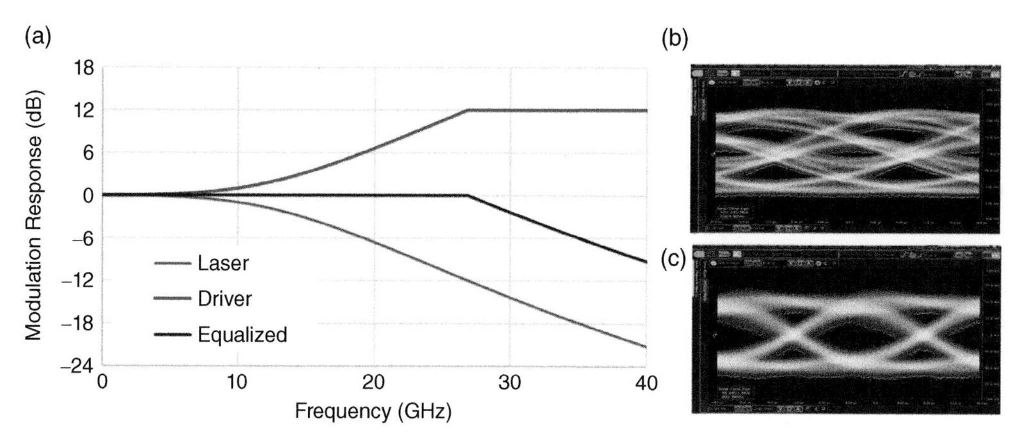

Figure 4.24 (a) Demonstration of equalization of the small signal response of a laser. (b) Unequalized optical eye diagram at 30 Gbps. (c) Resultant equalized eye diagram at 30 Gbps. Eye diagrams courtesy of S. Ralph at Georgia Tech.

VCSEL bandwidth when it is significantly underdamped, and this is why the large signal response simulated in Figure 4.23 shows lesser eye quality for the faster VCSEL with a less damped modulation response.

A second method of compensating the laser frequency response is to effectively reduce the required laser bandwidth by using raised cosine filters to modify the rise and fall times of the electrical signal. In this method, the rise and fall times are increased, but the opening at the center of the eye remains unchanged or can be improved significantly with the further use of equalization. The effect of the raised cosine pulse shaping is to reduce the frequency content of the required signal by introducing ringing in the signal at the bit period, and the result is a wider eye opening at the center. Figure 4.25a shows the frequency response of a rectangular waveform in the dashed line and the frequency response of a raised cosine filter with several different shape factors. The required bandwidth for transmission is reduced as the shape factor approaches unity, but it does

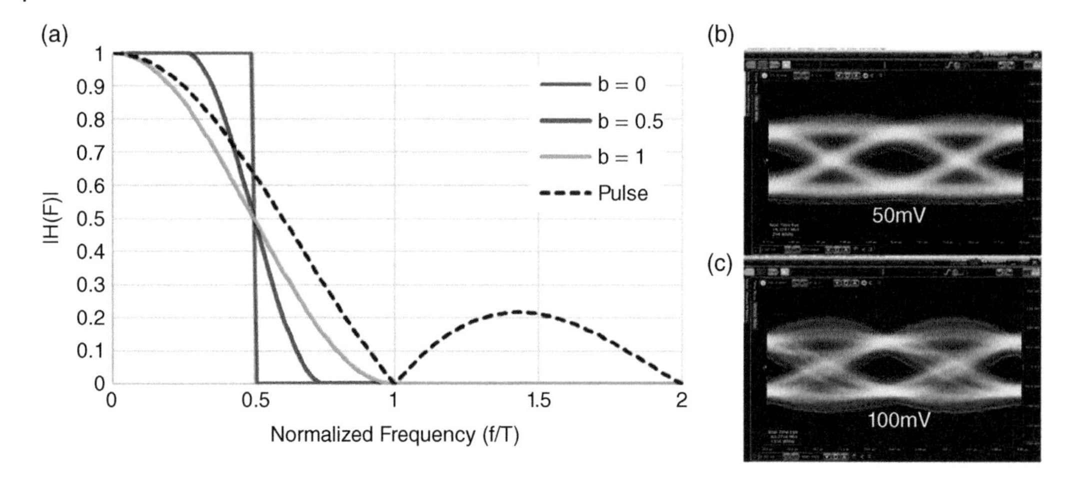

Figure 4.25 (a) Frequency response of a rectangular pulse (dashed line) and of raised cosine filters with several shape factors. (b) Optical eye diagram at 60 Gbps with equalization to 40 GHz showing eye opening of 50 mV. (c) Optical eye diagram with the same equalization used in (b) but with the addition of a raised cosine filter. The eye opening is nearly doubled. Eye diagrams courtesy of S. Ralph at Georgia Tech.

introduce more ringing in the waveform. The resulting optical eye diagram at 60 Gbps with equalization only is shown in Figure 4.25b. The signal was equalized to 40 GHz, and the eye opening is 50 mV. The result of adding an optimized raised cosine filter with the same equalization is shown in Figure 4.25c. The eye opening of the optical diagram has doubled to 100 mV. The effect of both equalization and raised cosine filtering is to dramatically improve the signal quality at the receiver and has become necessary to achieve higher data rates with bandwidth-limited VCSELs.

4.6.5 Forward Error Correction

The previous discussions in Section 4.6 have focused on how to increase the bandwidth or take better advantage of the available bandwidth of VCSELs. At speeds below 25 Gbps, the various components of the optical link had enough bandwidth to ensure error-free transmission. But as the bandwidth of the components has not kept pace with the desired data rates, inclusion of error-correcting codes has been adopted. FEC is the preferred method in most data communications systems. Forward error correction (FEC) works by adding extra information about the data being transmitted to aid the receiver in determining if the data is appropriate. The simplest error-correction codes (ECCs) may use one of the transmitted bits to inform what the next set of bits may contain. More complex codes contain polynomial encoding of the data, and Ethernet uses Reed-Solomon encoding. FEC can effectively recover random bit errors and increase the link distance or otherwise overcome bandwidth limitations. Because of the coding gain due to FEC, Ethernet receivers are now specified at an error rate of 5×10^{-5} instead of the traditional 1×10^{-12}, and this effectively adds multiple dB to the optical link budget. Details of coding gain are discussed in several references [28, 29].

4.6.6 Some Record Results

Figure 4.26 summarizes the achieved data rates for PAM-2 and PAM-4 VCSELs and the total carrying capacity of multi-mode fiber using wavelength division multiplexing [30]. Researchers continue to push the limits of current technology and set new records every year.

To our knowledge, the highest 3 dB bandwidth reported for a single traditional VCSSEL design is 35 GHz and was at 980 nm [23]. The reported bandwidth for a photonic resonance VCSEL is

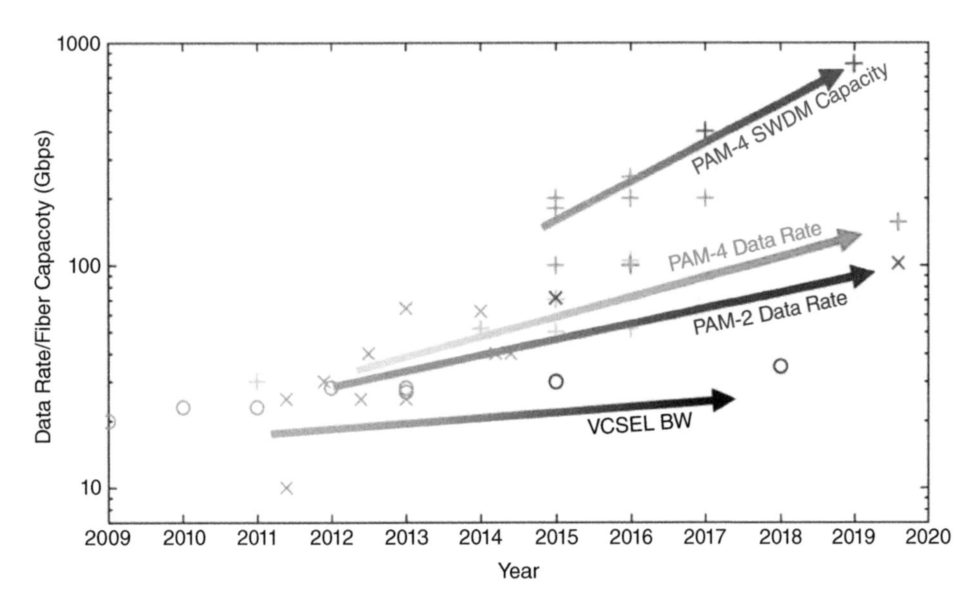

Figure 4.26 Summary of achieved data rates using PAM-2 and PAM-4 VCSELs [30]

45 GHz [31]. Two groups have reported 100 Gbps modulation with PAM-2 [22, 30]. The steadily increasing data rate has been made possible by the heavy use of equalization and pulse shaping to overcome the bandwidth limitation of VCSELs and other components. 160 Gbps using DMT modulation has also been reported [16]. Using PAM-4 modulation, 170 Gbps (85 GBd) has also been reported [32] and to our knowledge is currently the highest speed single-channel VCSEL link reported to date. For a single fiber transmission, several research groups have utilized wavelength division multiplexing to demonstrate more than 400 Gbps [33], while technically a single-fiber cable transmission at 240 Gbps has been reported on a multicore fiber [34]. While the results presented here are certainly impressive, continued increase in fundamental VCSEL bandwidth is needed to support the demand of even higher-speed data links [35].

4.7 Optical Link Impairments

When designing an optical communications link, it is necessary to consider all the possible sources of loss in the signal integrity. For purposes of this section the focus will be on the VCSELs and optical fiber. Photodiodes are discussed in Appendix J. A link budget describes the amount of loss in signal integrity in both amplitude and timing from one point to another. The budget will specify the amplitude and timing requirements of the laser transmitter and the optical receiver. The methodology of link specification has evolved over time from simple oscilloscope measurements of eye diagrams with a defined mask or exclusion zone to now include more complex mathematical analysis of the signals. Testing of the optical signals can be a complex and expensive proposition for TRX makers and has driven the adoption of AOCs to reduce test cost (the optical signal is not accessible to the user, and the link is tested only as a complete cable).

4.7.1 Transmitter Impairments

Consider the optical eye diagram in Figure 4.27, where many of the salient features are noted. The heavy black lines show a reference electrical driving signal, and the pseudocolor map is the resulting optical eye diagram. The amplitude-related features of the eye diagram are the power in the

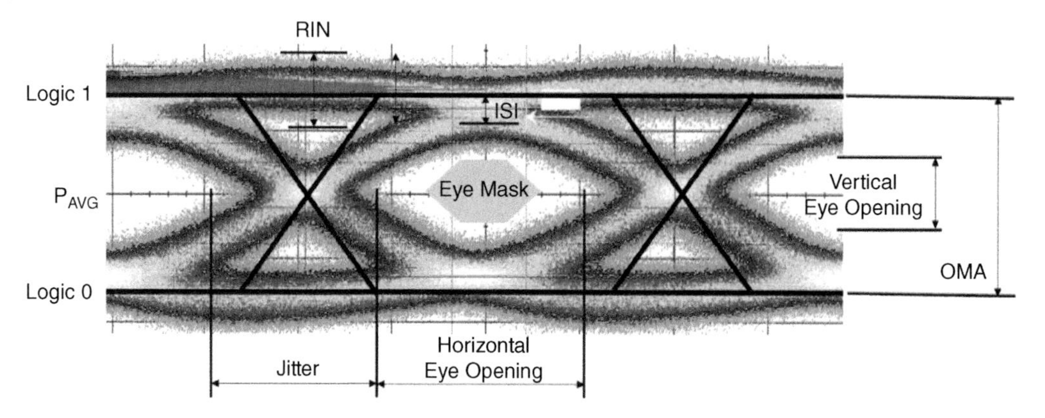

Figure 4.27 Reference electrical drive signal (solid lines) and the resulting optical eye diagram (pseudocolor) with the amplitude and timing definitions relevant for optical standards identified.

logical 1 and 0 levels, P_1 and P_0, the optical modulation amplitude, $OMA = P_1 - P_0$, the average power, $P_{AVG} = (P_1 + P_0)/2$, and the extinction ratio, $ER = P_1/P_0$. Note that ER is often expressed in decibel units, or $ER_{DB} = 10 \times \log(P_1/P_0)$. A useful conversion between ER and OMA is $OMA = 2P_{AVG}\,(ER - 1)/(ER + 1)$. The logic levels P_1 and P_0 are not perfect lines and have uncertainty in their exact level. The uncertainty in the levels are a result of RIN in the laser and is present on both P_1 and P_0. RIN is a fundamental laser parameter and was first described by Schawlow and Townes [36]. Another amplitude impairment is the intersymbol interference (ISI) caused by the limited analog bandwidth of the laser and is frequency dependent, meaning that a 1010101010 pattern will have more ISI than a 1111100000 bit sequence. In a bit stream, a single 1 level surrounded by a series of 0 levels will have a different amplitude than a bit that was preceded by a series of logic 1 bits. When the RIN and ISI are taken together, the result is called the vertical eye closure penalty (VECP) and when subtracted from the OMA, the result is the vertical eye opening. The laser transmitter will also have timing-related penalties in the optical eye diagram. These include the rise and fall times of the laser limited by its analog bandwidth and the jitter of the transition between logic levels. The total jitter (TJ) is a combination of the inherent random jitter (RJ) of the laser and deterministic jitter (DJ), which is a result of the bit pattern used to drive the laser, and TJ = DJ + RJ. The RJ and RIN are generally specified as the standard deviation of a normal distribution. The normal test method for 1 Gbps and 10 Gbps utilizes an eye mask, as depicted in Figure 4.27, and no part of the optical signal is allowed to impinge on this area. The transmitter specifications for 1 Gbps and 10 Gbps are similar in nature, and the Ethernet specifications are summarized in Table 4.4. The exact test methods for all of the parameters are described in IEEE 802.3 clause 38 and clause 52 for 1 and 10 Gbps, respectively [37]. The specifications for Fibre Channel speeds up to 8 Gbps are similarly defined. One final consideration for VCSEL link operation is mode partition noise (MPN) as described in [38]. MPN is a result of the power moving between transverse modes of a VCSEL during a pulse, and it interacts with the modal dispersion of the optical fiber and causes dispersion of the laser signal. In link modeling, the is commonly known as the k-factor. MPN can have a significant impact on link budgets and was initially set to 0.7 to be conservative, but as more careful measurements have been made, the generally accepted value is now 0.3 [38].

As optical standards have continued to increase in speed, limitations in the test methods described above have been noted, and a new measurement called transmitter and dispersion penalty (TDP) or more generally as transmitter and dispersion eye closure (TDEC) was introduced with 25 Gbps

Table 4.4 Transmitter-related specifications for IEEE 802.3 for 1Gbps (clause 38), 10Gbps (clause 52), 25Gbps (clause 92) and 50Gbps Clause (138) [37]

	1000BASE-SX		10GBASE-SX		100GBASE-SX4		50GBASE-SX		
Parameter	**Min**	**Max**	**Min**	**Max**	**Min**	**Max**	**Min**	**Max**	**Units**
Data rate[g]	1		10		25		50		Gbps
Wavelength	770	860	840	860	840	860	840	860	nm
Rise/fall time		260	Note ([a])		Note ([b])		Note ([f])		ps
RMS spectral width		0.85		0.45		0.6		0.6	nm
Average power	−9.5	0	−7.3		−8.4	2.4	−6.5	4	dBm
Extinction ratio	9		3		2		3		dB
Optical modulation amplitude[e]	−7.6		−4.3		−6.4	3	−4.5	3	dBm
Relative intensity noise		−117		−128	Note ([c])			−128	dB/Hz
Transmitter and dispersion eye closure						4.3 ([d])		4.5 ([d])	dBm
Length on OM3 fiber		550		300		70		70	m
Length on OM4/OM5 fiber						100		100	m

[a] Rise and fall times were removed and eye mask definition was used. Effective rise/fall time maximums are 40 ps. See clause 52 of [37].
[b] Effective maximum limit is 18 ps.
[c] Direct measurement of RIN was removed, but effective maximum limit is −128 dB/Hz.
[d] TDEC is measured by comparing to a golden transmitter and there are trade-offs allowed between OMA, RIN, and average power. See clause 95 of [37]. For 50GBASE-SX, the TDECQ is used and applies to both the outer and inner eye closure. A reference equalizer is used for compliance measurement. See clause 138 of [37].
[e] OMA for 50GBASE-SR applies to the outer eye.
[f] Rise and fall times are replaced by TDECQ specifications. Note the inner and outer eyes can have significantly different rise and fall times.
[g] 1G, 10G, 25G, and 50G are defined in clauses 38, 52, 92, and 138 respectively in [37].

Ethernet and carried forward to 50 Gbps Ethernet. This method represents what a real-world receiver would see and allows trade-offs in rise/fall times, RIN, and OMA and only specifies a maximum incurred penalty of the transmitter when compared to a golden reference receiver.

4.7.2 Fiber Impairments

MMF offers significant benefits for short-distance optical interconnects. The large core diameter, 50 μm, compared to the 9 μm diameter in SMF allows for looser mechanical tolerances and the use of plastic housing and epoxy assembly. The tolerance on SMF alignments requires metal housing and laser welding to manufacture. In addition, SMF applications often require some form of optical isolation to the laser to prevent feedback noise in the laser. The larger diameter of MMF also makes the connector manufacturing costs lower. MMF's principal drawback is its lower optical bandwidth, limited by both chromatic dispersion (CD) and modal dispersion (MD) [39]. The CD of MMF and SMF are essentially the same and predicted from the dispersion equation. The CD is partially responsible for the spectral bandwidth specification defined in the several communications standards. The more significant bandwidth restriction for MMF is the modal dispersion. MD results from the different paths of the modes within the optical fiber and is not present in SMF. The different paths taken by the individual modes result in different arrival times at the far end of the

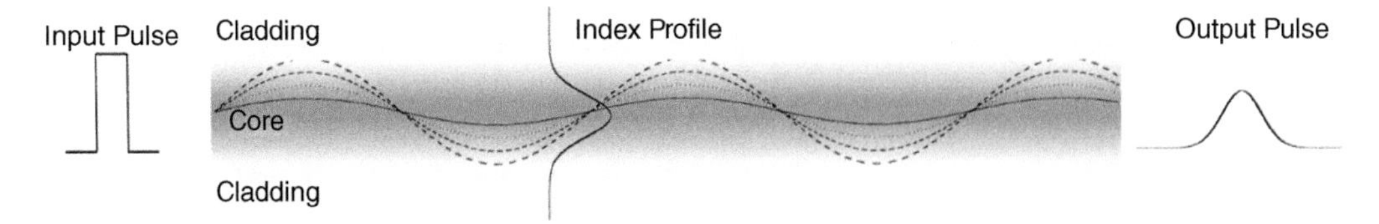

Figure 4.28 Schematic of light transmission through graded in optical fiber showing the effect of modal dispersion on the pulse shape after propagation.

Table 4.5 Classification of multi-mode optical fiber.

		Overfill modal bandwidth (MHz × km)		Effective laser bandwidth (MHz × km)
Fiber type	Core diameter (μm)	850 nm	1310 nm	850 nm
OM1	62.5	200	500	—
OM2	50	500	500	—
OM3	50	1500	500	2000
OM4	50	3500	500	4700
OM5	50	3500	500	4700

MMF. Figure 4.28 shows how an input pulse disperses as it traverses a multi-mode fiber as the different modes travel along different paths in the fiber.

Several standard types of MMF are used in data communication systems, and they are classified by both diameter and modal bandwidth, as shown in Table 4.5. There are two bandwidths specified. The overfill bandwidth is measured using an LED light source to overfill the fiber and excite all the available modes. However, this is not the case with laser coupling to the fiber where a more limited set of optical modes are excited. This was discovered during the development of the 1 Gbps Ethernet standard where testing of multi-mode fiber under single-mode launch conditions at 1310 nm revealed significant collapse in the bandwidth due to centerline defects in the index profile of the fiber. This led to defining of an offset patch cord to launch 1310 nm light into MMF and restrictions on the mode profile that can be launched using an 850 nm VCSEL source. Specifications for the launch profile into MMF was later standardized by the Telecommunications Industry Association (TIA) as FOTP455 [40]. The standard calls for a maximum percentage of light in the center of the fiber and a maximum radial extent of the light at the edge of the fiber and is known as the encircled flux (EF).

Figure 4.29 shows a schematic of the encircled flux. The left-hand side shows an MMF core and cladding, with the light intensity pattern in the center. The EF is measured by integrating the power along the radius and plotting as a function of the radius, as shown in the chart on the right-hand side. The EF specification is shown as the black boxes, which are exclusion zones. The maximum percentage of power allowed in the center 4.5 μm radius (9 μm diameter) is 30%, and more than 86.5% of the power must be inside a 19 μm radius (38 μm diameter).

When designing a laser transmitter, the optical coupling into the fiber is a key consideration. The coupling efficiency must be robust over the operating temperature range and pass the required mechanical tests for fiber side loading and retention. The EF specification must also be met over

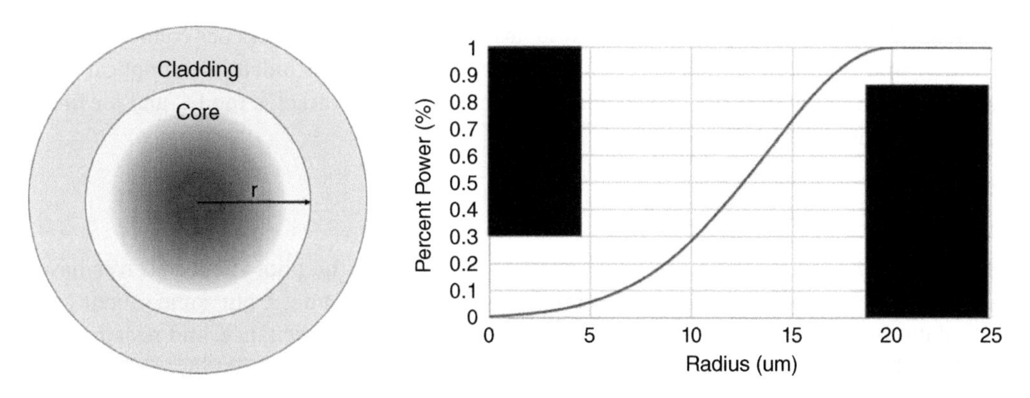

Figure 4.29 Schematic of power launch profile in a MMF and the resulting encircled flux calculation. The black areas in the chart are the exclusion zones for EF.

these conditions. The VCSEL numerical aperture (see Chapter 2) is typically 0.25, and the emission area is roughly 10 μm in diameter. A reasonable lens design will have a magnification near 2 and result in a spot size around 20 μm and a numerical aperture (NA) of 0.125. This will couple well to the MMF over environmental and mechanical stress and result in acceptable EF. When the EF is met, the MMF modal bandwidth is increased and is specified as the laser effective modal bandwidth (EMB) and is listed in Table 4.5. The EMB for OM4 and OM5 are the same at 850 nm, but OM5 was specifically designed to meet the laser EMB over a wider range of wavelengths, up to 940 nm, to allow for WDM solutions on MMF.

In addition to the modal bandwidth dispersion, the optical fiber also has chromatic dispersion. CD is the result of the small change in refractive index of the glass material in the fiber core. For the short distances covered by MMF links, CD is not typically an issue and only slightly modifies the effective bandwidth for single emitters. Figure 4.30 shows the calculated chromatic bandwidth of OM4 MMF as a function of the laser RMS spectral width. The complete fiber bandwidth (BW) includes the effects of both modal and chromatic dispersion and is expressed as $BW_{TOTAL} = (EMB^{-2} + BW_{CD}{}^{-2})^{-1/2}$.

In addition to the bandwidth impairments introduced by the MMF, there is also passive loss introduced. The attenuation of OM4 is specified as a maximum of 3.5 dB/km for OM1 and OM2, and 3.0 dB/km for OM3, OM4, and OM5. The passive loss is a result of absorption and scattering loss of the light as it propagates through the MMF. Additional passive loss occurs at fiber

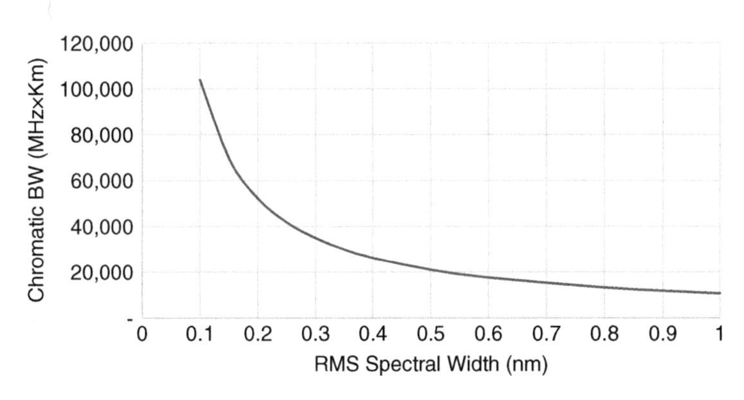

Figure 4.30 Chromatic dispersion bandwidth for OM4 MMF at 850 nm. The CD is nearly constant over the range of 850–1000 nm.

connections. Most standards specifications use an average value of 0.5 dB loss per connector. The total passive loss from attenuation and optical connectors must be included in the optical link budget. For single-fiber links, the most common connector in the market is the LC and for fiber arrays the MT connector.

4.7.3 Receiver Impairments

As the data rate has increased over the years, the limitations of the photodiode receiver have become apparent. Photodiodes are described in more detail in Appendix J, but some salient features for data communications are presented here. The bandwidth, capacitance, and responsivity of a photodiode all depend on the thickness of the intrinsic region where the photons are absorbed. The competing parameters need to be considered as part of the overall link budget. The capacitance is a function of both the size and thickness of the active region and can be expressed using the common formula of $C_{PD} = \varepsilon_0\varepsilon_r A/d$, where ε_0 is the permittivity of free space, ε_r is the relative permittivity of the material ($\varepsilon_r = 12$ for GaAs), A is the area of the PD, and d is the thickness of the intrinsic region. The bandwidth is determined by the transit time of the carriers generated by absorbing the incoming photons and is inversely proportional to d. The responsivity of the PD is directly proportional to d. The trade-off in design is the desire for a large active area for simplicity in alignment to capacitance versus the need for higher speed that reduces d at the expense of responsivity. The bandwidth needed in the PD is approximately 80% of the line rate (20 GHz for a 25 Gbps link). The responsivity directly effects the link power budget as an effective passive loss. For GaAs full responsivity is 0.62 A/W. The capacitance of the PD represents a noise source to the input transimpedance amplifier and can also reduce the overall bandwidth. Figure 4.31 demonstrates the triple trade-off of responsivity, capacitance, and bandwidth for a GaAs-based photodiode. In Figure 4.31a the capacitance as a function of the active area diameter is plotted for several different intrinsic layer thicknesses. For 25 Gbps operation, capacitance should be under 100 fF for the active area. For detector diameters less than 40 μm, the alignment of the PD using low-cost plastic components and maintaining operation over environmental and mechanical stresses is more difficult. Figure 4.31a shows the PD responsivity, intrinsic bandwidth, and the total bandwidth including capacitance for a 40 μm diameter detector. At 25 Gbps, a typical PD intrinsic thickness is approximately 2 μm, and the bandwidth is near 20 GHz with some degradation in optical responsivity. As speeds are scaled to 56 Gbps (PAM-2), the detector challenges become apparent. The PD bandwidth will need to increase to near 40 GHz, and this will significantly reduce the

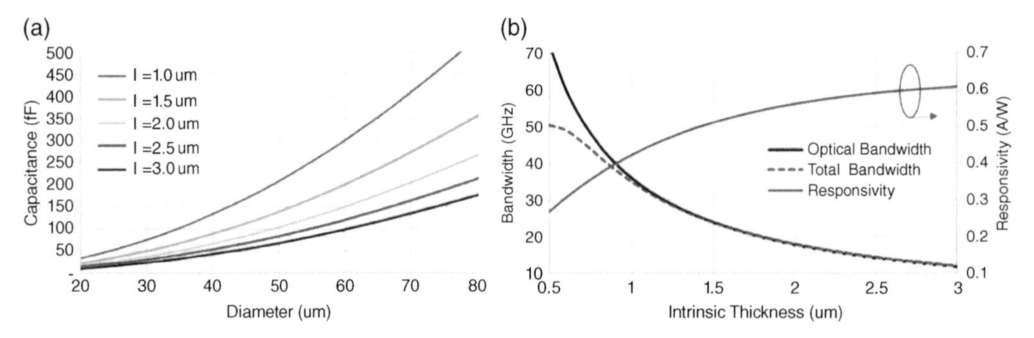

Figure 4.31 (a) Junction are capacitance of a GaAs photodiode as a function of the active diameter for several different intrinsic region thicknesses. (b) Transit time bandwidth and total bandwidth including the capacitance of the intrinsic region for a 40 μm diameter PD and the responsivity (right hand axis) as a function of the intrinsic region thickness.

responsivity. The capacitance requirement will drop to less than 80 fF and drive to smaller diameters making alignment more difficult. Improvement in PD designs are needed to make robust 50 Gbps PAM-2 links and maintain desired link distances.

4.8 Energy Efficient VCSELs

In 2005, the power consumption of the world's data centers was estimated to be 152.5 billion kW·h and is likely at least 10x higher today [41]. With an estimated 3% of global power consumption, data centers are one of the primary and rapidly growing consumers of power. The power consumption not only goes directly to the electronics but also to cooling. Figure 4.32 shows the breakdown of the overall power usage in a data center and within a typical server.

Cooling and power delivery are directly linked to the efficiency of the servers, data storage, and network environment. It is estimated for every watt of power reduction at the equipment level, a total to 3.5 W in overall data center power usage is realized. As the speed has increased, the power required per bit at the transceiver level (Figure 4.8) has not decreased dramatically and is approximately 10–20 pJ/bit. While most of the power consumption is in the laser driver and receiver amplifiers, it is incumbent upon all pieces of the transceiver to reduce power consumption as much as possible. Specifically looking at the VCSEL contribution, the required power to operate at 25 Gbps and 850 nm is approximately 2.2 V and 8 mA, or 16.6 mW of which 3 mW is emitted optical power, leaving 13.6 mW of heat generation. At 25 Gbps, this excess power per bit is approximately 500 fJ/bit ($13.6\ \text{mW}/25\times10^9$) So the question is how to reduce power lost to heat. The voltage can be reduced by moving to longer wavelengths by reducing the bandgap voltage. The bandgap voltage is related to the wavelength by $E_{BG} = 1.242/\lambda$, so moving to longer wavelength, say 980 nm, will decrease the voltage by nearly 200 mV. Keeping the same operating current and resistance decreases the heat loss by 1.76 mW. If the operating current can be reduced by half and maintain the same speed, the power can be further reduced by nearly 8 mW. Taken together, the power loss to heat can be reduced to under 5 mW. This leads to excess power dissipation of 200 fJ/bit, a significant reduction in the VCSEL contribution to the total power load of the transceiver. The benefits are amplified because the laser driver power dissipation is also reduced by running at a lower current and voltage. There has been active development by research groups to develop high-speed VCSELs at 980 nm by reducing the size of the active region and moving to longer wavelengths, and excess heat of 70 fJ/bit has been achieved [42, 43].

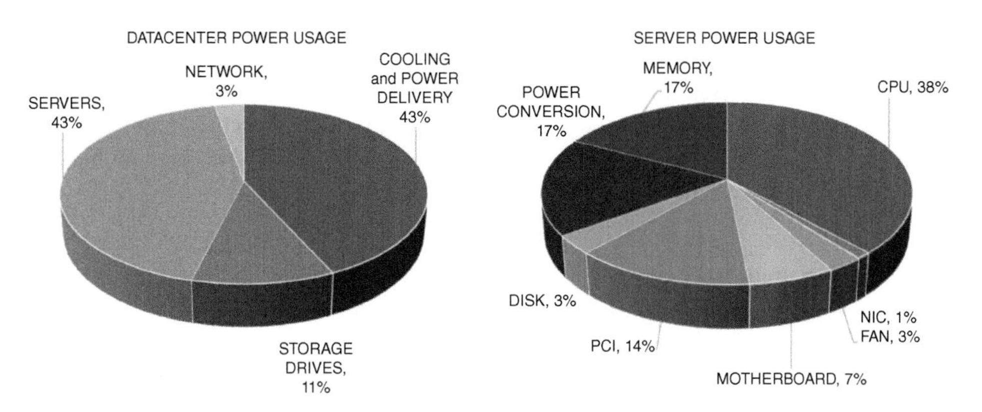

Figure 4.32 Breakdown of power usage in a data center and the breakdown of power usage within a single server.

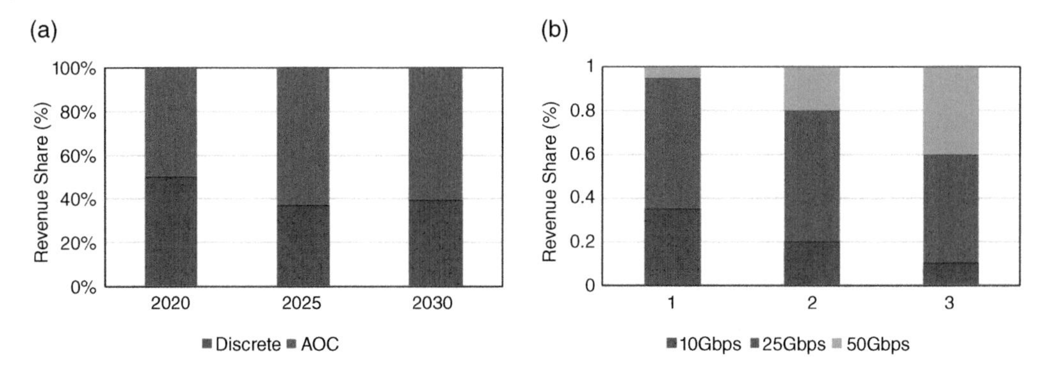

Figure 4.33 Breakdown of power usage in a data center and the breakdown of power usage within a single server.

4.9 Datacom Market

VCSEL-based transceivers, AOC, and OBO components now represent a multibillion-dollar industry and are poised to continue growing over $10 billion by 2025, with compound annual growth rate (CAGR) around 15%. Detailed market forecasts are available from multiple sources that detail the market by speed, reach, transceiver type, and applications. 10 Gbps transceivers are the highest volume type today and support a broad area of not only data centers but also consumer electronics components. Over the next several years, the volume share will transition to 25 Gbps TRXs. AOCs are expected to gain market share as they offer a lower-cost solution than traditional transceivers, especially as the link lengths decrease and they further supplant DACs. Higher-speed links will also command more market share over time as 25 Gbps becomes the dominant channel rate in the data center. Figure 4.33a shows the evolution of increased market revenue share for AOCs over discrete components and Figure 4.33b the revenue share over time by speed.

4.10 Summary

The data communications market was the first commercial driver for VCSELs and will continue to drive demand for the foreseeable future. The market will continue steady growth with a CAGR near 15% in revenue. Adoption of higher speeds within the network will drive more deployment of fiber connections over copper. The fundamental speed of the VCSEL has kept pace with data rates until now, but with 25 Gbps and higher speeds, more complex driver equalization and pulse shaping are needed. For now, increased connection bandwidth is addressed with more complex modulation formats in combination with either spatial or spectral multiplexing. Direct modulation is the lowest cost and power consumption solution for data communications, and VCSELs with higher fundamental speed will be needed to keep pace with competing technologies.

References

1 J. Guenter, B. Hawthorne, D. Granville, M. Hibbs-Brenner, R. Morgan, "Reliability of proton-implanted VCSELs for data communications," *Proc. SPIE* vol. **2683** (1996).

2 K. Choquette and H. Hou, "Vertical-cavity surface emitting lasers: moving from research to manufacturing," in *Proc. IEEE*, vol. **85**, no. 11 pp. 1730–1739 (1997).

3 B.M. Hawkins, R.A. Hawthorne III, J.K. Guenter, J.A. Tatum, J.R. Biard, "Reliability of various size oxide aperture VCSELs," *Proceedings of the 52nd Electronic Components and Technology Conference*, pp. 540–550, IEEE, Piscataway, NJ, (2002).
4 R. Herrick "Oxide VCSEL reliability qualification at Agilent Technologies," *Proc. SPIE* **4649**, (2002).
5 Data taken from Wikipedia https://en.wikipedia.org/wiki/Internet_traffic.
6 G. Moore, "Cramming more components onto integrated circuits," *Electron. Mag.* Vol. **38**, No. 8 pp. 119ff (1965).
7 D. Kuchta, "High Capacity VCSEL-based links," Optical Fiber Conference, 2017.
8 Data taken from https://phoenixnap.com/blog/data-center-tiers-classification.
9 The Top 500 project website, available at https://top500.org/statistics/perfdevel/.
10 "Introducing data center fabric, the next-generation Facebook data center network," available https://engineering.fb.com/2014/11/14/production-engineering/introducing-data-center-fabric-the-next-generation-facebook-data-center-network/.
11 Ethernet Alliance Roadmap available at https://ethernetalliance.org/technology/2020-roadmap/.
12 Fibre Channel Roadmap available at https://fibrechannel.org/roadmap/.
13 Infiniband Roadmap available at https://www.infinibandta.org/infiniband-roadmap/.
14 D. Mahgerefteh, C. Thompson, C. Cole, G. Denoyer, T. Nguyen, I. Lyubomirsky, C. Kocot, and J. Tatum, "Techno-economic comparison of silicon photonics and multimode VCSELs," *IEEE J. Light. Technol.*, vol. **34** no. 2 pp. 233–242 (2016).
15 S. Gebrewold, A. Josten, B. Baeuerle, M. Stubenrauch, S. Eitel, and J. Leuthold, "PAM-8 108 Gbit/s Transmission Using an 850nm Multi-Mode VCSEL," in 2017 European Conference on Lasers and Electro-Optics and European Quantum Electronics Conference, (2017).
16 C. Kottke, C. Caspar, V. Jungnickel, R. Freund, M. Agustin, and N. N. Ledentsov, "High Speed 160 Gb/s DMT VCSEL Transmission Using Pre-equalization," in Optical Fiber Communication Conference, (2017).
17 J. Hecht "*Future Photonics: 5G—Optics will be indispensable for 5G networks*," Laser Focus World, 2020.
18 T. Fanning, J. Wang, Z. Feng, M. Keever, C. Chu, A. Sridhara, C. Rigo, H. Yaun, T. Sale, G. Koh, R. Murty, S. Aboulhouda, and L. Giovane "28-Gbps 850-nm oxide VCSEL development and manufacturing progress at Avago," *Proc. SPIE* **9001** (2014).
19 J. Tatum, D. Gazula, L. Graham, J. Guenter, R. Johnson, J. King, C. Kocot, G. Landry, K. Lyubomirsky, A. MacInnes, E. Shaw, K. Balemarthy, R. Shubochkin, D. Vaidya, M. Yan, and F. Tang, "VCSEL-based interconnects for current and future data centers," *J. Lightwave Technol.*, vol. **33**, no. 4, pp. 727–732 (2015).
20 S. Healy, E. O'Reilly, J. Gustavsson, P. Westbergh, A. Haglund, A. Larsson and A. Joel, "Active region design for high-speed 850-nm VCSELs," *IEEE J. Quantum Electron.*, vol. **46** 4 (2010).
21 P. Westbergh, J. Gustavsson, B. Kogel, A. Haglund, and A. Larsson, "Impact of photon lifetime on high-speed VCSEL performance," *IEEE J. Sel. Top. Quantum Electron*, Vol. **17**, No. 6, pp. 1603–1613 (2011).
22 N. Ledentsov, Ł. Chorchos, V.A. Shchukin, V.P. Kalosha, J. P. Turkiewicz, and N. N. Ledentsov, "Development of VCSELs and VCSEL-based links for data communication beyond 50 Gb/s," OFC conference (2020).
23 N. Haghighi, G. Larisch, R. Rosales, M. Zorn, and J. Lott, "35 GHz bandwidth with directly current modulated 980 nm oxide aperture single cavity VCSELs," 2018 IEEE International Semiconductor Laser Conference (ISLC), (2018).
24 H. Dalir, A. Matsutani, M. Ahmed, A. Bakry, and F. Koyama, "High frequency modulation of transverse-coupled-cavity VCSELs for radio over fiber applications," *IEEE Photon. Technol. Lett.*, Vol. **26**, No. 4, pp. 281–284 (2014).

25 S. Fryslie, M. Siriani, D. Siriani, M. Johnson and K. Choquette, "37-GHz modulation via resonance tuning in single-mode coherent vertical-cavity laser arrays," *IEEE Photon. Technol. Lett.*, vol. **27**, no. 4 pp. 415–418 (2015).

26 H. Dave, P. Liao, S. Fryslie, Z. Gao, B. Thompson, A. Willner, and K. Choquette "Digital modulation of coherently-coupled 2 × 1 vertical-cavity surface-emitting laser arrays," *IEEE Photon. Technol. Lett.*, Vol. **31** pp. 173–176 (2019).

27 J. Tatum, "The Evolution 10 Gb/s 850 nm VCSEL to 25-56 Gb/s," Optical Fiber Communication Conference (2014).

28 "*Essentials of Error-Control Coding*," H. Imai, Elsevier Press, (1990).

29 J. Moreira and P. Farrell, "*Essentials of Error-Control Coding*," Wiley publishing, (2006).

30 J. Lavrencik, S. Varughese, V. Thomas, J. Gustavsson, E. Haglund, A. Larsson and S. Ralph, "102 Gbps PAM-2 over 50m OM5 Fiber using Multimode 850nm VCSEL," Post Deadline Paper, IEEE IPC San Antonio TX 3 Oct 2019.

31 E. Heidari, H. Dalir, M Ahmed, M. H. Teimourpour, V. Sorger and R. Chen, "Hexagonal transverse coupled cavity VCSEL: a promising platform for high speed communications," *Nanophotonics* vol. **9**, no. 16 pp. 4743–4748 (2020).

32 J. Lavrencik, S. Varughese, N. Ledentsov, Ł. Chorchos, N. N. Ledentsov, and S. E. Ralph, "168Gbps PAM-4 Multimode Fiber Transmission through 50m using 28GHz 850nm Multimode VCSELs," in Optical Fiber Communication Conference (2020).

33 B. Wang, W. Sorin, P. Rosenberg, L. Kiyama, S. Mathai, and M. Tan, "4×112 Gbps/fiber CWDM VCSEL arrays for co-packaged interconnects," *IEEE J. Light. Technol.*, vol. **38**, no. 23 pp. 3439–3444 (2020).

34 P. Westbergh, J. S. Gustavsson and A. Larsson, "VCSEL arrays for multicore fiber interconnects with an aggregate capacity of 240 Gb/s," *IEEE Photon. Technol. Lett.*, vol. **27**, no. 3, pp. 296–299 (2015).

35 A. Schawlow and C. Townes, "Infrared and optical masers". *Phys. Rev.* **112** pp. 1940–1949 (1958).

36 B. D. Padullaparthi, "Impact of Δn_{eff} of 850nm VCSEL cavity on low noise for 100G eSR4 transmission and its potential for ≥400G datacenter optical interconnects" *Proc. SPIE* **11704** 11704–24 (2021).

37 IEEE Standard for Ethernet, 802.3 (2020).

38 R. Murty, D. Cunningham, L. Giovane, J. Wang, Z. Feng, T. Fanning, "Mode partition noise characterization of 25 Gb/s VCSELs," *Proc. SPIE* vol. **9381** (2015).

39 Saleh, Bahaa E. A., and Malvin Carl Teich. *Fundamentals of Photonics*. 2. Hoboken, N.J.: Wiley-Interscience, 2007.

40 ANSI/TIA/EIA-455-203-2001, "Launched Power Distribution Measurement Procedure for Graded-Index Multimode Transmitters," (2001).

41 J.G. Koomey, "Worldwide electricity used in data centers," *Environ. Res. Lett.* **3**, 034008 (2008).

42 W. Hofmann, P. Moser, and D. Bimberg, "Energy-efficient VCSELs for interconnects," IEEE Photonics J., Vol. **4**, pp. 652–656 2012.

43 http://www.lr-link.com/product/NetworkingAccessories.html.

5

VCSELs for 3D Sensing and Computer Vision

Babu Dayal Padullaparthi

5.1 Optical Sensors in Consumer Electronics

In the past few years, VCSEL technology has matured and been adopted into high-volume market applications in consumer electronics such as IR illumination and imaging, optical mice and touchpads, biometrics, camera autofocus, gesture recognition (indoor navigation), proximity sensors, ranging, drones, robots, and machine vision (augmented and virtual reality) in human-machine interactions to detect the 3D objects surrounding them. These new applications of VCSELs in 3D sensing resulted in commercial success and created a big impact in the VCSEL industry. VCSEL-based sensors can be found in nearly every household today and have become an essential part of everyday life. When compared with LEDs and EELs, 2D arrays of VCSELs have several advantages in sensor applications, including low packaging cost, wavelength stability over temperature, narrow spectral width, and reliability. So VCSELs are becoming the ideal choice of optical sensors as light transmitters for providing a source of illumination in 3D imaging and sensing, a projected market of ~$9 billion by 2025.

5.1.1 3D Imaging Technologies

Following advances in high-definition (HD) video cameras for 3D movies in the film industry, image sensor and laser illuminator technologies, and pressing needs for higher levels of precise automation and image/process monitoring, 3D imaging (depth sensing) systems have rapidly evolved in global consumer electronics and machine vision markets. A camera system that can capture the depth information of a 3D object and reconstructs the real-world 3D (volume) viewing experience of the scene or object is called a 3D camera. With superior active focusing and improvement in machine interaction, 3D camera systems have completely replaced traditional 2D cameras. In 3D sensing applications as shown in Figure 5.1, there are three kinds of cameras used for object detection (sensing and imaging), namely (a) stereo vision, (b) time-of-flight (TOF), and (c) structured light.

Stereo vision does not use a light source, while both TOF and structured light use an integrated source of light such as LED or laser. So, laser-based TOF and structured light depth sensing technologies show superior performance over stereoscopic and IR cameras using LEDs and rely on software algorithms running on processors to interpret data. Details on 3D imaging (TOF and structured light) technologies can be found in the literature [1, 2, 3, 4, 5]. As TOF and structured

VCSEL Industry: Communication and Sensing, First Edition. Babu Dayal Padullaparthi, Jim A. Tatum and Kenichi Iga.

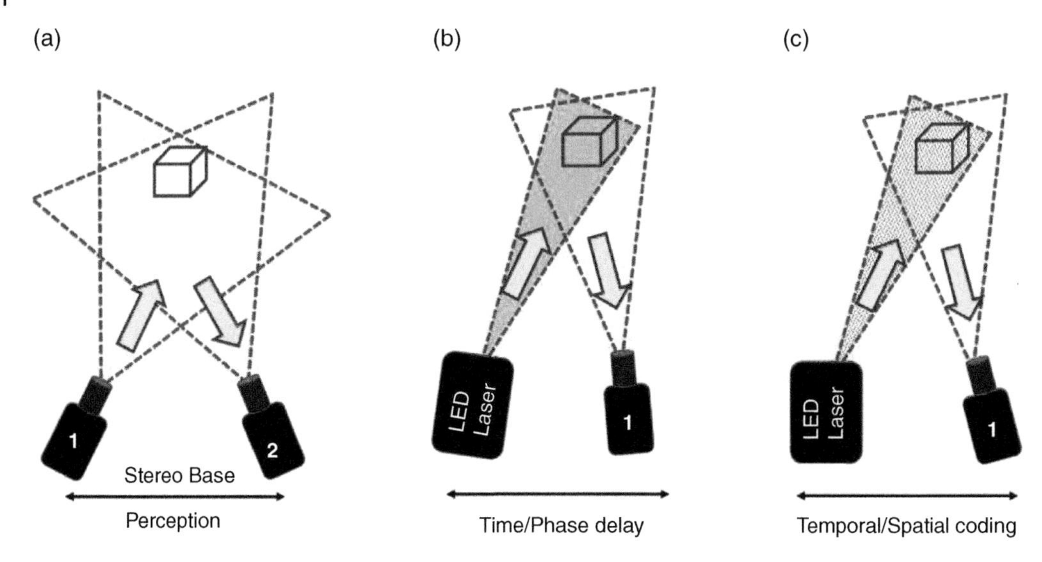

Figure 5.1 Three types of 3D sensing and imaging technologies object detection. *Source:* © Photonic Components DFM Ltd.

light have high-volume applications in illumination and sensing for short to medium distances (10 cm to 10 m or beyond), more details of these will be discussed in Sections 5.1.1.2 and 5.1.1.3. Detailed classification of 3D sensing and imaging technologies based on its function and specific product application is shown in Figure 5.2. The summary of all three 3D imaging technologies in today's consumer electronics era is shown in Table 5.1.

5.1.1.1 Stereo Vision

As there is no light source used in stereo vision, it is called passive imaging technique. The stereo vision technology mimics the binocular vision of human eyes, utilizing two cameras separated by a distance of roughly 60 mm (limited by phone size in mobile applications), which enable each of the cameras and lenses to capture slightly different visual perspectives of an object. The difference in perceptions is used to generate a depth map of the final image. Stereo vision camera system relies on ambient light for the imaging and is thus sensitive to the object reflection and lighting conditions. The resolution of a stereo vision system is limited by the physical separation of the cameras and disparity of the images on the cameras. Due to power and cost constraints, certain IR flood illuminator applications may use a light source in stereo vision technique, not suitable for mobile applications.

5.1.1.2 Time-of-Flight (TOF)

TOF technology is classified into pulse/flash, CW, pseudorandom number, and compressed sensing. A TOF sensor is composed of two essential elements, the IR light source and an optical detector. The return light is collected on each of the pixels of the receiving camera, and the arrival time of light to each pixel is determined. In other words, TOF technology calculates the direct acquisition of time of flight of photons between the object, camera, and reflected light from object that returns to sensors for image processing. TOF cameras have fast processing speed that depends on incident light pulse duration and reflected light capture (integration) time on to the receiver and are widely used in accurate measurement of 3D objects with short distances. The distance

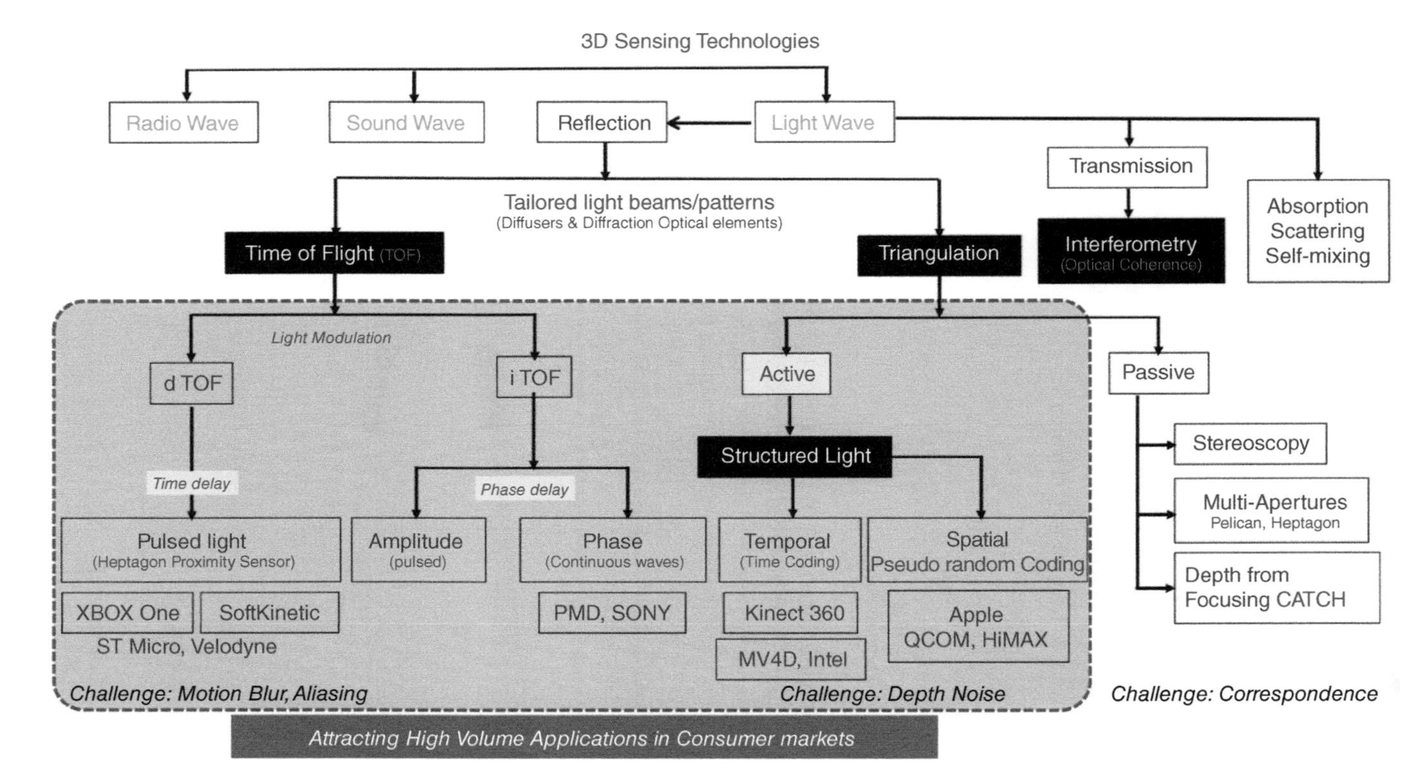

Figure 5.2 Various 3D sensing technologies, namely stereo vision, time-of-flight (TOF), and structured light used for 3D object detection. *Source:* © Photonic Components DFM Ltd.

Table 5.1 Difference between d-TOF and i-TOF.

Item	d-TOF	i-TOF
Concept	round-trip distance	round-trip phase
Measured parameter	time delay	phase shift
Detection range/accuracy	3–5 m (outdoor to indoor)/high	10–100s m/medium
Photodetector	standard CMOS image sensors	SPAD/APD/SiPM
Pixel density/count	small (VGA-640 × 480)	large (QQVGA[a] 160 × 120)
Packageability	medium	complex
Manufacturability	established	emerging
Interaction length with object	fast acquisition	long interaction
Interaction data volume (cal)	large (histogram based)	small (in-pixel)
Dynamic range	digital	analog
Illumination type	0.2–5 ns pulses	intensity modulation 20–100 MHz
Overall range solution	best for single/few points	best for 3D imaging

[*Source:* © Photonic Components DFM Ltd.]
[a] QQVGA- Quad Quad Video Graphics Array.

measurement in TOF is either direct (d-TOF) or indirect (i-TOF), as shown in Figure 5.3. In d-TOF, a complex and constraining time resolved apparatus is needed, where the light source emits a single, very short (<1 ns or ps) pulse of light and is limited by the rising edge of the IR light source, viz. the amount of time the receiver needs to blank to prevent reflections form the optical

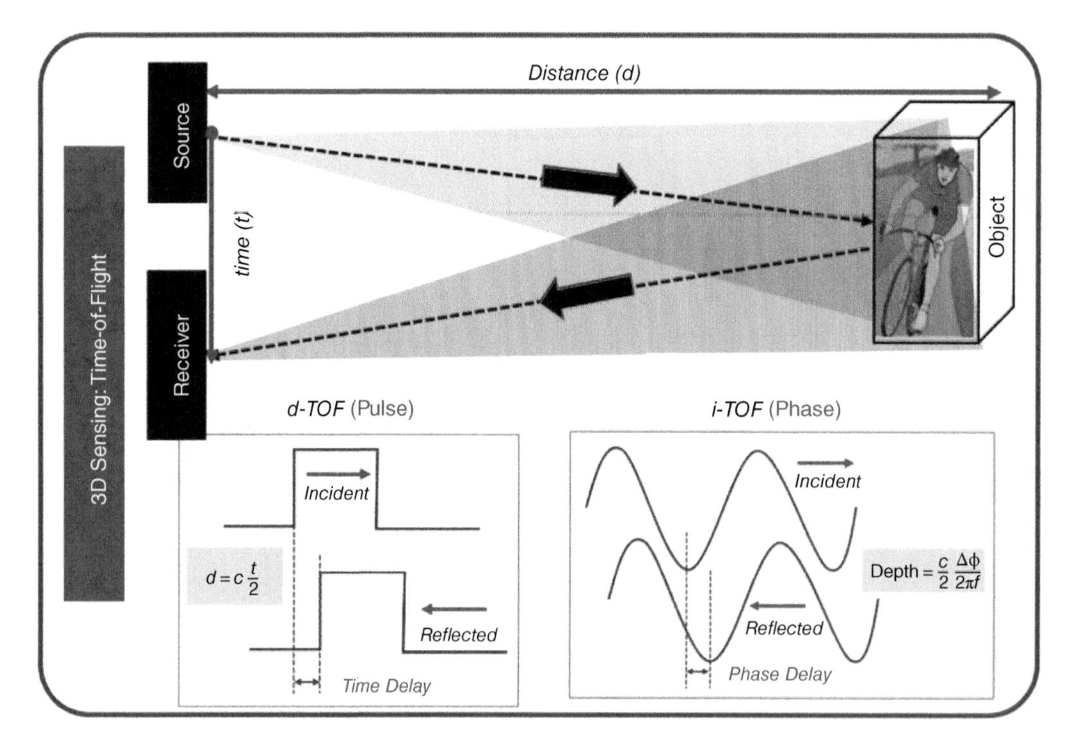

Figure 5.3 Schematic of d-TOF and i-TOF. *Source:* © Photonic Components DFM Ltd.

system and the timing and processing capability of the detector array. The range for d-TOF is only limited by the amount of return light and the wait time in the receiver.

Another method of reducing the speed requirements on the synchronized source and detector is known as indirect TOF (i-TOF). In i-TOF, the light source is modulated with a repetitive signal (square wave for example). The distance is then determined by the relative phase between the outgoing and returning light. As modulating laser source is easier than sending out short pulses, TOF systems based on modulated light source are good for low-cost TOF sensors. In other words, the entire scene is illuminated with modulated light rather than using a single laser pulse or beam to be scanned over the scene, to obtain 3D (depth) information. Using intelligent pixel array, each pixel can individually measure the turnaround time of the modulated light that is done by continuous modulation and measuring phase delay in each pixel. This is robust, and no scanning functions are needed. However, the principal limitation of i-TOF is the modulation period, which leads to uncertainty in phase (for example, a 100 MHz square wave has a period of ~10 ft, and this sets the maximum depth determination). Table 5.1 shows the differences between d-TOF and i-TOF.

Therefore, TOF systems are particularly advantageous as they overcome many challenges and provide a greater amount of flexibility with cost in consumer electronic applications. As of January 2021, as a result of pixel complexity and power consumption, most TOF commercial solutions remain limited in image solutions maximum to video graphics array (VGA).

5.1.1.3 Triangulation Technique and Structured Light

Triangulation is a 1D imaging technique based on scanning of the object distance and surface with the lasers field-of-view (FOV). High-resolution lasers with stable light intensity over temperature wavelength are needed to accurately track the position and displacement of objects; they have applications in logistics and the shipment industry in warehouse management. Here scanning is limited to short distances and use complex algorithms. Triangulation refers to a triangle formed by a light source (projector), an object (scene), and a light sensor (camera), as shown in Figure 5.4, and the distances and angles are geometrically related by triangulation principles as R = B (sin θ)/sin(α+θ), where R is the range between object and camera, θ and α are respectively projector and camera angles, and B is the correspondence distance between projector and camera. Triangulation-based 3D imaging differentiates a single light dot/spot from acquired image under a 2D projection pattern and measures the depth accurately. In triangulation, structured light adds high-precision texture and yields exact correspondences between a camera and projector; it can act as active stereo, unlike low precision in texture from passive stereo vision. The 3D imaging refers to acquiring data (point cloud) in terms of density (depth) volume as a function of its Cartesian coordinates (x, y, z), in particular for surface imaging the depth (z) map of an object as a function of position (x, y). For triangulation-based structured light applications, the measured signal is in the spatial domain, which requires uniform high performance across array emitters, so strict beam characteristics of individual emitters must be maintained.

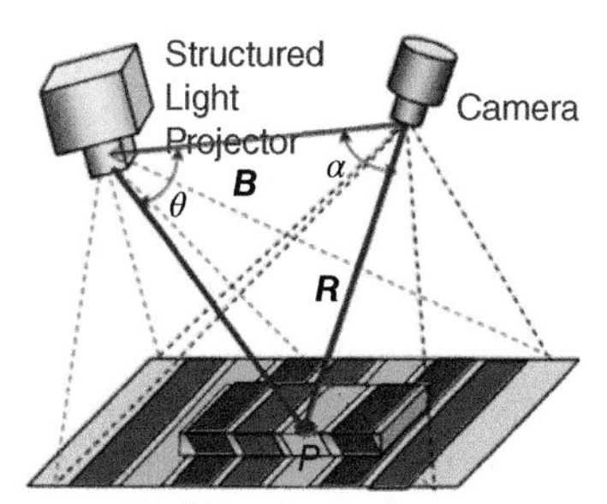

Figure 5.4 Principle of triangulation or structured light in 3D sensing/imaging system. *Source:* Reprinted with permission from Optica (formerly OSA) [3].

Structured light is essentially an active stereo replacing one of the cameras of the stereo vision system with a structured light projector, and the light patterns (mostly 2D) include spatial variations in all directions (for example, in intensity). As shown in Figure 5.4, in structured light technology a structured (distributed) light differentiates pattern changes of predetermined incident light and image depth information though distorted reflected light from the 3D object. The working principle includes the illumination of structured light pattern (projection) on an arbitrary 3D surface that has spatially varying color, establishing the correspondence based on the distortion of the image captured by the image sensor in comparison with undistorted patterns and accurately computes the 3D geometric shape (depth information). The segmented, or modulated, light points (distributed dots) prevent motion blur and create higher accuracy for short range with a depth resolution at sub-millimeter level. Structured light technology is robust against multi-path interfaces and has applications in indoor use for short-distance object navigation, such as face unlocking, but interferes with ambient sunlight outdoors. All of these important aspects of VCSEL arrays will be discussed in Section 5.2.

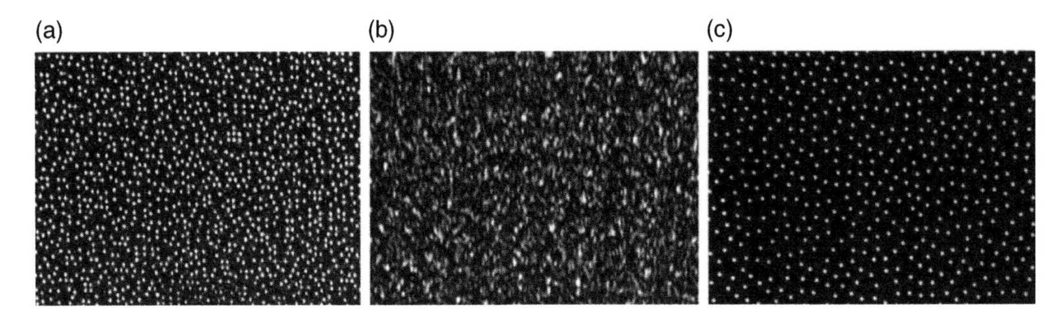

Figure 5.5 Random patterns projected by commercial 3D sensors. (a) Microsoft (Kinect V1); (b) Intel (RealSense R200); (c) Apple (iPhone X). *Source:* Reprinted with permission from Elsevier [4].

The ultimate goal of structured light technology is to robustly establish the correspondence between distorted (camera) and undistorted (projector) images using known structured patterns even for long distance ranging (>250 m). Structured light is particularly attractive for 3D optical shape acquisition for specific FOV, accuracy, or illumination wavelength by using a pattern coding method, and the encoded images are used to determine correspondence points to generate 3D scene points. This correspondence can be done through major codification strategies such as single-shot (statistical), multi-shot, binary, and gray-coding techniques that are used for high-speed 3D shape measurements and established in consumer electronics and medical fields [3]. If the object is static, a multi-shot technique can be used as it does not impose serious constraints on acquisition time, whereas when the object is in motion, the instantaneous position of the object can be acquired using a single-shot technique. Using a single-shot indexing technique, statically random coding patterns with multiple thousands of scene points with as little as one static pattern have been successfully employed to encode a scene in consumer products by Microsoft (Kinect V1), Intel (RealSense), and Apple (iPhone X), as shown in Figure 5.5. The advantages of these statistical or pseudorandom coding patterns include easy implementation and miniaturization, straightforward reconstruction of 3D image based on triangulation (although with limits on spatial resolution), depth noise, or denoising, to adverse weather conditions, and correspondence accuracy between projector and camera. The pros and cons of all three imaging techniques are presented in Table 5.2.

5.1.2 Apple's 3D Sensing Technology Breakthrough and its Impact

On celebrating the 10th anniversary of the iPhone (iOS) product series, Apple announced its latest generation of iPhone X in September 2017, which included a "TrueDepth" camera and optical array sensors. Apple's iPhone X has an impressive number of sensors integrated into a small space

Table 5.2 Differences among 3D cameras used in latest consumer electronic products[a].

Item	Stereo Vision	Time-of-Flight	Structured Light
Active illumination	no (LED or laser)	yes (LED or laser)	yes (laser only)
Distance and range	<2 m	0.5–5 m-SR and ~20–250 m LR	0.2–3 m
Detection speed	high	variable	~30 fps
Depth resolution/ accuracy	medium/medium	high/medium (max VGA)	medium/medium (SR)
Maturity	high	low	medium/very high (SR)
Computing power	high	low	medium
Application field	ranging, 3D construction, robotics, AR, 3D movies	AR, VR, autonomous vehicles, gaming	FR, GR, 3D scanning, gaming
Complexity (camera/ SW/sensors)	simple/high/complex	simple/low/complex	simple/high/complex
3D imaging capability	low	high	low
Low-light/outdoor response	weak/good	good / weak to good	good/weak
Alignment and calibration	complex (with baseline)	simple (no baseline)	complex (with baseline)
Latency	medium	low	medium
Size	medium	very compact-SR	medium
Cost	low	medium	high

[*Source:* © Photonic Components DFM Ltd.]

[a] SR: short range, LR: long range, SW: software, FR/GR: face/gesture recognition.

on the top of the screen. The so-called notch includes an IR camera, a VCSEL-based flood illuminator, a VCSEL-based dot (matrix) projector, a VCSEL-based proximity sensor, and an ambient light sensor. This event triggered shock waves in the Android-based mobile industry because Apple was able to successfully use both TOF and structured light technologies with a complex assembly of camera modules and light sources. The result is a fusion that uses the best features of two technologies, even though its combination is an expensive one! The development of 3D imaging using structured light technology in the iPhone X created a broad supply chain of using various passive and active optical components such as cameras, image sensors (Sony, STMicroelectronics, etc.) and VCSELs (Lumentum, Finisar, II-VI, AMS, etc.) to fulfill the industry demand.

After the iPhone X announcement, the 3D imaging and sensing market was forecasted to grow at a CAGR 44%, from $2.1 billion in 2017 to $18.5 billion in 2023, and the consumer, automotive, and other high-end markets were expected to experience double-digit growth in the same period. Massive events triggered by Apple's announcement include (i) Apple commitment to invest $380M in Finisar [6], (ii) Finisar merger with II-VI [7], (iii) AMS investment of $800M to expand its MFG plant in Singapore [8], (iv) UK government's investment in the Catapult Network project [9], Catapult partner IQE's Portland facility expansion with addition of 100 MOCVDs [10], and (v) Sanan building its mega-factory in Quanzhou [11]. Technologies such as image sensors, VCSELs, injection molded glass optics, lenses and cameras, DOEs, and semiconductor packaging are benefiting from the transition from imaging to sensing, and 3D sensing is becoming more universal.

The 3D sensing ecosystem is adopting a performance-based approach, and only a few players are expected to deliver components that meet high-quality and -volume market demands.

5.2 Why VCSELs for Smart Optical Sensors?

High-power 2D arrays of VCSELs as shown in Figure 5.6a can be easily scalable with emitter diameters as short as 10 μm to 50 μm, or larger, to get powers anywhere from 0.1 W to 100s of watts under pulse operation [12]. For 3D sensing applications, VCSELs will be driven under pulse operation to reduce junction temperature and get maximum pulse power controlled by its pulse width and duty cycle. The small pulse widths (typically a few nanoseconds) will result in at least 10x more power than their CW power, as shown in Figure 5.6b. This max pulse power from 2D arrays is needed to operate optical sensors at a particular distance (10 cm, 1–100 m, or above), which is critical for TOF applications, with the exception of near-CW power in structured-light-based applications. In consumer electronic applications such as smartphones, tablets, computers, and biometrics typically a distance of 10 cm to a few 100s of centimeters is needed. Above this distance, the authors will cover 3D object detection (ranging and imaging) in Chapter 6 on automotive LiDARs up to ~300 m.

5.2.1 Key Features of High-Power VCSEL Arrays

VCSELs have outstanding properties for low-cost mass manufacturing at 100 mm or 150 mm wafer diameters, similar to LEDs, but with unique features of wet oxidation process and deep etch mesa (~4 μm depth), as discussed in Chapter 3. The major aspects of rotationally symmetric (nearly Gaussian beam with M^2~1) with beam divergence of 20° (full width of $1/e^2$) allowing directionality, power scaling from 1 mW–100s W with CW power with power densities of 2–5 W/mm^2 (CW) or ~4 W/cm^2 (QCW), thermally isolated 2D array emitter configurations (independently of each other) with power conversion efficiency (PCE) >50%, and so forth [13, 14], all of which make them ideal for small- to large-scale arrays with excellent growth, process, and performance uniformities [15]. This has been evidenced by 100s of millions of VCSEL chips shipped by top-tiers in datacom and current demand of a few billions of VCSEL units in consumer electronics and automotive

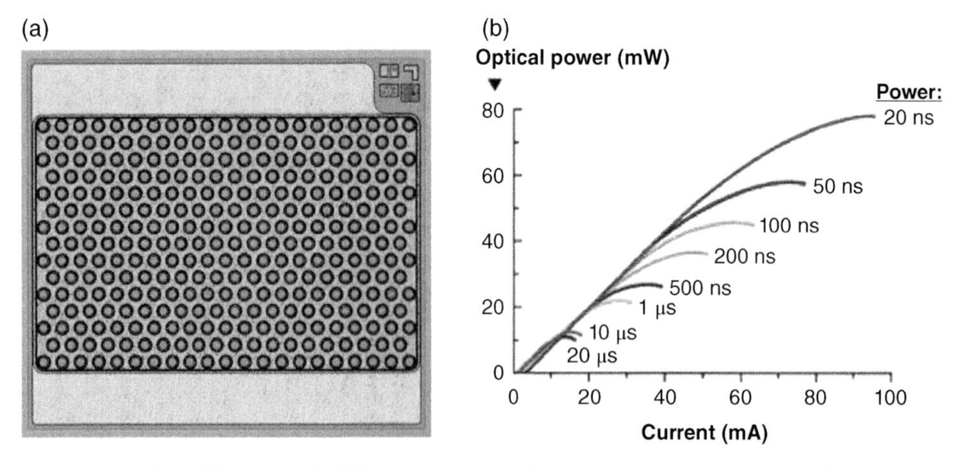

Figure 5.6 (a) A 3 W hexagonal VCSEL array chip for 3D sensing. *Source:* Reprinted with permission from II-VI Inc. (b) L-I characteristic of a typical VCSEL array under pulsed operation with decreasing pulse width. *Source:* Reprinted with permission from Laser Focus World.

transport applications by 2025. In the following sections, the authors look at some important figures of merit and challenges of VCSEL arrays that are critical for understanding and realizing efficient 3D sensor modules for consumer (mobile) use.

5.2.2 Figures of Merit of 2D VCSEL Arrays

5.2.2.1 Optimizing Losses: Slope Efficiency and Wall Plug Efficiency

Figure 5.7a shows a distribution of how the electrical input power to a VCSEL is used. A well-designed VCSEL has a PCE near 50%, which is the resultant optical output. The reduction in wall plug efficiency (WPE) from 100% is driven by Joule (self or internal) heating through the electrical resistance of the DBRs (~18%), material absorption (~13%), spontaneous emission (~12%), carrier leakage around oxide layer (~4%), and band alignment (~2%) [16]. An optimized and high PCE/WPE VCSEL design requires a careful balance between optical and electrical trade-offs in its epi-layer structure. The WPE can be engineered by (i) minimizing optical losses of the DBRs, reducing the operating voltage through band-gap engineering of the DBR mirrors (viz. optimum [modulation] doping schemes in graded index [GRIN] layers between DBR low- and high-index layers), optimizing the position of oxide layer at either a node or antinode of the standing-wave electric field, and reducing the leakage current and spontaneous emission by properly positioning the oxide layer [17], (ii) minimizing thermal impedance of DBRs through proper heat sinking in the device and process designs, etc., all of which have a strong impact on VCSEL array performances.

In 3D sensor applications, the desired SE is near 1 W/A. The higher the SE, the lower the absorptive optical losses from VCSELs. Together with lower optical loses, reduced thermal impedance (discussed in Section 5.2.3.1), and the low resistance of ohmic (electrical) contacts, the emission power increases, leading to an enhanced PCE/WPE. For fully optimized VCSELs, a PCE as high as 55% has been obtained. WPE directly affects the battery lifetime in a consumer device and is an important figure of merit when comparing VCSEL designs and vendors. Recently, novel methods of increasing PCE through use of multiple tunnel junctions that can result in PCE over 60% [18], and even 70% [19]. While the PCE increase is impressive, it does require a higher voltage to operate the stacked junctions, something that may not be available in all consumer electronics. Here both

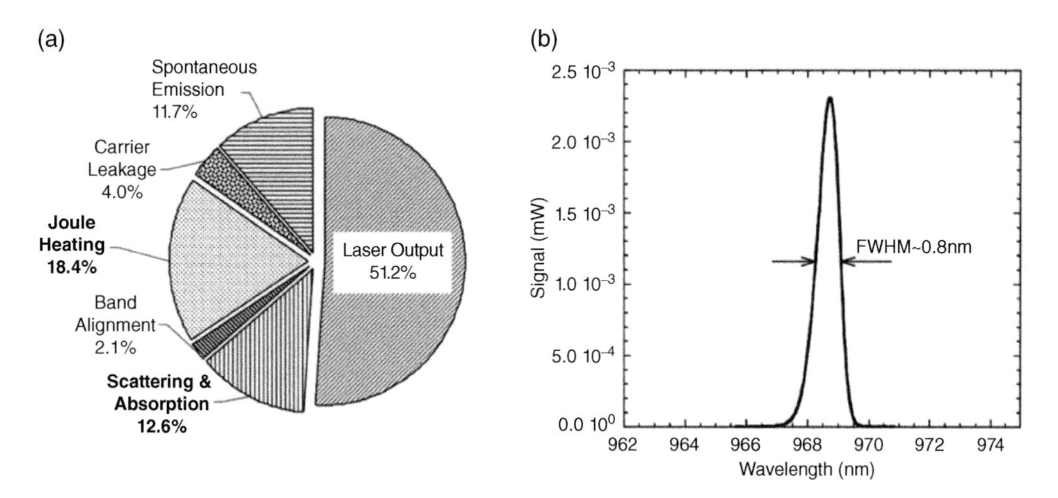

Figure 5.7 (a) Different contributions of losses in a VCSEL for reaching max PCE and (b) 976 nm, 230 W CW array spectral width as small as 0.8 nm. *Source:* Reprinted with permission from SPIE [16].

Table 5.3 Power scaling of VCSEL arrays for product applications.

Parameter	Unit	Single Emitter	N-Element Array
Example chip			
Dimensions	mm × mm	0.15 × 0.15	custom
Forward voltage	volt	2	2
Operating current	mA	10	N × 10
Series resistance	ohm	60	60/N
Optical power[a]	mW	10	N × 10
Application		Proximity	illumination/heating

Source: Table: © Photonic Components DFM Ltd. Inset Images: Reprinted with permission from Sino Semiconductor Photonics Integrated Circuit Co., Ltd.
[a] Pulse Driving: Up to 10X depending on DC and pulse width (~few nanoseconds).

SE and PCE serve as figures of merit for the performance of functional VCSEL arrays. Higher PCEs are important for next-generation 3D sensors (e.g., autonomous LiDARs, drones, and surveillance/night vision systems) that need high power densities, operate at longer distances, and consume even more power!

5.2.2.2 Fill Factor and Power Scaling

2D VCSEL arrays can be made with square, hexagonal, or random configurations (pitch) among them, and power can be simply scaled by physical dimension of the chip with geometrical arrangement (fill factor) among VCSEL emitters. Multiple emitters (mesas) will scale the power in a linear fashion, while voltage remains the same as for single mesa, implying that series resistance varies inversely with the number of mesas, as described in Table 5.3. Scaling power through a 2D array offers addressability of individual or groups of emitters with reduced coherence, tight control of wavelength, and improved heat dissipation. With active emitting area of VCSELs ~10 μm, VCSEL arrays with power as low as 100 mW to a few kW (six orders of magnitude under pulsed-driving conditions) can be realized. Thus VCSELs arrays have higher power densities ~4 kW/cm^2 against their EEL counterparts with 500 W/cm^2 due to low fill factors for efficient cooling. As VCSEL arrays have strong commercial applications in sensing, heating, and so forth, the magnitude of power density has paramount importance as a figure of merit for the working distance at which the product functions.

5.2.3 Key Challenges

5.2.3.1 Thermal Dissipation (Heat Sinking) and Packaging

As high power densities are needed for next-generation applications to increase the working distances to 10s and even 100s of meters, heat dissipation in the individual emitters and the overall die poses a major challenge. Heat dissipation / thermal crosstalk of VCSEL arrays can be optimized with cell geometry/symmetry and can be handled either internally or externally to the VCSEL die. Internally to the die, the VCSEL Joule heating can be reduced through selection of high thermal conductivity AlGaAs layers of DBRs [20], top or bottom GaAs heat sinks [21], thick optical cavities [22, 23], and so forth. External packaging also influences the VCSEL thermal impedance. External cooling measures such as microchannel coolers and high thermal conductivity submounts (SiC, AlN, diamond, etc.) with die-bonding of chips all help to reduce the overall heating of the VCSEL die. The packaging effects are most significant in ambient high temperature operation (>90°C) and high-duty cycle or CW operation. Thermal impedances as low as 0.15 K/W can be achieved with junction down heat removal processes using microchannel-based packages [24]. With all proper optical and thermal optimizations using external heat sinks, a max PCE/WPE well beyond 65% can be reached [25, 26].

5.2.3.2 Spectral Width, Wavelength Uniformity, and Beam Quality

Unlike EELs stronger dependence of gain-peak wavelength shift as a function of temperature, VCSELs operate only in a single longitudinal mode and their emission wavelength is highly stable against temperature and has minimal temperature dependence of optical cavity. NIR VCSELs based on a GaAs substrate show nearly 5x less sensitivity to temperature variation, with values ~0.07 nm/K for VCSELs against ~0.3 nm/K for EELs. With advancement of epitaxy growth methods in 4″ (100 mm) and 6″ (150 mm), VCSEL wafers offer emission wavelength uniformities within 2–3 nm from their center wavelength. This excellent growth uniformity together with a well-controlled oxidation process result in fabrication of 2D arrays with small wavelength changes among array emitters, with spectral width <1 nm (FWHM) in contrast to 3–5 nm for EELs. An example of emission spectrum with 100 W power from a 5 mm × 5 mm array in Figure 5.7b shows spectral width as small as 0.8 nm [16]. For short-distance (<10 m) 3D sensing applications, a 940 nm wavelength is preferred due to its compatibility with CMOS image sensors in NIR region and amount of background sunlight, which improves SNR. Further, both TOF and structured light techniques benefit from the narrow spectral width of VCSEL arrays and their low temperature dependence on wavelength shift. By a careful design of the epi-layer structure, VCSELs or their arrays can be made to emit single transverse mode with circular Gaussian beam (M^2~1), and this enormously reduces complex optics needed for coupling of light into lenses and image sensors. Besides, VCSELs with DBR reflectivity >99.5% are highly insensitive to optical feedback effects and exclude the need for expensive isolators or filters, in contrast to EELs.

5.2.3.3 Field-of-View (FOV) and Micro-Optic Illuminators

Unlike brightest LEDs, VCSELs possess the high brilliance (10–100x) and coherent length, and VCSEL arrays can be used for active illumination for a long-range 3D object detection in NIR camera applications such as surveillance, inspection, and night vision systems. Very uniform and tailored beams (with custom aspects ratios with controlled FOV) can be illuminated at longer distances (few 100 m) by the use of external diffractive optical elements (DOE) such as diffusers and encoders, as shown in Figure 5.8. The 3D (TOF and structured light) cameras generally operate with a near-uniform illumination across the desired FOV of the object, and different FOVs of the DOEs may be desired when used in face identification versus in cabin passenger monitoring. As consumer (mobile)-and automotive (LiDAR)-based electronic products demand narrow to wide focusing of light (−45° <FOV <120°), the FOV of laser light through optics and DOEs (engineered surfaces of diffusers and encoders) has a big impact on sensor module performance. Even though VCSEL emitters in 2D arrays are incoherent and with reduced speckle, they match well with DOEs. The large emitter count reduces the sensitivity to single-emitter failures, and often DOE surface structures are randomized to avoid aliasing effects. DOE and holography-based optical emitters for consumer markets show good growth projections of $1.038 billion and ~$0.7 billion, respectively, by 2027, according to OEMs [27].

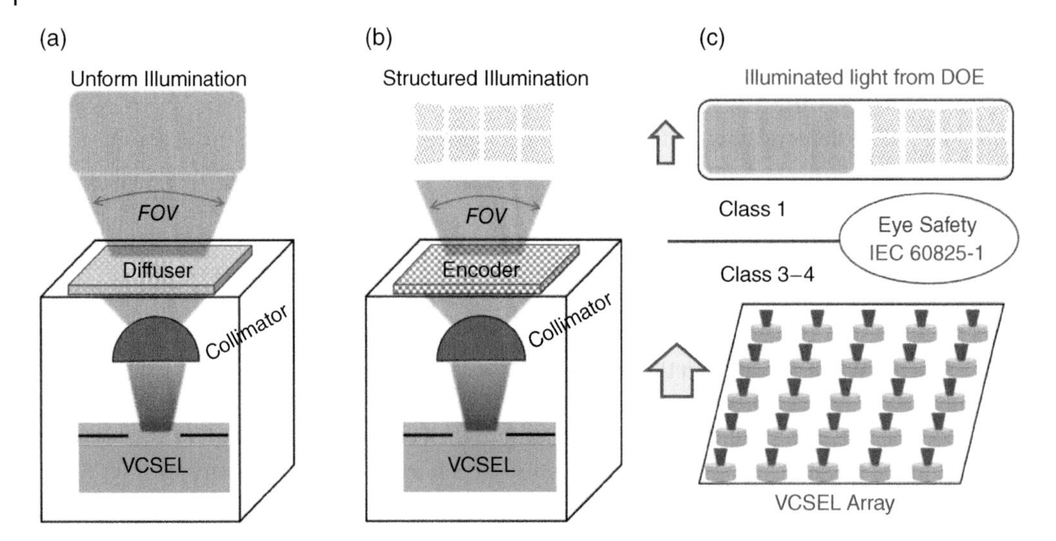

Figure 5.8 (a and b) Diffractive optical elements (DOE) diffusers and encoders used for uniform and structured light illuminations, respectively, and (c) eye-safety of illuminated light from DOEs. *Source:* © Photonic Components DFM Ltd.

Figure 5.9 VCSEL illumination module integrated with diffuser. *Source:* Reprinted with permission from Trumpf.

Consumer applications require uniform and controlled flat-hat light intensity distributions to be projected on the screen through diffusers [28], as shown in Figure 5.8a, whereas triangulation (structured-light) applications demand distributed light in terms of 100s–1000s of dots to be projected on the screen through encoders [5], as shown in Figure 5.8b and Figure 5.14a. As discussed in Section 5.1.1.2, single-shot (pseudorandom) projection cases, continuously varying structured light patterns use 1D (stripe indexing) or 2D (grid or dots indexing) or a combination of these encoding schemes to achieve the intended benefit of object depth tracking in 3D imaging systems [3]. Besides, it is also necessary for VCSEL arrays with Class 3 or Class 4 light intensity to integrate/co-package to convert into Class 1 intensity to meet eye-safety standards (such as IEC 60825-1), as shown in Figure 5.8c.

The illumination profile plays a key role in tracking the physical properties (size and depth) of the object at the projection scene, which are accomplished by passing the laser beam through engineered optical diffusers or encoders (to create a specified FOV) that scatters the light for the illumination. Typical illumination fields from DOEs are (30° H × 10° V) and can be different in the horizontal (H) and vertical (V) directions. The ability to engineer desired illumination patterns helps to reduce the amount of optical power needed and is carefully matched to the camera FOV. An example of VCSEL illumination module with diffuser from Philips Photonics (now Trumpf) is given in Figure 5.9.

5.2.3.4 Thermal Limits and Pulse Switching Times

Under CW operation at large current densities, VCSELs develop high junction temperatures (Tj) and quickly degrade the reliability of operation. As shown in Figure 5.2(b), when thermal rollover or saturation occurs, the intrinsic gain of the laser falls rapidly and the optical losses sharply increase. As shown in Eqs. (2.6) and (2.7), Tj exceeds 130°C and self-heating effect limits the emission of output power. In other words, when the VCSELs are driven with pulse operation shorter than the thermal time constant (time under pulse condition at which VCSELs' Tj significantly starts to increase) with low-duty cycle, significant high optical peak powers can be obtained. This implies that short-pulse or quasi-CW (QCW) operation restricts thermal limits, and the rollover or saturation point shifts to higher currents, as shown in Figure 5.2(b). The repetition of short pulses create an average heating effect, but this can be made smaller by a low-duty cycle. Typically, for pulses less than a few 10s of microseconds, the average heating is mostly localized and can reduce the thermal crosstalk among emitters. Thus, QCW or pulsed operations greatly reduce Joule heating and allow higher optical powers (~10x DC levels). For short-distance (<10 m) 3D sensing applications, both TOF and structured-light concepts demand high a PCE to minimize battery power consumption for long operating times.

For commercial TOF and structured-light applications, fast switching (rise and fall) times of a few nanoseconds are needed and can be simply obtained using VCSEL arrays. While MM VCSELs allow much faster switching times (~few 10s picoseconds in datacom devices), they are mostly limited by impendence matching (driver) circuits. Thus, for short-pulse illumination, MM VCSEL arrays are more efficient than their SM counterparts, and often the switching performance depends on laser dynamics (emission mode structure). The schematic shown in Figure 5.10 explains the combined effects of (a) far-field pattern (FFP) overlapped with radial LP mode structure and (b) Tj rise on pulse-switching behavior. The index-guiding structure of VCSELs due to wet thermal oxidation generates several modes within the oxide aperture area, and there is a mode competition between fundamental and higher-order modes. As both fundamental mode (LP01) and higher-order mode (say, LP02), shown in Figure 5.10a, have different FFP profiles, they represent different amounts of gain for given bias conditions. Further, when VCSELs are driven CW or pulses longer than their thermal time constant, thermal gradients appear in the active region, resulting in a dip at the center at the far-field emission pattern, as shown in Figure 5.10a. It is safe to apply current bias around the threshold, but it must be before the development of higher-order modes, which may take longer switch-on time and are critical for fast detection applications. Also, it is important to drive the laser roughly above the threshold to less than half the saturation or rollover point to ensure the availability of its 10x DC power with reasonable short pulse widths (few 10s ns). So when VCSELs are operated with short pulses, the duty cycle controls Tj at the beginning of the pulse, and then the pulse width controls Tj at the end of the pulse, as shown in the schematic in Figure 5.10b [29, 30]. Details of short-pulse operation are also described in Section 2.11.

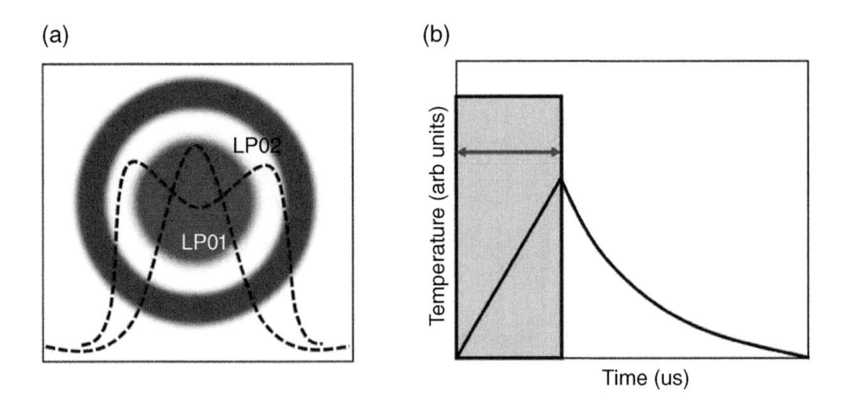

Figure 5.10 (a) Schematic of radial transverse mode patterns overlapped with far-filed pattern distributions, and (b) dependence of junction temperature (Tj) on current pulse. *Source:* © Photonic Components DFM Ltd.

5.3 3D Sensing (Mobile) Products

With Apple's breakthrough iPhone X announcement in September 2017 (explained in Section 5.1.2) for the use of VCSEL arrays as optical components of choice for potential 3D sensing products, the semiconductor industry has forecasted a demand of billions of VCSEL units for core product applications until 2025 (Figure 1.20). Since Apple's announcement, the smartphone industry supply chain players for iOS and Android operating systems have heavily invested in 3D sensing modules, and this has created a serious competition among players of high-volume manufacturing of VCSEL chips. The success of Apple in introducing iPhones and iPads integrated with 3D sensors based on TOF and structured light was mainly due to its innovative business model as the first to release these products while working with strong industry supply chain with diverse technologies. On the other hand, as fast followers, Android players such as Samsung, LG, Oppo, Xiaomi, Vivo, and Huawei have successfully demonstrated both TOF and structured-light technologies with maximum innovation in their smartphone models. For explaining various kinds of commercial products, a block diagram of 3D IR sensor module is shown in Figure 5.11.

5.3.1 Smartphones: iOS vs Android

Smartphones can be classified into two categories based on its operating system: iOS and Android. With the announcement of iOS-based iPhone by Apple using three kinds of VCSEL chips (proximity sensing, flood illuminator, and dot projector), a surge in demand of near 400M units of VCSEL chips by 2023 has been forecasted. iPhone's market share was 14% as of 2019, and it is also expected to grow. This massive quantity of chip demand was met by several VCSEL manufacturers including Lumentum, Philips Photonics (now Trumpf), and Finisar (now II-VI). On the other hand, Android players' (such as Samsung, Huawei, Xiaomi, Oppo, and Vivo and may include LG and Sony) strategies of manufacturing high-end phones with low volumes first and then quickly ramp up production holding a market share of ~86% [31] in a gigantic forecasted demand of ~2450M VCSEL chips [32]. This situation has created a fierce competition between iOS and Android players for VCSEL chips at very high volumes. Besides, with advancements in CMOS image sensors, cameras with high resolutions and smaller physical dimensions (including VCSEL chips), and so forth, the front-facing side (notch) of smartphones (as shown in Figure 3.4 and Figure 5.21) reduces

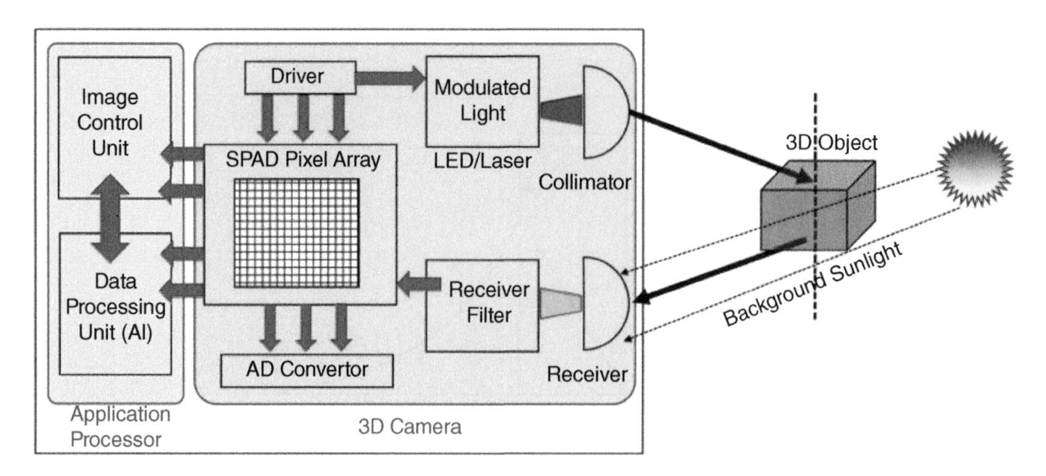

Figure 5.11 A block diagram of 3D IR sensor module with its essential elements. *Source:* © Photonic Components DFM Ltd.

nearly half of its original space, and the silicon content is also shrinking for environmental protection. In other words, Apple's September 2017 announcement set a clear threshold for market trend, and since then, every one to two quarters, a new announcement from Android players follows with a new smartphone model release with advanced functions, as described in Section 5.4.2.

5.3.2 TOF-Based Proximity Sensors

As shown in the 3D sensor module in Figure 5.11, a proximity sensor uses the reflection of IR light from 3D objects in proximity (few 10s cm) with a photodiode driven by an integrated VCSEL array. Light detection events occur with the interruption of light in the proximity of a sensor, which results in certain limit of threshold parameters and include a wide range of adjustments such as automatic subtraction of ambient light and compensating unwanted reflection at the sensor, all of which result in a higher SNR. Lenovo was the first company to use a TOF sensor in mobile phones (PHAB2 Pro 3D), using several optical components such as a high-resolution camera (Samsung), image sensor (Infineon), NIR VCSEL (Princeton Optronics), and a global shutter motion detector (Omni Vision). Quickly after this, Apple launched iPhone X, which uses optical components from STMicroelectronics, Philips Photonics, and Lumentum. The world's smallest TOF-based proximity sensor module (VL53L0X, with dimensions 4.4 × 2.4 × 1.0 mm) for 3D object detection at 2 m range with FOV 25° was first launched by STMicroelectronics [33]. Besides 3D object detection, it also acts as ambient light source and gesture recognition and is called a 3-in-1 function sensor. In particular it uses a 940 nm IR VCSEL chip as light transmitter and STMicroelectronic's patented Flight Sense™ technology with global shutter pixel from SPAD arrays, which has dynamic (very low) noise capability generating ultra-sharp images (avoiding motion blur aspect). This VL53L0X low cost and accurate TOF sensor has multiple applications in user detection in laptops, tablets, touch screens, obstacle detection in robotics, laser autofocus function, drones' take-off and landing phase assistance, ceiling detection, bathroom (automatic faucet water/soap disperser), and automation.

Figure 5.12 A four-emitter VCSEL transmitter used as proximity sensor in iPhone X. *Source:* Reprinted with permission from Trumpf.

Further, NIR VCSELs at 850 nm and 940 nm VCSEL (typically 3–6 emitters on small 0.0225 mm^2 or 0.04 mm^2 die area) have various applications in commercial products including smartphones, tablets, laptops, vacuum cleaners, automatic faucets and soap dispersers, auto-collision avoidance in drones, and smart lighting at homes and buildings. Figure 5.12 shows a four-emitter VCSEL transmitter used as proximity sensor in an iPhone X supplied by Philips Photonics (now Trumpf). Coincidentally iPhone X also uses an STMicroelectronics-based TOF sensor module for proximity sensing!

5.3.3 TOF-Based Illumination Sensors

After Apple's "TrueDepth" camera announcement with the use of VCSEL arrays as optical components (explained in Section 5.1.2), two kinds of VCSEL arrays, flood illuminator and dot projector, have emerged. An example of a 940 nm VCSEL array chip with a power of a few watts used as a TOF flood illuminator, for world-facing 3D sensing applications in consumer mobile applications is given in Figure 5.13. This hexagonal pitch array is made with flip-chip configuration for its direct mounting on PCB or COB without the need of wire bonding. The two pads on the left and right sides of the chip correspond to the anode and cathode, respectively. With high-power mode

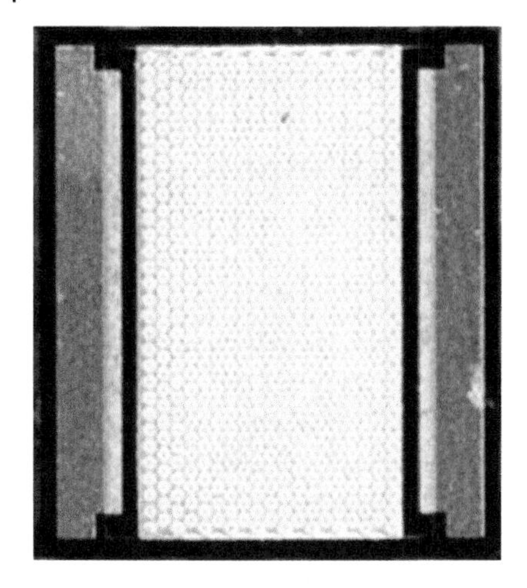

Figure 5.13 A flip-chip bonded 940 nm VCSEL array chip made for TOF flood illuminator applications. *Source:* Reprinted with permission from AMS AG.

operation, this can also be used in automotive vehicles for in-cabin experience/sensing (IVE) applications, a topic explained in detail in Chapter 6.

5.3.4 Structured-Light-Based Face Recognition Sensors

Similar to the flood illuminator described in Section 5.3.3, "TrueDepth" technology using a dot projector for a face-ID recognition (FR) functionality has emerged. This FR sensor module consists of a VCSEL array with active DOE encoder to create a semirandom structured or distributed (structured) light pattern to be projected on to the face of a human. Upon drive current, the VCSEL chip emits uniform light, then the encoder divides the light beam into 1000s of distributed dots, and then it is projected onto the 3D object face depth tracking. An example of distributed dots from VCSEL array and dots projected on a human face is shown in Figure 5.14a and b.

More advanced functions products emerged with both TOF and structured-light-based illumination for 3D sensing using cameras and micro-optical components (diffusers and encoders) with increasing resolutions, low costs, larger object detection ranges, and so forth. Figure 5.6a shows another such example of bare 3 W VCSEL chip (0.98 × 0.88 mm) used in TOF and structured-light sensing from II-VI Inc. Figure 5.15 shows first demonstrations of diffuser packaged 0.3 W 940 nm VCSEL chip (1.98 × 2.2 × 0.85 mm) used in TOF and structured-light sensing from OSRAM (from its subsidiary Vixar) that has application in face recognition.

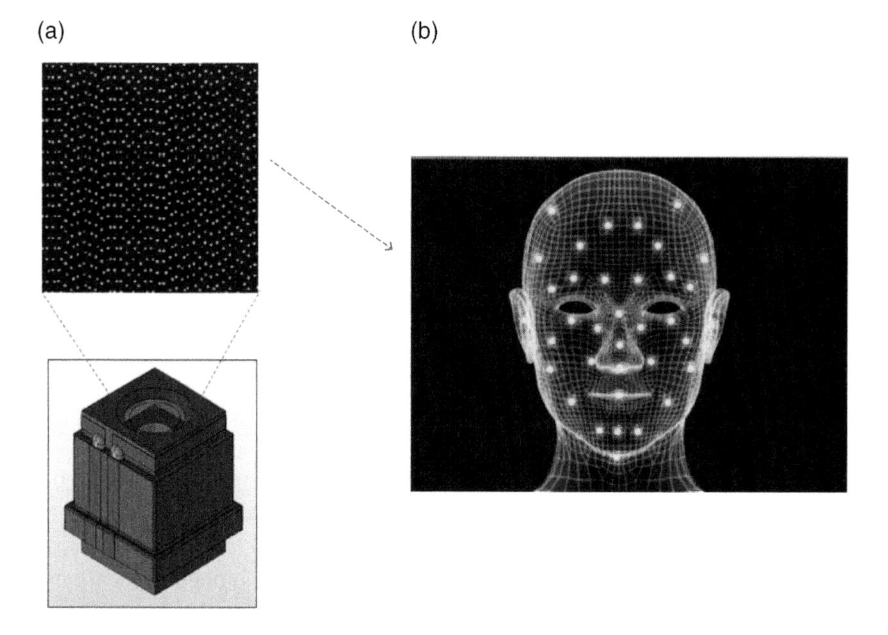

Figure 5.14 (a) Concept of generation of distributed light or dot matrix. *Source:* Reprinted with permission from AMS AG, and (b) its projection on a human face. *Source:* Adapted from IEEE Spectrum, [34].

Figure 5.15 A 0.3 W VCSEL packaged chip from OSRAM for TOF and structured-light-based 3D sensing applications. *Source:* Reprinted with permission from OSRAM.

Figure 5.16 A vacuum sweeper robot from iRobot Braava m6 using structured-light-based 3D sensors. *Source:* Adapted from IEEE [36].

5.3.5 Other Short-Range 3D Sensors

Advances in consumer electronic products further extended to short-distance home appliances (10 cm to <10 m) for vacuum sweeping or floor-mopping products, monitor cameras, and so forth. Outdoor long distance (8–60 m) applications such as drones also use IR-based proximity sensors and have high-volume markets for collision avoidance and obstacle detection applications. Figure 5.16 shows an iRobot Braava m6 product using three structured-light-based IR transmitters that can work even on a bumpy floor to detect the obstacles. Further, a VCSEL array–based LiDAR scanner for 5 m range detection was announced by Apple in its iPad 11 Pro, shown in [35].

5.4 Computer Vision and Virtual Reality

As described in previous sections on real-world (3D) sensing through optical components/modules and the importance of FOV for sensing objects, there exists an inherent relation between sensing and vision. Sensory cues or signals are a fundamental part of perception that can be extracted from the sensory input by a perceiver on how things appear. Visual cues are sensory cues received by the eyes in the form of light and processed by the dominant visual system in humans and provides a large source of information on how the world is perceived. In other words, the ability and accuracy of sensing comes from efficient FOV and better resolution of optical components. However, as shown in Figure 5.17, as human vision is limited to accurately perceive the objects nearby with

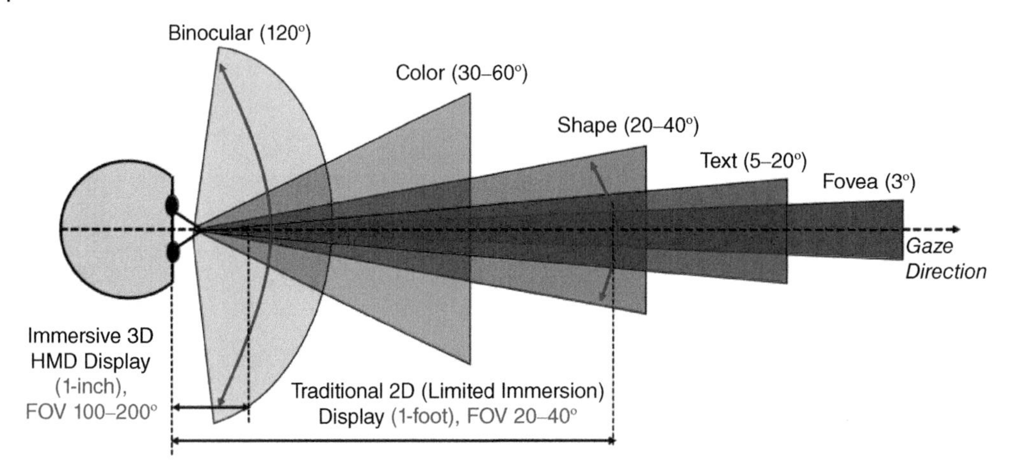

Figure 5.17 Limited human vision perceptions and FOV. *Source:* © Photonic Components DFM Ltd.

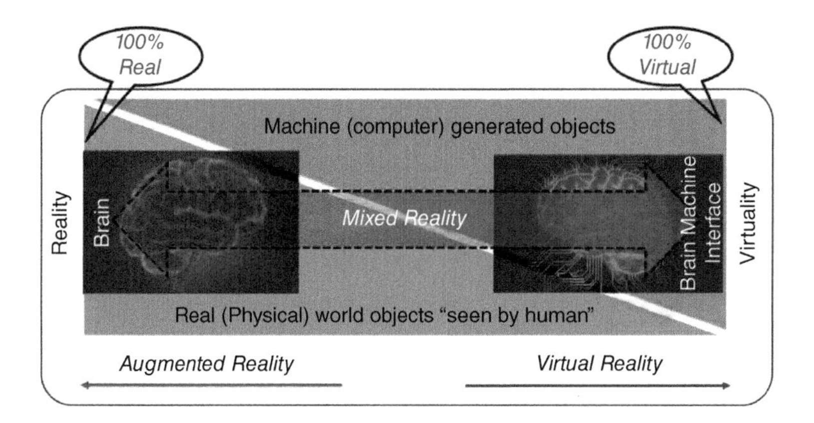

Figure 5.18 Schematic of reality-virtuality continuum. *Source:* © Photonic Components DFM Ltd.

limited FOV, the perception of the sensing its surroundings is limited. This section only focuses on visual cues in optical display applications (depth, motion, color, etc.), and auditory/aural (sound), haptic (touch), olfactory (smell), gustatory (taste), and environmental cues (temperature, food availability, scene statistics, etc.) are not covered.

From the beginning of the 1960s there has been tremendous effort to develop and enhance human perception through various concepts to describe real-world (3D) objects more accurately, using virtual electronic/mechanical tools [37, 38, 39]. Advances in optical sensing, computing power, and software algorithms, among others, have enabled a new field of human and machine interaction known as augmented and virtual reality. The level of immersion of virtual tools with real-world objects define these into augmented reality (AR), virtual reality (VR), and mixed reality (MR), collectively called extended reality (XR). Here the immersion does not depend on the device but on interaction and visualization of the design. Semi-immersive devices often use stereoscopic displays, whereas fully immersive displays surround the user with wide FOV with a virtual world. Figure 5.18 shows the modified concept of Milgram and Kishino's reality to virtuality continuum [40] that describes the extent of overlap among XR technologies, with examples of commercially available XR products.

5.4.1 Key Aspects of XR (AR, MR, VR)

Extended reality (XR) is based on the core concepts of immersion, interactivity, and presence of natural and virtual objects. The word *immersion* suggests that the real-world object is fully or partially surrounded by something so that all human senses (not limited to vision) perceive reality, and perception is mixed (both real and virtual together) both in physical and psychological space. The word *presence* signifies a physical or tele-presence with a meaning of sense of being somewhere. As we move away from the real world, real-world situations are immersed with virtual tools, and a machine (computer)-based simulation of real or virtual or both worlds is needed to navigate through the interaction. AR and MR show the presence of virtual objects without fully occluding the real-world objects (reality) implying the displays are *see-through*, so they utilize partially transparent display media. However with head mount displays (HMD), VR fully occludes the surroundings and displays the real world as video in the HMD. In MR the information requires a spatial computing element, whereas in AR the information is superimposed on a heads-up display (HUD) on the same spatial reference frame.

Thus, XR technologies represent a flatform for data-driven spatial computing real-world phenomena, enable in-situ visualizing, perceiving daily performance of spatial operations using AI (Boolean logic) and filling traditional gaps between traditional spaces of analysis, which often requires additional cognitive processing. Table 5.4 outlines the multiple key differences among XR technologies that are necessary to understand in order to develop new optical display products in the future.

Table 5.4 Difference among XR technologies.

Feature	AR	MR	VR
Digital content	overlays with real world	mixes with real world	replaces real with virtual world
Immersion	partly (addition)	moderate (interact)	full
Real-world visibility	high, enhanced	both real and virtual	totally occlude (block out)
Interaction/ computation	none	needed	fully simulative
Human perception	mostly real	mixed	fully artificial
Control /brain power	fully by human (hands)	mixed (hands and head)	totally by computer AI
System complexity	low end	high-end need	very high: cloud and 5€G
User presence at event	yes	yes and no	no
Real time / see through	yes/yes	yes/yes	sometimes/no
Applications	shopping, tourism, etc.	AI-based training, gaming	video games
Products	smartphone, smart glasses	head mount display	head mount display
Examples	ARKit iOS 11 (Apple) ARCore (Google) Pokomen GO Instagram filters	Magic Leap, HoloLens (Microsoft), Windows MR Headsets (Microsoft)	Oculus Go (Facebook), Oculus Rift, HTC Vive, Google Cardboard

Source: © Photonic Components DFM Ltd.

5.4.2 Augmented Reality (AR)

Augmented reality (AR) technology has become the future mobile computing platform and is revolutionizing the mobile industry and various consumer markets [41]. AR technology is offering unprecedented experiences with broad spectrum of technologies in (i) 3D reconstruction, (ii) object tracking, recognition, and registration, (iii) simultaneous localization and mapping (SLAM), (iv) virtual inertial odometry (VIO), and so forth, and is transforming human lives through interaction with the world with increased perception. AR industries include engineering, manufacturing, healthcare, education, military, marketing, and advertising. Some examples of daily-life scenarios that have changed due to AR include communication at personal (family) and professional (office work environments) gatherings, playing (gaming), tourism exploration (virtual tour guide), and sports (fitness centers). Even though AR technology is still at an early stage of growth, the level of immersion with quality vision, voice, and interactions with low-power consumption allow a seamless and efficient representation of real-world experiences with virtual objects. Besides accurately measuring 3D object dimensions, AR is also expanding human abilities and interreacting with several natural circumstances. They include GR (gesture recognition) and FR (face recognition), eye tracking, bringing efficient control of (life to) IOT devices, speech recognition and learning, and on-drive processing using 5G-LTE with wireless capacity through cloud-computing elements. Adding AI technologies is a key for AR adoption functions to make autonomous decisions (such as machine learning, security, and privacy) and help to meet real-world intuitive interactions with virtual sounds in airports and conference rooms. AR technology requires a powerful laser (VCSEL array) for auto-focusing function on the rear side of a smartphone, called a 3D scanning subsystem. The sensor works with the TOF technique described earlier to map the surrounding environment. This is similar to systems implemented in digital SLR cameras that used the concepts of phase shifting through autofocus and dubbed focus pixels. The phase detection system achieves accurate autofocus function by detecting and comparing multiple light rays that are reflected from an object onto its SPAD arrays.

Figure 5.19 shows two major AR-based emerging products, one embedded in a smartphone and one in smart glasses. The smartphone uses a 2D VCSEL array for identifying nearby 3D objects (10–15 m distance) with precise dimensions and finding applications in shopping in grocery stores, indoor navigation in railway stations and airports, outdoor navigation in tourism spots and restaurant finding in heavy traffic, photography, and so forth. A similar VCSEL-based 3D sensor from OSARAM, PLPVCQ 940, was released in December 2018 for machine vision and furniture

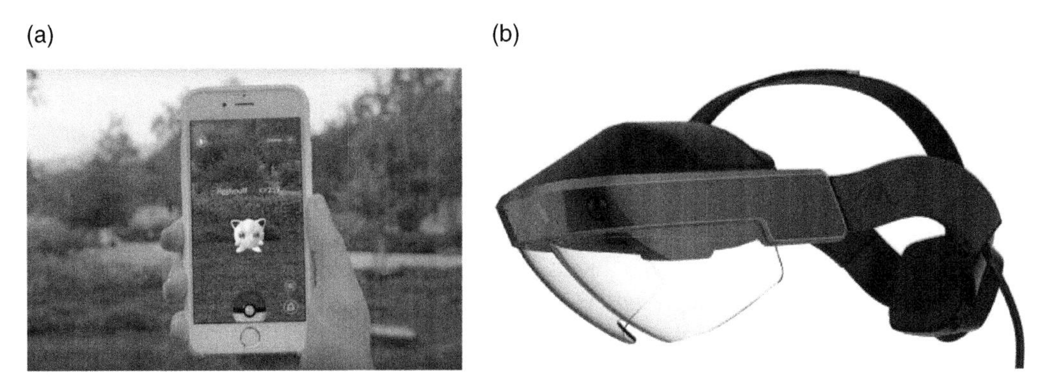

Figure 5.19 Examples of AR-based products: (a) smartphone and (b) smart glasses. *Source:* Adapted from IEEE Access [42].

Figure 5.20 A 2.0 W VCSEL chip packaged with microlens arrays for facial recognition and virtual positioning (Model PLPVCQ 940). *Source:* Reprinted with permission from OSRAM.

navigation [43], shown in Figure 5.20. Intel also released its AR-based smart glasses, called Vaunt, in February 2018, which used a low-powered VCSELs to support text and 400 × 150-pixel image directly on a person's retina, in contrast with other AR glasses that form images in front of the eyes [44]. Further, VishayOpto also released its VCNL36687S VCSEL-powered proximity sensing module in October 2018 for its use in smartphones, tablets, VR/AR headsets, and other battery-operated devices [45]. Interestingly the rumors of the use of AR technology during 2016–2017 [46] became true in October 2020 with the announcement of the iPhone 12 Pro and Max and iPad 13 Pro models, which use VCSEL arrays as short-range LiDARs [47, 48].

5.5 3D Sensing Mobile and Camera Industry Prospects (until 2025)

Since 2009, smartphones sales have dramatically increased from <200M to ~1.5B in 2016, and they are expected to steadily decrease to ~1.3B per year by 2024 Figure 5.21 [31, 49]. With advances and sophisticated functions such as proximity sensing, illumination, face unlocking, and autofocus for AR in smartphones (explained in Figure 3.4), it is very interesting to note that the number of VCSEL chips has increased from 3 (front facing) as of 2017 to at least 4 (front & world facing) in 2020. At the same time, the number of cameras in the front- and world-facing sides of smartphones is also expected to increase from 3 as of 2019 to 5 by 2025, according to a camera module forecast [50]. This point alone indicates a huge demand for VCSEL chips of 12B units and 16B cameras (for the cumulative 2020–2024 period 4B smartphones). This directly translates to at least a $12B market for VCSEL chips until 2024 (assuming $1 per die). Due to the trade war between the US and China, which resulted in a trade ban on Huawei, the VCSEL supply chain has seriously been affected, and market forecasts are now much lower than the previously estimated $10B.This is mainly because most of Android manufacturing fabs located in China and top VCSEL chips suppliers stopped supporting Huawei. The uncertainty in the smartphone forecast by IMF is given by (i) zero growth down-side scenario from its base line and (ii) 5G boost for upside scenario. Overall, the IMF forecast suggests that there is room for growth in global smartphone consumption with rapid innovations, and hence the same is true for VCSEL chip quantity in every model be it an iOS- or Android-based smartphone!

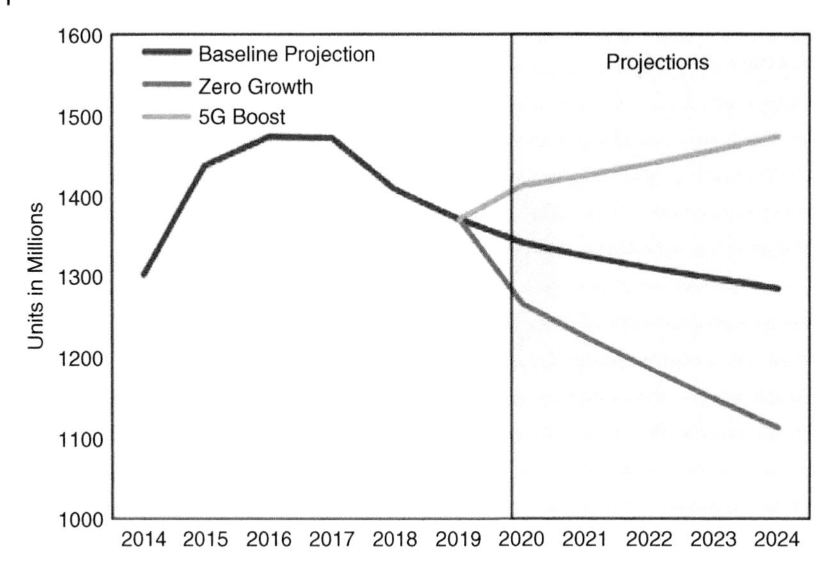

Figure 5.21 Market forecast for smartphones. *Source:* www.img.org, working paper No. 20/70, reprinted with permission.

5.6 Summary

In the past few years, the 3D sensing technology using 2D VCSEL arrays especially in smartphones has revolutionized the consumer electronics industry. This chapter has explained the basic principles and differences among 3D sensing and imaging technologies, namely stereo vision, time-of-flight (ToF), and structured light for short (< 10 m) distance applications. Insights into 2D VCSEL arrays with few commercial product examples, their key features for sensing functions, and some of the manufacturing challenges are highlighted. Author also introduced 'computer vision', a combination of real-world and virtual objects for enhanced human 3D vision, on augmented reality (AR) and intelligence with a brief note on 3D sensing (mobile and camera) market.

References

1 3D Imaging Technology - Time of Flight. AZoM, viewed 10 October 2020, https://www.azom.com/article.aspx?ArticleID=16003.

2 T. Ringbeck, and B. Hagebeuker, "A 3D TIME OF FLIGHT CAMERA FOR OBJECT DETECTION Optical 3-D Measurement Techniques," 09-12.07.2007 ETH Zurich Plenary Session 1: Range Imaging I https://www.ifm.com/obj/O1D_Paper-PMD.pdf.

3 J. Geng, "Structured-light 3D surface imaging: a tutorial," *Adv. Opt. Photon.* **3**, 126–160 (2011) doi:10.1364/AOP.3.000128.

4 S. Zhang, "High-speed 3D shape measurement with structured light methods: a review," *Opt. Lasers Eng.* **106**, 119–131 (2018) doi:10.1016/j.optlaseeng.2018.02.017.

5 B. Lyu, M. K. Tsai, and C. S. Chang, "Infrared structure light projector design for 3D sensing," *Proc. SPIE* **10690**, 1069020 (2018).

6 https://www.apple.com/newsroom/2017/12/apple-awards-finisar-390-million-from-its-advanced-manufacturing-fund.

7 https://www.ii-vi.com/news/ii-vi-incorporated-expands-compound-semiconductors-and-photonic-solutions-platforms-with-the-completion-of-the-finisar-acquisition.
8 https://www.eenewsanalog.com/news/ams-opens-singapore-site-under-200-million-plan.
9 https://assets.publishing.service.gov.uk/government/uploads/system/uploads/attachment_data/file/662319/catapult-programme-evaluation-framework.docx.pdf.
10 https://www.londonstockexchange.com/news-article/IQE/iqe-plc-finalresults/13572803?lang=en.
11 https://compoundsemiconductor.net/article/103086/Sanan_to_Establish_LED_Production_Base_in_Southeastern_China.
12 https://www.laserfocusworld.com/lasers-sources/article/16548239/verticalcavity-surfaceemitting-lasers-vcsel-arrays-provide-leadingedge-illumination-for-3d-sensing.
13 http://www.amstechnologies.com/fileadmin/amsmedia/downloads/4583_cwchiponsubmount2dvcselarrays.pdf.
14 H. Moench, R. Conrads, C. Deppe et al., "High power VCSEL systems and applications," *Proc. SPIE* **9348**, 93480W (2015).
15 https://xueqiu.com/9231373161/135486811.
16 J. F. Seurin, C. Ghosh, V. Khalfin et al., "High power high-efficiency 2D VCSEL arrays," *Proc. SPIE* **6908** 690808 (2008).
17 B. D. Padullaparthi, and F. Lin, "Consumer Semiconductor Laser," US 10, 283, 935 B1 (2019).
18 J. M. Maillard, E. Ruben, P. Thiagarajan et al., "Lasertel VCSEL development progress for automotive LiDAR," *Proc. SPIE* **11300**, 1130005 (2020).
19 J-F. Seurin, G. Xu, A. Miglo et al., "Multi-junction vertical-cavity surface-emitting lasers in the 800-1100nm wavelength range," *Proc. SPIE* **XXV**, 11704B (2021); doi:10.1117/12.2578767.
20 H. Motomura, N. Jikutani, and S. Sato, "Surface emitting laser element, surface emitting laser array, optical scanning apparatus, image forming apparatus, and optical communication system," US 2008/0056321 A1.
21 B. D. Padullaparthi, "Vertical cavity surface emitting laser device," US 9,014, 225 B2 (2015).
22 H. Li, "Temperature-Stable, Energy-Efficient, and High Bit-Rate 980 nm VCSELs," Ph. D. Thesis, TU Berlin, Chpater-3, **D-83**, (2015).
23 A. Mutig and D. Bimberg, "Progress on high-speed 980 nm VCSELs for short-reach optical interconnects," *Adv. Opt. Technol.* (2011), doi:10.1155/2011/290508.
24 J. F. Seurin, G. Y. Xu, V. Khlafin et al., "Progress in high power high-efficiency VCSEL arrays," *Proc. SPIE* **7229** 722903 (2009).
25 https://www.laserfocusworld.com/lasers-sources/article/16550121/vcsels-boast-634-pce.
26 N. Iwai, K. Takaki, H. Shimizu et al. www.furukawa.co.jp/review/fr037/fr37_01.pdf.
27 https://s3.i-micronews.com/uploads/2020/04/YDR20080-Displays-and-optics-for-AR-VR_Sample.pdf (Slide # 28).
28 H. Moench, M. Carpaij, P. Gerlach et al., "VCSEL based sensors for distance and velocity," *Proc. SPIE* **9766**, 97660A (2016).
29 https://optical.communications.ii-vi.com/sites/default/files/downloads/an-2109_high_power_vcsel_for_gesture_recognition.pdf.
30 H. Moench, S. Gronerborn, Xi Gu et al., "CSELs for short-pulse operation for time-of-flight applications," *Proc. SPIE* **10939**, 109380E (2019).
31 J. Mongardini and A. Radzikowski https://www.imf.org/en/Publications/WP/Issues/2020/05/29/Global-Smartphones-Sales-May-Have-Peaked-49361.
32 P. Boulay http://www.yole.fr/VCSEL_IndustryOverview.aspx#.X4JdIi1h0cg.
33 https://www.st.com/en/imaging-and-photonics-solutions/vl53l0x.html.
34 https://spectrum.ieee.org/transportation/sensors/how-3d-sensing-enables-mobile-face-recognition.

35 https://new.qq.com/omn/20200425/20200425A0L00900.html?pc.
36 E. Ackerman https://spectrum.ieee.org/automaton/robotics/home-robots/irobot-completely-redesigns-its-floor-care-robots-with-new-m6-and-s9.
37 DFW van Krevelen and R. Poleman, "A survey of augmented reality technologies, applications and limitations," *Int. J. Virtual Real.* **9**(2) 1–20 (2010).
38 A. Coltekin, I. Lochhead, M. Madden et al., "Extended reality in spatial sciences: a review of research challenges and future directions," *ISPRS Int. J. Geo-Inf.* **9**, 439 (2020); doi:10.3390/ijgi9070439.
39 S. M. Lavelle, Virtual Reality, Cambridge Univ. Press. http://vr.cs.uiuc.edu.
40 P. Milgram and F. Kishino, "A taxonomy of mixed realities visual displays," *IEICE Trans. INF & SYST.* **E77**(12) 1321–1329 (1994).
41 https://www.qualcomm.com/media/documents/files/the-mobile-future-of-augmented-reality.pdf.
42 A. Syberfeldt, O. Danielsson, and P. Gustavsson, "Augmented reality smart glasses in the smart factory: product evaluation guidelines and review of available products," *IEEE Access* doi:10.1109/ACCESS.2017.2703952.
43 https://www.osram.com/os/press/press-releases/osram-enters-the-3d-sensing-market-with-two-new-vcsels-plpvcq-850-and-plpvcq-940.jsp.
44 https://augmented.reality.news/news/intel-reveals-vaunt-smartglasses-normal-looking-glasses-work-with-ios-android-smartphones-0182626.
45 https://www.vishay.com/docs/84951/designingvcnl36687s.pdf.
46 https://appleinsider.com/articles/17/07/12/apples-iphone-8-to-feature-rear-facing-3d-laser-for-ar-and-faster-autofocus-report-says.
47 https://www.apple.com/hk/en/newsroom/2020/10/apple-introduces-iphone-12-pro-and-iphone-12-pro-max-with-5g.
48 https://www.systemplus.fr/wp-content/uploads/2020/06/SP20557-Apple-iPad-Pro-LiDAR-Module-sample.pdf.
49 https://www.gartner.com/en/newsroom/press-releases/2021-06-07-1q21-smartphone-market-share.
50 http://image-sensors-world.blogspot.com/2020/07/yole-forecasts-2020-25-camera-module.html.

6

Automotive LiDARs

Babu Dayal Padullaparthi

6.1 Introduction to LiDARs

Light detection and ranging (LiDAR) is a hot topic in the VCSEL industry and is potentially a game-changing technology in the automotive industry and others. LiDAR can provide high angular resolution (both vertical and horizontal field of view [FOV]) and depth resolution as small as a few millimeters for a given measured distance (~20 m SR, 40 m MR, ~120 m NV, and >120 m LR) between a source and an object. Automotive LiDAR is evolving as one of the most exciting applications providing premium safety, efficiency, and autonomy to vehicle operation. This market alone will create production demand of a few millions of units per year. Other LiDAR markets include mobility (urban transport through autonomous shuttles, robotic cars, grocery/medicine-delivering flying taxis), logistics (goods delivery taxi, warehouse and container terminal auto navigation in automated guided vehicles [AGV], forklifting and automated mobile robots [AMRs] with on-board intelligence and sophisticated environments, mining trucks), construction (static, mobile, and airborne LiDARs), smart buildings (office monitoring, surveillance, and security), manufacturing (human movement in harsh environments, reduced human presence through robotic operations, effective implementation of Industry 4.0 protocols, factory automation), and agriculture (landscaping, crop management), among many others. Implementation of LiDAR in consumer applications in smartphones for short-distance applications (<10 m) was explained in Chapter 5. The objective of this chapter is to describe the extent of VCSEL technologies in the automotive (in particular, autonomous vehicles [AV]) industry for 3D ranging (~250 m) and briefly compare VCSELs with edge-emitting lasers (EELs) in this application. In addition to LiDARs used outside the vehicle to a maximum of 400 m for cellular-based vehicle-to-everything (C-V2X), similar systems are being used inside the cabin to monitor driver awareness and improve the interface to in-vehicle communication/entertainment (IVE) and will be briefly addressed as part of the overall automotive market.

6.1.1 Classification of LiDARs

LiDARs can be classified into several categories as shown in Figure 6.1. To address the needs of the AV market, various LiDAR technologies are being developed categorized by the light source, scanning, and detection techniques, which are further discussed in Section 6.1.2 and Table 6.2. Objects

VCSEL Industry: Communication and Sensing, First Edition. Babu Dayal Padullaparthi, Jim A. Tatum and Kenichi Iga.

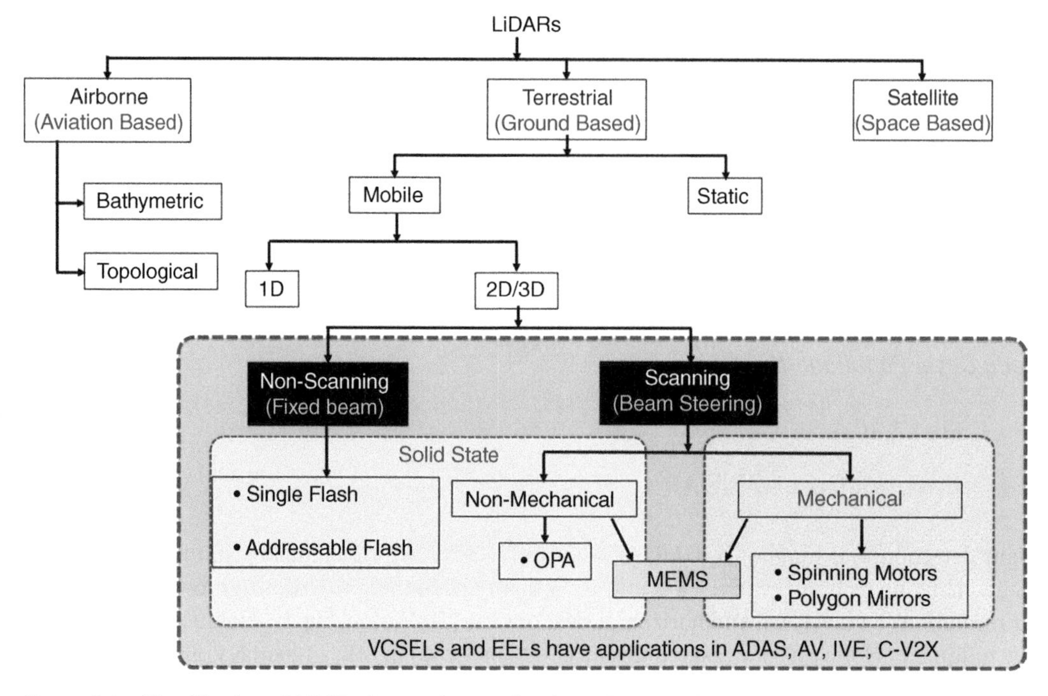

Figure 6.1 Classification of LiDARs for ranging applications. *Source:* © Photonic Components DFM Ltd.

can be detected in 1D, 2D, or 3D in simple one-time flash or sequential flashes. Sequential flash methods include individual emitters, matrix addressable row- or column-wise, and multi-zone emitters and may not need moving parts to perform the areal scan. LiDARs with moving parts steer the laser output using rotating micro/macro elements or electronically adjustable reflectors. VCSELs have high potential and practical applications in non-scanning mode LiDARs, while EELs may favor mechanical scanning-based applications, as described Section 6.3. Time-of-flight (TOF) and frequency-modulated continuous wave (FMCW)-based detection have evolved as popular techniques; their working principles are described in Figure 6.5.

6.1.2 Technologies and Sensor Fusion

Multiple technologies such as traditional digital cameras, sound navigation and ranging (SONAR), radio detection and ranging (RADAR), and LiDAR sensors show their promise for 3D object tracking and assistance in autonomous vehicle driving. Each technology has its own merits and demerits, which are described in Table 6.1. For example, camera merits include color vision and road sign recognition, radar provides for accurate measurement of speed in bad weather conditions, SONAR for object detection in adverse ambient conditions, and LiDAR for 3D object detection [1, 2, 3]. In other words, the technological ability to accurately detect and image objects in a wide range of angles (FOV) up to a few 100 m under background solar ambient or bad weather conditions has improved. As it is witnessed from Table 6.1, none of the aforementioned technologies alone provide all of the functionalities for advanced driver assistance systems (ADAS), as shown in Figure 6.2, and sensor fusion is needed for robust operation.

Optical sensor technologies based on near-infrared (NIR) and short-wavelength infrared (SWIR) emitters and detectors in 3D LiDAR systems are rapidly evolving to enable novel applications.

Table 6.1 Differences among 3D sensing technologies used in latest consumer electronic products[a].

Object/Parameter	Camera	RADAR	SONAR	LiDAR	Fusion
Detection, distance estimation, operation	◎	◉	◉	◉	◉
Resolution and accuracy	◉	◎	–	◉	◉
Operation in darkness	⊠	◉	◉	◉	◉
Operation in bad weather conditions	⊠	◉	◉	◎	◉
Range visibility	◎/⊠	◉	⊠	◉	◉
Lane tracking/read signs and colors	◉	⊠	⊠	⊠	◉
Speed measurement	⊠	◉	⊠	◎	◉
Environment analysis	⊠	⊠	◎	◉	◉
3D capability	⊠	⊠	–	◉	◉
Operation in bright sunlight	◎	◉	–	◉	◉
Interference effects	◉	◎	–	◉	◎
Cost	◉	◉	–	⊠	◎
Size	◉	◎	–	⊠	◉
HFOV/VFOV	◎/◉	◎/◎	⊠/⊠	◉/⊠	◉

Ability: ◉ high, ◎ poor, ⊠ inability, N/A.
Source: Babu Dayal Padullparthi © Photonic Components DFM Ltd.

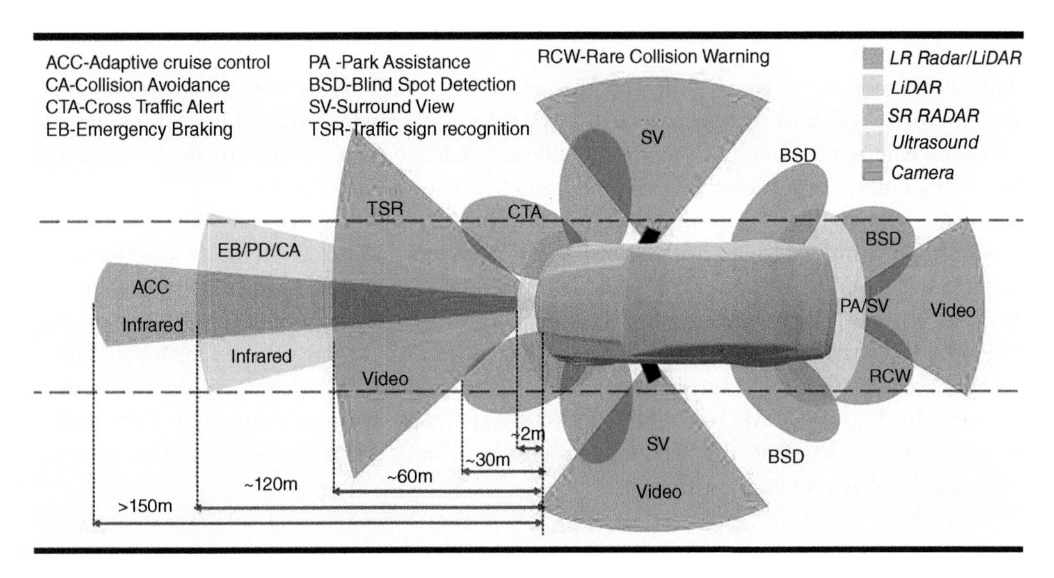

Figure 6.2 Concept of sensor fusion for advanced driver assistance systems (ADAS). *Source:* Babu Dayal Padullaparthi © Photonic Components DFM Ltd.

Scanning and non-scanning LiDARs, which were once traditionally confined to scientific and space explorations, are now effectively used in automotive transport with their specific features (detection range, sensitivity, resolution, frame rate, FOV, immunity in bad weather conditions, etc.), described in Section 6.1.2/Table 6.1; they have tremendous applications in various industries,

Table 6.2 Benchmarking of scanning and non-scanning LiDARs for applications in NIR and SWIR wavelength bands.

	Scanning: mechanical		Non-scanning: Flash		Scanning: MEMS		
Features	**NIR**[a]	**SWIR**[b]	**NIR**[a]	**SWIR**[b]	**NIR**[a]	**SWIR**[b]	**Non-scanning OPA**[b]
Speed	medium		fast		med–fast		fast
Range	long	very long	short–medium	very long	long	very long	long
Spatial resolution	high		low		high		medium
Performance on low R% target	good		low		good		good
Compactness	bulky		small	medium	small		very small
Eye safety	good	very good	good	very good	good	very good	very good
Bad weather cond.	poor	medium	poor	medium	poor	medium	medium
Maintenance	high		low		low		low–med
System cost	high	very high	low	med–high	low		low
Reliability	good		very good		good		very good
Systems per car	1		1–4 or more		1–4 More		

Source: Babu Dayal Padullparthi © Photonic Components DFM Ltd.
[a] VCSEL.
[b] F-P/DBR/EML EEL.

including automotive, transportation, robotics and drones, military, aerospace, smart building management, surveillance, and security. Currently there is no single or perfect LiDAR that meets all technical requirements, and new developments are needed to meet the industry needs. Table 6.2 shows a benchmarking of these LiDARs in NIR and SWIR wavelength bands

In LiDAR systems the maximum range (object tracking distance) critically depends on the amount of light emitted (power density) of laser transmitters and detection limit (sensitivity) of PDs (photodetectors). High-power GaAs and InP-based VCSELs and EELs are readily available at respective NIR and SWIR wavelengths. High peak powers (kW/mm^2) multi-mode- and multi-junction-based 2D arrays of VCSELs and 1D stacks of EELs are respectively used for high-quality (symmetric to asymmetric) beam shaping and narrow-to-wide beam divergences (FOV). On the other hand, the low-cost Si APD– or CMOS-based SPAD PDs use their maximum sensitivity in the NIR band (around 800–900 nm) but with expensive InGaAs APDs at 1550 nm. The authors summarize the above situation in Table 6.3.

6.1.3 Advanced Driver Assistance Systems (ADAS)

ADAS are electronic and optical sensing, as well as human-machine interfacing/interacting systems including vehicle-to-vehicle (V2V) and vehicle-to-infrastructure (V2I) communications, which assist driving and parking functions. The systems rely on inputs from the sensor fusion technologies described in Section 6.1.2. In other words, ADAS systems are developed to automate, adapt, and enhance vehicle systems for safety and better driving by minimizing driver errors, which cause most accidents, as described in Figure 6.2. These safety features are designed to avoid

Table 6.3 Laser transmitters and detectors for various LiDAR applications.

Applications	Range	Eye safety	Wavelength (λ) (nm)[a]	Laser transmitters	Laser detectors	Detection method
Autonomous vehicles Autonomous shuttles Robotic cars/taxicabs	SR MR LR	retinal damage (low MPE)	780, 850, 905940	VCSELs F-P EEL	APD SPAD	TOF
drones Flying taxicabs Logistics Building management	LR	eye safe (high MPE[c])	1310[b], 1550	F-P EEL DBR EEL EML, EDFA fiber laser	p-i-n APD	FMCW
Agriculture Industry 4.0 IoT	LR	moderate	808, 1064[b]	DPSS	SPAD	TOF

Source: Babu Dayal Padullparthi © Photonic Components DFM Ltd.
[a] NIR: 750–1200 nm, SWIR 1300–1550 nm.
[b] Eye safety between shortest-λ NIR and longest-λ SWIR.
[c] Maximum permissible exposure.

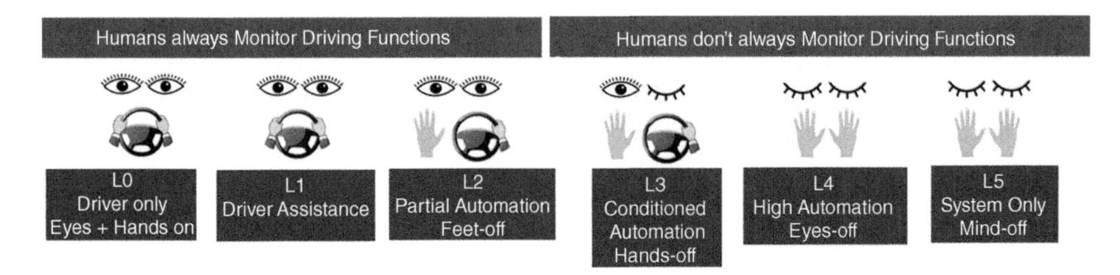

Figure 6.3 Levels of ADAS with key functions. *Source:* Modified from Ref [4, 5].

accidents and to take control of the vehicle, if necessary, by using on-board image processing and computer vision technologies.

Based on the amount of automation, the Society of Automotive Engineers (SAE) has provided a scale to divide ADAS into level 0 (L0) to level 5 (L5). These ADAS levels represent a scenario very similar to the virtual reality spectrum shown in Figure 5.4.2 and varies from full human control of vehicle to full control of vehicle by machine systems (virtual driving or driverless or self-driving)! In L0 ADAS cannot control the vehicle and can only provide information to the driver to interpret on their own. L1 and L2 are very similar in that they assist the driver to make the decision and provide information on parking sensors, surround view, traffic sign recognition, lane departure warning, night vision, blind spots, and so forth. Level L2 can take control over multiple functions to aid the driver differ from of L1. From L3 (highway chauffeur), L4 (automatic valet parking), to L5, the amount of control the vehicle has increases, L5 being where the vehicle is fully autonomous. Some of these systems have not yet been fully embedded in commercial vehicles. Figure 6.3 show the various function of ADAS from L0–L5 [4, 5].

As human binocular vision provides 200° HFOV, but only the central 45–50° FOV is high-resolution with maximum movement with color perception, it is necessary to have machine

Figure 6.4 Automotive LiDAR detection in the city center, where both light sources and objects are in motion. *Source:* Reprinted with permission from LeddarTech.

assistance during driving a vehicle for driver safety. This implies that the HFOV and VFOV play a key role to detect distant objects, and this warrants multiple LiDAR modules (example, 2–4 for short distance [~20 m] and 1–2 long distance [>100 m] in L3–L5 automation [6]). A case of automotive LiDAR detecting objects from its surroundings in the city center is shown in Figure 6.4 [7] where VCSELs and EELs are used in ADAS L0 to L5 applications.

6.2 Operating Principle of LiDARs

6.2.1 Time-Delay and Phase-Shift-Based Pulsed Light Detection

LiDAR works on the principle of TOF using an incident and reflected light pulses from a target. The TOF technique is the same as described in Section 5.1.1. In direct TOF (d-TOF), the distance is measured as a time delay (ΔT) between the transmitted and returned pulse, whereas indirect TOF (i-TOF) measures the phase difference between the transmitted and received signal (phase shift, $\Delta\phi$), as shown in Figure 6.5a and b. The d-TOF and i-TOF techniques will be effectively used to extract accurate information of the object with advanced (high-resolution) detectors and AI-based image processing techniques. The d-TOF method involves direct time measurement of light pulses with precise time differences from multiple reflection events registering at the sensitive (high pixel density) detectors and hence is a popular technique. On the other hand, the phase shift method uses CW using heterodyne detection and is more sensitive than d-TOF, but the maximum range is limited by phase unwrapping events [8, 9]. The range (R) is determined from the light reflected by an object (that returns to sensors for image processing) through Eqs. (6.2)–(6.4), given in Section 6.4.5.3.

Unlike short-distance sensing and imaging (<10 m) in consumer applications, in the AV industry both light sources and objects tend to be in motion. So large-scale image-processing resources are needed to accurately map the object over a function of time (multiple exposures of incident light and multiple captures of reflected light events), and hence computations are warranted using advanced algorithms.

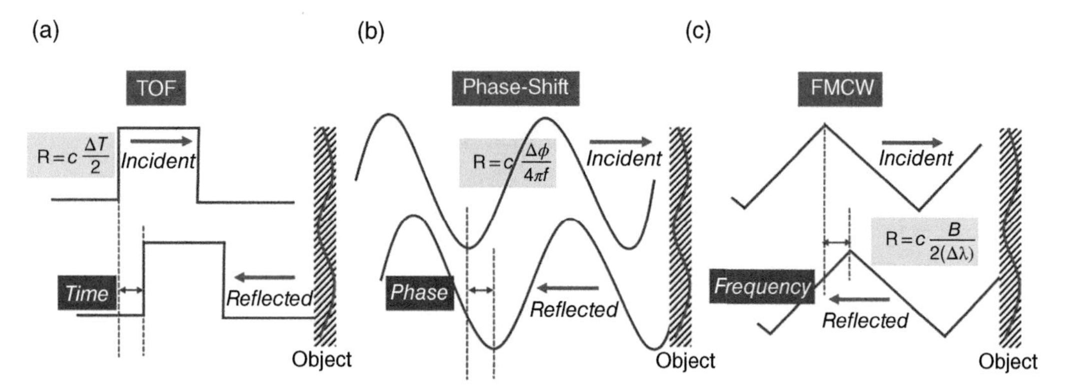

Figure 6.5 Working principles of TOF, phase shift, and FMCW measurement techniques. *Source:* Babu Dayal Padullparthi © Photonic Components DFM Ltd.

6.2.2 Frequency-Based Continuous Light Detection

The FMCW LiDAR is also emerging as a popular and promising, low-cost, chip-scale integration alternative detection technique to TOF technology to overcome range and measurement capabilities issues. Its working principle is the comparison of the frequency of the original and reflected light. As shown in Figure 6.5c, an FMCW LiDAR transmits a continuous beam of laser light that sweeps the object with different wavelengths, and hence the difference in wavelengths ($\Delta\lambda = 1/2\pi\Delta f$) are used to calculate the range using Eqs. (6.2)–(6.4), given in Section 6.4.5.3, with the help of a reference signal (local oscillator). In other words, when the incident light returns with the same frequency at the receiver (photon coincidence), the signal is amplified multiple (10–100x) times through heterodyne detection, and detection sensitivity is enhanced. FMCW LiDARs (also known as coherent detection laser for range finders) are more suitable for scanning applications (with no moving parts). With high power density and eye safety available at 1550 nm light sources, this technique offers direct measurement of the distance and the velocity through the Doppler effect and can reduce the interference effects from other laser sources and strong sunlight.

6.2.3 Light Transmitters in LiDARs

Several types of laser sources are used in LiDARs including EELs, VCSELs, and fiber lasers. GaAs- and InP-based EELs and VCSELs are well suited for LiDAR light sources in NIR (780–1200 nm) and SWIR (1300–1550 nm) bands. Even though fiber lasers generate high powers, they are bulky, expensive, and rarely used in LiDARs, and the same is true for DBR lasers with their mode-hopping characteristics. EELs offer much higher potential for their use in LiDARs with high power conversion efficiency (PCE), eye safety, and attainable longer ranges than lower-powered VCSELs but need high-cost complex manufacturing, testing, and packaging protocols (Table 1.3). VCSELs have sweeping characteristics of better reliability, low-cost manufacturing, easy packaging and assembly on to PCB or COB, 2D power scaling leading high-pulsed peak powers, and on-chip addressability of individual or group of emitters. Thus, VCSELs have totally surpassed EELs in terms of greater flexibility and becoming the preferred choices of LiDAR transmitters (850, 905, and 940 nm in NIR range). Further, with symmetric beam shapes, stability over temperature, and narrow linewidth of emission spectra (typically 1–3 nm), VCSEL arrays act as effective band-pass filters for

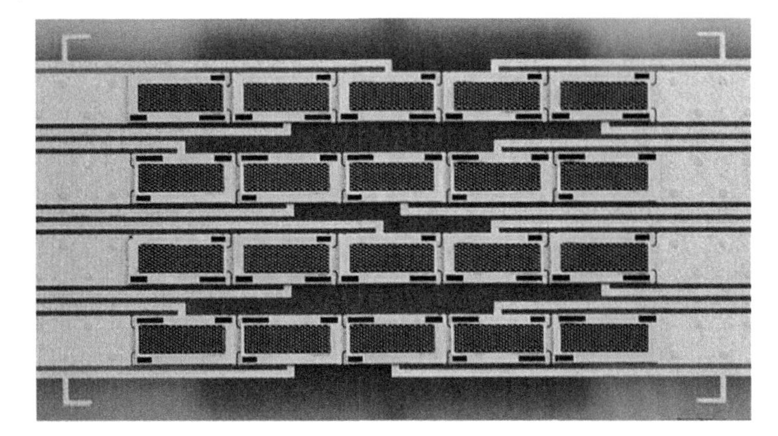

Figure 6.6 World first automotive standards qualified and individually addressable VCSEL array emitters in LiDAR (3000 emitters, 600 W, 300 m ranging. *Source:* Reprinted with permission from Trilumina Inc.

enhancing SNR at receivers that has paramount importance in object detection. Moreover, with large number of VCSEL emitters as few as 1000 to 10s of 1000s in LiDAR arrays, as shown in Figure 6.6, a single emitter failure may have a negligible impact on system performance. Several top-tier VCSEL makers are capitalizing on the LiDAR market opportunity and heavily investing in NIR VCSEL arrays for LiDARs [10]. Overall, EELs and VCSELs appear to compete with each other for LiDAR light sources, and there is no clear-cut direction on which one will be a major light source for future high-volume manufacturing of LiDAR units!

6.2.4 Light Detectors in LiDARs

PDs capture light from objects and their driver circuits register differences in time, amplitude, phase, or frequency of incident and reflected photons or waves. The high sensitivity of PDs in capturing pulses in the nano- or picosecond range time frames from tiny 3D (static or moving) objects in unfavorable weather conditions makes PDs indispensable optical components in long-range object detection and sensing. Details of PDs used for optical communication and sensing are described in Appendix J. Geiger mode PDs (APD, SPAD, and Si-PM) have high sensitivity (high bandwidth, gain and short response times) to detect CW and pulse photons to the extent of single photons and generate high SNR. SPADs are nearly noiseless due to their digital (electronic) threshold detection from quenching circuits and are extremely efficient in low background detection applications. Optimized SPAD arrays can be fully integrated with CMOS and used as TOF photon detectors in linear and image sensing applications in NIR band (780–1160 nm). Unlike Geiger mode PDs, linear mode (p-i-n-based Si and InGaAs) PDs without gain have long-distance ranging applications in the SWIR band (1400–1550 nm).

6.2.5 Lidar Module with Integrated System-on-Chip (SOC)

A system-on-chip (SOC) is a single substrate or microchip that has integration of several components to execute multiple functions. A SOC includes the electrical, electronic, optical, and mechanical components or software programming codes may contain separate CPU, graphics, memory, and storage interfaces, input/output ports, signal processing units, and USB connectivity through a central interfacing circuit board. A block diagram of a LiDAR SOC is shown in Figure 6.7a.

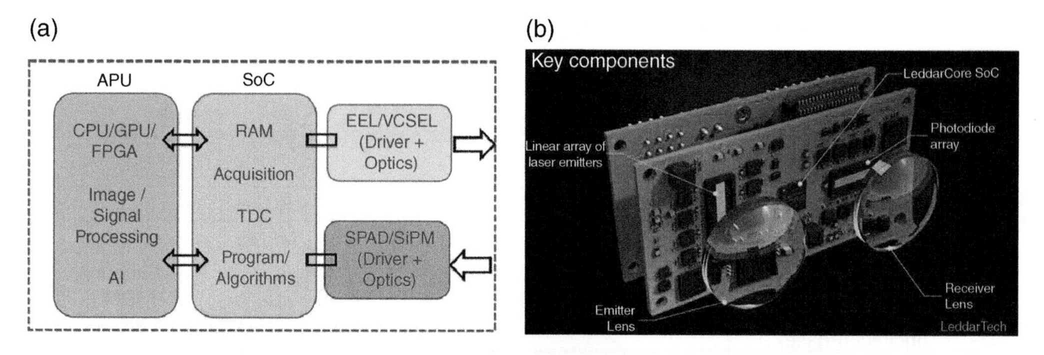

Figure 6.7 (a) Block diagram of LiDAR SOC. *Source:* © Photonic Components DFM Ltd. (b) A solid-state LiDAR module with integrated SOC. *Source:* Reprinted with permission from LeddarTech.

Recently in 2018, SOCs have emerged in the mobile (smartphones) and edge (data center) computing markets that can be viewed as embedded computing and hardware acceleration systems, in contrast to traditional motherboards with detachable or replaceable components. In August 2019, LeddarTech claimed the release of the world's first 3D solid-state LiDAR SOC with patented sensing technology, as shown in Figure 6.7b [11]. It uses a unique signal-processing method with proprietary algorithms to reduce signal noise and increase the sensitivity by lowering the detection threshold. This is achieved through sampling of the received (analog) signal, expanding the sampling rate and resolution, and analyzing the discrete-time signal to detect the distance of the objects in bad weather conditions. In this SOC LiDAR, one can see all of the key components: a linear array of laser transmitters, an array of photodetectors, beam optics at transmitter and receiver ends, a SOC, and a data processing IC behind it. LiDAR sensing involves storing a large amount of data over real-time events and analyzing and quick image processing through application processing units (APUs). This is handled with the use of FPGA along with VCSEL arrays and SPAD detectors for efficient and accurate detection of objects with high resolution and low cost [12, 13].

6.3 VCSELs in LiDAR Industry: Landscape and Direction

In the early 2000s, DARPA started the Grand Challenge program for ride sharing and movement automation with the vision of making LiDAR solutions commercially viable. In less than a decade, significant progress was made in key areas of automotive transport in obstacle avoidance, localization, and navigation. With 100s of LiDAR start-ups around the globe, Velodyne became the first LiDAR maker that garnered investment from the now ~$1.5B investment from VCs. As described in Table 6.4, LiDARs can offer key performance attributes in comparison to widely deployed cameras and RADARs.

Multiple laser sources, detectors, LiDAR types, wavelengths, object tracking schemes offering 100s of solutions to host mainstream and ancillary markets. LiDARs are built from their legacy laser and optical communication infrastructures and occupied nearly 50% of BOM (revenue from suppliers $0.7B in 2020 to a projected $3B by 2025). Lidar use in ADAS naturally led to fierce competition among tier-1 players (such as GM, Toyota, Bosch, Continental, Waymo, Argo, Aurora, Valeo, ZF, and Aptiv in the automotive space that forced them to invest in neighboring areas such as security, aerial mapping, drones, and industrial automations. Further, recently the German and French governments pumped heavy investments of €3B and €8B, respectively, to boost the automotive industries [14, 15]. With increasing demand from tier-1 LiDAR users that debuted in 2020

Table 6.4 Latest situation of LiDARs in the industry.

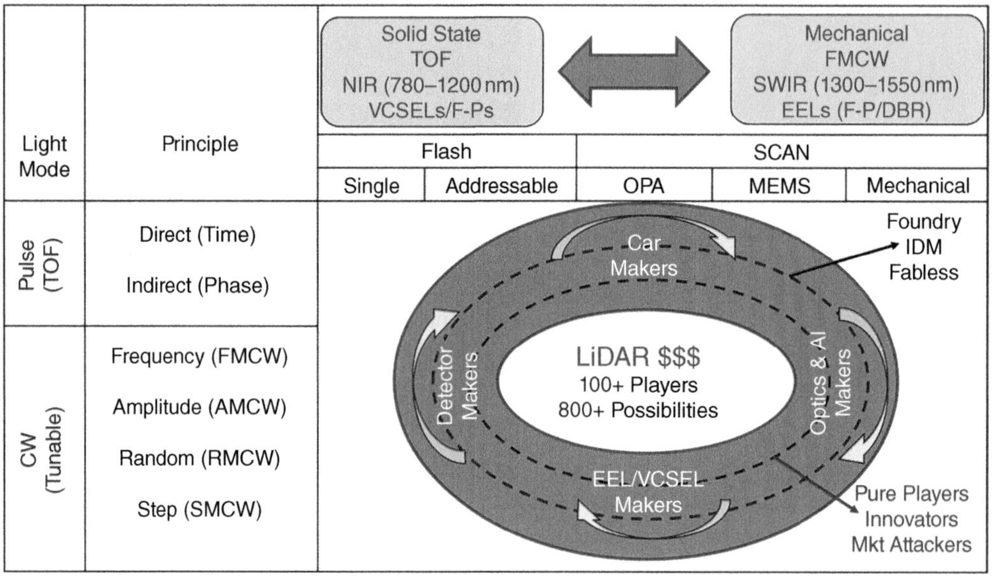

Light Mode	Principle	Solid State TOF NIR (780–1200 nm) VCSELs/F-Ps			Mechanical FMCW SWIR (1300–1550 nm) EELs (F-P/DBR)	
		Flash		SCAN		
		Single	Addressable	OPA	MEMS	Mechanical
Pulse (TOF)	Direct (Time) Indirect (Phase)	Car Makers Detector Makers LiDAR $$$ 100+ Players 800+ Possibilities Optics & AI Makers EEL/VCSEL Makers				Foundry IDM Fabless
CW (Tunable)	Frequency (FMCW) Amplitude (AMCW) Random (RMCW) Step (SMCW)					Pure Players Innovators Mkt Attackers

Source: Babu Dayal Padullparthi © Photonic Components DFM Ltd.

and the projected $118B TAM for LiDAR 2025 [16] and with successful Chang'e 5 lunar mission [17], the race is on for LiDAR opportunities from automotive to space transport.

The increase in market size of shared transportation from $30B to $1.5T by 2030 with 15–20% operating margin is attractive when compared to current ~5% growth in new car sales from OEMs and tier-1 players [18]. As 50–60% of total revenues goes to human drivers, driverless car projects are becoming attractive for ride-hailing companies to grow profitability. The number of driverless cars by 2025 is projected at 2.5M with total cars sales revenue of $175B [19]. With 100+ start-ups and several AV operators around the globe, LiDARs are getting public acceptance from USA and China. Tesla is a notable exception in LiDAR and is trying to promote non-LiDAR driverless cars! The range and scan point cloud are critical components for LiDARs. Range depends on laser power, sensitivity of PD, and object reflectivity, while point cloud density depends on pixel density and is a function of both FOV and frame rate. The target price for the widespread LiDAR adoption in the automotive industry is <$100 for 200 m range; 10% reflectivity and Class-1 eye safety are moving benchmark performance [20, 21].

The key evolving trend is that VCSELs are finding space on solid-state, TOF, NIR range with single/addressable flash options, while EELs find a major role in mechanical, TOF/FMCW, NIR/SWIR range with flash/scan options. There are currently more than 100 start-ups in LiDARs focused on NIR edge emitters with scan mechanisms, particularly solid-state scan mechanisms. As of January 2021, EELs appear to dominate the industry with major demonstrations by LiDAR makers and associated mass manufacturing plans by car makers. Nevertheless, VCSELs are actively penetrated into the LiDAR modules for long-distance ranging with the emergence of high peak powers, ease of 2D array manufacturing, low-cost optics, Si-CMOS-compatible SPADs, flash or addressable emitter options, and so forth.

Simultaneously, efforts are underway for using 1550 nm FWCM-based LiDARs with different random (RMCW) and step (SMCW) modulation schemes. VCSEL-based LiDARs are gaining traction, and recent announcements include (i) Apple's announcement of the iPad 11 Pro (March 2020)

Figure 6.8 Display of scanned image from flash LiDAR system with VCSEL transmitters. *Source:* Reprinted with Permission from Sense Photonics.

and iPhone 12 Pro (October 2020) with VCSEL-based LiDAR scanners [22, 23], (ii) VCSEL-based sequential flashes used for array row/column addressability from AMS/ZF (August 2020) [24], (iii) multi-zone "touch-on-focus" addressability of VCSEL-based LiDAR smart sensors from ST Microelectronics (October 2020) [25], and (iv) high peak powers bottom-emitting VCSEL arrays with multi-junction designs and integrated microlenses from Lumentum (October 2020) [26, 27]. These are indicative of VCSELs transitioning into LiDARs for short (~5 m) to long (250+ m) distance ranging and beginning to replace EELs. These are in addition to several VCSEL arrays and VCSEL-based LiDARs already demonstrated and explained [28, 29, 30]. Figure 6.8 shows an example of a scanned image of a scene by flash LiDAR using 940 nm VCSEL arrays from Sense Photonics.

6.3.1 Autonomous Shuttles: MaaS/ASaaS

The rapid advances in ADAS and their associated technologies have made it possible to implement AVs for public passenger transport such as autonomous shuttle bus (ASB), which navigate along predetermined, preprogrammed paths. ASBs provide an attractive and flexible solution to move people around; this is an example of mobility as a service (MaaS) [31]. Some countries and cities have already implemented ASBs at speeds up to 25–50 km/h in specific locations as university campuses, industrial areas, airports, and sports bases. They are also expanding MaaS to even connecting suburban neighborhood areas with main mass transit systems. ASBs as AVs will become foundations for the smart cities without traffic jams and rapidly increase the quality of life, and decrease energy consumption [32]. Two recent announcements of ASBs that have attracted global attention for MaaS, or autonomous shuttle as a service (ASaaS), are described in the references [33, 34]. ASB manufacturer NAVYA began operations in March 2018 from the small town of Neuhausen Rheinfall in Switzerland and appears to have beaten the rest of the world in the race to integrate an autonomous shuttle into their regular public transportation system [35, 36]. Further, in February 2020, Naganohara, a small town in Gunma Prefecture, Japan,

Figure 6.9 A flash LiDAR module integrated on rear side of autonomous shuttle buses (ASBs). *Source:* Reprinted with permission from LeddarTech.

also aims to be first at operating self-driving amphibious buses for local tourist attractions alongside the Yanba Dam and is supported primarily by the Nippon Foundation [37]. LiDAR sensors play a key role in ASBs and are emerging as active markets with a projected CAGR of more than 50% over the next 5 years.

Figure 6.9 shows an example of ASB with mechanical scanning and solid-state flash LiDARs (encircled in red). ASBs may use between 3 and 12 LiDARs, which can be either mechanical scanning or solid-state flash systems. They also include two essential CPUs for the main navigation system (robotic) and the security (or safety) systems that work redundantly in tandem to ensure complete collision avoidance and safety. Due to inherent sensor technology limitations, 2D or 3D solid-state flash LiDAR modules with wide FOV configurations are finding more deployment in detecting dead zones around the ASBs than alternate mechanical-scanning LiDARs [38, 39].

6.3.2 LiDARs in Drones, Robotics, etc.

Despite existing size, cost, power, and reliability issues, extremely accurate LiDARs have penetrated beyond ADAS into V2V, V2I, and V2X [40], which uses drones and robots, and are particularly useful in functions where humans can't see and understand. Multi-beam VCSEL arrays and SPAD arrays in combination with FPGA and AI are being innovatively used as flash LiDARs for providing scalable and flexible processing performances. An example of such a VCSEL-based drone was demonstrated by LiDAR maker Ouster, who partnered with Xilink, and is given in [41]. Flash LiDARs are also penetrating into robotic vehicle markets, which are seeing tremendous growth and are further discussed in Sections 6.5.1 and 6.5.2.

6.4 Key Aspects of LiDARs

The emitted light from a LiDAR travels through a transmission medium (normally air), then reflects back from a target surface, is received by a PD through optics, and a circuit creates an electrical signal. The amount of received power P_r by the PD originated from a pulsed laser source can be approximated by

$$P_r = E_p \frac{c\eta A r}{2r^2} \beta T_r, \tag{6.1}$$

where

E_p is the total energy of a transmitted pulse laser,
c is the speed of light in air,
A_r is the area of receiver aperture at distance r,
η is the overall system efficiency,
β $(= \Gamma/\pi)$ is the reflectance of the target's surface $(0 < \Gamma < 1)$, and
T_r denotes the transmission loss through the transmission medium.

P_r depends on both the surface properties and incident angle on the target as well as weather conditions and decreases quadratically with respect the distance r. E_P is limited in practice by eye-safety standards, and hence simply scaling the power of laser source may not be enough. The overall LiDAR system efficiency needs to be improved through optics, photodetectors, and more advanced signal processing algorithms [42]. In this section we focus on some of the key aspects and challenges of LiDAR for efficient use in product applications. The aspects on the laser side include wavelength, eye safety, and brightness while on the detector side the challenges include pixel density, background light rejection, single photon sensing, and range aliasing, all of which will also be discussed.

6.4.1 Measurement Techniques

Figure 6.10 shows the basic functionality of three kinds of LiDARs. Mechanical scanning LiDARs use pulsed laser light to illuminate a narrow FOV at a time, which is sequentially scanned across the wider FOV with a 360° rotating mirror, and the PD detects reflected photons from the object at the receiver. Similarly, solid-state scanning LiDARs use semiconductor MEMS-based solutions to scan the FOV across the target using pulsed laser light and work with the same principle as mechanical scanning LiDARs but without rotating mirrors. Contrary to both rotating or MEMS-based mirrors, a solid-state flash LiDAR illuminates the entire FOV at once using pulsed light and captures the reflected photons from the entire FOV simultaneously on its detector pixel array. Flash LiDARs are simple and offer low-cost solutions without the need of complex and expensive beam scanning methods. Single or sequential flash LiDARs based on the TOF enable real-time (3D depth) imaging of objects. The distance is determined by either the time or phase differences between and among reflected light at the receivers. Technology forecasts suggest that there is an

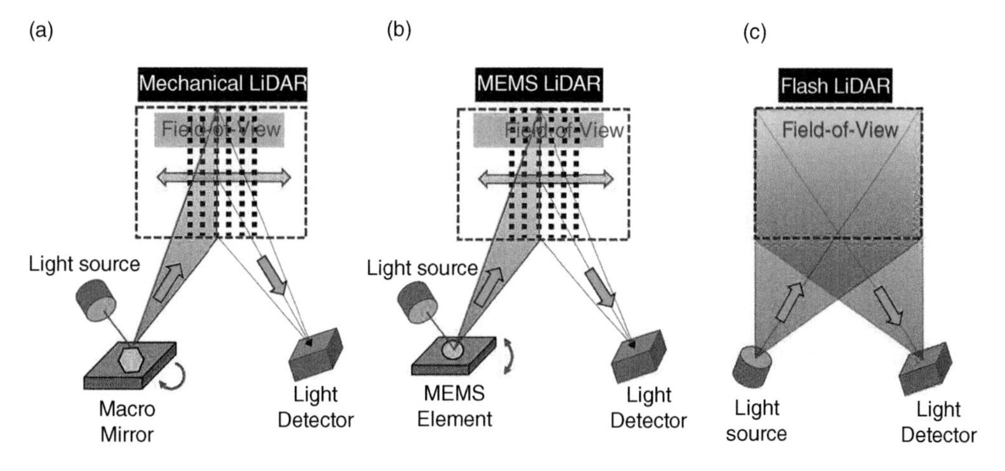

Figure 6.10 Function of three kinds of LiDARs. *Source:* © Photonic Components DFM Ltd.

increase in investments in both flash and MEMS-based scan LiDARs [6, 43]. Solid-state flash LiDARs have the significant advantage of 100% scene coverage in a single shot with FOV as high as 180°. Full 360° coverage can be obtained with multiple sensors to eliminate blind spots and require less data for point cloud processing with redundancy on the four corners of AV, which enhance perception robustness when compared to their mechanical scanning counterparts. High peak power NIR VCSEL arrays, typically at 850 nm [44], 905 nm [45], and 940 nm [46] are needed for such applications.

6.4.2 Wavelength

The choice of LiDAR wavelength is determined by many factors including atmospheric water absorption, background sunlight, and availability of suitable detectors to receive the reflected light. Light sources for LiDARs are chosen mostly in NIR band (780–1200 nm for GaAs-based lasers) and SWIR band (1310–1550 nm for InP-based lasers). Figure 6.11 shows the solar irradiance seen at the ground at 27°C from UV to MIR range [47]. It is evident from Figure 6.11 that at commercial wavelengths of LiDARs, that is, 808, 850, 905, 940, and 1550 nm, the solar irradiance is less when compared to other wavelengths. More specifically at 850 nm there is almost 2x more sunlight than at 905 nm, up to 10x more sunlight than at 940 nm, and up to 3x more sunlight than at 1550 nm. Besides, 808 and 850 nm wavelengths possess much lower water absorption coefficient than at 905, 940, and 1550 nm. Further, as described in Figure J 1.1.2b (Appendix J), the Si CMOS-based detectors are sensitive within NIR wavelengths due to their quantum efficiency/responsivity. For SR applications, 905 nm is a wavelength of choice due to its matching absorption with CMOS/CCD cameras and Si-APDs. A commonly accepted wavelength 940 nm comes from maximum sensitivity of Si-PIN detectors and reduced solar ambient (low noise) than at 905 nm. Due to higher quantum efficiency, designing a laser for LiDAR at 850 nm also facilitates to detect more reflected light back to detectors and can lead to long ranging and higher resolution [44]. For LR applications 1550 nm InP-based EELs with high peak powers are used because of eye-safety concerns, but they are very expensive and also have 100x higher water absorption coefficients than NIR wavelengths.

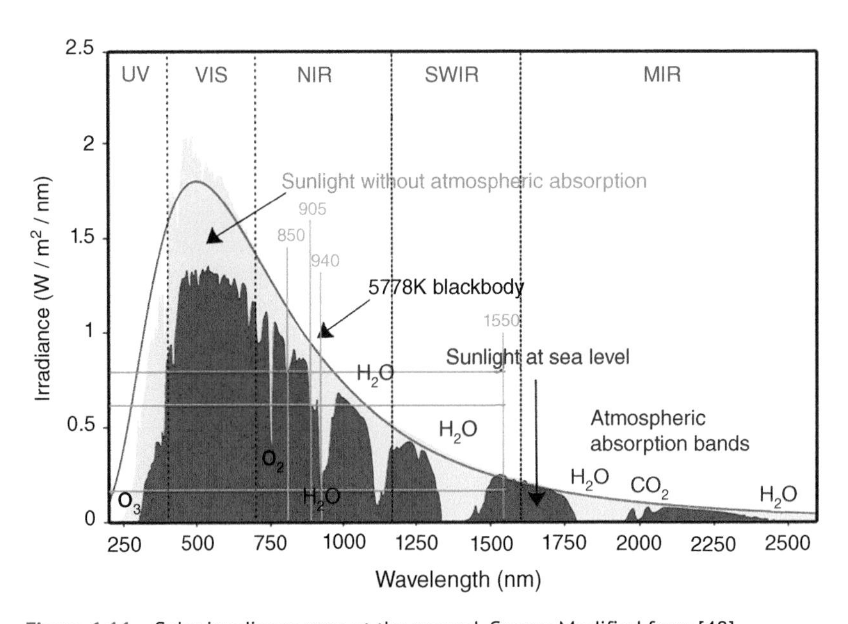

Figure 6.11 Solar irradiance seen at the ground. *Source:* Modified from [48].

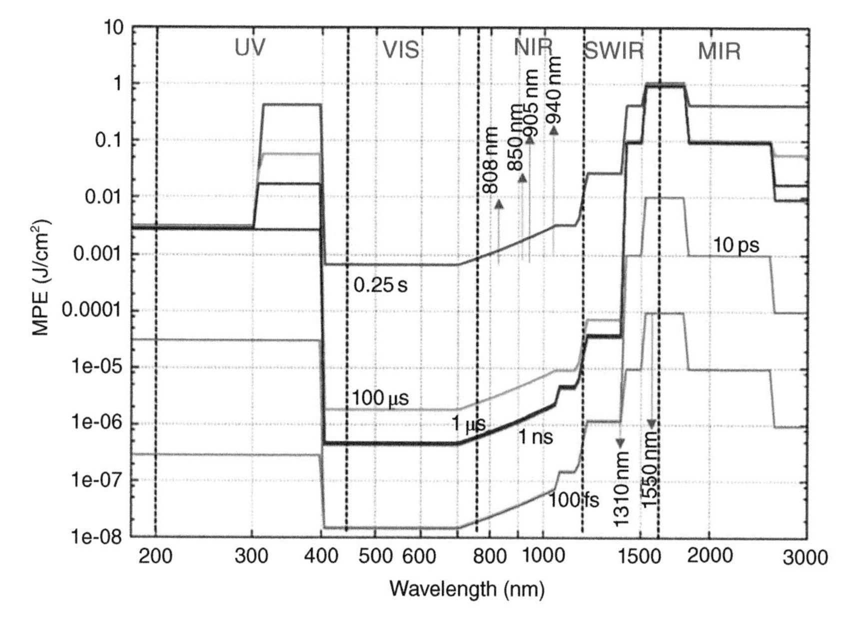

Figure 6.12 MPE as energy density versus wavelength for various exposure times (pulse durations). *Source:* Modified from [47].

6.4.3 Eye Safety

The maximum permissible exposure (MPE) is the highest average power or energy density (W/cm^2 or J/cm^2) considered to be eye safe (power density in the imaged area on the human retina) and depends on the wavelength of light [47]. As shown in Figure 6.12, MPE at the SWIR band is significantly higher than in the NIR band. This implies that for LR applications, 1550 nm LiDAR systems can offer more eye-safe power. The 905 nm wavelength lasers can penetrate through the front and the interior vitreous humor of the eye and reach the sensitive retina, and so eye-safety rules limit the LiDAR range to 100 m. The range depends on several factors from Eq. (6.1), and there is a trade-off. Light illumination from VCSEL arrays at 20–30 m has been found to be eye-safe, and the MPE for VCSEL-based illuminators are reported to be 30x larger than point source–based illuminators [49]. In all cases, the coherent light originates from distributed point light sources (VCSEL emitters) as a narrow pulses (ns) with low repetition rates, so eye safety is assured as it is totally different from brightness of narrowly focused point sources.

6.4.4 Laser Radiance and Perception

Unlike EELs, VCSELs show low radiance (power of light emitted into a solid angle with units of W/m^2/sr) and produce low to moderate powers in the mW level (typical 8 μm aperture). The power emitted from EELs and VCSELs are mainly due to their volume of active layers with high gain QW materials. With reduced physical chip areas and increasing fill factors, highly collimated laser powers can be designed with external optics to track the scenes at longer distances. Recent advances in epitaxial layer designs and manufacturing tools with multiple tunnel junction structures and integrated microlenses can provide peak powers necessary for long-distance ranging. Full wafer level fabrications have been reported [27] with powers as high as 1.0 kW/mm^2 at 1% DC (amounting to 100 kW/cm^2), SE >3 W/A, and PCE >60%. Besides these two, the ability of scaling to custom

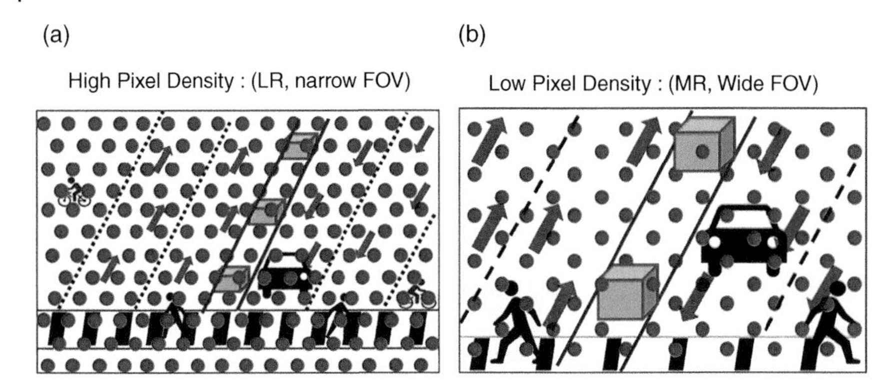

Figure 6.13 Structured light points from 2D-VCSEL arrays with (a) high and (b) low pixel densities to track long-range and medium-range objects, respectively. *Source:* © Photonic Components DFM Ltd.

2D arrays against 1D EEL bars is the other reason why VCSELs are becoming efficient light sources in LiDARs for long-distance (>200 m) ranging with low-cost manufacturing.

LiDAR perception is all about interpreting the environment in the vicinity or surroundings of the vehicle in the form of a point cloud mapping of hierarchical object descriptions, e.g., traffic lights, signs, and road markings, from multiple sensor outputs. It comprises a multi-stage localization of mapped data from object detection, object tracking, object recognition, and object motion tracking that can be made more precise by advanced deep learning technologies [42]. However, when it comes to the perception, matrix addressability unique to VCSEL arrays add an extra degree of freedom to LiDAR. As VCSELs appear to have high peak power densities of 100s W/cm^2, the structured light concept described in Section 5.1.1.3 for VCSEL array (light split into 1000s of points with DOEs) can be made to focus on distant targets with different pixel densities to track the precise size of the objects. A schematic in Figure 6.13 shows the object size and required pixel density. Small objects at longer distances need higher pixel density.

6.4.5 Challenges

As driver autonomy level expands from L1 to L5, more and more aspects need to be addressed simultaneously to ensure safety of the vehicle and its occupants. Firstly, at distances up to 250 m, AVs must detect objects quickly and with high accuracy as the vehicle can cover this distance in 9 seconds at a speed of 100 km/h. The ability to detect the signal is a function of the object reflectivity, and the target requirement is to acquire objects with >10% reflectivity at a distance of 250 m. Secondly, this drives the use of very high peak power optical pulses and very sensitive photon detectors. Thirdly, the angular resolution between pixels needs to improve from current (0.2° × 0.1° [for vertical and horizontal] <100 m] to 0.025° × 0.025° for distances of 250 m. This also linked with the desired LiDAR FOV of 360° vertical × 40° horizontal and frame rate greater than 20 fps to facilitate faster decisions by AV systems. Further care must be taken to ensure LiDARs operation in bright daytime sunlight, darkness at night, and potential interference from other LiDARs without affecting their performance. Traditional LiDARs address immunity issue with optical filters and pulse-encoding schemes (described in Section 6.4.5.3), while FWCM LiDARs are naturally immune by their default detection technique and offer additional velocity measurement of fast-moving objects [50]. These and other key issues are described below.

6.4.5.1 Background Light Rejection

Background light rejection in detectors is critically important to get high SNR to enable accurate object details. There are multiple ways to reduce background light including optical filters, multiple time measurements, and 1D scanning lasers [51]. In general optical band pass filters reduce most of the unwanted light, but often their performance depends on fabrication-induced variations and temperature-dependent wavelength shifts. Multiple time measurements offer event histograms and improve reliability but reduce frame rate. Using a point or 1D line scan source with a smaller FOV allows higher optical power density on the object, making it easier to distinguish laser and background light.

6.4.5.2 Single Photon Counting Using SPAD Arrays

In LiDAR systems, the received optical power of the reflected light from an object scales inversely with distance. With the upper limits on optical power driven by eye-safety restrictions, highly sensitive PDs are required to improve the SNR. single-photon avalanche detectors (SPAD, see Appendix I) utilize a prompt avalanche multiplication process to electronically multiply received photons and can enable near-single photon counting with time resolution in the picosecond range. Thus, SPAD are indispensable in LR automotive LiDAR applications. In addition to single photon counting, background light rejection and measurement at high ambient lights can simultaneously be made with combination of laser arrays with SPAD array detectors. By comparing temporal ~~(time)~~ changes through single photon detectors, false detections caused by ambient light can be reduced.

The detection technique can also be used to reduce the ambient light sensitivity. An example is the use of adaptive coincidence detection (ACD) which was demonstrated using 0.35 μm CMOS with 192 $\times$ 2-pixel dual line SPAD LiDAR, as reported in [51]. In ACD, the photon coincidence parameters are adjusted to the actual ambient light conditions to allow higher dynamic range for target reflectance. In other words, the photon coincident detection method easily distinguishes the time of arrival of laser pulses and ambient light signals through an AND gate integrated in the LiDAR module. Figure 6.14 shows the TOF detection with an AND gate and the probability of capturing background photons using varying target reflectances.

6.4.5.3 Range Aliasing

In TOF-based scanning LiDARs, the laser is operated with a series of short pulses with a pulse repetition period (PRP). When the PRP is smaller than the time difference between the reflections,

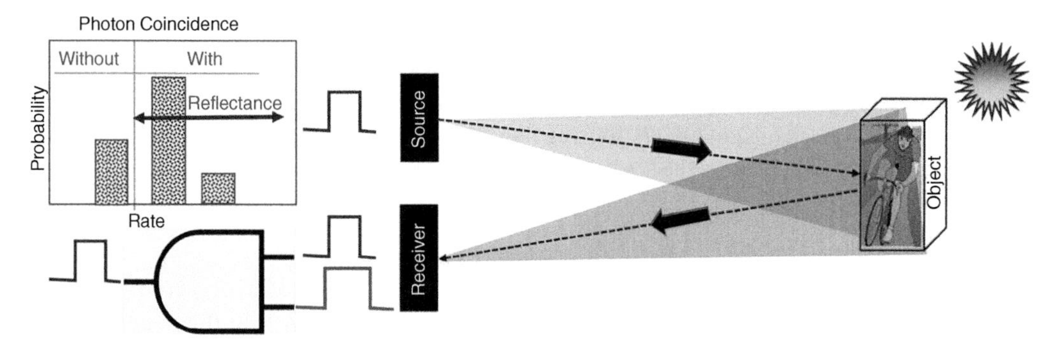

Figure 6.14 TOF detection with AND gate and probability of capturing background photons using varying target reflectance. *Source:* Modified (Fig 1 and Fig. 3) from Ref [51].

range aliasing (ambiguity) occurs. The maximum measurable range R for TOF and FMCW, as explained in Figure 6.10, depends on the measurement technique as follows:

$$\mathrm{R} = c\frac{\Delta T}{2} \text{ for pulsed LiDAR,} \tag{6.2}$$

$$\mathrm{R} = c\frac{\Delta\phi}{4\pi f} \text{ for AMCW LiDAR,} \tag{6.3}$$

$$\text{and } \mathrm{R} = c\frac{B}{2\Delta\lambda} \text{ for FMCW LiDAR,} \tag{6.4}$$

where c is the speed of light,

T is the PRP of the pulsed LiDAR,
ϕ is the phase shift between the transmitted and reflected signal,
f is the modulation frequency of the AMCW LIDAR,
B is the frequency sweep bandwidth of FMCW LiDAR, and
λ $(1/2\pi f)$ is the frequency shift per unit of time.

Unlike RF (coherent) systems where the amplitude, frequency, and phase of the carrier signal are modulated, by using the intensity-modulated direct detection (IM/DD) optical code-division multiple access (OCDMA) (i.e., unipolar signaling of optical wireless communication system); the problem of range aliasing can be solved by distinguishing the orthogonal reflected light pulses. In a pulsed scanning LiDAR system, OCDMA coding increases the PRP and hence the idle listening time between the laser pulse transmission and reception. This reduces the number of measurement points, frame refresh rate, and angular resolution per second and decreases the range ambiguity by increasing the maximum distance. Often multiple pulses are used to enhance the accuracy of the distance measurement, and the peak power of each pulse is distributed across several pulses so that the energy allocated to a pulse decreases in inverse proportion to the number of pulses and the time required to transmit pulses at a given measurement point is proportional to the number of pulses [52].

6.4.5.4 Power Consumption and System Integration

There are other challenges down the LiDAR development road, namely power consumption and components integration. Efficient heat dissipation and optimized laser designs with proper high thermal conductivity heat sinks will reduce the internal Joule heating removal from LiDAR assemblies. Some approaches separate the active optical components (lasers, PDs) from the mechanical scanning components and control/computing electronics. Placement of the LiDAR systems is also critical to thermal management. Consider that the heat flow from a vehicle radiator increases 6°C per minute compared to lasers wavelength shifts (0.062 nm/°C for VCSELs and 0.3 nm/°C for EELs)! For better thermal management and efficient integration, the simple flash optical engines can be detached from the rotational mechanism [30]. Optimizing these two challenges to enable robust LiDARs to operate in rugged environments is the key to move the technology toward mass production.

6.5 Examples of VCSEL- and EEL-Based LiDARs

Light source types are split among applications, as described in Table 6.3. In this section some significant examples of VCSEL- and EEL-based LiDARs or systems are presented with innovative technologies for future automotive transport (ADAS) applications. As described in Figure 6.1, two major types of solid-state and mechanical LiDARs are presented. The solid-state system refers to a flash or scanning system without moving parts (with the exception of integrated micro-mirrors for beam steering) and a mechanical system with moving parts (spinning or rotating bulky elements with beam steering options for scanning).

6.5.1 Solid-State Flash LiDAR

At the Consumer Electronics Show in 2020, LiDAR start-up Ouster demonstrated its OS2-128, the first long-range 850 nm VCSEL arrays–based LiDAR system and received an Honoree Innovation Award from CES [53]. Ouster claims that the sensor packs 128 individual channels to be offered at the scale, price, and reliability necessary to unlock commercially viable economics for AVs while maintaining high-volume availability and automotive-grade quality. Ouster has been developing 850 nm based LiDARs for several years [44] and now is among the fastest-growing LiDAR manufacturers. Their major customers include robotaxi providers, autonomous trucking companies, global mining companies, and automotive OEMs. It has been reported that Ouster sensors were on Postmates' robots on the sidewalks of Los Angeles [54], Kodiak's trucks on the highways of Texas [55], and on drones in the DARPA SubT challenge coal mines of Pennsylvania in 2019 [56]. Recently in 2020, NVIDIA partnered with Ouster to bring the OS2 to their autonomous trucking platform and their OEM partners to build LiDAR systems for commercial purposes [57, 58]. Another LiDAR start-up, Sense Photonics, used structured light from 940 nm VCSEL arrays that are transfer-printed on flexible (curved) substrates for 200 m ranging [59], as shown in Figure 6.15.

TriLumina Inc. (acquired by Lumentum in December 2020) also showcased its innovative back-emitting, flip-chip bonded 940 nm VCSEL illumination arrays and low-cost VCSEL illumination modules for LiDAR using a novel solid-state 3D LiDAR system in CES 2018 [60]. Its high-density VCSEL arrays offers (i) high-performance laser sources for LiDARs and (ii) monolithic micro-lenses, which significantly reduces the size and cost in TOF or SL applications. LeddarTech's LCA2 3D flash LiDAR module chip (with patented DSP algorithms using SoC concept [described in Section 6.1.5]) combined with VCSEL illuminators generates 245 000 waveforms per second and processes nearly 1.3 billion samples per second to generate the range data needed for ADAS applications [61]. LeddarTech also showcased its LCA3 technology at CES-2020 to target 350 m range and a resolution as low as 0.01° [62]. In November 2020, ROHM Semiconductor announced its new VCSEL module technology that helps to increase the reflected light from objects by 30% over conventional modules. Rohm claimed that its single optimized package containing VCSELs and

Figure 6.15 Majestic display of LiDAR transmitter made from 940 nm VCSEL arrays printed on a flexible curved substrate. *Source:* Reprinted with Permission from Sense Photonics.

MOSFETs with reduced parasitics (inductance from wire bonds) to modulate light in TOF systems results in high performance spatial recognition in both AGV and industrial inspection systems [63]. These novel LiDAR system developments will accelerate high-volume market prospects for automotive OEMs for industrial applications in AVs and ADAS with reliable, high-performance, low-cost solid-state LiDAR solutions.

6.5.2 Solid-State Addressable-Flash LiDARs

Solid-state scanning LiDARs have the advantage of no moving parts (mechanical or moving mirrors), and a robust beam-steering device with scanning functions is a key element for various sensing and imaging applications. These solutions offer significant benefits in terms of reliability and packaging flexibility, and in modern LiDAR systems, large-scale 2D VCSEL arrays with specific addressability or custom multiplexing of the emitter patterns are being developed. In the recent past some significant developments took place in terms of individually addressable emitters (rows, columns, zones, etc.), also called "electronic scanning," as shown in Figure 6.16. Reports indicate that the Austrian-based LiDAR module maker AMS (partnered with Ibeo) developed a 2D array of 940 nm VCSELs as IR light sources with 128 columns and 100 rows in sequential flashes (different from point-by-point scan functions) with an aim to generate a point cloud from reflected rays to calculate the image of the surroundings [64, 65, 66]. This could represent a true 3D (depth) model of the environment that can recognize different objects such as crash barriers, road markings, nearby cars, cyclists, and pedestrians and their position and movement accurately.

In October 2020, Ouster also announced its latest ES2 product with long-distance solid-state digital LiDAR that uses an electronic scanning mechanism to sequentially emit light from an array of VCSELs fabricated on a single chip. The ES2 is a true solid-state sensor with no moving parts that could be a game-changer to reach the range, FOV, and resolution targets using 880 nm VCSEL arrays and leverages standard CMOS manufacturing with SPAD detectors. It is the same as the core contactless slip ring technology used in its OS series spinning LiDAR sensors [30, 57]. Similar matrix-addressable emitter patterns in VCSEL arrays are used in numerous consumer devices for short-distance ranging (up to 5 m) as LiDAR sensors, described in Section 5.3.

6.5.3 MEMS Scanning LiDAR

Though not technically fully solid-state, micro electro-mechanical systems (MEMS) mirrors with scanning functions (dynamic FOV and scanning paths) allows to focus on the precise mapping of objects. There have been serious efforts from industry to push the MEMS-based LiDARs for object identification at longer distances. The MEMS mirrors are embedded in a chip that use Lorentz

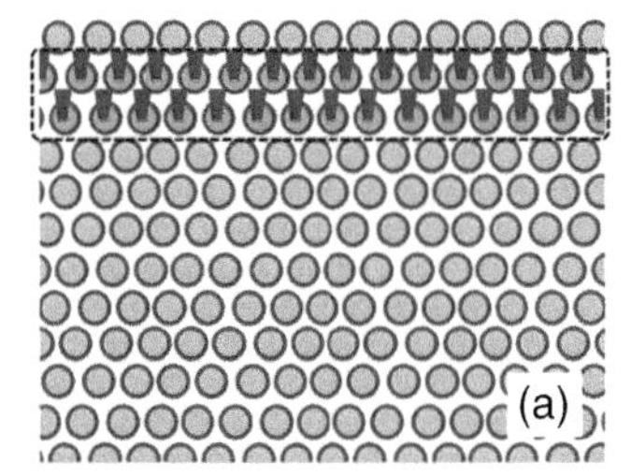

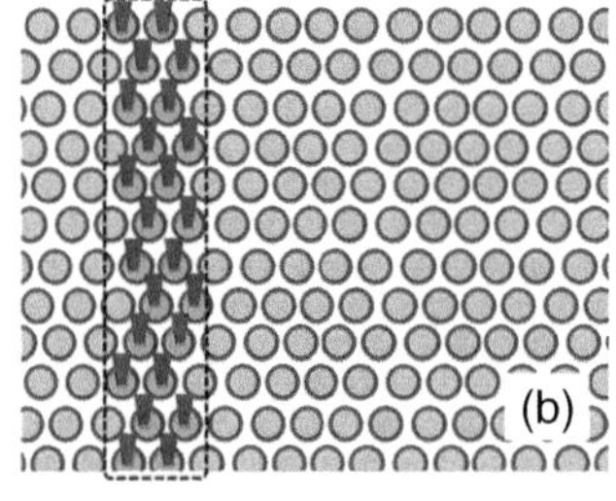

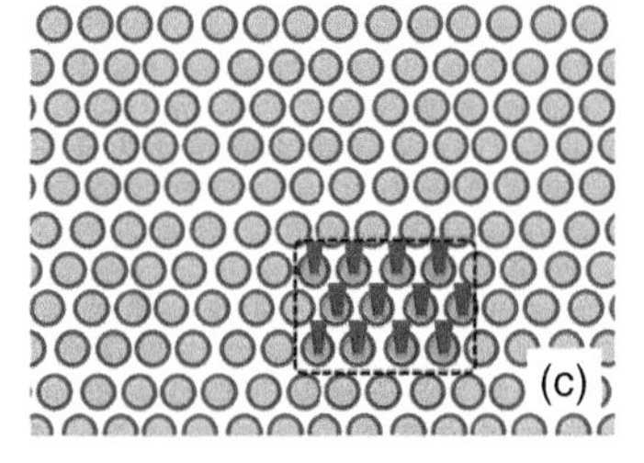

Figure 6.16 Individual (a) row-, (b) column-, and (c) zone-wise addressable or multiplexed patterns in 2D VCSEL arrays. (Electrical pads to drive selective emitters controlled by software programs are not shown intentionally.) *Source:* © Photonic Components DFM Ltd.

forces to deflect the axis of rotations either in single axis (1D) or dual axis (2D) in resonant and non-resonant (fixed programmed trajectory) modes. Leveraging the Si IC fabrication techniques, MEMS-based scan LiDARs may be able to meet the strict cost requirements in the industry. Other advantages of MEMS LiDAR scanners are (i) dynamic adjustment of the scanning pattern to a smaller angular step (to focus on objects of particular interest), (ii) handling of the variety of laser wavelengths and powers using metal coated Si mirrors, and (iii) they can perform 2D scan over wide FOV in short time intervals facilitating real-time operation of a LiDAR [42].

Innoviz, an Israeli start-up, demonstrated a 905 nm Fabry-Pérot EEL-based TOF LiDAR with 120° × 25° FOV scanning using MEMS mirrors. Operation at 150 m ranging appears more mature in the industry in terms of integration and vehicle deployments [67]. In addition to reliability and longevity challenges, Innoviz focused on efficient thermal management designs with power consumption ~20 W to cover auto-grade temperatures −40 to +125°C. This has attracted automaker BMW for its L3, with highly automated driving, to be deployed in iNEXT electric cars by 2021 [30]. Mitsubishi Electric Corporation (MEC) has also announced its MEMS-based LiDAR [68], but details about the light source are not disclosed. The LiDAR solution uses a dual axis (±15° HFOV and ±3.4° VFOV) with the industry's largest MEMS mirror (7 mm × 5 mm) to collect the maximum amount of reflected light from the objects. The optimized, extra-wide VFOV >25° allows the detection of vehicles, traffic lights, traffic signs, roadside obstacles, and pedestrians crossing the road even in close proximity, which contributes to safe and secure autonomous driving.

A 3D mapping project called the mobile mapping system (MMS) [Figure 6.17] is under development with a consortium of 15 Japanese auto (system and component) makers for cars including Toyota, Honda, and Nissan and is led by MEC itself [69]. This MMS is designed to take advantage of sensor fusion technologies along with a GPS antenna that aims to create detailed HD 3D maps to improve the safety of ADAS operation. The project is supported by Japan's government, whose objective is to have driverless (L4–L5) vehicles on the road for the Tokyo Olympics in 2020 (event finished)! These L4–L5 vehicles cruising at 40 km/h are expected to capture objects that are 7 m away with resolution of 10 cm using a laser-scanning technique and create 3D positioning data in the form of points to map roadside objects [70]. With MEC's established MEMS-based laser scan system that collects 27 100 data points per second, the information-rich 3D MMS has enough technological strength to improve the safety of ADAS. It is quite exciting to see the soaring private investments ~$1.45 billion by 2019 in all these scanning (mechanical, MEMS) and non-scanning (flash) LiDARs [71].

Figure 6.17 A car loaded with MEMS-based LiDARs for MMS project. *Source:* Adapted from IEEE Spectrum 2016.

6.5.4 Mechanical Scanning LiDAR

Mechanical scanning systems are currently the popular LiDAR solution heading for mass production. These scanning solutions offer high SNR and nearly 360° FOV. The spinning LiDARs direct the laser beams through rotating assemblies such as nodding/polygon mirrors, prisms, and so forth. They are controlled by motors, which often pose bulky integration problems inside the harsh vehicle environment. Multiple players are competing in both the NIR region and a few in the SWIR region with products at the final stages of volume production. Most of the players use the TOF LiDAR concept, and a few others use the FMCW approach, as shown in Figure 6.10. F-P EELs are the main laser sources from LiDAR makers Valeo, Velodyne, and Ibeo in the NIR band, while Luminar uses coherent tunable DBR lasers in the SWIR band. Examples include Audi's use of Valeo's Scala, a four-layer mechanical spinning LiDAR in its luxury sedan A8 for L3 automated driving functions. Audi claims to offer the first commercially available vehicle carrying auto-grade LiDAR in the world [72]. Velodyne HDL64 uses arrays of FP-EELs and photodiodes to increase point cloud densities and reduce movable mechanisms [73]. Ibeo's Lux series also uses 905 nm EELs [74]. The SWIR wavelength range offers a higher eye-safety limit, with an MPE 6x higher than the NIR region and has been adopted in some cases. Examples include Luminar collaboration with automaker giants Volvo and Toyota to supply LiDARs with 200 m range to detect objects with a mere 10% reflectivity using its mechanical mirror scanning technology [75]. A VCSEL-based mechanical scanning LiDAR system has not yet appeared in the market.

6.5.5 FMCW LiDARs

As coherent detection–based FWCM LiDARs require high-quality CW and tunable lasers with long coherence length and narrow linewidths, 1550 nm EELs in SWIR band used in fiber optics communications networks may be suitable for developing long-range eye-safe LiDARs. One of the drawback of this wavelength is lack of low-cost detectors and use of expensive InGaAs APDs. Lumentum has a long history of providing high-quality and reliable 1550 nm DBR-EELs and is active in the automotive market. An example of a LiDAR optical subassembly (LOSA module) made from its 1550 nm laser is shown in Figure 6.18 [76]. Companies such as Cruise and Aurora are investing in this area (respectively acquired Strobe and Blackmore), and details of their technologies have not been disclosed. In CW mode, operating techniques such as random modulation (RMCW) and step modulation (SMCW) improve object detection capabilities, while RMCW uses pseudorandom binary sequence modulated signals to filter random noise out to improve SNR and

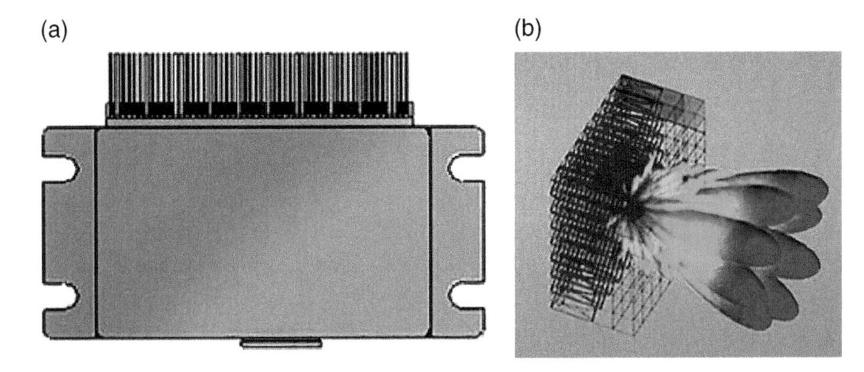

Figure 6.18 (a) Lumentum FWCM LOSA using 1550 nm DBR laser. *Source:* Reprinted with permission from Lumentum. (b) Far-field patterns from phased arrays. *Source:* Adapted from IEEE Spectrum 2016.

the precision of distance measurement [77]. The SMCW method uses phase unwrapping of scenes to improve resolution of mapped objects/targets [78].

Baraja demonstrated a 1550 nm LiDAR by using RMCW with a prism scanning approach routed through SM POF cables. It has smart integration with partitioning of its active and passive sections to manage heat, is acoustically (<−50 dB) and electronically (EMC/EMI) quiet, and appears in a more compact form than traditional polygon mirrors [20, 79].

6.5.6 Optical Phased Array (OPA) and Si-Photonics-Based LiDARs

Another truly solid-state LiDAR is the optical phased array (OPA). It works on the principle of phased array (radar antenna) by steering of the laser beams through various types of phase modulators. In OPA, the optical wavefront is passed through different phase modulators to steer the beam in different directions (1D or 2D), which allows control of the steering angles, as shown in Figure 6.18b. A typical OPA consists of a number of lasers arranged in a 1D configuration to allow beam steering in one plane only. It appears to be a promising technology with prototype products demonstrated and tested by Quanergy for autonomous vehicles and flying cars [80]. The OPA LiDARs use photonic ICs (Si-Photonics-based components lasers, waveguides, couplers, etc.) with CMOS SPADs to cover ~150 m distances with a wide FOV originating from diffracted wavefronts. The OPA minimizes blind spots with high resolution tracking of objects to the level of <5 cm at a distance of 100 m. Details are reviewed in [81, 82].

6.5.7 VCSELs for in-Cabin Sensing

LiDARs are also being developed for in-cabin sensing much like their use in smartphones, gaming, and tablets. Automotive customers are more interested in interacting with in-vehicle experience (IVE) infotainment systems during the ride and for driver awareness systems for their safety. New driver assist functions include more high-resolution visual displays and projections on different surfaces; gesture, speech, and voice recognition systems; lane departure and crossing, collision avoidance warning and blind spot detections, and parking assistance, all of which are gathering momentum [13]. In order to use lasers in automotive applications, auto-grade and high-performance VCSELs operating at −40°C to 125°C with high reliability are needed. Figure 6.19a and b show examples of packaged 940 nm VCSEL chip for L2–L4 in-cabin monitoring (IVE).

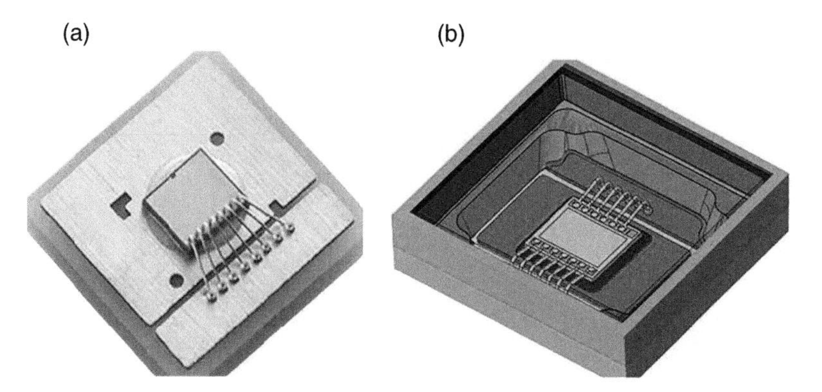

Figure 6.19 An example of highly reliable 940 nm VCSEL (a) bare and (b) packaged array for L2–L4 in-cabin monitoring (IVE) applications. *Source:* Reprinted with permission from Lumentum.

6.6 Automotive Communication: IVE (Infotainment) and C-V2X

In order to overcome the human errors and to meet destination zero to have no accidents, driver assistance or driverless cars make sense for the future of the automotive industry. Automotive innovation is accelerating at an unprecedented pace, and as shown in Figure 6.20, new built-in infotainment systems are in high demand for next generation AVs. Increasing innovations in sensor fusion, wireless (Wi-Fi), HD 3D mapping, powerful signal processing technologies, and AI for enhanced safety and improved connectivity in V2V, V2N, V2I, V2P, V2U, and V2X systems are fueling the demand. In the above technologies all electronic components need to handle large amounts of data (5 Tb to 20 Tb per day) with many different operating speeds (100 kbps to 70 Mbps). Short-range communication in ADAS and AVs need high bandwidth and low latency. New protocols are being developed by IEEE and other standards developing organizations to cater to the needs of the automotive industry, which include transmitting high-speed data reliably and with low latency [83].

In this direction, automakers are developing the Automotive Ethernet (AEnet), a wired and diagnostic network, as an emerging solution for dedicated IVE and V2X experiences. The AEnet concept is a multivendor and licensed-based solution introduced by Broadcom to use 100 Mbps proprietary PHY layer structures. It uses PAM3 signaling for data transmission up to 1 Gbps (1000BASE-T1) over a single pair in both directions. AEnet is adopted and regulated by OPEN (One-Pair Ethernet) Alliance (IEEE and OPEN 802.3 bp and 802.1 groups). Faster system communication groups have been established (802.3ch for 2.5/5/10Gb/s), and new task force (IEEE 802.3cz: Optical Multi Gig Ethernet) is also looking for 25/50Gb/s options [83, 84].

Traditional automotive serial bus technologies such as CAN, LVDS, LIN, MOST, FlexRay, and CAN-FD have limitations and can't provide enough speed or bandwidth for next-generation vehicles. To overcome the above legacy buses' limitations (for example a LiDAR requirement to transmit 70 Mbps), automakers are simply looking for AEnet to enable ADAS and AVs as a new reality. Figure 6.21 depicts the historical evolution of serial bus technologies, which helps to understand the needs of AEnet [83, 84, 86, 87]. Serial bus technologies use traditional local and domain-distributed architectures, whereas AEnet uses zonal architectures that require high bandwidth 10G+ links between zones (dashed squares in Figure 6.21). A host of auto giants including Audi, General Motors, Jaguar, Land Rover, Toyota, Volkswagen, and Volvo have publicly announced the implementation of zonal architectures in their ADAS vehicles [84].

Figure 6.20 An infotainment system in the cockpit of an AV. *Source:* Adapted from [85].

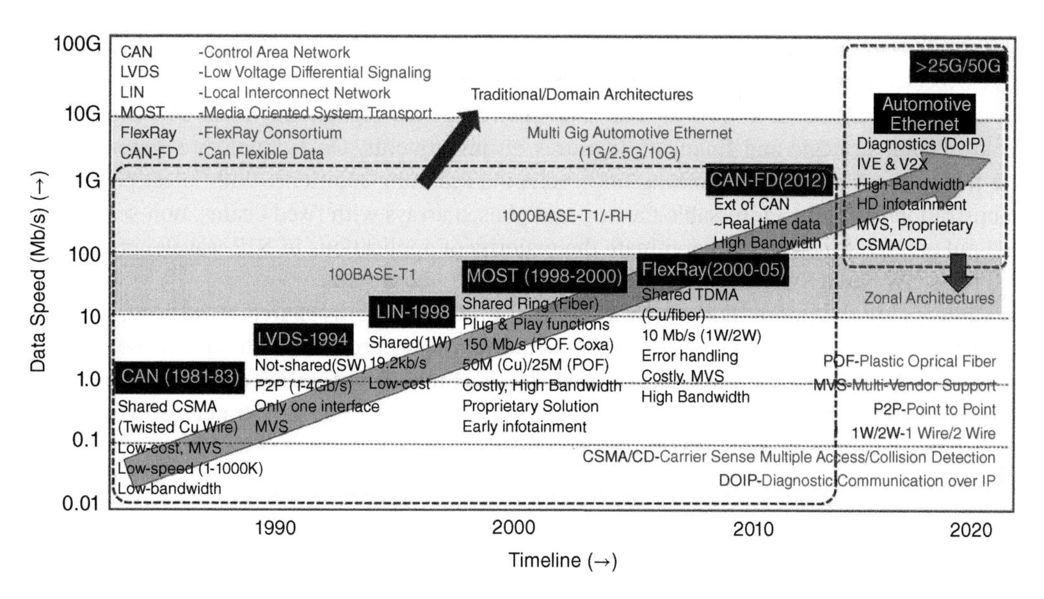

Figure 6.21 Historical evolution of serial bus technologies. *Source:* © Photonic Components DFM Ltd.

As the vendor-dependent car architectures are getting more complex with increased sensing elements and interfaces (say, L4–L5 require 20 radars, 6 cameras, and V2X), AEnet serves as the common standard for all communication and reduces cable connectivity cost by ~80% and cable weight by ~30%. Further, while transitioning from MOST to AEnet, LEDs will be replaced by laser transmitters (VCSELs and EELs), and stringent auto-grade reliability requirements have to be met by all of the component makers [88]. AEnet has the potential to offer advanced capabilities of in-vehicle high-speed communication among electronic systems, replacing traditional serial buses. The expected CAGR for the market exceeds 21% for 2019–2026 [89] and car OEM test vehicles are already using enterprise class Ethernet devices with 25 Gbps or 50 Gbps; even looking for more variants before volume production begins in model year 2025!

In order to connect form vehicle to vehicle (V2V), a cellular-based V2X (C-V2X) unified platform is emerging to enhance the functionalities of ADAS with next generation vehicle connectivity experience. C-V2X uses 3GPP, 4GLTE, and 5G ecosystems to communicate in- and out-of-vehicle longer reaction (response) times. For example, when a vehicle travels at 140 km/h, the 802.11P standard for 225 m covers 3.3 seconds, while for the C-V2X platform for 450 m the reaction time will be 9.2 seconds. Qualcomm's Snapdragon is an example for future C-V2X technologies for next level infotainment, and details are given in reference [90].

6.7 Market Summary

For the sake of market briefing, LiDAR applications can be categorized into multiple sectors such as automotive, robotic, topological, and industrial. According to market analysists, the combined market forecasted is $1.6 billion in 2020 and is expected to reach $3.8 billion in2025, with a CAGR of 19% [91]. Though automotive and robotic applications are the main drivers for LiDAR growth in the next five years, topography and industrial applications are strongly emerging, as described in Section 6.1. As the LiDAR component supplier ecosystem (makers of light transmitters, photodetectors, optical elements, LiDAR sub-systems/modules, systems, and ICs) [92] is fully crowded with 100+ players, and there is anticipation of industry consolidation. LiDAR and car manufacturers are expected to adopt ADAS mostly into robots and smart (infrastructural) facilities. However, with the market effects of COVID-19 and strict environmental regulations on carbon emissions, they are facing strategic and financial pressures on investments. On technology segmentation, more than 80% of players are working on developing scanning solutions. With latest prototype demonstrations in flash, addressable flashes, and phased arrays with fixed beams, non-scanning-based pulse LiDARs continue to dominate the majority of applications in NIR wavelengths compared to its CW-based SWIR counterparts [93].

It is interesting that more than 60% of the global LiDAR market is shared by just four players (Trimble, Hexagon AB, Sick AG, and Topcon) with their heavy product portfolio in airborne and topological LiDARs besides traditional auto industry. Further, current LiDAR market appears to be dominated by a mix of traditional established players and new entrants with strategic investments. Established Chinese firms are expected to penetrate deep into this market with low unit costs [94]. With increasing numbers of driverless cars/robotic vehicles hitting the road, a record number of shipments of automotive LiDARs of 3.2M units by 2024 have been forecasted [95].

Despite all the above prospects, there is some twist on ADAS, being treated beyond hype for vision zero or destination zero (meaning zero accidents by 2020) ideas for automated driving (beyond L2++). There are some reports of robo-taxis that use massive computing powers becoming a reality and disrupting ADAS [96]!

References

1 https://resource.lumentum.com/s3fs-public/technical-library-items/lidar3dsensing_wp_cl_ae.pdf.
2 OSRAM Licht AG, "LiDAR Teach-In" Slate 20 June 2018 https://www.osramgroup.com/~/media/Files/O/Osram/Investor%20Relations/Publications_other/2018/presentation-investor-analyst-call.pdf.
3 https://www.swissphotonics.net/libraries.files/Cochard.pdf.
4 https://www.osramgroup.com/~/media/Files/O/Osram/Investor%20Relations/Publications_other/2018/presentation-investor-analyst-call.pdf.
5 https://www.sae.org/news/press-room/2018/12/sae-international-releases-updated-visual-chart-for-its-"levels-of-driving-automation"-standard-for-self-driving-vehicles.
6 A. Debray., https://www.i-micronews.com/epic-meeting-on-lidar-technologies-for-automotive-at-anteryon.
7 https://www.ces.tech/Innovation-Awards/Honorees/2020/Honorees/L/Leddar-PIXELL-The-Cocoon-LiDAR-for-Autonomous-Veh.aspx.
8 R. Mellors, E. Lindsey, X. Xu et al., https://www.unavco.org/education/professional-development/short-courses/course-materials/insar/2017-insar-gmtsar-course-materials/08_2017_unwrap_v6.pdf.
9 J. Mei, A. Kirmani, A. Colaco et al., https://citeseerx.ist.psu.edu/viewdoc/download?doi=10.1.1.640.8420&rep=rep1&type=pdf.
10 https://optics.org/news/11/8/16.
11 https://www.allaboutcircuits.com/news/leddartech-first-3D-solid-state-LiDAR-system-on-chip-autonomous-vehicles.
12 https://www.forbes.com/sites/tiriasresearch/2020/08/24/ouster-and-xilinx-partner-to-bring-cost-effective-lidar-to-autonomous-vehicles/#c9a5a0f1daf3.
13 https://www.intel.com/content/dam/www/programmable/us/en/pdfs/literature/wp/wp-01167-custom-arm-soc.pdf.
14 https://www.dw.com/en/germany-to-pump-additional-3billion-in-ailing-automotive-industry/a-55641102.
15 https://www.dw.com/en/france-unveils-stimulus-plan-worth-8-billion-for-car-industry/a-53578294.

16 https://www.nanalyze.com/2020/12/four-lidar-stocks.

17 https://www.scmp.com/news/china/science/article/3114252/change-5-lunar-mission-returns-earth-after-collecting-moon.

18 https://www.mckinsey.com/~/media/mckinsey/industries/automotive%20and%20assembly/our%20insights/disruptive%20trends%20that%20will%20transform%20the%20auto%20industry/auto%202030%20report%20jan%202016.pdf.

19 https://www.photonicsonline.com/doc/the-driverless-car-revolution-creates-massive-opportunities-for-optics-0001.

20 https://www.forbes.com/sites/sabbirrangwala/2020/06/15/the-big-bend--theory-and-beyond--are-we-there-yet/?sh=1c43d3591edc; https://www.forbes.com/sites/sabbirrangwala/2020/09/22/money-for-somethin/?sh=1f70b667511a.

21 https://www.forbes.com/sites/sabbirrangwala/2020/11/23/money-for-everythin/?sh=22d9b6dc4e1c.

22 https://www.apple.com/hk/en/newsroom/2020/10/apple-introduces-iphone-12-pro-and-iphone-12-pro-max-with-5g.

23 https://www.systemplus.fr/wp-content/uploads/2020/06/SP20557-Apple-iPad-pro-Lidar-Module_flyer.pdf.

24 https://ams.com/-/ibeo-lidar http://epic-events.eu/epic/2019/lidar2019/191030_EPIC_LIDAR2019_P31.pdf https://www.epic-assoc.com/wp-content/uploads/2020/07/Hanno-Holzhuter_ibeo.pdf.

25 https://www.st.com/en/imaging-and-photonics-solutions/vl53l5.html.

26 https://www.lumentum.com/en/products/multi-junction-vcsel-array.

27 https://www.lumentum.com/en/media-room/videos/autosens-2020-high-power-vcsels-automotive-applications.

28 https://www.forbes.com/sites/sabbirrangwala/2020/04/12/there-must-be-860-ways-to-build-an-av-lidarpart-1/?sh=3dc1a120545c.

29 https://www.forbes.com/sites/sabbirrangwala/2020/05/16/there-must-be-860-ways-to-build-an-av-lidarpart-2/?sh=1a3240f929bd.

30 https://www.forbes.com/sites/sabbirrangwala/2020/07/17/lidarnow-you-see-me-soon-you-wont/?sh=4a42d68e4384.

31 https://innovationatwork.ieee.org/mobility-as-a-service-maas.

32 C. Iclodean, N. Cordos, and B. O. Varga, "Autonomous shuttle bus for public transportation: a review" *Energies* **13**, 2917 (2020); doi:10.3390/en13112917.

33 https://www.intelligenttransport.com/transport-articles/70339/autonomous-shuttle-regular-public-transport/.

34 A. Bucchiarone, S. Battisti, A. Marconi et al., "Autonomous Shuttle-as-a-Service (ASaaS): Challenges, Opportunities, and Social Implications," *IEEE Transactions on Intelligence Transport Systems* (s3-eu-west-http://1.amazonaws.com › ASaaS_Vision2020).

35 https://navya.tech/en/solutions/moving-people/self-driving-shuttle-for-passenger-transportation.

36 https://navya.tech/wp-content/uploads/documents/Brochure-Autonom-Shuttle-Evo-EN.pdf.

37 https://spectrum.ieee.org/cars-that-think/transportation/self-driving/small-japanese-town-to-test-first-autonomous-amphibious-bus.

38 https://leddartech.com/market/autonomous-shuttles.

39 B. D. Padullaparthi, T. Tamanuki and D. Bimberg., Field-of- view control of segmented NIR VCSEL arrays for next-generation flash LiDARs," *Proc. SPIE* **11262**, 112620Z (2020); doi:10.1117/12.2564489.

40 https://www.forbes.com/sites/tiriasresearch/2020/08/24/ouster-and-xilinx-partner-to-bring-cost-effective-lidar-to-autonomous-vehicles/#c9a5a0f1daf3.

41 https://aithority.com/internet-of-things/ouster-announces-os-1-128-channel-lidar-sensor.

42 Y. Li and J. I. Guzman, "Lidar for autonomous driving: the principles, challenges, and trends for automotive Lidar and perception systems," *IEEE Signal Process. Mag.*, **37**, 50–61 (2020).

43 https://www.idtechex.com/zh/research-article/lidar-making-sense-of-the-complicated-technology-landscape/18795.

44 V. Koifman, http://image-sensors-world.blogspot.com/2018/11/ouster-discusses-its-lidar-principles.html.

45 R. F. Carson, M. E. Warren, P. Dacha et al., "Progress in high-power, high-speed VCSEL arrays," *Proc. SPIE* **9766**, 97660B 1-13 (2016).

46 https://www.laserfocusworld.com/lasers-sources/article/16548239/verticalcavity-surfaceemitting-lasers-vcsel-arrays-provide-leadingedge-illumination-for-3d-sensing.

47 https://en.wikipedia.org/wiki/Laser_safety.

48 https://en.wikipedia.org/wiki/Solar_irradiance#/media/File:Solar_spectrum_en.svg.

49 D. Zhou, J. F. Seurin, G. Xu et al., "Progress on vertical-cavity surface-emitting laser arrays for infrared illumination applications," *Proc. SPIE* **9001**, 90010E (2014) doi:10.1117/12.2040429.

50 G. Smolka https://www.photonics.com/Articles/Market_Trends_in_Automotive_Lidar/a65302.

51 M. Beer, J. F. Haase, J. Ruskowski et al., "Background light rejection in SPAD-based LiDAR sensors by adaptive photon coincidence detection," *Sensors*, **18**, 4338 (2018) ; doi:10.3390/s18124338.

52 G. Kim and Y. Park, "Suitable combination of direct intensity modulation and spreading sequence for LIDAR with pulse coding," *Sensors*, **18**, 4201 (2018); doi:10.3390/s18124201.

53 https://www.ces.tech/Innovation-Awards/Honorees/2020/Honorees/O/OS2-128-Long-range-lidar-sensor.aspx.

54 https://www.roboticsbusinessreview.com/news/postmates-ouster-team-up-for-lidar-enabled-delivery-robots.

55 https://money.yahoo.com/ouster-lidar-raises-42-million-120000035.html.

56 https://medium.com/penn-engineering/penn-engineers-to-send-robots-underground-in-darpa-subterranean-challenge-836fa6e3001a.

57 https://ouster.com/products/es2-solid-state-lidar-sensor.

58 https://ouster.com/blog/the-es2-the-first-true-solid-state-high-performance-digital-lidar.

59 https://sensephotonics.com/photons-flood-the-scene.

60 https://www.trilumina.com/latest-news/press-releases/trilumina-to-demonstrate-3d-solid-state-lidar-using-its-3d-sensing-vcsel-illumination-solutions-at-ces-2018.

61 https://innovation-destination.com/2018/01/18/3d-lidar-delivers-automotive-solutions.

62 https://leddartech.com/leddartech-showcase-first-3d-solid-state-lidar-ic-autonomous-driving-ces-2018.

63 https://www.globenewswire.com/news-release/2020/11/30/2136947/0/en/ROHM-Announces-New-VCSEL-Module-Technology-Increasing-the-Output-of-Spatial-Recognition-and-Ranging-Systems-by-30.html.

64 https://optics.org/news/11/8/16.

65 https://www.photonics.com/Articles/Lidar_A_Technology_at_Use_in_Daily_Life/a65437.

66 https://www.eejournal.com/industry_news/ibeo-and-ams-solid-state-lidar-technology-used-by-prominent-automotive-manufacturer-great-wall-motor-to-enable-future-autonomous-driving-vehicles.

67 https://www.prnewswire.com/news-releases/innoviz-technologies-introduces-its-next-generation-innoviztwo-which-will-accelerate-the-industry-towards-autonomous-driving-enabling-safe-mobility-to-all-301151672.html.

68 https://emea.mitsubishielectric.com/en/news/releases/global/2020/0312-a/index.html.

69 https://spectrum.ieee.org/cars-that-think/transportation/self-driving/japans-upgraded-mobile-mapping-technology-aims-to-make-autonomous-driving-safer.

70 S. Chen, B. Liu, C. Feng et al., "3D Point Cloud Processing and Learning for Autonomous Driving," TR2020-066 June 03, 2020 (https://www.merl.com/publications/docs/TR2020-066.pdf).

71 http://epic-events.eu/epic/2019/lidar2019/191030_EPIC_LIDAR2019_P11.pdf.

72 https://europe.autonews.com/article/20170804/COPY/308049989/lidar-is-here-valeo-expects-price-to-drop-over-next-5-years.

73 https://velodynelidar.com/products/hdl-64e.

74 https://www.ibeo-as.com/en/products/sensoren/IbeoLUX.

75 C. Rablau, “LIDAR – a new (self-driving) vehicle for introducing optics to broader engineering and non-engineering audiences,” *Proc. SPIE* **11143**, 111430C (2019) doi:10.1117/12.2523863.

76 https://www.lumentum.com/en/fmcw-lidar.

77 http://oa.upm.es/42613/1/INVE_MEM_2015_230461.pdf.

78 R. Whyte, L. Streeter, M. J. Cree et al., “Application of lidar techniques to time-of-flight range imaging, ”
54, 9654–9664 (2015).

79 https://www.baraja.com/wp-content/uploads/2020/01/2020-02-06-Baraja-Immunity-CES-Press-Release.pdf.

80 https://www.imaging.org/Site/PDFS/Conferences/ElectronicImaging/EI2019/AVMKeynotes/LouayEldada_Quanergy_AVM%202019.pdf.

81 X. C. Sun, L. X. Zhang, Q. H. Zhang et al, “Si photonics for practical LiDAR solutions,” *Appl. Sci.* **9**, 4225 (2019); doi:10.3390/app9204225.

82 J. Sun, E. Timurdogan, A. Yaacobi et al., “Large-scale silicon photonic circuits for optical phased arrays,” *IEEE J. Sel. Top. Quantum Electron.*, **20**, 8201115 (2014).

83 https://www.keysight.com/hk/en/assets/7018-06381/white-papers/5992-3430.pdf.

84 https://www.ieee802.org/3/cz/P802d3cz_objectives_01_0520.pdf

85 https://spectrum.ieee.org/transportation/self-driving/accelerating-autonomous-vehicle-technology.

86 https://talks.navixy.com/reviews/most-wireless-can-flexray-and-automotive-ethernet.

87 T. Kibler, S. Poferl, G. Bock et al., “Optical data buses for automotive applications,” *J. Lightwave Technol.* **22**, 2184 (2004), doi:10.1109/JLT.2004.833784.

88 https://www.ruetz-system-solutions.com/uploads/RUETZ-SYSTEM-SOLUTIONS-MOST-Forum-2013-conference-presentation-VCSEL.pdf.

89 https://www.globenewswire.com/news-release/2020/01/21/1973389/0/en/The-automotive-Ethernet-market-is-projected-to-grow-from-USD-1-6-billion-in-2019-to-USD-4-4-billion-by-2024-at-a-Compound-Annual-Growth-Rate-CAGR-of-21-7-from-2019-to-2024.html.

90 https://www.qualcomm.com/media/documents/files/cellular-vehicle-to-everything-c-v2x-technologies.pdf.

91 http://www.yole.fr/iso_upload/News/2020/PR_LiDAR_MarketUpdate_LIVOX_LiDAR_YOLE_SYSTEMPLUSCONSULTING_August2020.pdf.

92 https://s3.i-micronews.com/uploads/2020/08/YDR20114-LiDAR-for-Automotive-and-Industrial-App-Sample.pdf.

93 http://epic-events.eu/epic/2019/lidar2019/191030_EPIC_LIDAR2019_P11.pdf.

94 https://s3.i-micronews.com/uploads/2020/08/YDR20114-LiDAR-for-Automotive-and-Industrial-App-Sample.pdf.

95 http://epic-events.eu/epic/2019/lidar2019/191030_EPIC_LIDAR2019_P11.pdf.

96 https://www.i-micronews.com/the-rise-and-fall-of-the-adas-promise-now-disrupted-by-avs/?cn-reloaded=1.

7

Illumination, Night Vision, and Industrial Heating

Jim Tatum

7.1 Introduction

Infrared illumination sources are commonly used in machine vision, night vision, security, and surveillance systems. The illumination is not visible to the human eye and does not distract or alert the viewer or the object under illumination. LEDs (light-emitting diodes) have been the primary light source in these applications, but they are beginning to be replaced with vertical-cavity surface-emitting lasers (VCSELs) to improve the efficiency and brightness of the source. Two-dimensional VCSEL arrays offer significantly more usable optical power than an LED, and the narrow spectral bandwidth and low wavelength tuning with temperature enable the use of narrower background optical filters to further increase the signal-to-noise ratio (SNR). One possible drawback to laser sources is the potential for speckle in the illumination profile, but these effects are reduced with 2D arrays of VCSELs. VCSEL arrays can be scaled almost infinitely, and individual die can be co-packaged to achieve kilowatt power levels, making them good candidates for large-area terrain illumination and industrial heating applications. In this chapter, the optical properties of VCSELs in these application areas are compared to edge-emitting lasers (EELs) and LEDs, and a number of commercial applications are presented.

7.2 Optical Properties of Illumination Sources

As described in Chapter 2, VCSELs have circularly symmetric output beams with relatively low optical divergence, making the output easy to collimate and otherwise manipulate. Figure 7.1 compares the illumination profile from a VCSEL, LED, and EEL. Both VCSELs and EELs can be collimated to illuminate a small area or can be used with engineered optical diffusers to create a prescribed divergence profile, while LEDs are generally used as broad area illuminators.

One potential limitation of laser-based illuminators is speckle, or non-uniformity of the laser intensity. Speckle results from different path lengths taken by coherent light to an object or detection system. The coherence length, $L_{COHERENCE}$, is the distance over which the emitted photons from a light source maintain a phase relationship and can thus constructively or destructively interfere, resulting in speckle patterns. The coherence length is defined by the wavelength and the spectral width of the source and can be written as

$$L_{COHERENCE} = \frac{c}{\pi \Delta v} = \frac{\lambda^2}{\pi \, \Delta\lambda}$$

VCSEL Industry: Communication and Sensing, First Edition. Babu Dayal Padullaparthi, Jim A. Tatum and Kenichi Iga.

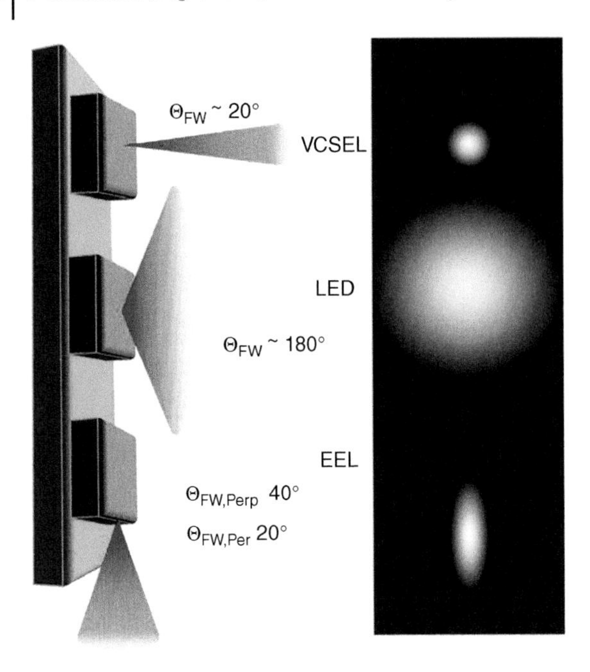

Figure 7.1 Output beam characteristics of VCSEL, LED, and EELs.

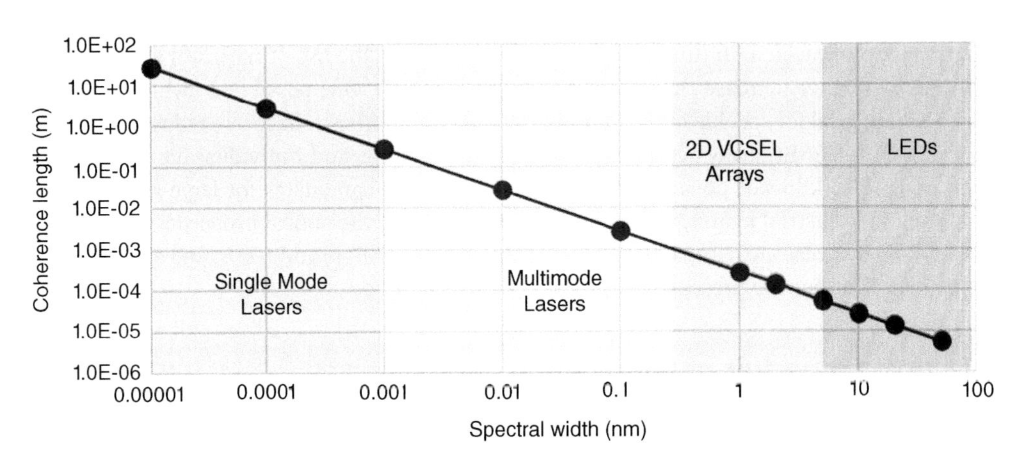

Figure 7.2 Coherence length of an illuminator as a function of the spectral width. The highlighted regions show the different light source technologies.

where c is the speed of light, $\Delta\nu$ is the spectral width in frequency space, λ is the wavelength, and $\Delta\lambda$ is the spectral width in wavelength space. Figure 7.2 is a plot of the coherence length as a function of the spectral bandwidth in wavelength space. Ranges for single-mode lasers, multi-mode lasers, 2D VCSEL arrays, and LEDs are highlighted on the graph. A conservative design target for speckle-free illumination is to operate at a minimum distance of $10 \times L_{COHERENCE}$, as shown in Figure 7.2.

Another advantage to VCSELs in illumination is the relatively small change in the spectral emission across temperature when compared to EELs and LEDs. The normalized spectral output of the three source types is shown in Figure 7.3. (Note: For simplicity, a Gaussian emission profile has been assumed for all the sources.) As can be seen from the figure, if a spectral filter is used in front of the detector to improve the SNR by reducing any ambient light, an LED would need a filter

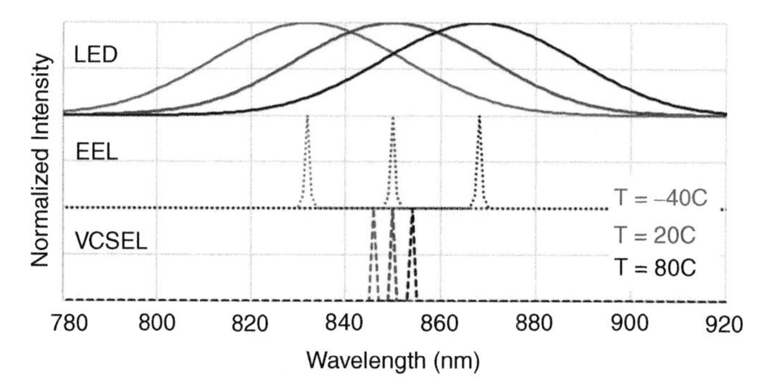

Figure 7.3 Relative optical emission spectrum of LEDs, EELs, and VCSELs at −40, 20, and 80 °C.

nearly 200 nm wide, an EEL would need a filter 50 nm wide, and a VCSEL would require only 20 nm. The amount of SNR improvement from ambient light is nearly 10x using a matched spectral filter with a VCSEL compared to an LED. The SNR improvement can be used to either increase the range of the sensor or reduce the required illumination power.

7.3 Commercial Examples of VCSEL Illuminators

The first high-volume commercial introduction of a VCSEL as an illuminator was in an optical mouse. The VCSEL replaced the LED commonly used and was initially targeted in-high end gaming and drafting use [1, 2]. The VCSEL source allowed operation on a wider variety of surfaces and offered 2000 counts per inch at speeds up to 45 in. per second, acceleration up to 20 g, and operated at 7000 frames per second. The movement was tracked using image correlation where movement is registered as the image shifts on the camera detector (same as an LED-based mouse). The components of the optical mouse are shown in Figure 7.4.

Other VCSEL-based optical mice have been developed that use the VCSEL coherence to track movement. One example was developed by Philips and used self-mixing of the VCSEL with light reflected from a surface [3]. Another example used speckle reflections from a metal ball to map the optical motion in conjunction with the self-mixing [4]. All of these VCSELs were low power (~1 mW) single-emitter sources. The properties that made VCSELs attractive in this application

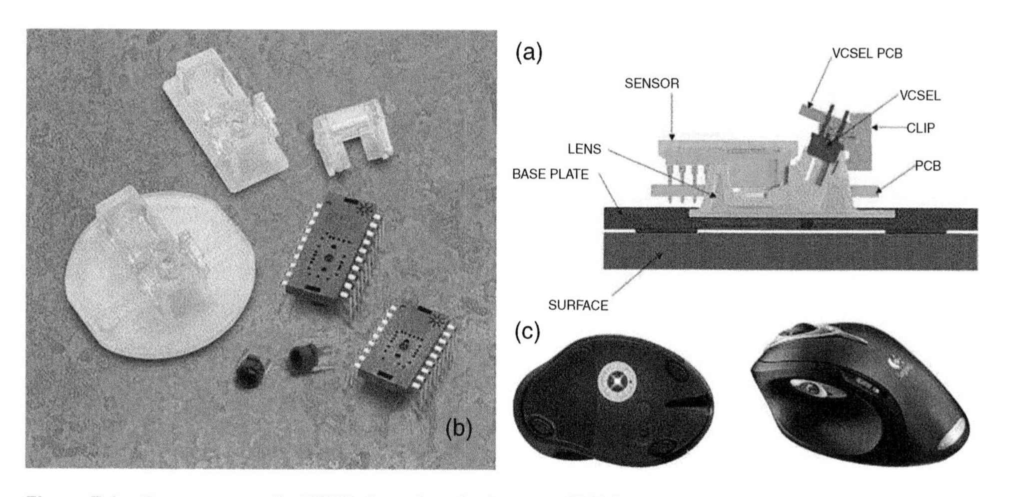

Figure 7.4 Components of a VCSEL-based optical mouse [12]. Image copyright from Elsevier.

Figure 7.5 VCSEL-based infrared flashlight available on www.alibaba.com

were the low power consumption and simple packaging schemes. But this really did not take advantage of one of the major VCSEL attributes.

The mobile phone revolution of the last 10 years has driven the availability of low-cost cameras, which coupled with the availability of high-speed Internet access, created a new market segment in home security. Along with this market came the desire for low-cost night vision camera equipment to monitor nighttime and low-light environments. In the home and industrial security market, infrared cameras are used to survey areas where ambient white light may be intrusive to the user or otherwise alert an intruder to its presence. These cameras are used in many places, including home and building security systems and recreational uses such as hunting. The example shown in Figure 7.5 is a handheld VCSEL-based flashlight that is designed to be used with infrared night vision equipment.

Another example of a VCSEL illumination source is a night vision security camera with IR illumination. Again, the infrared light is not visible to humans, so it is a good choice when surveillance is desired but ambient lighting is poor or is undesirable. A good example of this might be outside of a bedroom window, where lighting might disrupt the occupants. Figure 7.6. shows a VCSEL-based illumination ring that is housed within a so-called bullet camera.

One of the VCSEL aspects that enables rapid deployment of VCSELs to replace infrared LEDs in some applications is the ability to package the VCSELs identically to the existing LEDs. For example, plastic surface mount components, leadframes, and others can handle VCSEL emitters without changes. These applications continued to use single-emitter VCSELs packaged in traditional methods and did not take advantage of the true VCSEL differentiation—the ability to make 2D arrays of emitters on a single die. Making 2D VCSEL arrays means that the output power is essentially scalable with the area of the VCSEL array. This offers several significant advantages in the illumination application. For example, one driver in the illumination market is military, which drove some of the innovation in high-power VCSELs. A VCSEL illuminator with more than 30 W peak optical power and collimated to enable directional illumination to several kilometers is shown in Figure 7.7.

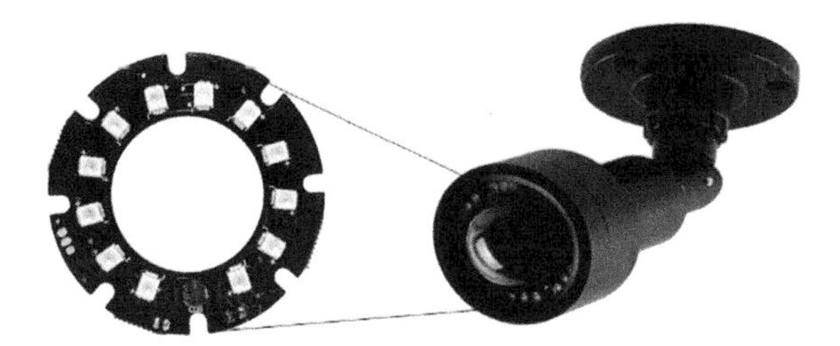

Figure 7.6 Infrared night vision camera with a VCSEL illuminator available on www.alibaba.com

Figure 7.7 Illuminator with 30 W of peak power and capable of direction illumination of more than 1 km. Photo courtesy for FLIR Inc. (www.flir.com)

By spreading the emission to many individual apertures, the failure of any one VCSEL element has a limited effect on the overall power and emission pattern. This enables significantly higher reliability than other optical sources (Appendix E) and helps in meeting eye-safety challenges (Appendix F). A common value for the CW power emission of a 10 μm aperture VCSEL is ~10 mW and increases to more than 100 mW when driven with short pulses and low-duty cycles (see Section 2.12). The total power emission from a 2D array depends on the geometry of the emitters and the driving conditions. Figure 7.8 shows the thermal transit time as a function of the spacing between nearest emitters and the transit time through the substrate for 100, 150, and 200 μm thick substrates. The thermal diffusivity of the DBR mirror is about three to five times smaller than that of bulk GaAs and must be considered in the overall thermal response [5].

For emitter spacings more than 50 μm, and with the substrate thinned to 100 μm, the emitters are nearly independent from each other (similar transit time of heat to the neighbor as the substrate). This leads to the question of how much optical power (number of emitters) can be packed into a VCSEL die area. The densest packing is a hexagonal pattern; with emitters spaced at a distance d in mm, the number of emitters per square mm is $2/(d^2\sqrt{3})$. For a emitter with 50 μm spacing and 10 mW per emitter 4 W/mm^2 is achievable CW, and for short pulse applications with densely packed emitters, more than 150 W/mm^2 of peak power is attainable. If a 150 mm wafer was fully patterned and all the VCSELs turned on CW, more than 50 kW of optical power would be generated and more than 1 MW of peak power in a short pulse. The scalability of power has led to some very impressive illumination system demonstrations. For example, a system with > 1 kW of power for military terrain illumination was discussed with many tens of thousands of VCSELs in [6, 7]. Figure 7.9 shows a series of illuminators fabricated by Princeton Optronics (now AMS) capable of (a) ~5 W, (b) ~1 kW, and (c) ~100 kW CW power.

Figure 7.10 shows several images taken with VCSEL illumination, with distances of several hundred meters being realized with less than 10 W of optical power.

These modules demonstrate the extreme scalability of VCSEL arrays but also highlight the practicality of having so many VCSEL emitters in parallel, which essentially eliminates the sensitivity of the system to failure of any single element. The larger chip area and relatively low power per facet of the VCSELs make it possible to air-cool many of these modules. The applications presented in Section 7.3 have centered on illumination and vision, both far field, or relatively distant applications. The most ubiquitous example of a VCSEL illuminator is in the face ID application and associated 3D depth sensors available in a wide range of consumer devices, as described in Chapters 5 and 6.

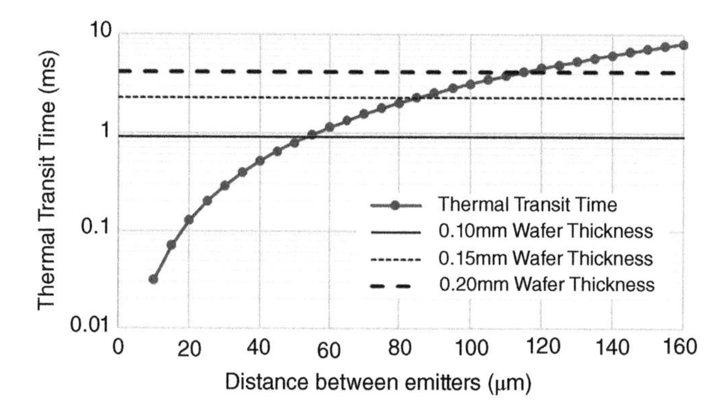

Figure 7.8 Thermal transit time as a function of distance between the emitters (red line) and transit time through the substrate for 100 μm (solid line), 150 μm (dotted line), and 200 μm (dashed line).

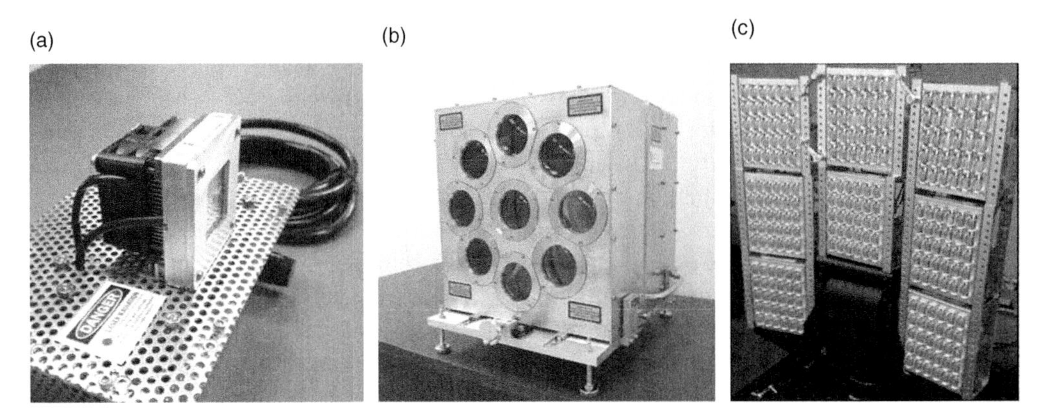

Figure 7.9 Illuminator modules produced by Princeton Optronics (now AMS) capable of (a) ~5 W, (b) ~1 kW, and (c) 100 kW of CW optical power [6, 7].

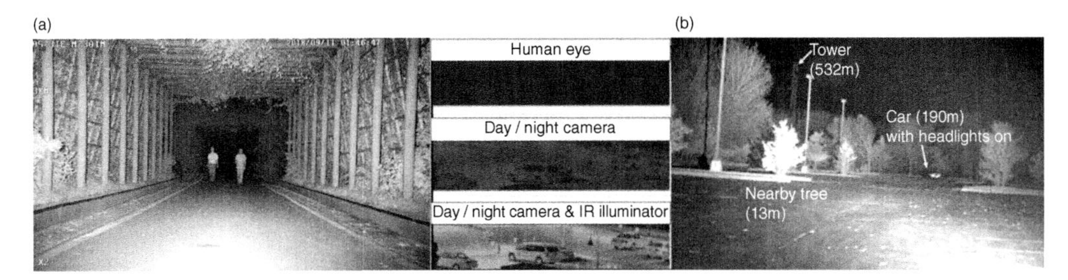

Figure 7.10 Night vision images collected with VCSEL illuminators [6]. Note the images are speckle free.

7.4 VCSEL-Based Industrial Heating

Like the VCSEL-based illuminators described in Section 7.3, the principal advantage of VCSELs in heating applications is the ability to scale the optical power with the emission area. One critical point is that the overall wallplug power conversion efficiency (see Chapter 2) stays nearly constant with the power scaling and is generally over 40%. Figure 7.11 shows how a single VCSEL emitter is scaled into a 2D array of individual chips, which are then mounted on a heat sink and connected in series to provide very high-power assemblies. Figure 7.12 shows a final assembly that delivers more than 5 kW of optical power and contains over 3 000 000 individual VCSEL emitters. This assembly method works well in broad area heating but does not lend itself to fiber coupling or high radiance applications traditionally addressed with lasers.

With heating modules assembled using many individual laser die, several new degrees of freedom are available: the ability to create custom temperature profiles, to create specific temporal profiles, and to dynamically adjust the profile and sequence. This can be very beneficial in shaping certain plastics and other materials. The VCSELs can be made to emit at wavelengths from 800 to 1000 nm, and wavelengths may even be mixed in the assembly. In some cases where a high-power density is required, lenses can be used with these systems to increase the power density. To understand the range of heating applications that can be addressed with these VCSEL solutions, the parameter space can be portioned by power density and pulse width. A common figure of merit for optical sources in heating is the beam parameter product (BPP), which is defined as the product of the beam spot size at focus and the divergence and is expressed in mm × mrad. Figure 7.13 shows the BPP and power requirements for a wide range of applications.

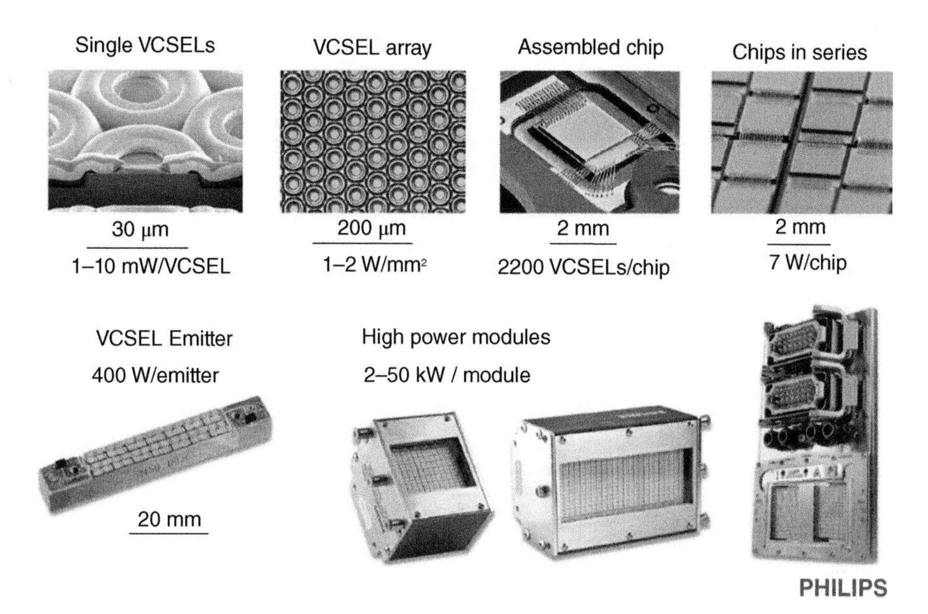

Figure 7.11 Scaling of VCSEL power from a single emitter at 10 mW to systems with 10s of kilowatts. Image courtesy of Philips.

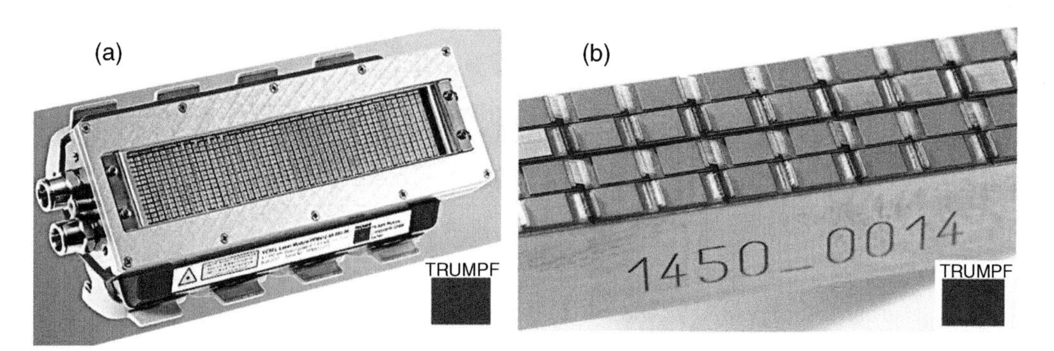

Figure 7.12 (a) A 10 kW heating system with over 3M VCSEL emitters, and (b) packaging of VCSEL array chips in a heating module. *Source:* Reprinted with permission from Trumpf, Germany.

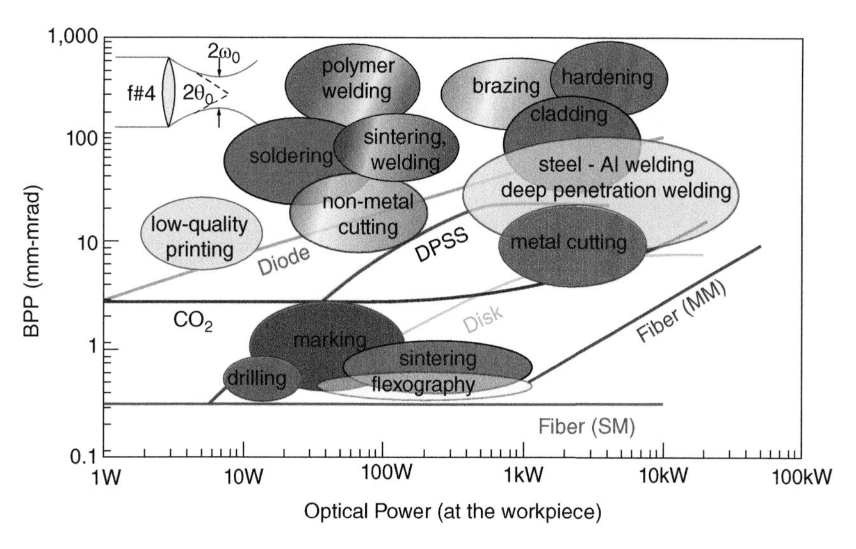

Figure 7.13 Map of heating applications and the required BPP and total optical power. The solid lines represent available laser heating solutions [13].

VCSEL-based heating systems tend to be above the diode line when used without an external lens. Another figure of merit that can be mapped into the available application space is the laser pulse width and the power density, as shown in Figure 7.14. The VCSEL sources discussed here tend to have pulse widths > 1 μs and power densities < 1 MW/mm^2 when focused. Taken with the space mapped in Figure 7.13, the addressable applications are typically thermal responses over relatively larger areas.

The unique features of the VCSEL-based illuminator are further illustrated in Figure 7.15. In the upper part of Figure 7.15, a traditional heating profile form a flashlamp or carbon filament is shown on material traveling under the illumination, and the lower portion shows the profile form

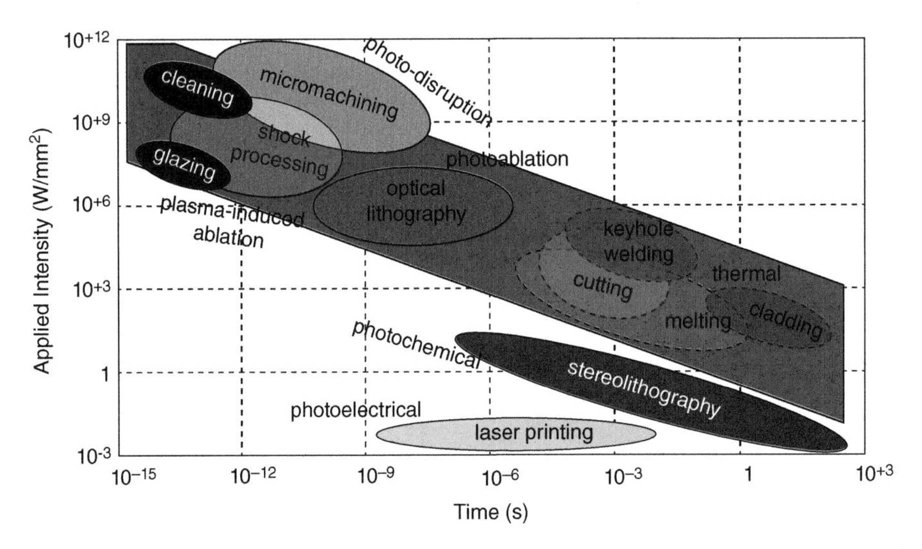

Figure 7.14 Map of heating applications and the required pulsewidth and power density [13].

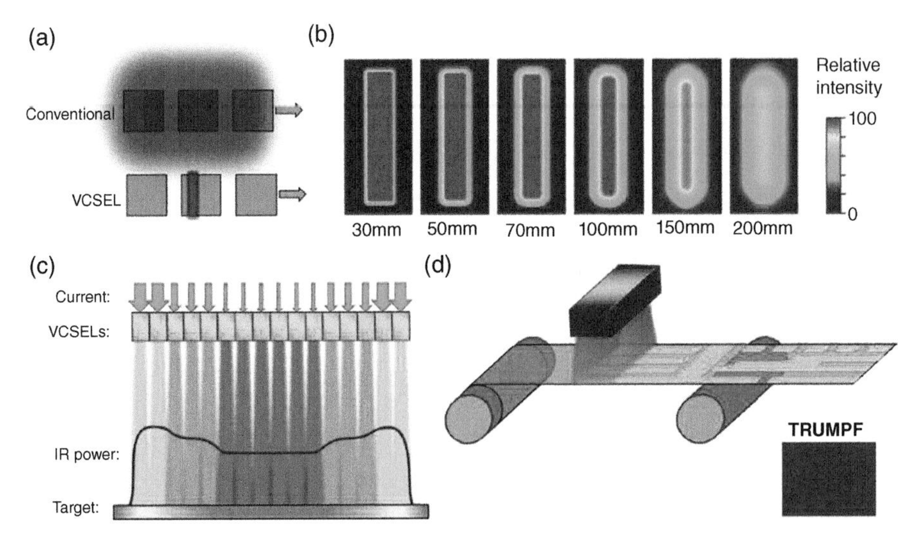

Figure 7.15 (a) Difference between flash lamp or carbon filament heaters and VCSEL heating profiles. (b) Power density as a function of distance from the VCSEL heater. (c) Illustration of a custom heating profile engineered by adjusting the current to each element. (d) Illustration of a custom heating solution that combines the spatial and temporal aspects of a VCSEL heating solution. Images courtesy of Trumpf.

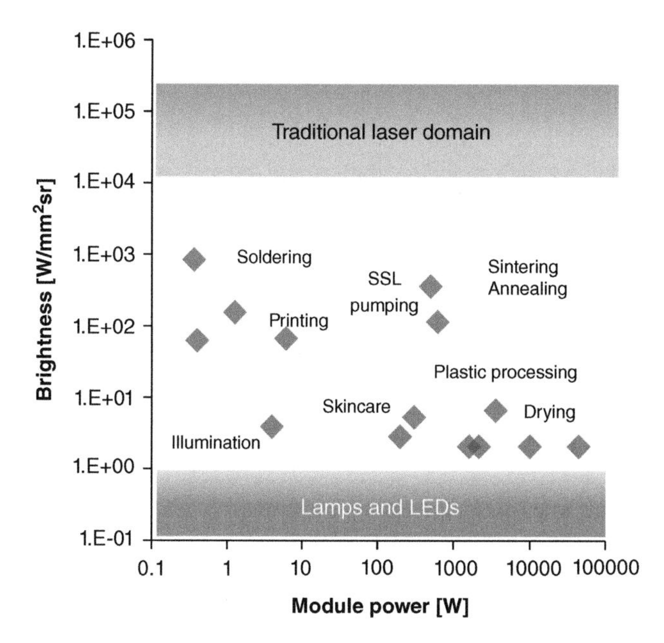

Figure 7.16 Map of heating applications and the required pulsewidth and power density [8].

a VCSEL illuminator. Note the localized heating area, which reduces power consumption and footprint within a factory. In Figure 7.15b, the intensity profile of the VCSEL heater is mapped, and the variation in intensity is shown for several distances form the illuminator. The available working distance can be many millimeters and maintains excellent thermal uniformity. Note that these profiles are shown without lenses. Figure 7.15c demonstrates the ability to control the spatial profile of the infrared source using the current injected into each VCSEL subassembly, as described in Figure 7.11. Figure 7.15d shows that when used in conjunction with a conveyor belt, the heating profile can be adjusted in time, giving the ability to spatially program the illumination, and thus 2D heating patterns can be achieved. Finally, the wavelength of the heating system can be tailored to the application at the complete system level or at the individual die level, as described in Figure 7.11. The complete flexibility of wavelength, spatial profile, and temporal adjustment make VCSEL-based industrial heaters a highly flexible manufacturing tool.

Some of the applications demonstrated with VCSEL-based heating are mapped in Figure 7.16 [8] and further described in [9, 10, 11]. Some notable achievements in these applications are: (a) heating rate of more than 100°C/s in solar panel metal sintering, (b) rapid drying of inks in a printing application, (c) bonding of carbon fibers in critical aerospace components at over 300°C and 30 m/min, (d) forming and welding of plastic sheets and components, (e) localized heating of 1.5 mm thick steel to > 900°C to facilitate cold roll processing, (f) welding of many foam layers, and (g) thermal application and activation of epoxies to adhere metal edging to laminate at more than 45 in. per minute.

7.5 Summary

Illumination, night vision, and industrial heating applications take advantage of the ability to scale the optical power by simply scaling the number of emitters in 2D arrays and stacking these chips together in clever ways. By distributing the total power to many individual emitters, the sensitivity to power failures in any single element is greatly reduced. Additionally, the distributed nature of the emission means the system may be considered as an extended source, as described in Appendix E. This makes eye-safety considerations potentially more manageable when compared to other solutions. The principal drawback of the 2D scaling of emitters is that the assemblies may not lend themselves well to fiber coupling and high radiance applications typical of other laser sources. The addressable applications for VCSEL illuminators and heaters fall between those addressed with sources such as LEDs and flashlamps and high-radiance laser solutions.

References

1 "Agilent launches industry's first laser-based optical mouse sensors," in EE Times, (2005).

2 M. Grabherr, H. Moench, and A. Pruijmboom, "VCSELs for Optical Mice and Sensing", in VCSELs, vol. 166, (2013).

3 A. Pruijmboom, M. Schemmann, J. Hellmig, J. Schutte, H. Moench, J. Pankert "VCSEL-based miniature laser-doppler interferometer," *Springer Ser. Opt. Sci.* **166** (2013).

4 "Pen with VCSEL replaces computer mouse," Laser Focus World (2002).

5 G. Chen, C. Tien, X. Wu, J. Smith, "Thermal diffusivity measurement of GaAs/AlGaAs thin-film structures," *J. Heat Transfer*, vol. **116**, no. 2, (1994).

6 J Seurin, G. Xu, B. Guo, A. Miglo, Q. Wang, P. Pradhan, J. Wynn, V. Khalfin, W. Zou, C. Ghosh, and R. Van Leeuwen, "Efficient vertical-cavity surface-emitting lasers for infrared illumination applications," SPIE vol. 7592 (2011).

7 D. Zhou, J. Seurin, G. Xu, A. Miglo, D. Li, Q. Wang, M. Sundaresh, S. Wilton, J. Matheussen, and C. Ghosh, "Progress on vertical-cavity surface-emitting laser arrays for infrared illumination applications," Proc. SPIE vol. 9001 (2014).

8 A. Pruijmboom, R. Conrads, C Deppe, G. Derra, S. Gronenborn, X. Gu, G. Heusler, J. Kolb, M. Miller, H. Moench, F. Ogiewa, P. Pekarski, J. Pollmann-Retsch, U. Weichmann, "Near-infrared digital heating solutions with power VCSEL arrays," 2014 IEEE Photonics Conference, (2014).

9 H. Moench and G. Derra, "High power VCSEL systems: a tool for digital thermal processing," *Laser Tech. J.*, pp. 43–47, (2014).

10 H. Moench, R. Conrads, C. Deppe, G. Derra, S. Gronenborn, X. Gu, G. Heusler, J. Kolb, M. Miller, P. Pekarski, J. Pollman-Retsch, A. Puijmboom and U. Weichmann "High-Power Diode Laser Technology and Applications," Proc. SPIE vol. 9348, (2015).

11 A. Pruijmboom, R. Apetz, R. Conrads, C. Deppe, G. Derra, S. Gronenborn, X. Gu, J. Kolb, M. Miller, H. Moench, F. Ogiewa, P. Pekarski, J. Pollmann-Retsch, U. Weichmann, "VCSEL arrays expanding the range of high-power laser systems and applications," ICALEO, (2015).

12 R. Szweda, "VCSEL applications diversify as technology matures," III-Vs Review Volume 19, Issue 1, (2006).

13 A. E. Willner, R. L. Byer, C. J. Chang-Hasnain, S. R. Forrest, H. Kressel, H. Kogelnik, G. J. Tearney, C. H. Townes, "Optics and photonics: key enabling technologies," *Proc. IEEE* **100**, 1604–1643 (2012).

8

Single-Mode VCSELs for Sensing Applications

Kenichi Iga and Jim Tatum

8.1 Introduction

The previous chapters have focused mostly on multi-mode VCSELs and their applications. Chapter 8 will describe single-mode VCSELs and their applications in several sensors, and Chapter 9 will describe single-mode VCSELs and applications in communications. The basic description of single-mode VCSELs in the following sections applies to both chapters. The material systems previously described in Chapters 2, 3, and 4 apply to both multi- and single-mode VCSELs (SM-VCSELs). Figure 8.1. shows some of the applications and wavelengths for single-mode VCSELs covered in the next two chapters. The VCSEL-based sensor market is forecasted to exceed $3 billion in 2025, as shown in Figure 1.21 [1, 2].

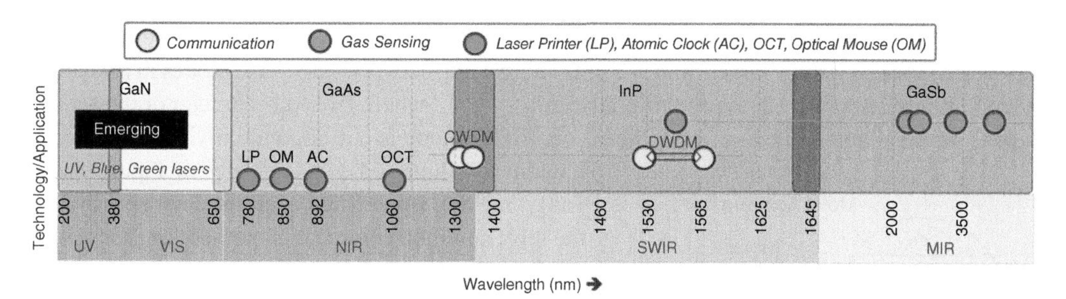

Figure 8.1 Schematic of InP-, GaSb-, and GaAs-based VCSEL technologies used for single-mode communication and sensing. *Source:* Figure by Babu Dayal Padullaparthi © Photonic Components DFM Ltd.

8.2 Single-Mode VCSELs

Semiconductor lasers oscillate in different modes (power radiation patterns) that depend on the dimensions of the optical resonator. As described in Figure 1.16 and in more detail in Section 2.7, several kinds of modes may be present in a laser cavity, including longitudinal and transverse, along with different polarization modes of each. The optical spectrum and the far-field beam profiles are ways of quantifying the modal behavior. In some applications, the VCSEL can be single mode, which means that it is operating in exactly one mode, but in other cases single mode may simply mean a Gaussian-like divergence pattern but have more than one spectral or polarization

VCSEL Industry: Communication and Sensing, First Edition. Babu Dayal Padullaparthi, Jim A. Tatum and Kenichi Iga.

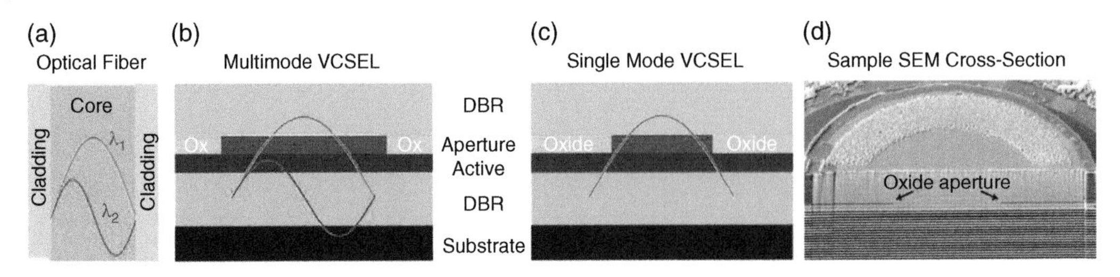

Figure 8.2 (a) Analogy of a VCSEL to an optical fiber showing multiple transverse modes. (b) Schematic of a multi-mode VCSEL where the diameter is large enough to support multiple transvers modes. (c) Schematic of a single-mode VCSEL where the aperture diameter has been decreased to cut off the higher-order mode. (d) SEM cross section of a VCSEL [4].

mode. In this chapter, the focus will be on true single-mode lasers, meaning operation in a single polarization and single spectral mode.

8.2.1 Spatial Mode Control

There are several methods for making single-mode VCSELs, all of which involve increasing the loss of higher-order modes of the cavity. A VCSEL structure can be modeled as a cylindrical waveguide, and the modal solutions are the well-known Laguerre-Gaussian profiles [3]. For simplicity, the electric field of the two lowest order modes are shown in Figure 8.2a. Also note that a similar mathematical treatment applies to the modes of a step index optical fiber, and the solutions are generally expressed as Bessel functions of the first kind instead of Laguerre-Gaussians. The most common approach to making single-mode VCSELs is to reduce the diameter of the aperture: values are less than 5 μm and more typically less than 3 μm. Figure 8.2b shows a simple cross-section schematic of a multi-mode VCSEL. Here the aperture diameter is large enough to support more than one transverse mode, generally larger than 3–5 μm. The spacing between transverse modes in a weakly guided cylindrical waveguide is proportional to $\Delta\lambda_T \sim \lambda^2/D_A$, where D_A is the diameter of the active region. As the diameter is decreased, the second-order mode is cut off, and only a single mode remains, as indicated schematically in Figure 8.2c. Today, making small oxide aperture devices is by far the most common commercial method for producing single-mode VCSELs. An SEM cross section of a typical VCSEL is shown in Figure 8.2d [4].

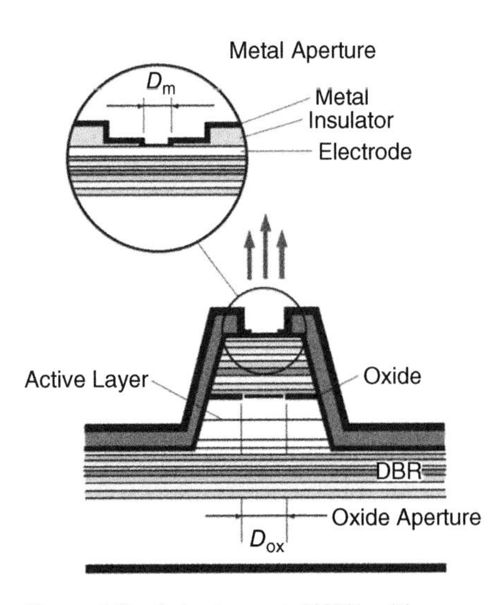

Figure 8.3 A single-mode VCSEL with a small metal aperture used in laser printers [9, 10].

The reduction of the oxide aperture diameter increases the loss of the higher-order mode and causes it not to reach the threshold. There are other methods to increase the loss of the higher-order modes. Early VCSELs used a small metal aperture on top of the VCSEL, as shown in Figure 8.3. This method was also applied to VCSELs with proton implants for controlling the current injection instead of oxide apertures. These VCSELs were widely used in optical mice, as will be discussed in Section 8.3.1. Another method of making single-mode VCSELs uses a surface etch into the top contact layers of the VCSEL to make an index contrast that increases the loss of higher-order modes [5]. Figure 8.4a shows a schematic cross

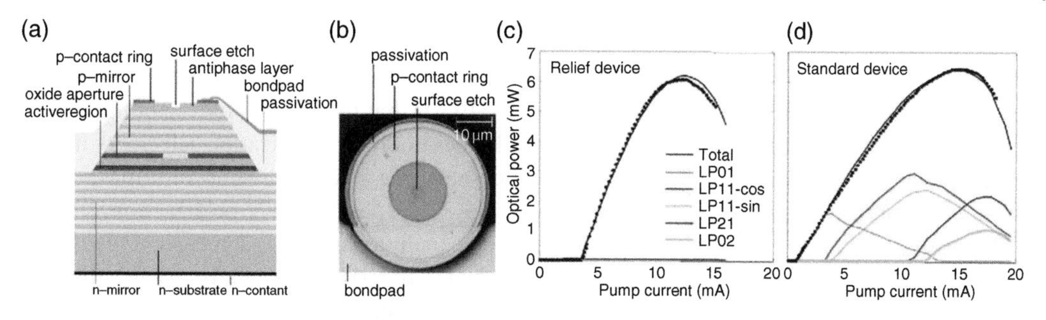

Figure 8.4 (a) Cross section of a VCSEL with surface relief etch. (b) SEM image of the top surface of the VCSEL showing the surface etch. (c) and (d) show the light output as a function of current showing the total light output (black lines) and the contributions to the total power of the higher order modes of the cavity with and without the surface etch feature, respectively [5].

section of the VCSEL with the shallow surface relief, and an SEM cross section of the device is shown in Figure 8.4b. The light output as a function of current is shown in Figure 8.4c and d for a VCSEL with and without the surface relief feature, respectively. The figures also show the contribution of each of the LP modes of the cavity to the overall light output characteristic. The mode stability of SM VCSELs using shallow surface relief was also comprehensively studied by the Chalmers group [6] and found stable SM operation for surface relief diameters nearly half of the of oxide aperture diameters [7, 8].

Another method of achieving single-mode operation uses a photonic crystal bandgap structure, or PC-VCSEL. In this case, the index guide is formed by patterning and etching holes into the top DBR structure to form an index guide [11]. Figure 8.5a shows a schematic cross section of the PC-VCSEL. Current guiding is accomplished using proton implantation, and the index guide is formed by the array of etched holes, as shown in Figure 8.5b. The photonic crystal parameters can be adjusted to force operation in a single mode, as shown in Figure 8.5c.

The spatial-mode dynamics in VCSELs can be quite complex and are influenced by multiple factors, including the index modifications described previously. Some special cases of SM VCSELs include: (i) the theoretical investigation of the weakest index guiding by shifting the position of oxide layer much away from QW active region (3rd node in DBR outside the optical cavity) [12] and (ii) experimental demonstration of high-index contrast 1D grating with grating area (12 μm × 12 μm) with oxide aperture diameters as large as 10 μm [13]. With increasing current, the gain profile can change as carrier and thermal profiles influence the optical mode and ultimately may cause to become multi-mode. Controlling the modal-dependent gain and loss over environmental and operating conditions is critical to maintaining robust single-mode VCSELs.

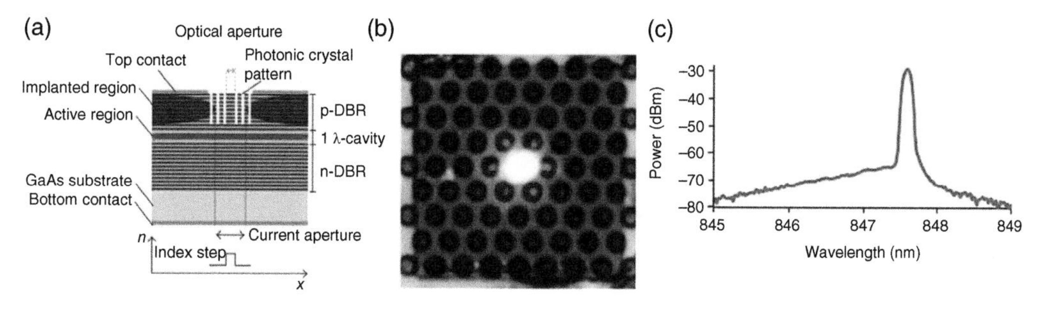

Figure 8.5 (a) Cross-section schematic of a photonic crystal VCSEL. (b) Top-view image of the PC-VCSEL showing the lasing mode. (c) Optical spectrum of a single-mode PC-VCSEL [11].

8.2.2 Polarization Control

The methods described in Section 8.2.1 focused on controlling the loss of higher-order transverse modes of the VCSEL cavity. One feature of VCSELs is that the circular nature of the cavity and the orientation of the light to the quantum wells does not have polarization selectivity in the gain. This can lead to polarization degeneracy in the transverse modes and allow multiple polarizations to operate simultaneously within the VCSEL. To control the polarization, either polarization-dependent gain or loss methods need to be included in the design. Several methods have been shown to be effective, including the use of 1D and 2D surface gratings, asymmetric (non-circular) emission areas, miscut or off-axis substrates, and quantum wire or quantum dot active regions [14].

Early attempts to control the polarization of VCSELs focused on introducing anisotropy in the gain of the polarization modes by using off-axis cut or inclined semiconductor substrates to grow the VCSEL structure. These are high-index surface substrates and off substrates with surfaces slightly offset from the exact angle of the (hkl) surface with respect to the *(100)* substrate. Figure 8.6a shows the crystallographic orientation of a GaAs crystal and the polarization of the light output. GaAs substrates are available in several cuts of the substrates, as shown in Figure 8.6b, including slight offcuts from the [100] plane (not shown). Figure 8.6c shows the relationship between some high-index planes and the crystal structure. Here, the Miller indices A and B indicate that A is a group III atom and B is terminated by a group V atom. There are atomic steps on the surface of the actual off-substrate. In the case of the (n11) substrate, the (100) plane and the (111) plane appear alternately, and the lengths of the respective planes are different. The larger area is called the terrace, and the smaller area is called the step. The inclined substrate has a crystal structure that is asymmetric with respect to the in-plane direction. As a result, the internal band structure changes, and it has different characteristics from the (100) substrate that can be used to control polarization, including: (a) strained quantum wells on the inclined substrate generate higher optical gain than the (100) substrate, (b) optical anisotropy appears in the quantum well in the in-plane direction, and (c) an internal electric field is induced by the piezoelectric effect in a crystal with lattice mismatch.

VCSELs grown on inclined substrates have demonstrated good polarization behavior. Figure 8.7a shows a VCSEL grown on a [311]B substrate, and Figure 8.7b shows the optical power output as a function of polarizer angle for several different current levels. More than 20 dB difference in the polarization-resolved optical power was obtained [15, 16].

Another approach to locking the polarization of a VCSEL uses asymmetric (non-circular) apertures. Many different geometries such as ellipses and dumbbells have been shown to have polarization stability [17]. However, the polarization locking is relatively weak and can lead to instabilities

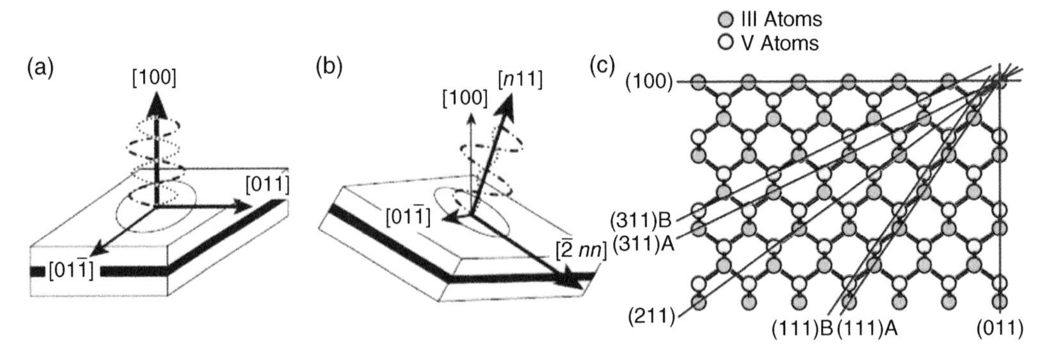

Figure 8.6 (a) Orientation of [100] GaAs substrate and (b) [n11] GaAs substrates with the polarization of the optical output. (c) Crystallographic planes of a GaAs crystal showing off-axis or inclined cut directions. (*Source:* After Nobuhiko Nishiyama and Kenichi Iga)

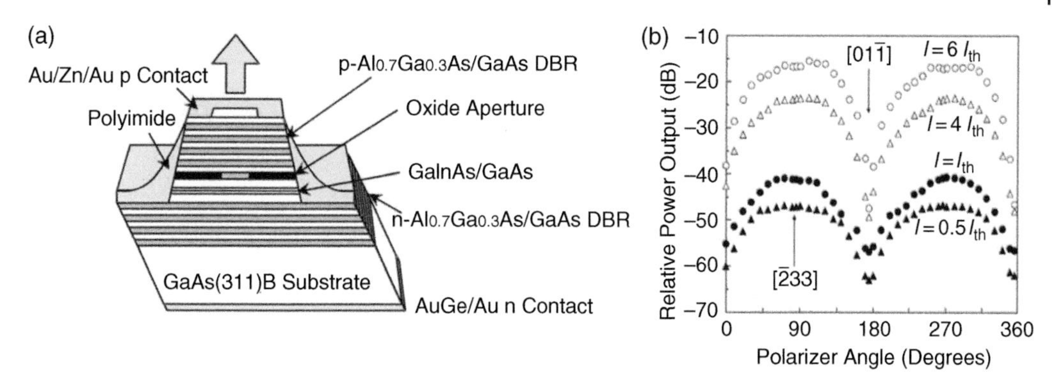

Figure 8.7 (a) A VCSEL structure grown on a [311]B substrate and (b) the optical output as a function of the polarization angle for several laser currents [15, 16]. (*Source:* After Nobuhiko Nishiyama and Kenichi Iga)

over environmental and operating conditions. Further, the output beam characteristics are generally higher order and not as attractive as a simple Gaussian beam. Photonic crystal structures can also generate polarization stability [18]. However, due to the relative difficulty of manufacturing, these devices have not been widely commercialized.

By far the most common method of polarization control in commercial VCSELs use surface gratings [19]. In this method, a grating is formed on the surface of the VCSEL and can be fabricated in semiconductor, metal, or in dielectrics. The grating pitch, Λ, is on the order of the wavelength of emission ($\Lambda \sim \lambda$). The grating enhances the reflectivity of the polarization parallel to the grating and reduces the reflectivity of the polarization perpendicular to the grating. Figure 8.8a shows a picture of the top surface of a VCSEL with a 1D surface grating [19]. Typically, the grating is aligned to the natural cleavage planes of the GaAs substrate. Figure 8.8b shows an atomic force microscope image of the surface grating, showing the pitch L of 0.8 μm, very close to the emission wavelength of 0.85 μm. The total light output, and the polarization-resolved output as a function of current are shown in Figure 8.8c and d with and without the surface grating, respectively. The optical polarization extinction ratio ($PER = 10 \times \log(P_{POL1}/P_{POL2})$) is also shown in Figure 8.8c and d, and more than 15 dB is achieved.

Gratings with pitch substantially less than the wavelength ($\Lambda << \lambda$) have also been attracting attention as a method of controlling the polarization, as shown in Figure 8.9 [20]. These have been termed high-contrast gratings (HCGs). One unique feature of HCGs is that they can create very high reflectivities, can replace a large portion of the DBR mirror structure, and are particularly attractive in long-wavelength single-mode VCSELs, described in Chapter 9. Figure 8.9 shows polarization control results using an HCG [20]. The most effective HCGs have $\Lambda \sim 0.5\lambda$ and are thus more difficult to manufacture using standard lithography and nanoimprinting techniques than other gratings. This has led to limited commercial applications to date.

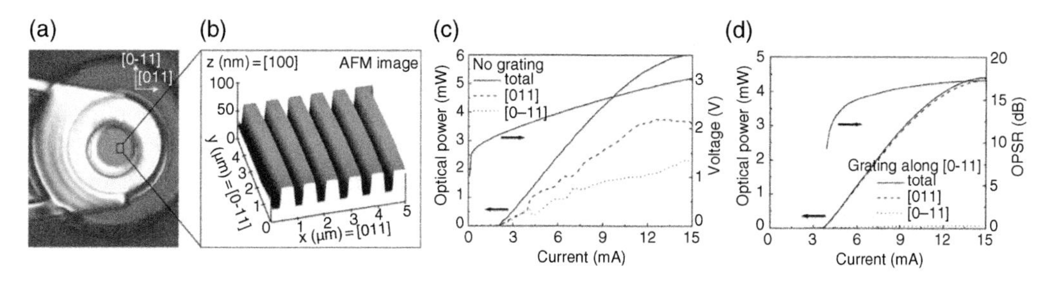

Figure 8.8 (a) Top-view image of a VCSEL with a surface grating. (b) AFM image of the surface grating. (c) and (d) show the polarization-resolved output of a VCSEL with and without a surface grating, respectively [19].

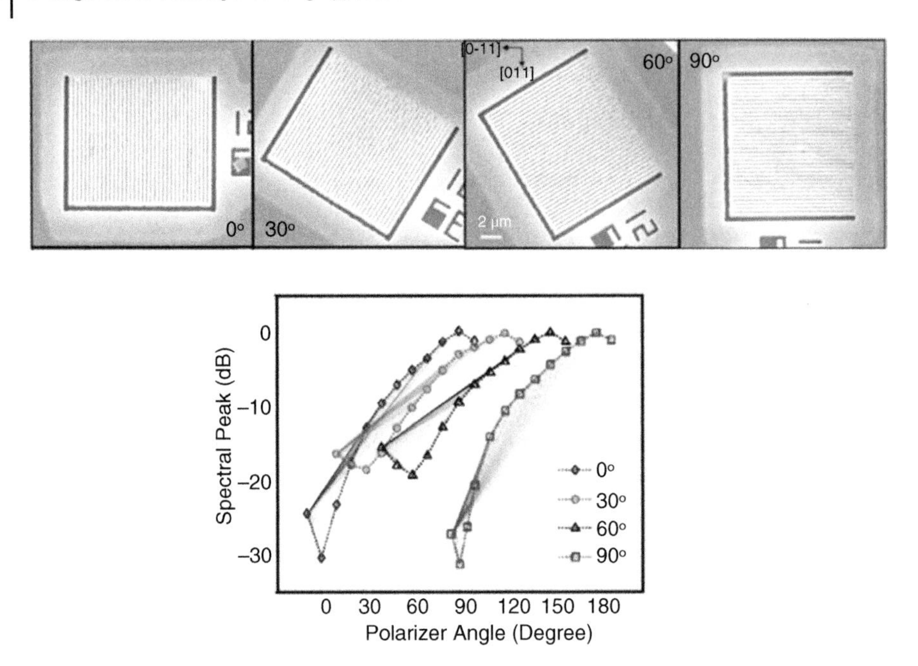

Figure 8.9 Polarization-controlled VCSEL using an HCG. *Source:* Adapted from IEEE [20].

8.2.3 Wavelength Tuning Principles

The ability to adjust the wavelength of a laser, particularly of single-mode lasers, is very attractive in many applications. For example, in a data communications system with wavelength division multiplexing (WDM), the ability to precisely tune the output to a specific wavelength is needed. Other applications such as gas sensing or atomic clocks require very precise wavelengths to operate. Other devices such as optical coherence tomography (OCT) use the wavelength change to measure depth. As previously described in Section 2.7, the VCSEL wavelength changes with both temperature and current but only at a modest rate and over a limited range of a few nanometers. The ability to tune over many tens of nanometers opens many new possibilities for VCSELs. Some of these applications are discussed in the following sections, but first a description and examples of broadly tunable lasers is presented.

In a laser with a cavity length of d, the longitudinal mode spacing is $\Delta\lambda = \lambda^2/(2nd)$, where n is the refractive index of the material and λ is wavelength. This is often referred to as the free spectral range (FSR). In an edge-emitting laser, where the cavity length is hundreds of microns, the FSR is less than 1 nm, and the laser hops between longitudinal modes as the environmental conditions change. In a VCSEL, with very short cavities of a few microns, the FSR can be many tens of nanometers, as previously described in Section 2.7. The wide FSR can be used to create a tunable wavelength VCSEL if the mirror spacing can be adjusted. This is shown schematically in Figure 8.10a where the VCSEL bottom mirror and top mirror are epitaxially grown, and a separate mirror is brought into close proximity. The cavity length d is the sum of the air gap distance, d_{AIR}, and the semiconductor optical thickness nd_1. The FSR as a function of the total distance d is shown in Figure 8.10b. The output wavelength can then be continuously adjusted across the entire FSR by changing the air gap distance.

In practice, the wavelength tuning range is limited by the reflectivity bandwidth of the mirrors, the gain spectrum of the quantum wells, and the ability to adjust the air gap. The first realizations

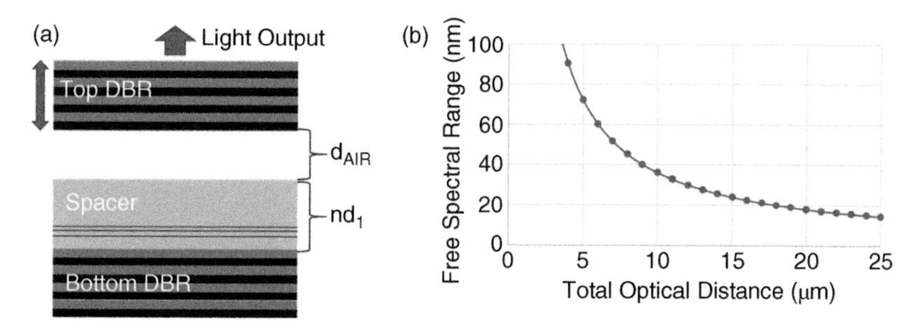

Figure 8.10 (a) Simplified schematic of a VCSEL with a widely tunable wavelength. (b) FSR as a function of the total optical cavity length.

of a tunable VCSEL structure used an external mirror placed in close proximity to a partial VCSEL structure, and the air gap was mechanically adjusted with a translation stage. The device operated at liquid nitrogen temperature, and tuning of 4 nm was demonstrated [21]. In the following years, development has focused on integrating the tabletop demonstration to the chip scale. The integration of the top DBR movement into in a mechanical system to adjust the air gap was first conceived with the mirror on a cantilever that is electrostatically adjusted. Other embodiments use a dielectric mirror and more recently, integration with a high-contrast grating. Figure 8.11 shows a schematic of several structures that have been developed [22].

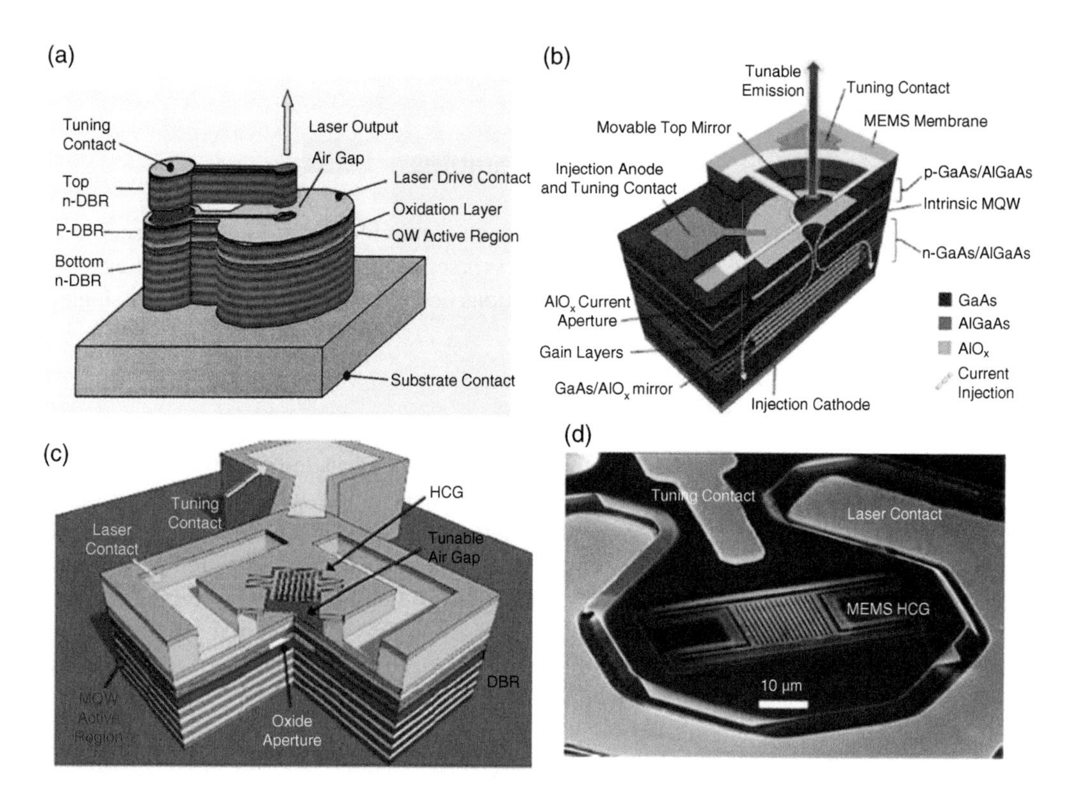

Figure 8.11 (a) Simplified schematic of a tunable VCSEL with a cantilever DBR. (b) Schematic cross section of a tunable VCSEL using a dielectric DBR. (c) Cross section of a tunable VCSEL with an HCG mirror. (d) SEM image of a tunable HCG-based VCSEL [22].

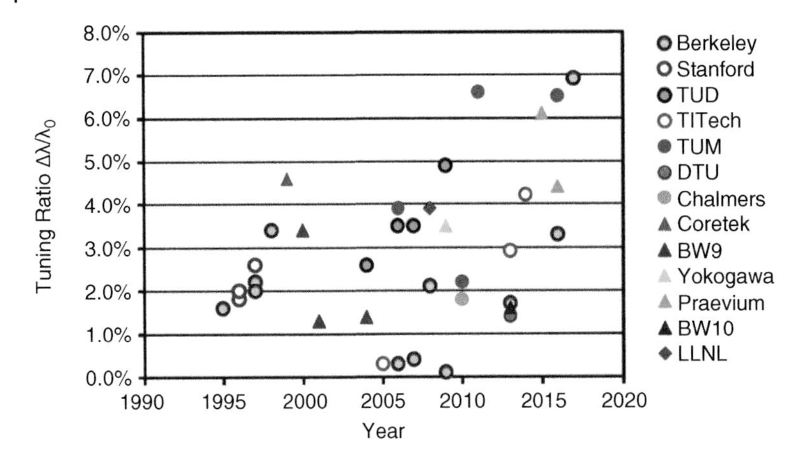

Figure 8.12 Evolution of the tuning ratio as a function of time [22].

With the major advances in MEMS capability and HCGs in the recent years, tunable VCSELs are emerging as a new market area. To compare tunable VCSELs at different wavelengths, the normalized tunability, expressed as $\Delta\lambda/\lambda_{NOM}$ in percent, as a function of time is shown in Figure 8.12. One of the unique features of a MEMS-tunable VCSEL is the tuning rate. Because of the small size and electrostatic actuation, the entire wavelength range can be scanned in microseconds (> 100 kHz). This feature enables applications like OCT. Today there are multiple companies with MEMS-tunable VCSELs offering products in the 1060 nm and 1300 nm regions.

8.3 Single-Mode VCSEL Application Examples

Many of the applications for SM-VCSELs have a unique requirement that forces the users to a single-mode device. Because the manufacturing process is somewhat more complicated, they tend to be somewhat more expensive. Each of the applications described below are enabled by single-mode VCSELs.

8.3.1 Laser Mouse and Finger Navigation

Optical mice with LEDs have been widely available for nearly two decades, so why are laser mice a large market for VCSELs? The working principle of an optical mouse, whether VCSEL or LED, is the same and was briefly introduced as an illuminator in Section 7.3.1. A light source is used to illuminate a surface, and a camera takes a picture of the reflected and scattered light. Each successive picture is spatially correlated to the previous one, and motion is determined. A simplified schematic is shown in Figure 8.13a for both an LED and VCSEL and in a 3D representation in Figure 8.13b. The VCSEL mouse was first commercialized in 2003 by Agilent Technologies in the Logitech MX1000 laser mouse, shown in Figure 8.13c, and later offered the complete kit of components (VCSEL, camera, lens, driver) to manufacture the mouse [23]. The mouse had excellent reviews and enjoyed considerable commercial success [23, 24]. One subtle feature is that the VCSEL can work on specular reflection from the surface, and the image on the sensor can have speckle, as described in Section 7.2, which can form a much higher resolution than

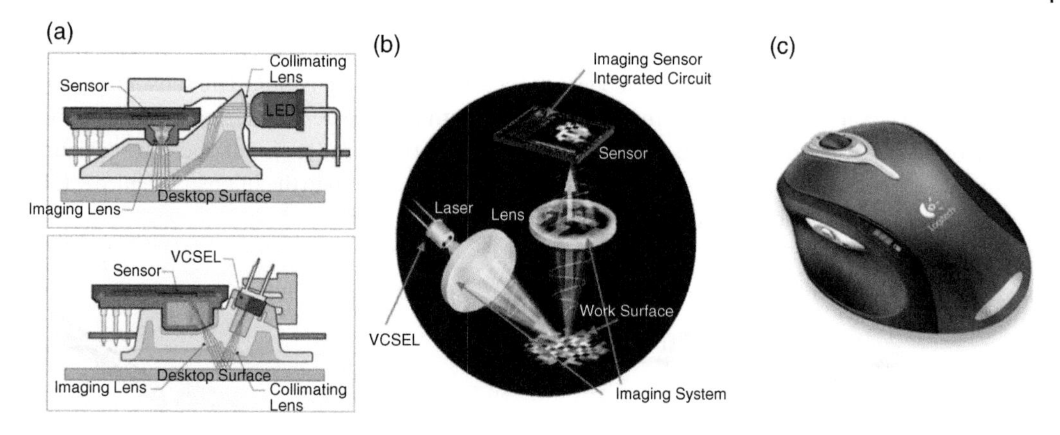

Figure 8.13 (a) Schematic of an LED- and-VCSEL based mouse [26]. (b) 3D schematic of a VCSEL-based mouse operation [23, 24]. (c) A mouse appearance. *Source:* Courtesy of W. Ishak ©Logitech MX1000 the world's first VCSEL-based mouse [23].

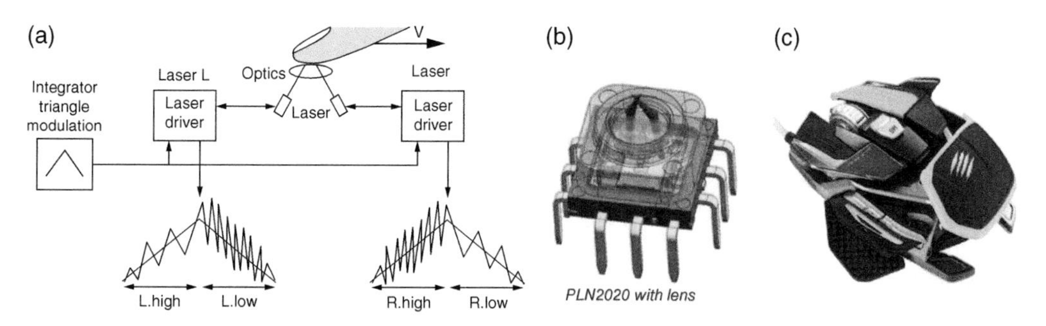

Figure 8.14 (a) Schematic of self-mixing [28]. (b) Philips Twin-Eye™ sensor [28]. (c) Philips Mad-Catz™ mouse using self-mixing to track movement (available at www.amazon.com).

scattered light imaging. The primary advantages of a VCSEL illuminator are: (a) the ability to work on a wider range of surface, especially glass and cloth, (b) the coherent light gives a higher speckle contrast in the image sensor, (c) a smaller focus spot increases the resolution of the sensor, and (d) reduction in electrical power consumption (longer battery life) [25]. The VCSEL is single mode, which increases the spectral contrast in the image and does not have changes in the far field as a result of polarization switches or transverse mode dynamics that would lead to false movement in the mouse. Today it is estimated that nearly 1B VCSEL-based optical mice have been produced [23].

Another approach to laser mice or more generally optical guidance was developed by Philips, based on the principle of self-mixing, as shown in Figure 8.14a [27]. The sensor works by scattered light being coupled back into the laser and creating a mixing signal that is proportional to the velocity of the object. Both finger tracking and optical mice were made using a so-called Twin-Eye sensor, had the advantage of much higher resolution and speed tracking, and was targeted at the demanding gaming users [28]. The Twin-Eye sensor is shown in Figure 8.14b, and a high-end gaming mouse by Philips is shown in Figure 8.14c.

8.3.2 Optical Encoders

Lasers have long been used as optical encoders to measure absolute position. The sensors use the reflection of a coherent light source from a diffraction grating to create an interference pattern on a detector array, as shown in Figure 8.15a. The resolution of the encoders is enhanced by measuring quadrature (relative phase) of the light intensity pattern in addition to the intensity. Figure 8.15b and c show several products based on this technique.

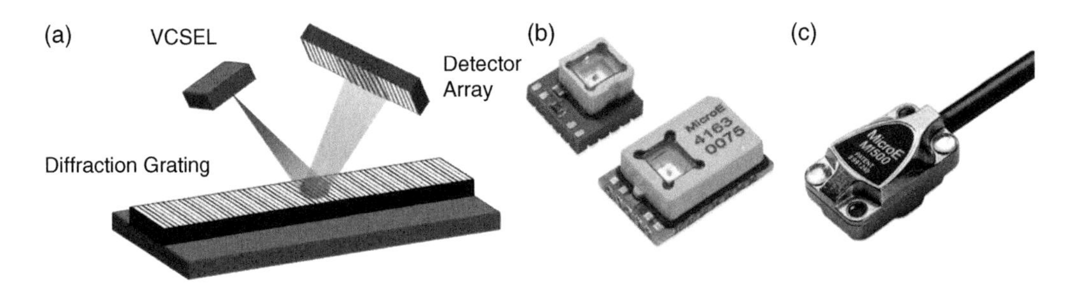

Figure 8.15 Optical Encoder. (a) Schematic of an optical encoder. (b) Chip scale encoder package. (c) Complete optical encoder. *Source:* Photos courtesy of Novanta (www.novanta.com).

8.3.3 Laser Printers

One of the most ubiquitous applications of lasers is in printing, where nearly every office and many homes have at least one laser printer. For everyday home and light business use, a traditional EEL-based printer running a few pages per minute is sufficient, but in high-volume printing, the ability to run hundreds of pages per minute is needed, and the ability to make one- and two-dimensional arrays of VCSELs is a natural fit. Today, printing companies such as Fuji Film Business Innovation, Ricoh, and Sony/Canon have introduced industrial printing machines using VCSELs; these companies have posted more than $1 billion in system sales and are the market share leaders today [29,30]. The initial idea by Bob Thornton at Xerox was to use a single row of VCSELs to fit across an entire page, more than 6000 individual emitters, but this was not practical to implement. Others, including one of the authors, were investigating a two-dimensional array of emitters. Among the many challenges was making a reliable 780 nm single-mode VCSEL. Ultimately, the VCSEL design shown in Figure 8.3 with inclined substrates and elongated emission area for polarization control was employed in a 2D array and successfully introduced to product by multiple companies. To understand the benefit of VCSELs in this application, Figure 8.16 shows a schematic of a laser printer using multi-element VCSEL array as shown in Figure 8.17a. Each element of the 2D array is addressed individually, and with the aid of a scanning polygon mirror and lens,

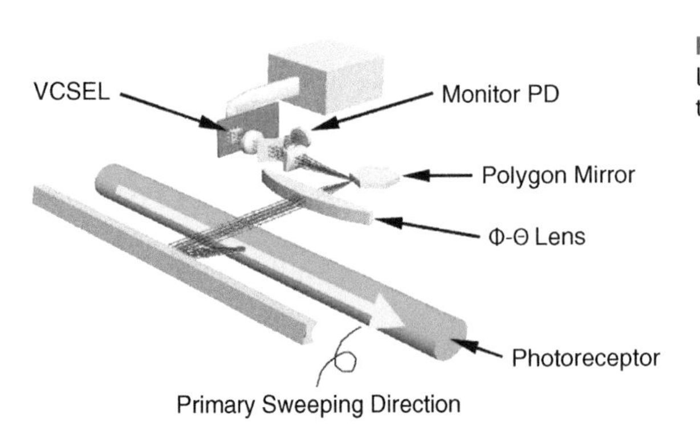

Figure 8.16 Schematic of a VCSEL-based laser printing engine. *Source:* Fuji Film technical report [10].

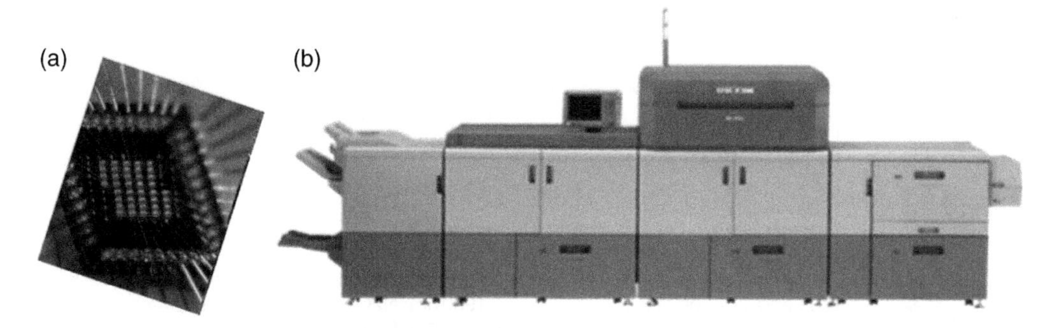

Figure 8.17 (a) 40 element VCSEL array used in an industrial laser printer, as shown in (b) [31, 32, 33].

the beams can write multiple lines at a time on the toner [10]. The highest resolution achieved by Ricoh was 4800 dots per inch and at very high speeds with a machine as shown in Figure 8.17 [31] [32][33]. The industry forecast is for continued multi-billion-dollar sales of industrial laser printers in the next few years.

8.3.4 Gas Sensors

Gas sensors work by tuning a laser to the atomic absorption spectrum of the subject molecule. Many harmful, toxic, and green-house gases have absorption spectrum in the near and mid infrared and are specific to each molecule's atomic content and energy band structure. Figure 8.18 shows absorption lines for a wide variety of gaseous species that are important for monitoring in either safety or chemical processes. Laser-based gas detection can be quite sensitive with detection levels in the few parts per billion depending on the other species present in the gas. Laser absorption is measured in a sample by tuning the laser across the absorption band of the gas. Narrow wavelength tuning can be done either thermally or by control of the laser current. In wideband applications, a tunable VCSEL, as described in Section 8.2.3, can be used. The technique is known as tunable diode laser absorption spectroscopy (TDLAS). The large FSR of a VCSEL makes it an ideal laser source for this measurement because it does not experience mode-hopping, which is present in Fabry-Pérot edge-emitting lasers and in some DFB/DBR lasers [34].

Most of the gases shown in Figure 8.18 have absorption lines beyond the reach of GaAs-based VCSELs, and there has been effort by many universities and companies to develop VCSELs at these

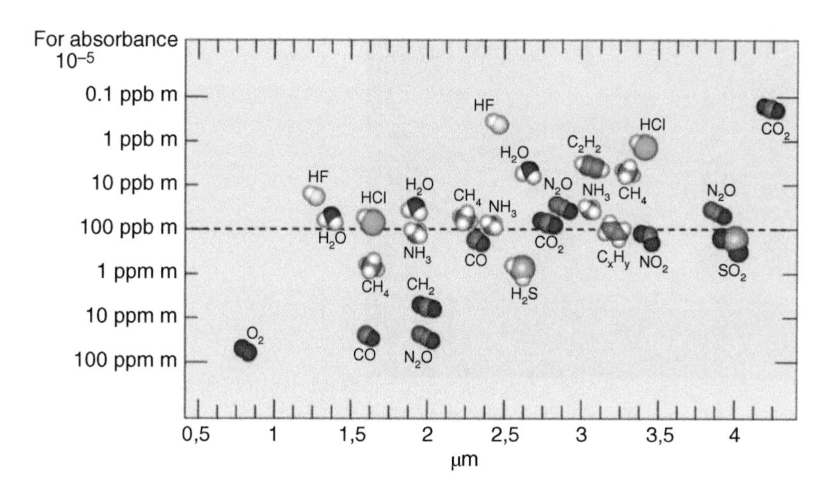

Figure 8.18 Absorption spectrum and detectivity level of some important gases [35].

VCSEL Products for NIR Gas Analysis (TDLAS)

Standard Wavelengths

- 1280 nm: HF
- 1392 nm: H2O
- 1512 nm: NH3
- 1564 nm
- 1579 nm: H2S
- 1590 nm: H2S
- 1654 nm: Methane, CH4
- 1680 nm: Combustables
- 1730 and 1742 nm: HCl
- 1854 nm: H2O
- 2004 nm and 2008 nm: CO2

Application Specific Wavelengths

- Any wavelength from 1.3 μm to 2.3 μm
- Examples:

  – 1560 nm: CO
  – 1645 nm: Ethylen Oxide
  – 1800 nm, 1877 nm : H2O
  – 1960 nm: N2O
  – 2012 nm: CO2

Figure 8.19 VCSEL wavelengths available for gas sensing form Vertilas (www.vertilas.com).

longer wavelengths. The efforts at 1310 and 1550 nm will be deferred to Chapter 9 as these are important for data communications. To extend operating to beyond 1.55 μm, InP or GaSb substrate–based materials are needed [34]. Figure 8.19 shows a list of wavelengths commercially available from Vertilas GmbH.

Mid infrared VCSELs are reported using GaInAsSb QWs, AlAsSb/GaSb DBRs, and dielectric mirrors between 2.3 and 4.0 μm with different process variations such as (i) buried tunnel junction (BTJ) and substrate removal [36, 37], (ii) bipolar and cascaded heterostructure laser structures [38], and non-cascaded heterostructures with wafer bonding [39]. The material systems for these VCSELs are not ideal for light emission, and due to the lack of type-I quantum well (QW) materials, this band uses type-II QWs, which permits optical transitions to neighboring bands in contrast to the direct band-to-band transitions in type-I QWs. The probability of matching wave functions in conduction and valence bands is lower than that of type-I QWs. Several companies have been working to develop VCSELs at this wavelength including Vertilas GmbH, Brolis Sensor Technology, and Vixar (now AMS). One such VCSEL structure is shown in Figure 8.20.

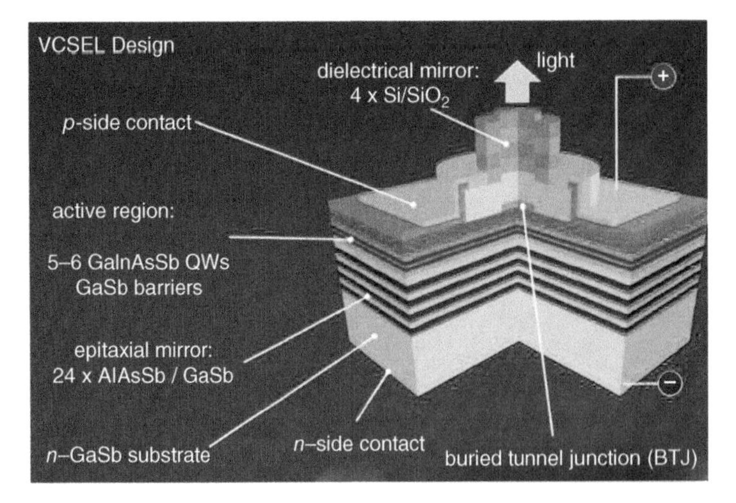

Figure 8.20 GaSb VCSEL designed to operate in the mid-infrared region. *Source:* Image courtesy of the late M.C. Amann, TU Munich and Walter Schottky Institute, Germany.

8.3.5 Atomic Clocks and Magnetometers

Atomic clocks (ACs) are used to create timing standards in telecommunications systems and for precise synchronous detection of events such as in oil field exploration. The VCSEL-based ACs operate on the principle of coherent population trapping (CPT) [40, 41], shown schematically in Figure 8.21, and generally use Cs or Rb atoms at 894 nm and 795 nm, respectively. In CPT, the laser is amplitude modulated with a sine wave, which creates a splitting of the VCSEL emission wavelength, or sidebands, as shown in Figure 8.21b. The frequency is adjusted to match the fine splitting of the atomic levels of the Cs or Rb atoms. When the frequency is matched, the transmission is maximized, and a precise frequency standard is determined. The accuracy of the clock standard is determined by the VCSEL linewidth (typically < 50 MHz) and intensity noise.

Magnetometers are used to measure the earth's magnetic field and in particular minor local variation caused by the presence of a ferrous substance or other electromagnetic fields.

Figure 8.22a shows the schematic of an atomic clock physics package with the VCSEL and a Cs vapor cell inside a magnetically shielded and hermetic package, as shown in Figure 8.22b and c [43, 44], developed by Kenichi Okada's group in Tokyo Institute of Technology. They obtained an Allan variance of 2.2×10^{-12} with an averaging time of 10^5 seconds. VCSELs have been commercialized into chip-scale atomic clocks (CSAC) by Microchip (formerly Symmetricom). The physical package concept is shown in Figure 8.22d, the complete package assembly is shown in Figure 8.22e, and the final clock assembly in Figure 8.22f. Figure 8.23 shows the oscillator module for an atomic clock using a modulated VCSEL. (The VCSEL and gas cell part are not shown here due to copyright restrictions.)

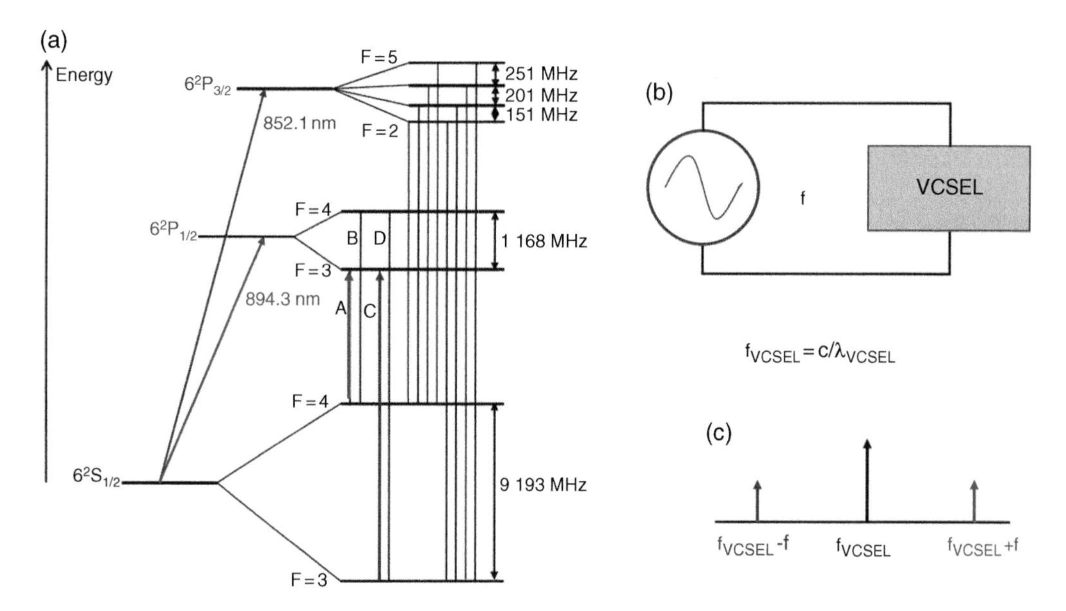

Figure 8.21 (a) Portion of the atomic spectrum of Cs showing the relevant VCSEL transitions at 894.3 nm and 852.1 nm and the fine splitting of the atomic levels used for CPT [42]. (b) Amplitude modulation of the VCSEL results in splitting of the VCSEL optical spectrum, as shown in (c).

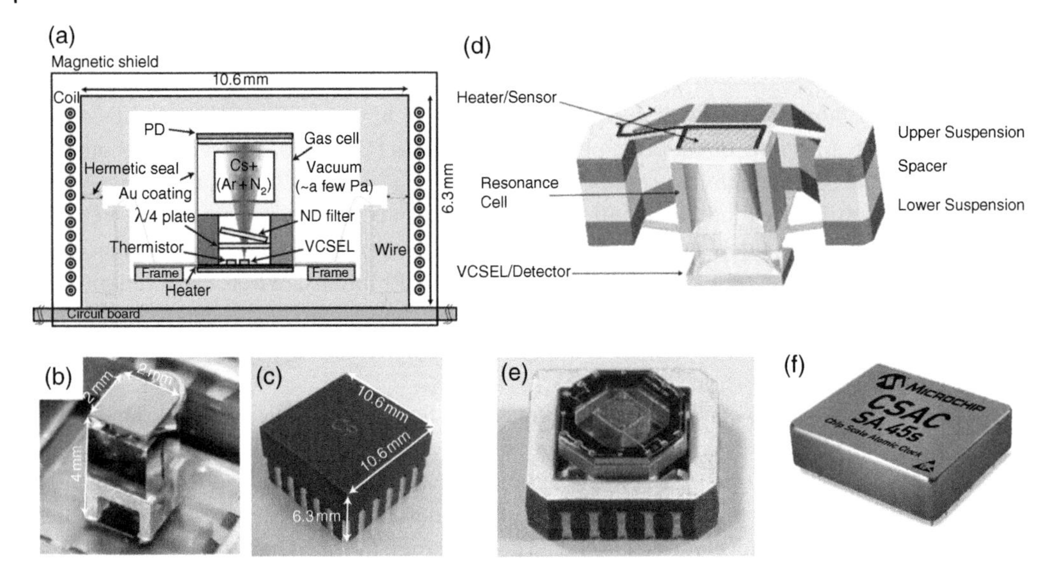

Figure 8.22 (a) Schematic of a physics package for a VCSEL-based atomic clock. (b) The interior package realization of (a). (c) The final sealed package of (a) [43]. *Source:* Photos courtesy of K. Okada at Tokyo Institute of Technology. (d) Schematic of the physics package of the Microchip atomic clock. (e) The assembled physics package. (f) The complete CSAC (Microchip) (www.microchip.com).

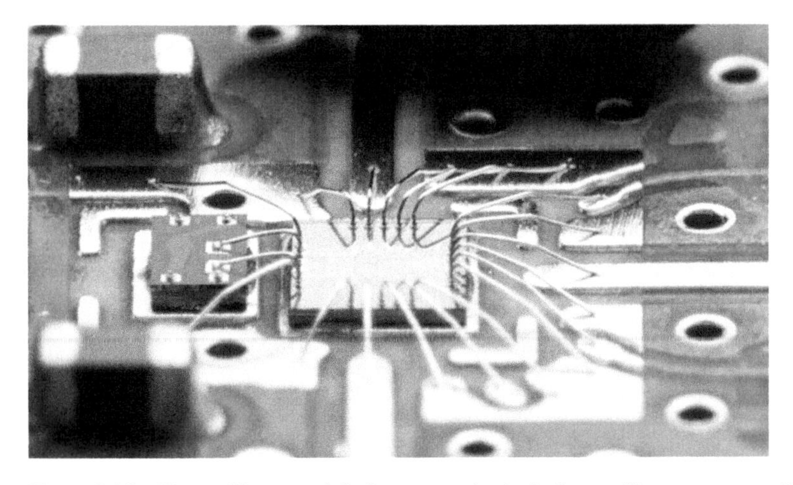

Figure 8.23 The oscillator module for an atomic clock. *Source:* Photo courtesy of M. Hara of NICT.

8.3.6 Optical Coherence Tomography

Optical coherence tomography (OCT) is a method of obtaining a cross-sectional image by interference in the frequency domain of light [45]. According to the literature [46], OCT was first patented in 1990 by Naohiro Tanno of Yamagata University [47]. In 1991, James Fujimoto of MIT made an independent proposal [48], which was linked to practical applications pioneered by Carl Zeiss. Tunable VCSELs provide an optical frequency swept-source OCT (SS-OCT) with the advantage of sweeping speeds in the hundreds of kHz range. This enables real-time imaging with a very large number of samples to improve the signal level and accuracy. The optical frequency is swept, and the interference signal is Fourier transformed to obtain imaging in the depth direction.

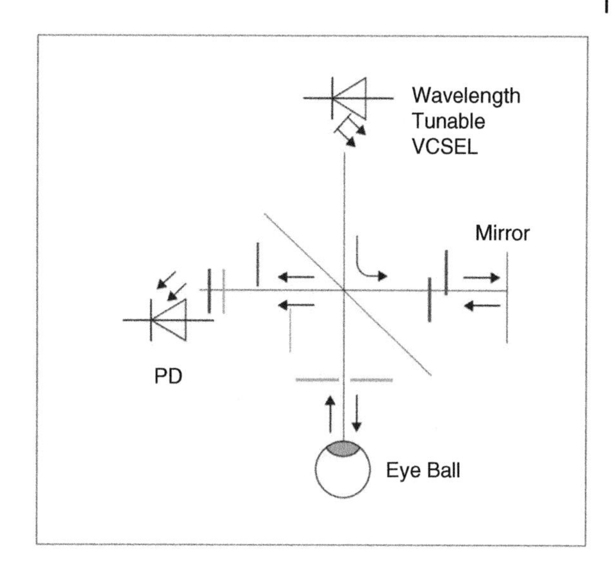

Figure 8.24 The principle of SS-OCT. *Source:* Genichi Hatakoshi [16].

The principle of this SS-OCT is shown in Figure 8.24 using a Michelson interferometer. Here the OCT is used to image a human eye, an upcoming application for OCT.

The light from the tunable VCSEL is split to a reference mirror and to the object (eye), then the reflected light from each is directed to a photodiode, and an interference signal is created. The amplitude of the interference is related to the relative intensity from the mirror and the object. Signals from every position in the test body undergo a slightly different path length, ΔL, (amplitude is proportional to the reflectance coefficient) and thus overlap. When the signal is Fourier transformed, the distribution on the frequency axis is transformed into the distribution in the depth direction. This is the basic principle of SS-OCT. In other words, as the wavelength of the laser is swept, the signal obtained on the time axis is Fourier transformed, the interference signals of different frequencies are expanded in the frequency domain, and the signal strength for a plurality of continuous distances ΔL is obtained. Reflections from the refractive index distribution of the layer are automatically visualized. In an actual SS-OCT device, FFT (fast Fourier transform) computation is performed, and the distance spectrum and intensity of ΔL are displayed. Using electrically pumped MEMS-VCSEL, SS-OCT wide-field imaging has been carried out *in vivo* on a human retina, as shown in figure 8.25a, b, c [49].

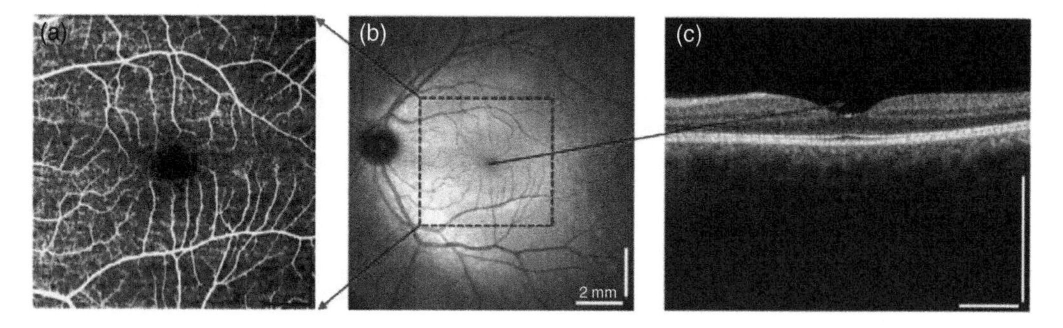

Figure 8.25 Images of OCT: (a) Vasculature and capillaries (red and yellow) of OCT fundus (12 mm × 12 mm). (b) OCT angiography of the macula (6 mm × 6 mm). (c) OCT cross-sectional image consisting average of 5 repeated B-scans (5 mm × 2 mm). *Source:* Adapted from IEEE [49].

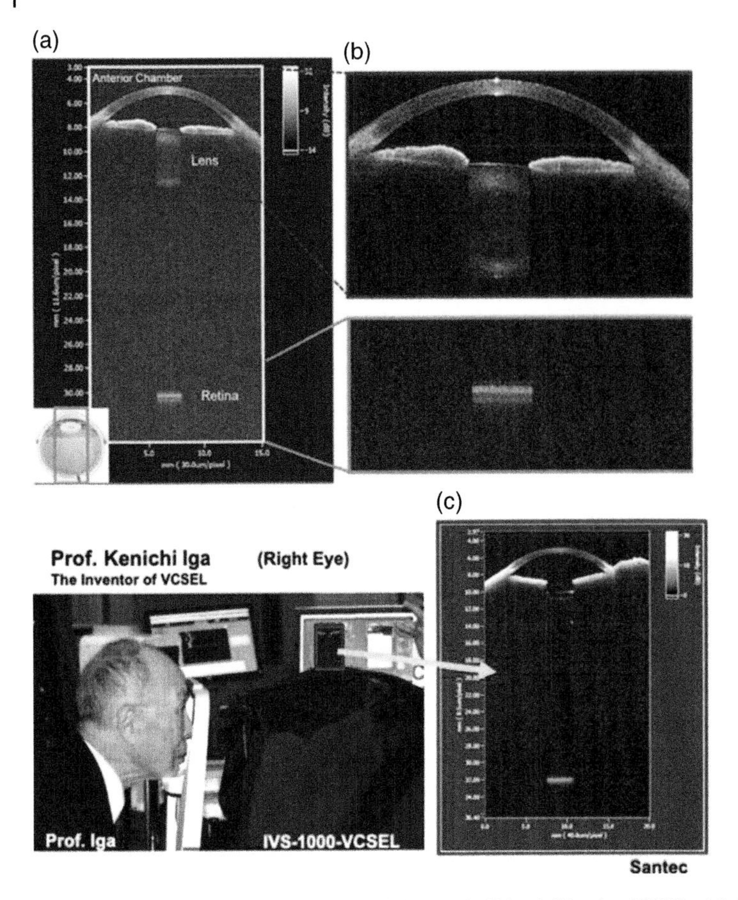

Figure 8.26 Cross-sectional image of the eyeball by OCT using VCSEL: (a) OCT image in the depth direction of the eyeball (upper side is the inner corner of the eye). (b) Upper: upper cornea, lower center lens. (b) Lower: retina. (c) Actual measurement (subject is Kenichi Iga). *Source:* Photo: Courtesy by President Chung Daiko of Santec Corporation.

Further, this MEMS-tunable VCSEL concept using 1060 nm wavelength was practically implemented in ophthalmic imaging by Santec Corp. as a swept-source optical coherence tomography (SS-OCT) imaging with a record dynamic tuning bandwidth of 63.8 nm. Figure 8.26 shows an OCT ocular tomographic photograph using a wavelength tunable VCSEL. With a laser with a large free spectral range (FSR) such as a VCSEL, the coherence length is maintained and a wide range in the depth direction can be measured because there is no mode jump even if the wavelength sweep width is increased. In 2020, an OCT system with a high-speed sweep of 1–2 MHz using MEMS-VCSEL was developed and its applications expanded to include not only ophthalmology but also dermatology, digestive organs, gynecology, orthopedics, surgery, and neurosurgery [50]. Figure 8.26 shows Professor Kenichi Iga undergoing a retinal scan with a VCSEL-based OCT appliance.

8.3.7 Other Emerging Applications

During the last several years, GaN-based VCSELs in 280–650 nm (UV–VIS) band of the electromagnetic spectrum are emerging for diverse applications replacing LEDs, as described in Figure 8.1. These include a mixture of both SM and multimode for (i) retinal scanning displays as smart glasses (AR/VR), (ii) adaptive headlights, (iii) underwater visible communications, (iv) portable mobile projectors and heads-up display as smart automotive lighting, (v) high-resolution color printing and laser TVs, (vi) medical therapy and point-of-care testing, (viii) green houses and horticulture, (ix) atomic clocks, (x) UV-c light for curing and disinfections, and (xi) biomedical sensing. Several prototypes are being commercialized, and the market forecasts are not known yet. Some of the latest developments that may have high commercial value are described in Appendices G and J.

8.4 Summary

The single-mode VCSEL market is characterized by a large number of diverse applications. Many sensor technologies have been enabled with the availability of a low-cost, polarization-stable, and wavelength-tunable devices. Unlike large area 2D arrays discussed in Chapters 5 and 6, these applications will not likely drive a large volumes of wafer manufacturing, but they have become indispensable in many applications.

References

1 https://www.globenewswire.com/news-release/2020/11/11/2124454/0/en/569-Million-Tunable-Diode-Laser-Analyzer-TDLA-Market-Global-Forecast-to-2025-Opportunities-in-Modernization-and-Migration-Services.html.

2 https://www.technavio.com/report/laser-printer-market-industry-analysis.

3 R. Michalzik and K. Ebeling "Operating principles of VCSELs," In: Li H.E., Iga K. (eds) *Vertical-Cavity Surface-Emitting Laser Devices.* Springer Series in Photonics, vol **6**. Springer, Berlin, Heidelberg. (2003).

4 E. Haglund, "Vertical-Cavity Surface-Emitting Lasers: Large Signal Dynamics and Silicon Photonics Integration," Chalmers Univ. Press, (2016). Available at https://publications.lib.chalmers.se/records/fulltext/236337/236337.pdf.

5 P. Debernardi, A. Kroner, F. Rinaldi and R. Michalzik. "Surface relief versus standard VCSELs: a comparison between experimental and hot-cavity model results," *IEEE J. Sel. Top. Quantum Electron.*, vol. **15** (2009).

6 A Larsson, "Advances in VCSELs for communication and sensing," *IEEE JSTQE* **17**, 1552 (2011).

7 Å. Haglund, J. S. Gustavsson, P. Modh, and A. Larsson, "Dynamic mode stability analysis of surface relief VCSELs under strong RF modulation," *IEEE Photon. Technol. Lett.* **7**, 1602 (2005).

8 E. Söderberg, J. S. Gustavsson, P Modh, A. Larsson, Z.Z. Zhang, J Berggren, and M. Hammar, "Suppression of higher order transverse and oxide modes in 1.3-um InGaAs VCSELs by an inverted surface relief," *IEEE Photon. Technol. Lett.* **19**, 327 (2007).

9 N. Ueki, A. Sakamoto, T. Nakamura, H. Nakayama, J. Sakurai, H. Otoma, Y. Miyamoto, M. Yoshikawa, and M. Fuse: "Single-transverse-mode 3.4-mW emission of oxide-confined 780-nm VCSEL's," *IEEE Photon. Technol. Lett.*, vol. **11**, no. 12, pp. 1539–1541 (1999).

10 N. Ueki, J. Ichikawa, S. Ikeda, H. Tezuka, and A. Ota, "Exposure equipment using surface emitting semiconductor laser array elements," *Fuji Xerox Technical Report*, no.**16**, pp. 11–19 (2006).

11 D. Siriani, M. P. Tan, A. M. Kasten, A.C.L. Harren, P. Leisher, J. D. Sulkin, J. Raftery, A. Danner, A. Giannopoulos and K. D. Choquette. "Mode control in photonic crystal vertical-cavity surface-emitting lasers and coherent arrays," *IEEE J. Sel. Top. Quantum Electron.*, Vol. **15** (2009).

12 V. P. Kalosha, N. N. Ledentsov, D. Bimberg, "Design considerations for large-aperture single-mode oxide-confined VCSELs," *Proc. NUSOD* 91–92 (2013) doi: 10.1109/NUSOD.2013.6633139.

13 M. C. Y. Huang, Y. Zhou, and C. J. Chang-Hasnain, "Single mode high- contrast subwavelength grating vertical cavity," *Appl. Phys. Lett.*, vol. **92**, pp. 171108-1–171108-3, 2008.

14 B. D. Padullaparthi and F. Koyama, "A review on polarization control of vertical-cavity surface-emitting semiconductor lasers," *Recent Pat. Electr. Eng.* **4**, 81 (2011) DOI: 10.2174/1874476111104020081.

15 N. Nishiyama: "Surface emitting laser and polarization control using inclined substrate," *Basics and Applications of Surface Emitting Laser* (Edited by K. Iga and F. Koyama), Chapter 7, Kyoritsu Publishing (1999).

16 K. Iga and G. Hatakoshi "The principle and application systems of vertical cavity surface emitting lasers," ,"Adcom-Media, September 2020.

17 K. D. Choquette and R. E. Leibenguth, "Control of vertical-cavity laser polarization with anisotropic transverse cavity geometries," *IEEE Photon. Technol. Lett.*, vol. **8**, no. 1 (1994).

18 D. S. Song, Y.-J. Lee, H.-W. Choi, and Y.-H. Lee, "Polarization-con-trolled, single-transverse-mode, photonic-crystal, vertical-cavity surface-emitting lasers," *Appl. Phys. Lett.*, vol. **82**, (2003).

19 P. Debernardi, J. Ostermann, M. Feneberg, C. Jalics, and R. Michalzik, "Reliable polarization control of VCSELs through monolithically integrated surface gratings: a comparative theoretical and experimental study," *IEEE J. Sel. Top. Quantum Electron.*, Vol. **11**, No. 1 (2005).

20 C. J. Chang-Hasnain, Y. Zhou, M. C. Y. Huang, and C. Chase, "High-contrast grating VCSELs," *IEEE J. Sel. Top. Quantum Electron.*, **15**, 869 (2009).

21 N. Yokouchi, T. Miyamoto, T. Uchida, Y. Inaba, F. Koyama, and K. Iga, "40 Å continuous tuning of a GaInAsP/InP vertical-cavity surface-emitting laser using an external mirror," *IEEE Photon. Technol. Lett.*, vol. **4**, no. 7, (1992).

22 P. Qiao, K. Cook, K. Li, and C. Chang-Hasnain, "Wavelength-swept VCSELs," *IEEE J. Sel. Top. Quantum Electron.*, Vol. **23**, No., 6 (2017).

23 W. Ishak: "My Story with VCSEL at HP/Agilent/Avago," Symposium of 30th Anniversary of the Invention of the VCSEL, Tokyo, Japan 12–17, (2007).

24 https://phys.org/news/2004-09-logitech-unveils-world180s-laser-mouse.amp.

25 D. Baney: "Oral history of Doug Baney," Interview by Gunter Steinbach in Mountain View, CA August 23, 2017. https://www.youtube.com/watch?v=SpUos4r3sSU. (Be advised to start at 24:35.)

26 D. Moy, "Optical mouse technology: Here to stay, still evolving," Electronics Products, April 2011.

27 A. Pruijmboom, S. Booij, M. Schemmann, K. Werner, P. Hoeven, H van Limpt, S. Intemann, R. Jordan, T. Fritzsch, H. Oppermann, M. Barge, "VCSEL-based miniature laser-self-mixing interferometer with integrated optical and electronic components," *Proc. SPIE* **7221**, 72210S (2009) doi: 10.1117/12.808810.

28 M. Grabherr, H. Moench, and A. Pruijmboom, "VCSELs for Optical Mice and Sensing," in VCSELs, Springer Series in Optical Sciences Vol. 166, pp. 521–538 (2013).

29 K. Iga: "40 Years of vertical-cavity surface-emitting laser: Invention and innovation," Japan. J. Appl. Phys., Special Issue of MOC2017, vol. 57, no. 8S2, pp. 1–7 Aug. (2018).

30 K. Iga "VCSEL Odyssey," 2nd Ed., e-book, Optronics Co. Ltd., (2021). http://optronics-ebook.com/products/detail.php?product_id=168.

31 Shunichi Sato: OITDA Optnews vol. 14, No. 1, pp. 2–6 (2017).

32 N. Akiya, T. Hara, A. Ito, H. Shoko, M. Uenishi, H. Motomura, K. Harasaka, S. Sugawara, S. Sato: Ricoh Technical Report, No. 37, pp. 74–80 (2011).

33 S Sato: "Front Runner Interview," https://jp.ricoh.com/technology/rd/f_runner/fr11.

34 M. Ortsiefer, C. Neumeyr, J. Rosskopf et al., "GaSb and InP -based VCSELs at 2.3um emission wavelength for tuneable diode laser spectroscopy of carbon monoxide," *Proc. SPIE* **7945**, 794509 (2011).

35 M. Ortsiefer, J. Rosskopf, C. Neumeyr, T. Gründl, C. Grasse, J. Chen, A. Hangauer, R. Strzoda, C. Gierl, P. Meissner, F. Küppers, M.-C. Amann, "Long-wavelength VCSELs for sensing applications," *Proc. SPIE* **8276**, 82760A (2012).

36 A. B. Ikyo, I. P. Marko, K. Hild, A. R. Adams, S. Arafin, M.-C. Amann & S. J. Sweeney, "Temperature stable mid-infrared GaInAsSb/GaSb vertical cavity surface emitting lasers (VCSELs)," *Sci. Rep.* **6**, 19595; doi: 10.1038/srep19595 (2016).

37 G. Veerabathran, S. Sprengel, A. Andrejew, and M.-C. Amann, "Room-temperature vertical-cavity surface-emitting lasers at 4μm with GaSb-based type-II quantum wells," *Appl. Phys. Lett.* **110**, 071104 (2017).

38 D. Sanchez, L. Cerutti, and E. Tournié. "Mid-IR GaSb-Based Bipolar Cascade VCSELs," *IEEE Photon Technol. Lett.* **25**, pp. 882–884 (2013).

39 V. Jayaraman, S. Segal, K. Lascola, C. Burgner, F. Towner, A. Cazabat, G.D. Cole, D. Follman, P. Heu, C. Deutsch., "Room temperature continuous wave mid-infrared VCSEL operating at 3.35um," Proceedings SPIE 10552, 10552xx (2018).

40 M. Hara, Y. Yano, M. Kajita, H. Nishino, Y. Ibata, M. Toda, S. Hara, A. Kasamatsu, H. Ito, T. Ono, and T. Ido: "Microwave oscillator using piezoelectric thin-film resonator aiming for ultra-miniaturization of atomic clock," *Rev. Sci. Instrum.* vol. **89**, no. 10, p. 105002, 2018.

41 M. Hara: "Scenario for the realization of atomic clock portable in smart phon," Nikkei Electronics, no. 9, pp. 85–93 September 2020.

42 D. Serkland, G. Peake, K. Geib, R. Lutwak, R. Garvey, M. Varghese, and M. Mescher "VCSELs for atomic clocks," Proc. SPIE 6132, Vertical-Cavity Surface-Emitting Lasers X (2006).

43 H. Zhang, H. Herdian, A.T. Narayanam, A. Shirane, M. Suzuki, K. Harasaka, K. Adachi, S. Goka, S. Yanagimachi, and K. Okada, "ULPAC: a miniaturized ultralow-power atomic clock," *IEEE J. Solid State Circuits*, **54**, 3135 (2019).

44 H. Zhang, H. Herdian, A. T. Narayanan, A. Shirane, M. Suzuki, K. Harasaka, K. Adachi, S. Goka, S. Yanagimachi, and K. Okada: "ULPAC: a miniaturized ultralow-power atomic clock," *IEEE Solid-State Circuits*, vol. **54**, no. 11, pp. 3135–3148, Nov. 2019.

45 M. Haruna, "Optical coherence tomography," *J. IEICE Japan*, vol. **90**, no. 3, pp. 226–231 (2007).

46 Y. Oota: "The review of optical coherence tomography," J. Effect of Economy/ Medical Photonics, No. 5, pp. 47–52, (2011). Also, https://eu.wikipedia.org/wiki/Lankide:**光干渉断層像**.

47 Naohiro Tanno, Tsutomu Ichikawa, and Akio Saeki: "Light wave reflection image measuring device," Japanese patent No. 2010042 (1990).

48 D. Huang, E. A. Swanson, C. P. Lin, J. S. Schuman, W.G. Stinson, W. Chang, M. R. Hee, T. Flotte, K. Gregory, C. A. Puliafito, J. G. Fujimoto, "Optical coherence tomography," *Science*, **254** (5035), (1991).

49 D. D. John et al., "Wideband electrically pumped 1050-nm MEMS-Tunable VCSEL for ophthalmic imaging," *J. Lightwave Technol.* Vol. **33** 3461 (2015).

50 https://www.laserfocusworld.com/detectors-imaging/article/14174348/new-directions-in-biomedical-oct.

9

Single-Mode VCSELs for Communications Applications

Kenichi Iga and Jim Tatum

9.1 Introduction

As described in Chapter 4, the size of data centers has continued to grow, and the need for longer distance fiber optic links has become evident. VCSEL-based multi-mode fiber (MMF) links are restricted to a few hundred meters by the bandwidth of the optical fiber, and more of these links are being addressed with active optical cables. However, when cabling infrastructure is embedded in the building infrastructure, many companies are choosing to use single-mode fiber (SMF) because the bandwidth of the fiber is not a limiting factor. In fact, with the size of many data centers today, longer links even between equipment racks can be a challenge for MMF interconnects. It is clear today that MMF solutions continue to be the lower-cost alternative. For many years, the promise of VCSELs compatible with SMF has been an attractive alternative to traditional FP and DFB lasers. The primary advantages of VCSELs are lower power consumption and presumably lower cost. The factors that have limited adoption have been the difficulty of producing highly reliable VCSELs, the limited optical power, and the overall cost of the transmitter optical subassembly (TOSA). Many of the factors that drive cost of the TOSA, as described in Chapter 4, are not addressed by simply replacing the DFB laser with a VCSEL, i.e., the cost of materials and alignment are significant compared to the cost of the laser itself. As a result, there has been limited commercial success for VCSELs in the 1310 and 1550 nm band traditionally used with SMF. Despite these drawbacks, there has been considerable research and development activity, and some excellent transmission results have been obtained. Some of the more recent results include 40 Gbps NRZ and 28GBd PAM4 modulation [1] with laser showing 22 GHz small signal bandwidth, 56 Gbps NRZ [2], 180 Gbps PAM4 over 300 m [3], 84 Gbps PAM-4 over 1.6 km SSMF [4], and 726.7 Gbps at 1.5 μm [5, 6, 7]. Using discrete multi-tone transmission, 726.7 Gbps over 2.5 km has been achieved [8], and 1550 nm 56 Gb/s, PAM-4, 2.0 km [9]. Studies on ultra-high-speed transmission using advanced modulation formats are underway [10].

Also as described in Chapter 4, one approach to increasing the fundamental VCSEL speed is to decrease the active volume, and single-mode VCSELs at longer wavelengths have been developed. In this chapter the challenges to manufacture long-wavelength VCSELs (LW-VCSELs) are reviewed.

VCSEL Industry: Communication and Sensing, First Edition. Babu Dayal Padullaparthi, Jim A. Tatum and Kenichi Iga.

9.2 LW-VCSEL Design and Manufacturing

Industry standard SMF has a cutoff wavelength near 1260 nm where a second optical mode is supported. The second mode has a large significantly different propagation constant and can lead to significant mode dispersion. For this reason, lasers operating at cutoff wavelength are almost exclusively used with SMF. The traditional wavelength bands in SMF are near 1310 nm, where the chromatic dispersion is minimized, or at 1550 nm, where the power attenuation is minimized. The following discussions will focus on these two bands and the associated materials, as depicted in Table 2.2 and Figure 8.1. The first section describes some of the unique wafer fabrication processes for LW-VCSELs, focusing on the mirror-related issues. The active materials are described in the following sections, and some relevant results on transmission experiments are reported.

9.2.1 LW-VCSEL Structures

The principal issues with LW-VCSELs are material related, especially for InP lattice matched structures. These include large inter-valence band absorption, non-radiative Auger recombination, current leakage in the quantum wells, low index contrast of lattice matched semiconductor materials for the DBR, very thick epitaxial structures, and the lack of a well-defined method for current confinement (no oxidation process). The challenges of the DBR structure apply to all LW-VCSEL structures and have led to more exotic methods of epitaxial growth and subsequent fabrication process. One of the principal advantages of VCSELs is the ability to realize the structure in a single epitaxial growth and relatively simple wafer processing requirements. Figure 9.1 shows some of the relevant VCSEL structures for circumventing the DBR problem. The conventional VCSEL structure shown in 9.1a is grown in a single epitaxial growth and is the most common type of commercial device. Figure 9.1b shows a technique of wafer-bonding a suitable active region to GaAs-based semiconductor mirrors [11]. In this structure, the AlGaAs-based n and p DBRs are grown on separate GaAs substrates, and the active region is grown on a lattice matched InP substrate. The active region is thermo-compression-bonded to the n-type DBR structure, and the InP substrate is removed. The p-DBR is then thermo-compression-bonded to the active region, and the substrate is removed. The wafer is then ready for more conventional VCSEL processing. In Figure 9.1c a dielectric DBR is used for the VCSEL mirrors. In this case, the active region is grown

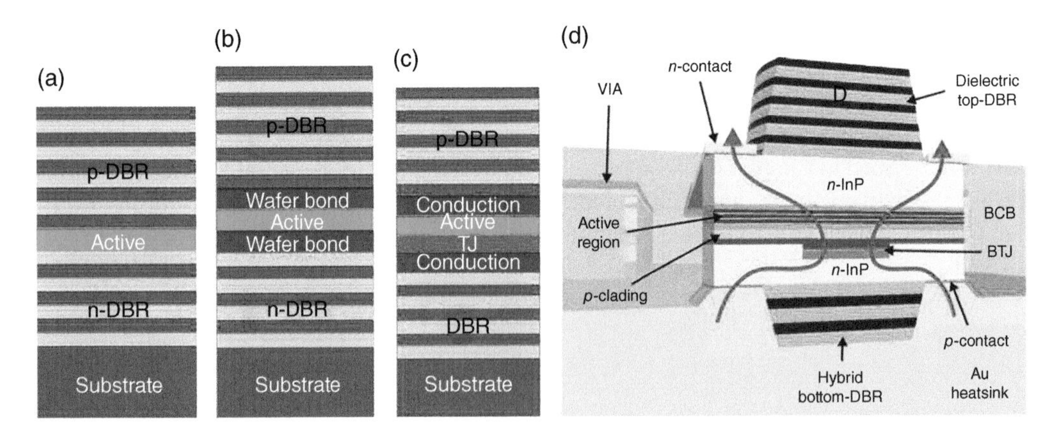

Figure 9.1 VCSEL structures for LW-VCSEL.

on lattice matched InP substrates with a tunnel junction to steer the current; such a device is depicted in Figure 9.1d. During the wafer fabrication process the substrate is removed and a dielectric mirror is deposited.

9.2.2 1310 nm VCSEL

The first 1310 nm VCSEL to operate continuously at room temperature borrowed a buried heterostructure from EELs and used an InGaAsP active region on InP substrates [12]. In this approach, the active region current is confined by multiple regrowth steps and intermediate laser processing. The device operation is limited by the holes and crystal defects during the regrowth process as a result of the circular aperture.

Subsequent developments using InGaAsP/InP and InGaAlAs/InP materials focused on more complex structures including the use of tunnel junctions (TJ) and both semiconductor and dielectric mirrors. Transmission to more than 50 Gbps has been demonstrated using this structure, and commercial products are available [13].

Wafer-bonding disparate semiconductor materials has also been widely attempted as a commercial path for 1310 nm VCSELs and reasonable reliability results have been reported [14]. In this process, undoped AlGaAs mirrors were thermo-compression wafer-bonded to an InGaAlAs/InP active region, and device performance to more than 10 Gbps was achieved with reliability of approximately 20 years at 70°C.

VCSELs at 1310 nm have been developed as traditional VCSEL structures using InGaAsN (or GaInNAs) as the active region. The lasers are generally grown in MBE systems, or a hybrid approach is taken where the mirrors are grown by MOCVD, and the active region grown by MBE. Promising results on performance and reliability have been reported [15, 16, 17]. These VCSELs use AlGaAs-based mirrors and are grown on GaAs substrates. It is difficult to incorporate sufficient N in the growths to push the wavelength to 1310 nm, and the VCSELs typically operate near 1290 nm, still within the wavelength limits of the optical standards and standard SMF. Achieving long term reliability from high quality epitaxial growth is challenging because of the strain introduced by N replacing As atoms in the lattice [18].

Despite several successful demonstrations of both performance and reliability, 1310 nm VCSELs have not been widely adopted for data communications.

9.2.3 VCSELs in the 1550 nm Band

At 1550 nm, and in particular the InGaAsP/InP material system, the effects of valence band absorption and Auger non-recombination transition become remarkable because of its band structure. For this reason, it is even more important to reduce the loss of the DBR mirrors. The difference in refractive index between InGaAsP and InP is larger compared to the 1300 nm band, so the InGaAsP/InP semiconductor DBR can employed [19]. The principal advantage to using the semiconductor DBRs is that absorption in the cladding layer can be reduced and the VCSEL can be made without removing the substrate. In 1550 nm VCSELs, a hybrid structure using an n-doped mirror is used, where free carrier and Auger recombination is small compared to a p-type material, and a dielectric second mirror is sometimes employed. The primary drawback to a full semiconductor DBR is the thickness, driven by the wavelength and the low index contrast. To achieve > 99.5% reflectivity, more than 50 pairs are required, and this represents more than 10 μm of epitaxial growth just for one mirror. (For reference, an entire 850 nm VCSEL is approximately 10 μm in total thickness.) The thick structure drives impractically long growth times and stress within the semiconductor material. Another approach is to use a composite reflector structure that

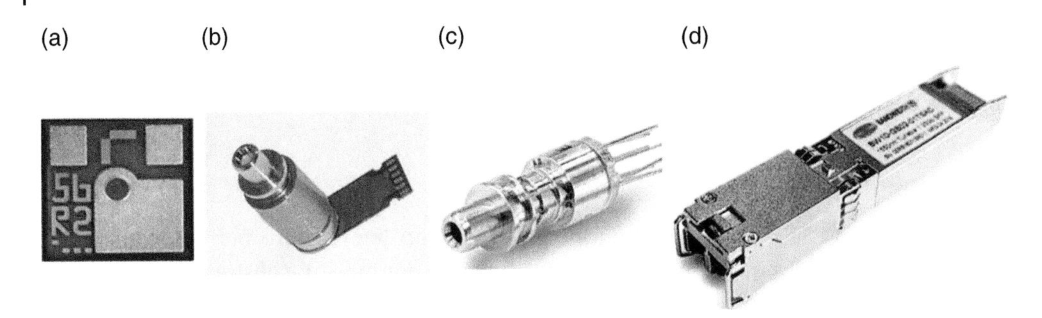

Figure 9.2 (a) LW-VCSEL die. (b) LW-VCSEL in an LC fiber TOSA. (c) Tunable VCSEL in LW TOSA. (d) Tunable VCSEL-based optical transceiver operating at 10 Gbps. *Source:* a, b are courtesy of Vertilas, www.vertilas.com, and c, d are courtesy of Bandwidth 10, www.bandwidth10.com.

combines a dielectric DBR with the semiconductor DBR. There is an example of a surface-emitting laser realized by using a compound reflector in the n side. Eight pairs of n-type InGaAsP/InP DBR and four pairs of Si/SiO2 DBR deposited by the chemical beam epitaxial growth method resulted in a structure that operated at room temperature under pulse operation [20].

Many of the structures aimed at addressing the material issues have been applied to 1.55 μm VCSELs as well. Today there are multiple (small) companies providing VCSELs in this wavelength band [13]. In the 1550 nm band, tunable sources are also of interest, which use high-contrast grating, as described in Section 8.2.3 and specifically Figure 8.11c. The use of HCG at 1310 nm band has been studied but does not a have high value in the data communications market. Similarly, 1550 nm VCSELs development at 1550 nm, shown in Figure 8.11, has led to some commercial products, but speed has been limited to 10 Gbps [21].

Like VCSELs at 1310 nm, the primary disadvantages to 1550 nm VCSELs are the relatively low optical power and the high cost of packaging. As estimated in Section 4.4.5, the cost of the laser die in a single-mode optical assembly is relatively small in comparison to the total cost. For VCSELs the packaging cost is made even worse by the relatively low power that drives even tighter alignment tolerances. Some examples of LW-VCSEL assemblies are shown in Figure 9.2.

The structure described at the 1300 nm VCSEL, shown in Figure 8.11, was also extended to 1550 nm and exhibited similar characteristics to those in the 1300 nm band. It also has drawbacks such as low thermal conductivity of a quaternary crystal InGaAsP. In order to solve these problems, the development of GaAlSbAs-based materials that can obtain a large difference in refractive index while lattice-matching with InP and the wafer-bonding method were studied [22]. In addition, as device technology, new initiatives such as spin injection type can be seen in 1550 nm devices [23].

It should be mentioned that there have been several efforts for long-wavelength VCSELs—especially (i) Coldren's group and Bower's group (UCSB), (ii) Amann's group (Vertilas Company), (iii) Eli Kapon (Beam express), (v) QD VCSELs (Bimberg's group), (vi) Chang-Hasnain's group (Bandwidth), JDSU, Seiko Epson, Raycon, among others—that have had significant industrial impact.

9.2.4 Other Wavelengths for Data Communications

As previously described in Chapter 4, one method to increase the laser bandwidth is to decrease the size of the optical cavity. This has been proposed and demonstrated at wavelengths of 850 nm [24], 980 nm [25, 26], and 1060 nm [27]. It is important to note that the optical standards such as Ethernet do not exclude the use of single-mode VCSELs, but due to low power and potential reliability issues, they have not found widespread utility. At longer wavelengths such as 980 nm and 1060 nm, reliability may be less of a concern because of the lower photon energy. By increasing the In content in InGaAs quantum wells, emission to nearly 1200 nm can be obtained and remain lattice-matched to GaAs [28, 29]. This allows for the use of AlGaAs-based DBR mirrors and conventional device fabrication. The use of wavelengths in this region has been hampered by the lack of standards and specifications and being caught between two fiber specifications. As previously stated, standard SMF has a cutoff wavelength around 1260 nm, and below that the fiber will support more than one optical mode. This makes the modal dispersion quite high and limits the utility in SMF. MMF is generally specified at 850 nm, or more recently to 970 nm in the case of OM5. So many of these VCSELs are caught between the traditional transmission windows.

9.3 Quantum Communications

While not strictly data communications, the use of VCSELs as a single photon source, or source of photon pairs is an attractive approach to quantum cryptography. A quantum optical source is needed to transmit along with data and serve as the key for encryption. Often this may be in an optical fiber, and thus LW-VCSELs may be a possible solution. In this section we quickly highlight the technology and requirements in quantum communications.

Quantum information technology (QIT) uses states of light in superposition in the form of 0s and 1s with multiple combinations called quantum bits, or qubits. QIT based on laser sources and detectors has tremendous applications in telecommunication, sensing and computing, and finance, where encryption plays a critical role. In quantum communication, the word *entanglement* is often used to describe photon states. In quantum theory, superposition and linear combinations are used to express the mixed state of the energy of one quantum. The entanglement referred to here is, for example, when two or more quanta are interacting with each other, they are random when viewed from each other, but when both are measured, they are intertwined in a correlated manner. The concept of entanglement is often used in the field of quantum information processing. Quantum communication or teleportation uses this quantum entanglement to transfer the quantum state. "Teleportation" is originally a teleportation to a distant place. In quantum communications, transmission of a series of single photons occurs in one light pulse, or entangled photons. To accomplish this, a single-photon emitter and single-photon detector are needed. To understand how this is possible, it is necessary to understand the concept of "photon." The second quantization of light is based on the similarity with the harmonic oscillator. Now consider a resonator as shown in Figure 9.3. It is assumed that the light is in a single horizontal (transverse) mode and vertical (longitudinal) mode resonance state, respectively. The field can be described as a set of quantum harmonic oscillators. In this case, the harmonic oscillator must follow Bose-Einstein statistics. Then, the energy of the light can be expressed by the number states such as the number of harmonic oscillators. Then, its eigenvalue of energy in number state is expressed as:

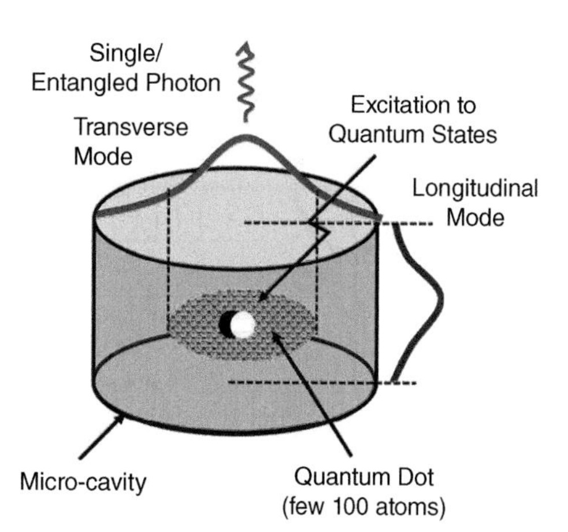

Figure 9.3 Light field and single-photon emission [30].

$$E_n = \left(n + \frac{1}{2}\right)\hbar\omega.$$

Here, ω is angular frequency of light, $\hbar$ is h/2π (*h* is Planck's constant), and *n* is the number of photons and is represented by an integer. The state of a single photon is the number state where $n = 1$.

A single-photon emitter can be made by placing a single quantum dot in the cavity and exciting it with either an optical pulse or current to produce an electron and hole pair, or exciton. By a careful design of the cavity and excitation condition, there exists the probability of having one single photon in one pulse. This is a single-photon emitter. The especially designed micro-cavity that comes from VCSEL resonator may have an opportunity to function as a single-photon emitter in the future. Single photon ($n = 1$) or entangled photons with controlled phase, polarization, and angular momentum states generated from lasers or VCSELs may play a critical role in quantum communication and sensing. Lasers with on-demand short optical pulses are used as light sources to prepare atoms to initiate quantum states to read quantum information. Further, with advances in EUV light lithography, < 5 nm gate length CMOS processes feasible [31], in the future it might be possible to fabricate scalable quantum dot or quantum well lasers with single or entangled photon emission in the form of controlled and uniform 2D arrays [32]. If this happens, there will be a tremendous opportunity for high-volume manufacturing of quantum light sources.

Quantum cryptography refers to the delivery of encoding keys based on quantum theory that are required to decrypt the transmitted information. Passwords have come to be used in computers, and three-dimensional face authentication has come to be used in smartphones. However, unlike these authentications, when it is necessary to send information to a distant place, it must be encrypted so that it cannot be intercepted and read, and an encryption key must be sent so that (only) the intended recipient can read it. With classical cryptography and cryptographic key distribution, there is a risk of being intercepted or cracked by someone other than the recipient. That's where the quantum cryptography key distribution method (or simply, quantum cryptography) may offer superior security. Quantum cryptography was proposed by Stefan Weizner in 1970 and was rediscovered by Charles H. Bennet and Giles Brassard in 1984 [33]. The proposed protocol is called BB84. A different B92 protocol with two non-orthogonal linearly polarized states in 850 nm VCSEL cavity was proposed in 2005 [34]. A polarization-controlled VCSEL may take a potential role for its realization [35].

The quantum transmission is considered a secure way of sending cryptography due to the so-called no-cloning theorem in quantum information technology [36][37]. It says that a single quantum cannot be cloned, if somebody tries to intercept the information.

The quantum communication shown in Figure 9.4 implies a one-to-one communication method that was further developed from quantum cryptography key distribution or quantum key

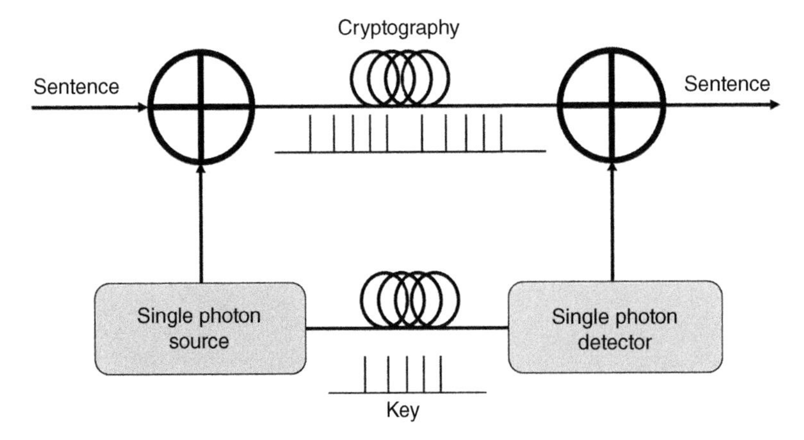

Figure 9.4 Concept of quantum encryption (K. Iga).

distribution (QKD). A quantum network is a group of quantum communications that started based on the delivery of quantum cryptography and is attracting attention as a communication network to increase the security of the network. Quantum networks are highly attractive methods of transmitting highly confidential information such as national secrets, financial information, and sensitive personal information such as genetics.

In this era information is sent and received via the Internet, but it is difficult to protect confidentiality with classical encryption. Therefore, the importance of quantum networks is increasing. As for the scale of delivery of quantum cryptography, international networks using optical fiber networks and high-frequency networks using satellite communication networks have been tested. Already, an international quantum communication network is about to be formed from tabletop experiments.

In 2020, there were notable experiments in different parts of the globe with off-the-shelf components to teleport information in optical fibers and free space through 10s to 100s of miles, with the ultimate objective of a developing a quantum Internet [38]. Once developed, it will be the most secure method for transmitting and receiving sensitive data. Several experiments are underway by companies such as Intel, Apple, Microsoft, IBM, Toshiba to develop quantum technologies for future secure communications and transactions [39].

9.4 Summary

The commercial success of VCSELs in data communications on SMF has been limited, but the promise of low-power devices continues to drive research. The material challenges and complex manufacturing processes have been difficult to overcome at the same scale as shorter wavelength VCSELs.

References

1 F. Karinou, C. Prodaniuc, N. Stojanovic, M. Ortsiefer, A. Daly, R. Hohenleitner; B. Kögel, C. Neumeyr, "Directly PAM-4 modulated 1530-nm VCSEL enabling 56 Gb/s/ λ data-center interconnects", *IEEE Photon. Technol. Lett.*, Vol. **27**, No. 17, 1872–1875 (2015).

2 D. Kuchta, T. Huynh, F. Doany, L. Schares, C. Baks, C. Neumeyr, A. Daly, B. Kögel, J. Rosskopf, M. Ortsiefer: "Error-free 56 Gb/s NRZ modulation of a 1530-nm VCSEL link," *IEEE J. Lightwave Technol.*, Vol. **34**, No. 14, (2016).

3 S. M. R. Motaghiannezam, I. Lyubomirsky, H. Daghighian, C. Kocot, T. Gray, J. Tatum, A. Amezcua-Correa, M. Bigot-Astruc, D. Molin, F. Achten, and P. Sillard: "180 Gbps PAM4 VCSEL Transmission over 300m Wideband OM4 Fibre," Optical Fiber Communication Conference OSA Technical Digest (online) (Optical Society of America, 2016), paper Th3G.2.

4 N. Eiselt, H. Griesser, J. Wei, R. Hohenleitner, A. Dochhan, M. Ortsiefer, M. H. Eiselt, C. Neumeyr, J. Olmos, and I. Monroy, "Experimental demonstration of 84 Gb/s PAM-4 over up to 1.6 km SSMF using a 20-GHz VCSEL at 1525 nm," *J. Lightwave Technol.* **35**, 1342–1349 (2017).

5 A. Malacarne, V. Sorianello, A. Daly, B. Kögel, M. Ortsiefer, C. Neumeyr, M. Romagnoli, A. Bogoni: "Performance analysis of 40-Gb/s transmission based on directly modulated high-speed 1530-nm VCSEL," *IEEE Photon. Technol. Lett.*, Vol. **28**, No. 16, 1735–1738 (2016).

6 S. Moon and E. Choi: "VCSEL-based swept source for low-cost optical coherence tomography," *Biomed. Opt.Express* Vol. **8**, Issue 2, 1110-1121 (2017).

7 S. Spiga, W. Soenen, A. Andrejew, D. Schoke, X. Yin, J. Bauwelinck, G. Boehm, M. Amann: "Single-mode high-speed 1.5-μm VCSELs," *IEEE J. Lightwave Technol.*, Vol. **35**, No. 4 (2017).

8 J. Van Kerrebrouck, L. Zhang, R. Lin, X. Pang, A. Udalcovs, O. Ozolins, S. Spiga, M. C. Amann, G. Van Steenberge, L. Gan, M. Tang, S. Fu, R. Schatz, S. Popov, D. Liu, W. Tong, S. Xiao, G. Torfs, J. Chen, J. Bauwelinck, and X. Yin: "726.7-Gb/s 1.5-μm Single-Mode VCSEL Discrete Multi-Tone Transmission over 2.5-km Multicore Fiber," Optical Fiber Communication Conference (2018).

9 G. Kanakis, N. Iliadis, W. Soenen, B. Moeneclaey, N. Argyris, D. Kalavrouziotis, S. Spiga, P. Bakopoulos, H. Avramopoulos, "High-speed VCSEL-based transceiver for 200 GbE short-reach intra-datacenter optical interconnects," *Appl. Sci.* 2019, **9**, 2488. doi:10.3390/app9122488.

10 J. Van Kerrebrouck, X. Pang, O. Ozolins, R. Lin; A. Udalcovs, L. Zhang, H. Li, S.a Spiga, M. Amann, L. Gan, M. Tang, S. Fu, R. Schatz, G. Jacobsen, S. Popov, D. Liu, W. Tong, G. Torfs, J. Bauwelinck, J. Chen: "High-speed PAM4-based optical SDM interconnects with directly modulated long-wavelength VCSEL," *IEEE J. Lightwave Technol.*, Vol. **37**, No. 2, 356–362 (2019).

11 Y. Okuno, J. Geske, K-G. Gan, Y-J. Chiu, S. DenBaars, and J. Bowers, "1.3 μm wavelength vertical cavity surface emitting laser fabricated by orientation-mismatched wafer bonding: a prospect for polarization control," *Appl. Phys. Lett.*, vol. **82**, no. 15 2377–2379 (2003).

12 T. Baba, K. Suzuki, Y. Yogo K. Iga, and F. Koyama: "Threshold reduction of 1.3μm GaInAsP/InP surface emitting laser by a maskless circular planar buried heterostructure regrowth," *Electron. Lett.*, vol. **29**, no. 4, pp. 331–332 (1993).

13 M. Verplaetse, L. Breyne, C. Neumeyr, T. De Keulenaer, W. Soenen, X. Yin, P. Ossieur, G. Torfs, and J. Bauwelinck, "DSP-free and real-time NRZ transmission of 50Gb/s over 15km SSMF and 64Gb/s back-to-back with a 1.3μm VCSEL," in Optical Fiber Communication Conference, Postdeadline Papers, (2018).

14 A. Sirbu, G. Suruceanu, I. Vladimir, A. Mereuta, Z. Mickovic, A. Caliman and E. Kapon, "Reliability of 1310 nm wafer fused VCSELs, *IEEE Photon. Technol. Lett.* vol. **25**, no. 16, 1555–1558 (2013).

15 L. Graham, J. Jewell, K. Maranowski, M. Crom, S. Feld, J. Smith, J. Beltran, T. Fanning, M. Schnoes, M. Gray, D. Droege, V. Koleva, M. Dudek, J. Fiers and R. Patterson "LW VCSELs for SFP+ applications," *Proc. SPIE* vol **6908** 6908–02 (2008).

16 L. A. Graham, M. Schnoes, K. Maranowski, T. Fanning, M. Crom, S. A. Feld, M. A. Gray, K. Bowers, S. L. Silva, and K. Cook, "New developments in 850 and 1300nm VCSELs at JDSU," *Proc. SPIE* **7229** 7229–0B (2009).

17 "Alight and ULM partner on volume VCSEL production," Published in Semiconductor Today, (2009).

18 T Nishida, M Takaya, S Kakinuma, and T Kaneko, "4.2-mW GaInNAs long-wavelength VCSEL grown by Metalorganic chemical vapor deposition," IEEE *JSTQE* **11**, 958 (2005).

19 N. Nishiyama; C. Caneau; B. Hall; G. Guryanov; M.H. Hu; X.S. Liu; M.-J. Li; R. Bhat; C.E. Zah: "Long-wavelength vertical-cavity surface-emitting lasers on InP with lattice matched AlGaInAs–InP DBR grown by MOCVD," *IEEE J. Sel. Top. Quantum Electron.*, vol. **11**, no. 5, pp. 990–998, Sept.-Oct. 2005.

20 T. Miyamoto; T. Uchida; N. Yokouchi; K. Iga: "Surface emitting lasers grown by chemical beam epitaxy," *J. Crystal Growth*, vol. **136**, no. 1-4, pp. 210–215, 1994.

21 C. Chase, Y. Rao, M. Huang, and C. Chang-Hasnain "Tunable 1550nm VCSELs using high-contrast grating for next-generation networks," Proc. SPIE 9008, (2014).

22 N. Nishiyama; C. Caneau; B. Hall; G. Guryanov; M.H. Hu; X.S. Liu; M.-J. Li; R. Bhat; C.E. Zah: "Long-wavelength vertical-cavity surface-emitting lasers on InP with lattice matched AlGaInAs–InP DBR grown by MOCVD," *IEEE J. Sel. Top. Quantum Electron.*, vol. **11**, no. 5, pp. 990–998, Sept.-Oct. 2005.

23 N. Yokota, R. Takeuchi, H. Yasaka, K. Ikeda: "Lasing polarization characteristics in 1.55- μm spin-injected VCSELs," *IEEE Photon. Technol. Lett.*, Vol. **29**, No. 9 711–714 (2017)..

24 N. Ledentsov, Ł. Chorchos, V.A. Shchukin, V.P. Kalosha, J. P. Turkiewicz, and N. N. Ledentsov, "Development of VCSELs and VCSEL-based Links for Data Communication beyond 50Gb/s," OFC conference (2020).

25 G. Larisch, P. Moser, J. A. Lott, and D. Bimberg, "Impact of photon lifetime on the temperature stability of 50 Gb/s 980 nm VCSELs," *IEEE Photon. Technol. Lett.* **28**, 2327–2330 (2016)..

26 N. Haghighi, G. Larisch, R. Rosales, M. Zorn, and J. Lott, "35 GHz bandwidth with directly current modulated 980 nm oxide aperture single cavity VCSELs," 2018 IEEE International Semiconductor Laser Conference (ISLC), (2018).

27 E. Simpanen, J. S. Gustavsson, E. Haglund, E. P. Haglund, A. Larsson, W. V. Sorin, S. Mathai, and M. R. Tan, "1060 nm single-mode vertical-cavity surface-emitting laser operating at 50 Gbit/s data rate," *Electron. Lett.*, vol. **53**, no. 13, (2017).

28 Keishi Takakia, Norihiro Iwaia, Shinichi Kamiyab, Hitoshi Shimizua, Koji Hiraiwaa, Suguru Imaia, Yasumasa Kawakitaa, Tomohiro Takagia, Takuya Ishikawab, Naoki Tsukijia and Akihiko Kasukawaa: "Experimental demonstration of low jitter performance and high reliable 1060nm VCSEL arrays for 10Gbpsx12ch optical interconnection," Proc. Of SPIE Vol. 7615, 761502. © 2010 SPIE CCC code: 0277-786X/10/$18 doi: 10.1117/12.840130.

29 Suguru Imai, Keishi Takaki, Shinichi Kamiya, Hitoshi Shimizu, Junji Yoshida, Yasumasa Kawakita, Tomohiro Takagi, Koji Hiraiwa, Hiroshi Shimizu, Toshihito Suzuki, Norihiro Iwai, Takuya Ishikawa, Naoki Tsukiji, and Akihiko Kasukawa: "Recorded low power dissipation in highly reliable 1060-nm VCSELs for "green optical interconnection," *IEEE J. Sel. Top. Quantum Electron.*, vol. **17**, no. 6, Pp. 1614–1620, November/December 2011.

30 K. Iga and G. Hatakoshi "The principle and application systems of vertical cavity surface emitting lasers," Adcom-Media, September 2020.

31 https://www.iws.fraunhofer.de/en/pressandmedia/news/2020-1124_news_euv_zukunftspreis.html.

32 R. Uppu, F. Pedersen, Y. Wang, C. Olesen, C. Papon, x. Zhou, L. Midolo, S. Scholz, Sven, A. Wick, A. Ludwig, and P. Lodahl, "Scalable integrated single-photon source," *Sci. Adv.*, Vol. **6**, No. 50, eabc8268(1-6) (2020).

33 C. H. Bennett and G. Brassard: "Quantum cryptography: Public key distribution and coin tossing," Proc. IEEE Int. Conf. on Computers, Systems and Signal Processing, pp. 175–179, Bangalore (1984).

34 V. Fernandez, K. Gordon, R. Collins, P. Townsend, S. Cova, I. Rech, G. Buller, "Quantum key distribution in a multi-user network at gigahertz clock rates," *Proc. SPIE* vol. **5840** (2005) https://doi.org/10.1117/12.608650.

35 Á. Schranz and E. Udvary, "Transmitter Design Proposal for the BB84 Quantum Key Distribution Protocol using Polarization Modulated Vertical Cavity Surface-emitting Lasers," Proceedings of the 6th International Conference on Photonics, Optics and Laser Technology (PHOTOPTICS), (2018).

36 W. Wootters and W. Zurek, "A Single Quantum Cannot be Cloned," Nature **299**, 802–803 (1982).

37 D. D. Dieks, "Communication by EPR devices," Physics Letters A, **92**, No. 6, 271–272 (1982).

38 R. Valivarthi, S. Davis, C. Pena, S. Xie, N. Lauk, L. Narvaez, J. Allmaras, A. Beyer, Y. Gim, M. Hussein, G. Iskander, H. Kim, B. Korzh, A. Mueller, M. Rominsky, M. Shaw, D. Tang, E. Wollman, C. Simon, P. Spentzouris, D. Oblak, N. Sinclair, and M. Spiropulu, "Teleportation systems toward a quantum internet," *PRX Quantum*, **1**, 020317 (2020).

39 https://s3.i-micronews.com/uploads/2020/01/YDR20062-Quantum-Technologies-2020-Yole-Développement-Sample.pdf.

10

Future Prospects

Babu Dayal Padullaparthi, Kenichi Iga, and Jim Tatum

The previous chapters have described many of the current applications of VCSELs, ranging from high-speed data communications and high-power LiDAR systems to single-mode sensors. In the nearly 44 years since the modern VCSEL was conceived and 25 years since the VCSEL was commercialized, the VCSEL has been a true economic and technical success. Many aspects of everyday life are enhanced by VCSELs, whether keeping people safer in driver awareness systems, helping with secured transactions with face ID, or simply browsing the web. The photonics market in general is growing rapidly, and VCSELs will be a significant part of the future. In this chapter some of the future needs, directions, and possibilities are discussed that could end up in consumer and industrial products.

10.1 VCSEL Industry

- Many of the traditional applications of VCSELs previously described will continue to grow in their own right. The average forecasted cumulative annual growth rate (CAGR) of these markets is 17% for the core segments of data communications and optical sensing. LiDAR is still an emerging application and has the potential to drive tremendous growth in the VCSEL industry. The global photonics market is forecasted to reach $ 1.2 trillion by 2030, and VCSEL chips alone will be more than $12 billion (B). The combined estimates of core and edge markets at module level comes out to be $40 billion and $4.8 billion by 2025, respectively. Figure 10.1 shows the market projections for the photonics industry overall and the VCSEL segment in particular.
- The overall revenue growth places massive demands on the existing supply chain infrastructure. Over 5B VCSEL chips will be needed in the next five years alone in the core and edge markets. With the strong emergence of high-power density 2D MM VCSEL array-based LiDARs for autonomous vehicles and low power SM and single-photon-based applications, a larger demand for VCSEL chips is expected between 2025 and 2030 and even beyond.
- To meet future demand, the VCSEL industry has continued to invest considerably in both epitaxy and fabrication capacity. Work has begun to transition VCSEL manufacturing from 150 mm wafers to 200 mm and even 300 mm substrates. Recent work has shown that in some applications, VCSELs can be grown on Ge [1] and even Si wafers [2]. With the modest demands on the photolithography tolerances in VCSELs, leveraging the traditional LED and Si-CMOS

VCSEL Industry: Communication and Sensing, First Edition. Babu Dayal Padullaparthi, Jim A. Tatum and Kenichi Iga.

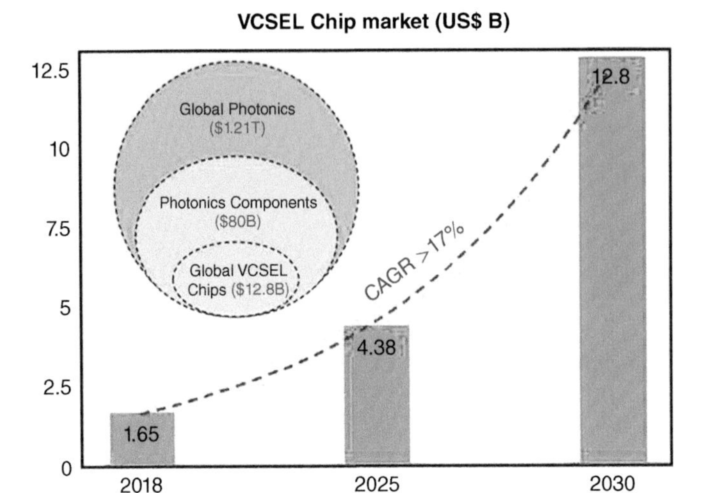

Figure 10.1 Estimated future VCSEL chip markets. *Sources:* Babu Dayal Padullaparthi © Photonic Components DFM Ltd., Global newswire 2020, IPSR-2020, Business Wire (Dublin), and Laser Photonics Conference.

technology infrastructures may be possible [3]. Today there is an overcapacity in the industry, but further adoption of consumer application such as LiDAR could rapidly use that capacity. Wet thermal oxidation is a critical process step, and the industry is working to address that step in larger wafer tools.

10.2 Datacom VCSELs

- The data communications industry has experienced steady growth over the last two decades and will continue to grow. The demand for high-speed Internet access will be driven by existing applications such as video on demand and further penetration of high-speed networks around the world. Today VCSELs are routinely deployed at 25 Gbps, with standards being developed for 50 Gbps and even 100 Gbps in the near future. VCSELs will be a large part of the connectivity within the data centers and high-speed computers of the future.
- High transmission speeds are needed for large-scale infrastructure such as data centers, super computers, and high-performance computing. The race to exaflop computing is on, and low-latency high-speed interconnects are needed to support that effort. Intense research and development activities on higher speed VCSELs is happening today for the networks of tomorrow. The challenges to increase the data rates past 100 Gbps/channel are many, and some revolutionary ideas are needed.

10.3 VCSEL Arrays for 3D Sensing (Short Distance)

- VCSELs and high-density two-dimensional VCSEL arrays with power levels in the few W/mm^2 that are capable of fast rise and fall times (a few ps) are needed for short-distance (~10 m) 3D sensing applications. Examples include, gesture recognition, face recognition, illumination, proximity sensing, autofocus for augmented and virtual reality, and smart home appliances.
- Efficiency improvements are always welcome in consumer electronics to increase the battery lifetime. In addition, increasing the power density and chip size are key drivers to reducing costs.

Smaller chips that may require better heat dissipation are needed for next generation sensors in consumer gadgets.

- 3D sensing technologies that acquire position and depth of an object such as computerized tomography scan and magnetic resonance imaging could use VCSELs as volumetric pixels (voxels). The volumetric 3D display is needed to understand the massive amount of data being analyzed using AI analytics to identify the internal structure of a projected target.
- From COVID-19 experiences, the desire for contactless sensing surrounding an objects is increasing. Toward this end, the development of high-performance components (with long range, low power consumption, increased SNR [reduced interferences], and lower thermal budgets) with cost-effective packaging options (lower die dimensions and higher emitter densities, low-cost optics [DOE/diffusers, reduced parasitics, including drive currents and voltages in driver ICs]) is also increasing.
- Further integration of face ID technology using VCSELs may offer potentially secure solutions for access control in buildings, criminal identification in large crowd gatherings or at public places, smart locks and video doorbells, secure shopping at payment kiosks, and so forth.

10.4 VCSEL Arrays for 3D Sensing and Imaging (Long Distance)

- There is tremendous activity in high power density VCSEL arrays using multi-junction cavities. High power densities (up to a few 100s W/mm^2 of pulse power) and fast rise and fall times (few ps) are possible. These VCSEL emitters in NIR eye-safety wavelengths are needed for long-range (<400 m) 3D imaging applications to meet C-V2X platforms. Examples include, flash and electronic scan LiDARs in autonomous vehicles, drones, and robotics.
- In this direction it is very interesting and exciting to see companies such as Ouster using VCSELs for high-performance digital LiDAR solutions with proprietary SPAD architectures, expanding their broad product portfolios into robotics and automotive markets, merging with Colannade and having an IPO [4, 5]. This point alone underscores that VCSEL are going to stay until 2030 or even beyond.
- In January 2021, Huawei officially released their automotive-grade high-performance LiDARs for intelligent heating and intelligent cleaning in passenger cars [6]. They could be VCSELs/EELs as light sources at NIR/UV range; details are not yet known.
- Further, in January 2021, a breakthrough in flash LiDAR was reported from Sense Photonics for 200 m based on 940 nm VCSEL arrays and 3D global shutter SPADs that can detect targets with 10% reflectance [7]. Thus, VCSELs can a play critical role in replacing EELs, especially in LiDARs with no moving parts for flash and electronic scan projections.

10.5 kW-Level VCSEL Arrays for Industrial and Night Vision

- More investments are needed to mass produce high-power (kW) level VCSEL arrays to make modules and panels with millions of emitters for industry applications. Efforts are needed to integrate photonics and electronics to make VCSEL array panels for several applications such as interior heating, material processing, 3D printing, and additive manufacturing.
- With the projected growth in the VCSEL industry, more technicians, engineers, scientists, and managers are needed. These needs will be acute in the future as VCSEL manufacturing continues to transition to more of a foundry model.

10.6 Single-Mode VCSELs for Communication and Sensing

- Single mode VCSELs will continue to have applications in some of the edge markets described in Chapter 8. Atomic clocks are needed to synchronize 5G networks and further increase the usable bandwidth. Encoders will become more common in a world with ever-increasing robotics.
- Long-Wavelength VCSELs for data communications continues to be a market of interest. The challenges are high, but the rewards are potentially huge.
- Swept wavelength sources will become ubiquitous and will be found in many medical offices in the future, particularly in the area of ophthalmology.

10.7 Quantum Technologies

- Quantum information technology (quantum IT) follows the laws of quantum physics, which allow data to transmit through optical fiber cables or free space. Quantum IT uses superposition states of light in the form of 0s and 1s with multiple combinations called quantum bits, or qubits. Quantum IT based on laser sources and detectors has tremendous applications in telecommunication, sensing and computing, healthcare, transport, oil and gas, finance, and so forth. These technologies are at the early stage of their prototype developments and may create a market in the 2025–2035 time frame that is on par with the supercomputing market ($50 billion) [8].
- Lasers with on-demand short optical pulses are used as light sources to prepare atoms to initiate quantum states to read quantum information. Single or entangled photons with controlled phase, polarization, and angular momentum states generated from lasers play a critical role in quantum communication and sensing. Besides external-cavity lasers, DPSS, DBR, and DFB lasers; QD-based VCSELs also find some interesting space in quantum key distributions (QKD) alongside their classical (digital) communication. Compact and rugged, narrow linewidth, stable output, and high peak powers with fast repetition rate are needed as quantum light sources. Single-mode VCSELs are possible candidates for quantum light sources. QDs embedded in photonic waveguides and N_2 vacancy centers in diamond are another candidate light source for single-photon emission. If suitable light sources are found for quantum applications, this may further open a window of opportunity for their use even after 2030.
- In 2020, there has been a tremendous activity in different parts of the globe with off the-shelf components to teleport information through optical fibers and free space over distances of 10–100s of miles with the goal of creating a quantum Internet, which would be primarily used for secure and sensitive data sharing. Several costly experiments are underway with big giants such as Intel, Apple, Microsoft, and IBM to develop sustainable quantum technology for future secure communication/transactions [9].
- Further, with the advancements in EUV light lithography that made < 5 nm gate length CMOS processes feasible [10], it is possible that future QD- or QW-based lasers with single or entangled photon emission in the form of controlled and uniform 2D arrays will be feasible. If this happens, there will be a tremendous opportunity for semiconductor lasers to open a big window for high-volume manufacturing of quantum light sources.

10.8 Neuromorphic/Neurophotonic Technologies

- Besides the core markets of datacom and consumer sensing & imaging, VCSEL-based systems are finding a great potential for new developments in large-scale intelligent data processing such as neuromorphic computing and sensing [11]. There has been a strong interest in neuromorphic computing for data-intensive processing tasks in brain-inspired artificial neural networks (ANNs) due to VCSELs' unique advantages of ultrafast performance, large bandwidths, low cross talk, and high parallelism. This means, laser-enabled operation of brain-inspired photonic systems with speeds up to nine orders of magnitude faster than biological neurons and—crucially—up to six orders of magnitude faster than electronic approaches, which follow Moore's law [12]. There will be a bright future for photonic ANN hardware architectures based on spiking VCSEL-neurons for ultrafast AI and neuromorphic computing platforms for performing functional tasks such as coincidence detection and pattern recognition at ultrafast sub-nanosecond (~100 ps) signal rates.

10.9 Biomedical/Bio-Photonic Applications

- The controlled low power, low beam divergence, and narrow linewidth of light from VSCELs are attractive in medical applications of imaging, analytics, sensing, and laser surgery (processing and curing treatments). Physicians, veterinarians, and specialists are using laser-based tools for in vivo diagnostic and image systems such as surface, see-through and depth imaging, and oximetry measurements. This also extends to in vitro biomedical analysis in laboratories, clinics, and hospitals in optical microscopy, and analytical systems such as sequencing, cytometry, spectrometry, and bio sensors. Hence the scope of VCSELs use is enormous and ever expanding in bio-photonics.

10.10 New Directions of VCSEL Technologies (as of March 2021)

- **VCSELs on Ge substrates:** Epi-wafer growth foundry IQE has successfully demonstrated 940 nm VCSEL growth and processing on 150 mm Ge substrates as IQGeVCSEL™ [1]. Contrary to large bowing (~100–200 μm) from VCSEL epi-structures grown on traditional GaAs substrates, Ge substrates offer nearly 10x more flatness (bowing ~10–20 μm) and are truly defect free (zero EPD). This breakthrough approach can open a large window for VCSELs to have similar performance on Ge substrates and pushes wafer sizes to 200 mm or even to 300 mm, which will fully utilize advanced planar CMOS processing tools at par with the high-volume LED industry.
- **High-power VCSELs with multiple tunnel junctions:** There is a clear trend for high-power VCSELs with multiple tunnel junctions for use in consumer and automotive LiDARs. It is rather unusual to see VCSELs with slope efficiencies > 3.0 W/A, and even up to 6 W/A, for 3–6 cascaded junctions monolithically grown VCSEL structures. This has led to power densities in excess of 1 kW/mm^2 under short pulse operations and array power conversion efficiencies exceeding 55% [13, 14, 15, 16]. This reflects the maturity of VCSEL technologies that can meet the needs of the consumer (~20 W/mm^2) and automotive industries (> 500 W/mm^2). These super-high performance 2D VCSEL arrays can potentially replace F-P EELs in the NIR band (750-1100 nm) in consumer and automotive applications.
- The ability to make power scaling from 10s of mW/mm^2 to 100s of W/mm^2 by (i) adjusting a number of tunnel junctions with multiple optical cavities and area of emission, (ii) controlling the emission wavelength through epitaxial growth, and (iii) individually addressable emitter zones with different FOV for flash and e-scanning in auto-grade reliable and environmentally robust 940 nm and 905 nm VCSEL transmitters opened tremendous opportunities for structured light, i-TOF, d-TOF, and LiDAR-based sensing/imaging applications [13, 17, 18]. These include both in-cabin (driver monitoring systems, seat occupancy monitoring systems, gesture recognitions, etc.) and outside-car sensing applications, such as long-range LiDARs.

- In particular, the superior performance of 905 nm VCSEL arrays with 6 tunnel junctions with huge power densities of 1500 W/mm^2 and their individually addressable zones almost kicked EELs out of the auto-transmitter race for long-range imaging LiDARs. Further, with high beam quality and ease of monolithic growth of epitaxial structures and leveraging standard high-volume manufacturing tools, these 2D VCSEL arrays are inching toward highly advanced Autonomy of Things (AoT) applications in robotics mobility (shuttles, robotaxis), a massive parcel delivery market ($665 billion by 2030), last-mile delivery market (which includes the Starship Robot, Amazon Scout, the Postmates Serve robot, and the Nuro robot and has a projected value of $33 billion by 2026), AGV for factory automation/warehouse material handling ($3.14 billion by 2026), and so forth [17]. It is also forecasted that robotaxis and shuttles may exceed 1M units by 2030 and logistics applications using more than 175K units by 2031 [19].
- Despite material challenges in development of VCSELs in DUV, UV, and visible (blue and green) wavelength bands (280–560 nm), several advanced performances of GaN-based semiconductor lasers and receivers have been reported for VCSELs [20, 21, 22], and high reflective nanopore mirrors [23], polarization behavior [24], 2D arrays of PDs [25] with moderate to high power densities that have emerging high end applications in bio-sensing (water purification), high resolution (3D) printing, retinal scanning displays for AR/VR, adaptive headlights in automotive cars, underwater visible communication, atomic clocks, single-photon light sources, and so forth.

10.11 Concluding Remarks

Overall, from Figure 10.2, the digital era has rapidly replaced analog components especially in the consumer mobile and automotive sectors. It is also evident that the current VCSEL industry is transitioning from rapid growth to full digital maturity and attracting further markets in new

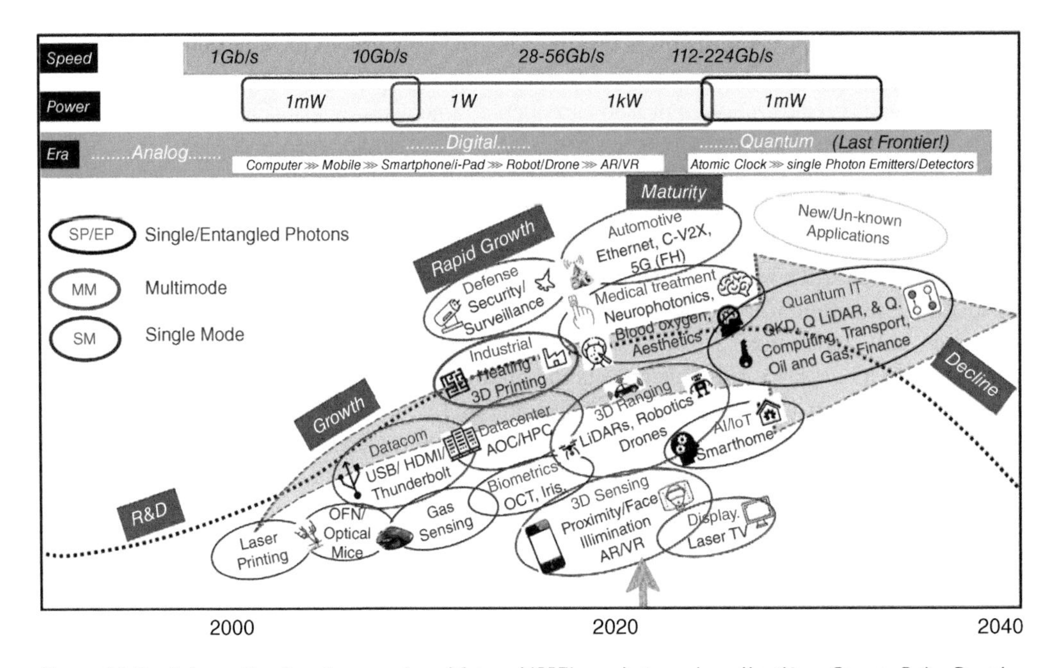

Figure 10.2 Schematic of past, present, and future VCSEL markets and applications. *Source:* Babu Dayal Padullaparthi © Photonic Components DFM Ltd.

applications with higher speeds and a variety of powers. If VCSEL arrays are standardized in applications such as long-range LiDARs and VCSEL-based single- and entangled-photon light sources in quantum applications, the VCSEL industry will continue to expand and occupy a space in the history of semiconductor lasers at par with the transistor or the solar cell!

References

1 A. Johnson, A. Joel, A. Clark, D. Pearce, M. Geen, W. Wang, R. Pelzel, S W Lim., "High performance 940nm VCSELs on large area germanium substrates: the ideal substrate for volume manufacture," SPIE Photonics West 11704–1 (2021) (Digital Forum On-demand Video Presentation).

2 H. K. Sahoo, T. Ansbæk, L. Ottaviano, E. Semenova, F. Zubov, O. Hansen, and K. Yvind., "Tunable MEMS VCSEL on silicon substrate," *IEEE J. Sel. Top. Quantum Electron.* **25**, 1700707 (2019).

3 2020 Integrated Photonic Systems Roadmap - International (IPSR-I) April 2020 https://photonicsmanufacturing.org/events/2020-ipsr-i-integrated-systems-roadmap-0.

4 https://www.businesswire.com/news/home/20201222005187/en/Ouster-a-Leading-Provider-of-High-Performance-Digital-Lidar-Sensors-to-Combine-With-Colonnade-Acquisition-Corp.-to-Accelerate-Digital-Lidar-Adoption-in-Industrial-Smart-Infrastructure-Robotics-and-Automotive-Markets.

5 https://data.ouster.io/downloads/Ouster-investor-presentation_12222020.pdf?__hstc=34987006.4fdcc41d3d47f45e52aff24bfcdc6fe7.1610105712531.1610105712531.1610105712531.1&__hssc=34987006.1.1610105712532&__hsfp=3240560975.

6 https://consumer.huawei.com/ae-en/community/details/TECHNOLOGYHuawei-releases-a-new-range-of-advanced-lidar-smart-car-solutions/topicId_132785.

7 https://www.reuters.com/article/us-sense-photonics-lidar-autonomous/sense-photonics-jumps-into-self-driving-fray-with-new-sensor-technology-idUSKBN29D2NT.

8 https://www.osapublishing.org/abstract.cfm?uri=OIDA-2020-3.

9 https://s3.i-micronews.com/uploads/2020/01/YDR20062-Quantum-Technologies-2020-Yole-Développement-Sample.pdf.

10 https://www.iws.fraunhofer.de/en/pressandmedia/news/2020-11-24_news_euv_zukunftspreis.html.

11 https://www.i-micronews.com/products/neuromorphic-sensing-and-computing-2019.

12 J. Robertson, M. Hejda, J. Bueno et al., "Ultrafast optical integration and pattern classification for neuromorphic photonics based on spiking VCSEL neurons," *Nat. Sci. Rep. (*2020*)* **10**:6098 doi:10.1038/s41598-020-62945-5.

13 G. Zhao, J. Yang, E. Hegblom, A. Barve, B. Kesler, M. Tashima, Z. Bian, S. Xie, A. Robit, M. Peters, J. Skidmore., "Multi-junction VCSEL arrays with high performance and reliability for mobile and automotive 3D sensing applications," SPIE Photonics West 11704–7 (2021) (Digital Forum On-demand Video Presentation).

14 J-F. Seurin, G. Xu, A. Miglo, R. V. Leeuwen, H. Othman, S. Okur, Y. Zhang, A. Amtout, H. Bian, H. Yuan, E. M. Shaw., "Multi-junction vertical-cavity surface-emitting lasers in the 800-1100nm wavelength range," SPIE Photonics West 11704–8 (2021) (Digital Forum On-demand Video Presentation).

15 Y. Xiao, S. Y. Tan, J. Wang, H. Liu, "High efficiency and high power density VCSEL chip for 3D sensing," SPIE Photonics West 11704–4 (2021) (Digital Forum On-demand Video Presentation).

16 M. Dummer, K. Johnson, S. Rothwell, K. Tatah, M. K. Hibbs-Brenner., "The role of VCSELs in 3D sensing and LiDAR," SPIE Photonics West 11692–9 (2021) (Digital Forum On-demand Video Presentation).

17 A. Wong, "VCSELs and Their Use in Emerging Applications," Industry Event on Applications of sensing and imaging solutions, SPIE Photonics West, 11 March 2021, (Digital Forum On-demand Live Presentation).

18 J. S. Eng, "VCSEL-based 3D Sensing: Market and Technology," Industry Event on Applications of sensing and imaging solutions, SPIE Photonics West, 11 March 2021, (Digital Forum On-demand Live Presentation).

19 P. Cambou, "Sensors for robotic mobility and good transportation the upcoming market disruption," Industry Event on Applications of sensing and imaging solutions, SPIE Photonics West, 11 March 2021, (Digital Forum On-demand Live Presentation).

20 F. Hjort, J. Enslin, M. Cobet, M. A. Bergmann, J. Ciers, G. Cardinali, N. Prokop, T. Kolbe, F. Nippert, M. R. Wagner, J. S. Gustavsson, T. Wernicke, M. Kneissl, Haglund, "Advances in ultraviolet-emitting vertical-cavity surface-emitting lasers," SPIE Photonics West 11686–2 (2021) (Digital Forum On-demand Video Presentation).

21 T. Takeuchi, S. Kamiyama, M. Iwaya, I. Akasaki, "Developments of GaN-based VCSELs with epitaxially grown DBRs," SPIE Photonics West 11686–3 (2021) (Digital Forum On-demand Video Presentation).

22 K. Terao, H. Nagai, D. Morita, S. Masui, T. Yanamoto, S. Nagahama., "Blue and green GaN-based vertical-cavity surface-emitting lasers with AlInN/GaN DBR," SPIE Photonics West 11686–1 (2021) (Digital Forum On-demand Video Presentation).

23 J-H. Kang, R. T. Elafandy, J. Han., "Development of highly reflective mirrors for III-nitrides from green to UV," Proc. SPIE Photonics West 11686–5 (2021) (Digital Forum On-demand Video Presentation).

24 T-C. Lu, T-C. Chang, L-R. Chen, K-B. Hong, S-Y. Kuo., "Polarization and transverse mode control in GaN-based vertical-cavity surface-emitting lasers," SPIE Photonics West 11686–4 (2021) (Digital Forum On-demand Video Presentation).

25 M. B. Noodeh, M. Cho, Z. Xu, H. Jeong, N. Otte, S-C Shen, T. Detchprohm, A. K. Sood, J. W. Zeller, P. Ghuman, S. Babu, R. D. Dupuis., "Demonstration of uniform 6x6 GaN p-i-n UV avalanche photodiode arrays," SPIE Photonics West 11686–27 (2021) (Digital Forum On-demand Video Presentation).

Appendix A

VCSELs Design Engineering

Babu Dayal Padullaparthi

A.1 Background

There are numerous reports on the design of VCSELs for various device/product applications [1, 2, 3]. This appendix aims to highlight some of the key design aspects of oxide-confined MM VCSELs related to high-volume production in core markets of communication and sensing applications described in Chapter 3.

A.1.1 Key Stages of VCSEL Mass Production

VCSEL production has several stages, as described in the Figure A.1. The foremost part is epi-layer design, the second part is the epi-layer growth, the third part is wafer processes, the fourth part is testing and characterization, and the fifth part is reliability and chip qualification. All these aspects will be systematically explained in Appendices A, B, C, D, and E respectively.

A.1.2 Basic Design Elements

Figure A.2a shows a generic top-emitting, oxide-confined VCSEL structure for mass production. It consists of a substrate, bottom DBR, active optical cavity (active region [AR], or gain medium), oxide window (index guiding for current confinement), top-DBR, and top surface. Optimization of this structure requires detailed understanding of the optical, electrical, thermal, and mechanical properties of the material system and the inevitable design trade-offs.

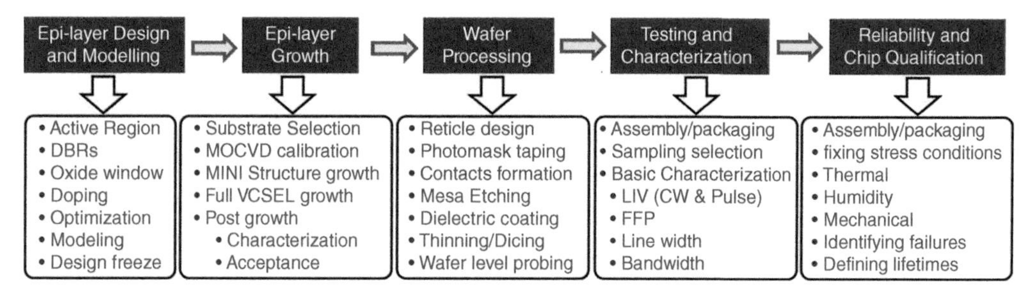

Figure A.1 Key stages of VCSEL development for mass production. *Source:* Babu Dayal Padullaparthi © Photonic Components DFM Ltd.

VCSEL Industry: Communication and Sensing, First Edition. Babu Dayal Padullaparthi, Jim A. Tatum and Kenichi Iga.

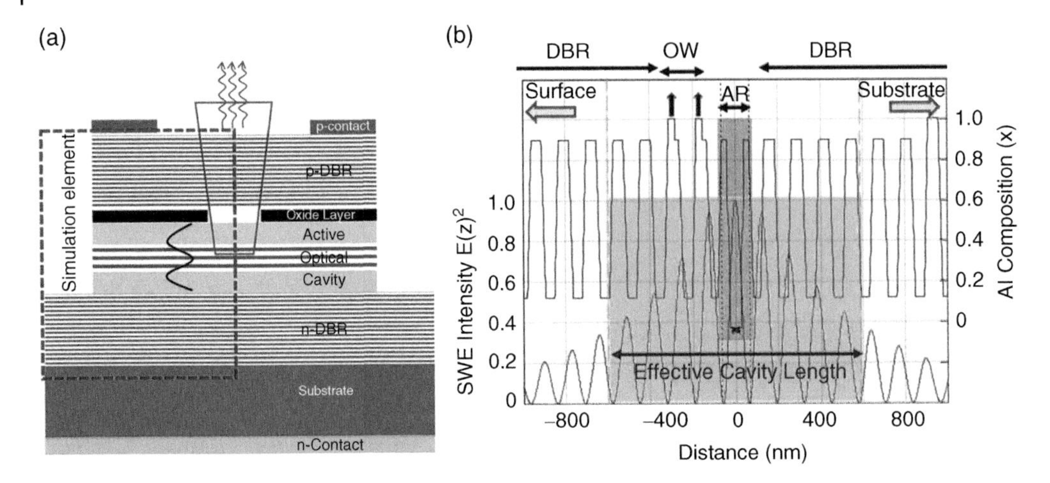

Figure A.2 (a) A generic top-emitting, oxide-confined VCSEL structure for mass production and (b) a schematic of Mathcad-simulated dual oxide (upward black arrows) VCSEL epi-layer structure showing Al(x) composition overlapped with a standing-wave electric field. *Source:* Babu Dayal Padullaparthi © Photonic Components DFM Ltd.

A.1.2.1 Active Region (Optical Cavity)

As described in Sections 2.3 and 2.4, the AR consists of low-bandgap and high refractive index materials with thickness of a few nanometers (also called quantum wells, or QWs) sandwiched by higher-bandgap and low refractive index materials with thickness of a few nanometers (also called barrier layers, or barriers). These QWs and barrier materials form band (energy) discontinuities and confine the injected carriers for recombination process to generate light by following the laws of quantum physics. As QWs are light-emitting or gain materials, the material gain (α) is a figure of merit for this AR. The photoluminescence (PL) linewidth of MQW is another key parameter that reflects the quality of epitaxial wafer to be described in Appendix B. Adjacent to both (substrate [bottom] and surface [top]) sides of barrier materials, a separate confinement heterostructure (SCH) layer is defined to confine injection carriers and control the optical field. Often this SCH is made with a graded index layer called GRIN-SCH and offers reduced resistance to carrier flow for electrical and optical confinement. All these elements are shown in Figure A.3a. The distance between surface and substrate sides of GRIN-SCH defines optical cavity, and the cavity length L_c is often an integer multiple of half a wavelength (λ), $Lc = \frac{m\lambda}{2}$ where m = 1,2,3... and λ is emission wavelength. In practice $m = 1$ or 2 are the most common values. A half-wave cavity may have increased mode overlap with the gain, as shown in Section A.1.4.1.2. One important aspect of the AR is that the PL wavelength shifts with 0.3 nm/K, and this is different than the wavelength shift of the optical cavity of about 0.06 nm/K. This difference requires the gain peak to be offset from the cavity resonance for wide temperature operation. When the gain peak and cavity wavelength are aligned, the threshold current is generally minimized. This behavior leads to the parabolic dependence of threshold current on temperature shown in inset of Figure A.3 and Section 2.10.

A.1.2.2 Distributed Bragg Reflectors (DBRs) and GRIN Layers

In VCSELs, distributed Bragg reflector (DBR) structures are needed to provide feedback to the AR and stimulate lasing activity, as described in Section A.1.2.1. The DBR together with the AR function as an optical resonator with a high-reflectivity band, or *stopband*, that is approximately

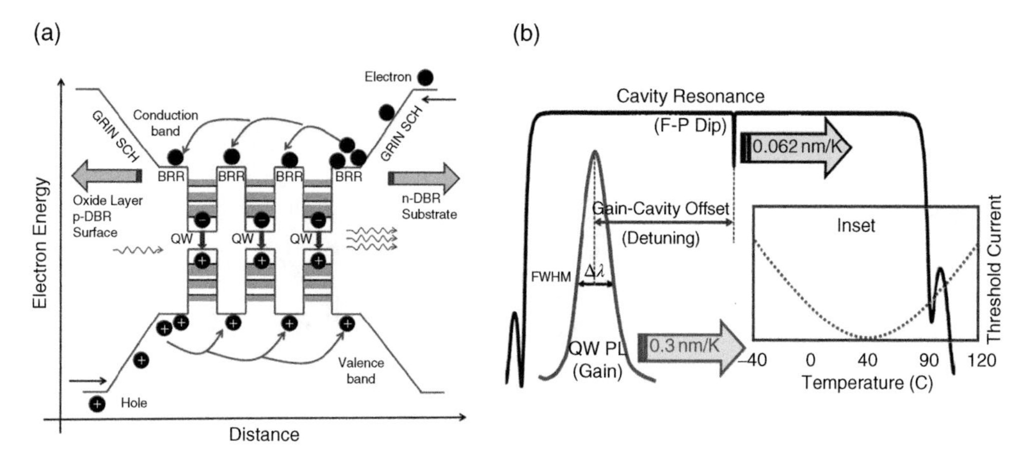

Figure A.3 (a) Basic elements in MQW-based active region. (b) Example of photoluminescence spectrum and cavity resonance (F-P dip) in a VCSEL structure. *Source:* Babu Dayal Padullaparthi © Photonic Components DFM Ltd.

100 nm wide. Light generated from the AR is reflected by the DBRs back to the active region. Lasing is achieved when the round trip phase of the optical field is a multiple of 2π, and there is sufficient gain to overcome the losses. This results in an emission wavelength at the Fabry-Pérot (F-P) resonance (dip) in the reflectivity spectrum, as shown in Figure A.3b. Maintaining PL and F-P dip tolerance and uniformity across epitaxial wafers are major challenges for VCSEL growth, as described in Appendix B and in Section 3.2.3. It is common to use a thin (10–20 nm) graded composition between the high and low index materials of the DBR. The graded composition is used to reduce the electrical resistance of the mirror while maintaining acceptable optical performance [3].

A.1.2.3 Oxide Window and Current Confinement

The size and location of the oxide layer in a VCSEL strongly influences all of the VCSEL operating characteristics. The oxide layer confines the electrical carriers and the optical mode and thus affects the threshold current, slope efficiency, resistance, spectral width, transverse mode structure, and so forth. Typically, $Al_xGa_{1-x}As$ with $0.96 < x < 1$ is used to make the oxide layer. The oxidation rate is strongly dependent on the Al mole fraction, and there can be >10x difference in oxidation rate just in a few percent change in mole fraction. As described in Section 3.2.2.4, the thickness and composition of Al(x) in $Al_xGa_{1-x}As$ needs to be carefully chosen to maintain a stable and repeatable wet thermal oxidation process. The thickness and Al composition of layers below and above the oxide layer also play a significant role in forming the tip of the oxide layer. All these layers constitute an oxide window, as seen in Figure A.4a, that must be carefully designed to yield good device performance and reliability. Consider the VCSEL with the standing-wave electric field (SWE) shown in Figure 2.8. To minimize the optical scattering loss due to the oxidation layer, it is generally placed at a field minimum (node) of the optical standing wave in the VCSEL cavity. Other designs may place the oxide layer at a peak (antinode) of the standing wave. Examples of an oxide window (OW), anti-node, and node positions of SWE for a typical single oxide layer are shown in Figure A.4. The position and shape of the oxide layer strongly affects the optical emission profile through effective index guiding of the SWE [4] and strongly influence the key characteristics of VCSELs such as spectral linewidth and far-field, to be described in Section A.1.4.1.5.

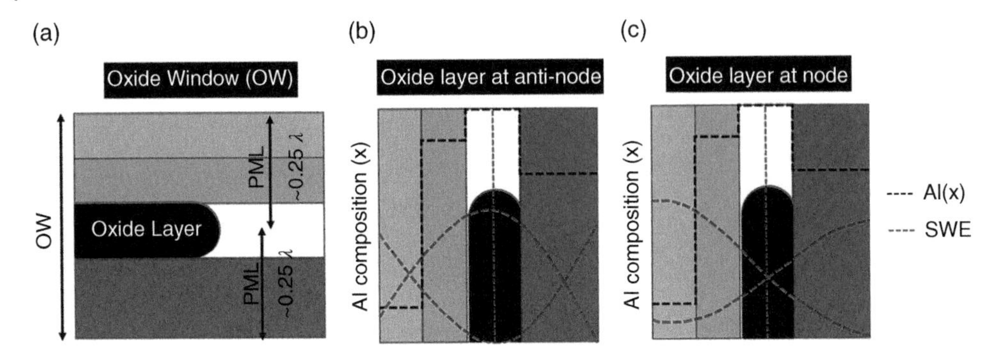

Figure A.4 Structure of (a) oxide window (OW) and tilted OW at (b) node and (c) anti-node. *Source:* Babu Dayal Padullaparthi © Photonic Components DFM Ltd.

A.1.2.4 Heat Management and Thermal Limitations

Thermal management is of paramount importance in both electronic and photonic devices to deliver stable device performance and reliability. In VCSELs most of the heat is generated in the DBR layers as Joule heating, a result of electrical resistance [5]. The heat flow from a VCSEL goes through the DBR mirror structure (Figure 2.1), and the trade-offs for electrical, optical, and thermal considerations must be considered in the DBR design. For an $Al_xGa_{1-x}As$ system, the thermal conductivity varies nearly an order of magnitude over composition, as shown in Figure A.5. The selection of DBR mirror materials depends on emission wavelength. The lower end of the Al composition is determined by optical absorption, and the upper end of the Al composition is determined by the amount of inherent oxidation allowed in the layers adjacent to the intended oxidation layer. For a standard 850 nm datacom VCSEL, $0.15< x <0.9$ can be used without compromising the optical losses. For longer wavelengths, the lower end of the x composition can be reduced, and for wavelengths greater than 900 nm, it may be possible to utilize GaAs. The DBR mirror does not have to be a single design in an epitaxial structure, and it is common to use different compositions away from the oxidation layer to maximize thermal conductivity. For example, in

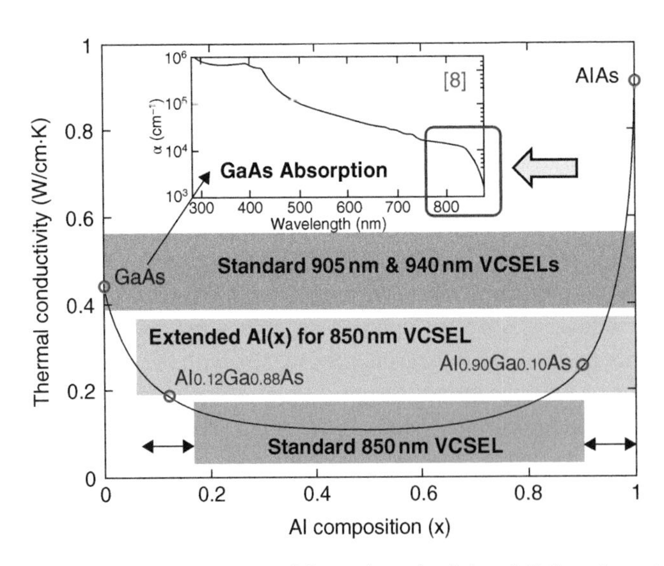

Figure A.5 The concepts of thermal conductivity of $Al_xGa_{1-x}As$ systems and increased heat dissipation using integral multiple of AlAs stacks were used. *Source:* Modified from [7]. Inset: Absorption coefficient of GaAs. *Source:* Reproduced by courtesy of the Electromagnetics Academy [8].

a 940 nm VCSEL, a combination of (i) Al(x) from 0.0 to 0.92 for top-DBRs and (ii) Al(x) from 0.0 to 1.0 for bottom-DBRs (called binary DBRs) can be used for efficient heat dissipation. In most of the cases the heat is dissipated downward to the substrate and out through a metal heat sink. Some improvement in thermal performance may be realized by removing heat from the top surface as well through the use of a thick GaAs window as part of the DBR [6]. Heat removal may be further enhanced with thick metal coatings on the top surface.

A.1.3 Device Modeling and Commercial Software

Modeling of VCSEL layer structures is needed to determine device operation and performance prior to committing the resources for epitaxial growth and device fabrication. VCSEL modeling can be classified into three types: (i) analytical, (ii) semi-analytical, and (iii) numerical. Analytical modeling requires extensive mathematical manipulations with necessary realistic approximations, but these are limited in the range of applications. In semi-analytical modeling, the solutions are accurate within the limit of approximations and often use spectral index and free-space radiation modes to solve the problem. Both of these methods involve approximations to the governing equations to improve computational speed at the expense of ultimate accuracy. In numerical modeling, variables are discretized to represent them on a computer with limited simplifying assumptions, and the electric fields are represented at the computation nodes to confirm Maxwell's equations. Numerical modeling doesn't generally yield exact solutions but converges toward them with increasing levels of discretization. The discretization (mesh) nature of numerical modeling of a section or slice of the system/device is far more conducive to the transfer of information between the various physical models than attempting to solve the coupled optical, electrical, and thermal problem of VCSEL design with single-step analytical or semi-analytical models.

Several commercial software tools are available to model both active and passive semiconductor optoelectronic devices. Depending on the discretization of the models, numerical methods can be further classified into two types, the finite difference (FDM/FDTD) and finite element (FEM), as shown in Figure A.6a and b. As the density of the mesh is high, FEM methods are generally more accurate than FDTD methods. Both solve appropriate differential equations for the AR and bulk semiconductor lasers and include the electrical, optical, and thermal behavior of the semiconductor material in 2D and 3D cross sections. For semiconductor lasers, models must simultaneously solve

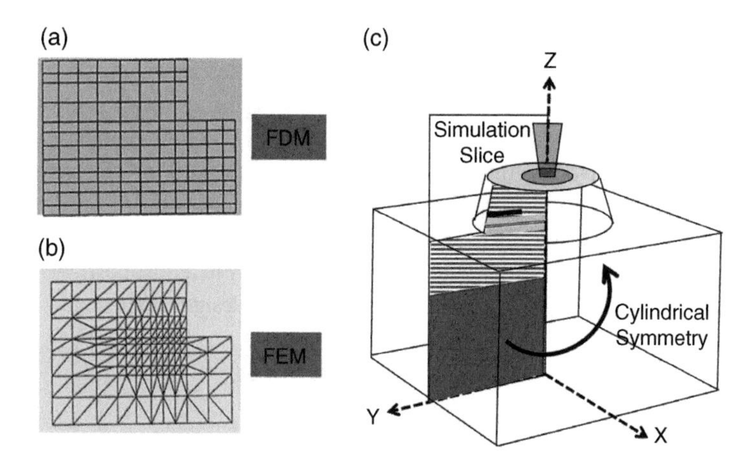

Figure A.6 Discretization of device into (a) finite difference (FDM/FDTD) and (b) finite elements (FEM) models and (c) creation of a VCSEL slice for modeling. *Source:* Babu Dayal Padullaparthi © Photonic Components DFM Ltd.

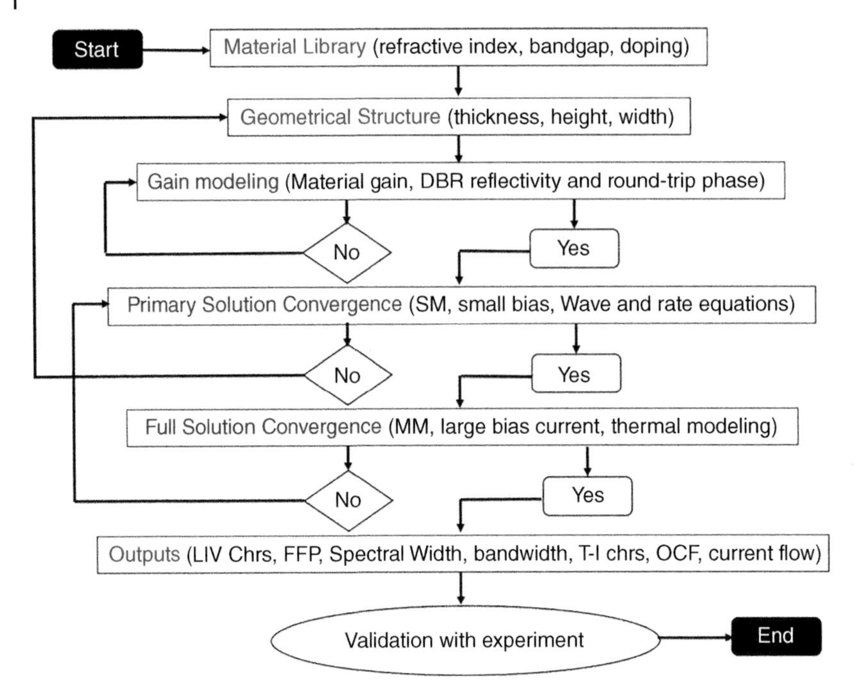

Figure A.7 Flowchart of VCSEL device in numerical modeling. *Source:* Babu Dayal Padullaparthi © Photonic Components DFM Ltd.

Poisson's equations with current continuity, drift-diffusion of carriers, optical gain models, Maxwell's and Schrödinger equations for the electromagnetic field, rate equations to account for the interaction of the carriers and the optical mode, and finally thermal condition (junction temperature evolution) [9, 10]. A flowchart for modeling the complex physics of a VCSEL structure is shown Figure A.7.

An example of discretization of a cylindrically symmetric VCSEL structure is shown in Figure A.6c. Once the fields are calculated for one section (slice) of VCSEL, the rotational symmetry will automatically calculate the field properties of a fully 3D structure of the VCSEL. Commercial software to model VCSELs is available from multiple vendors including Synopsys/Rsoft (LaserMod) [11, 12], Photon Design (FIMMPROP), Optiwave, Lumerical (MQW), COMSOL (MultiPhysics) [13], and CrossLight-PICS 3D/TCAD [14]. A majority are FDTD based, a few are FEM based, and some only offer passive photonic and active Si-photonic component simulations. Still many designers choose to design VCSELs with a more specialized set of tools for each of the problems instead of using a fully integrated model. In this case, the design process might begin with defining the QW active region, designing the DBR mirrors and resonant cavity, determining the placement of the oxide layer, and calculating the total mirror loss to determine the required optical gain for lasing. Electrical impedance and thermal properties can also be estimated. Regardless of the chosen design method, the results should be compared to actual device performance, and it is often necessary to adjust portions of the models to match device performance. With all the capability for modeling, designers are generally happy with less than 10% error in predicting device performance.

A.1.4 High-Performance Designs for Communication and Sensing

It has been evident from the previous sections that VCSEL design is a complex interplay of electrical, optical, thermal, structural, and mechanical properties of materials, as shown in Figure A.8. An in-depth understanding of semiconductor, dielectric and metallic materials,

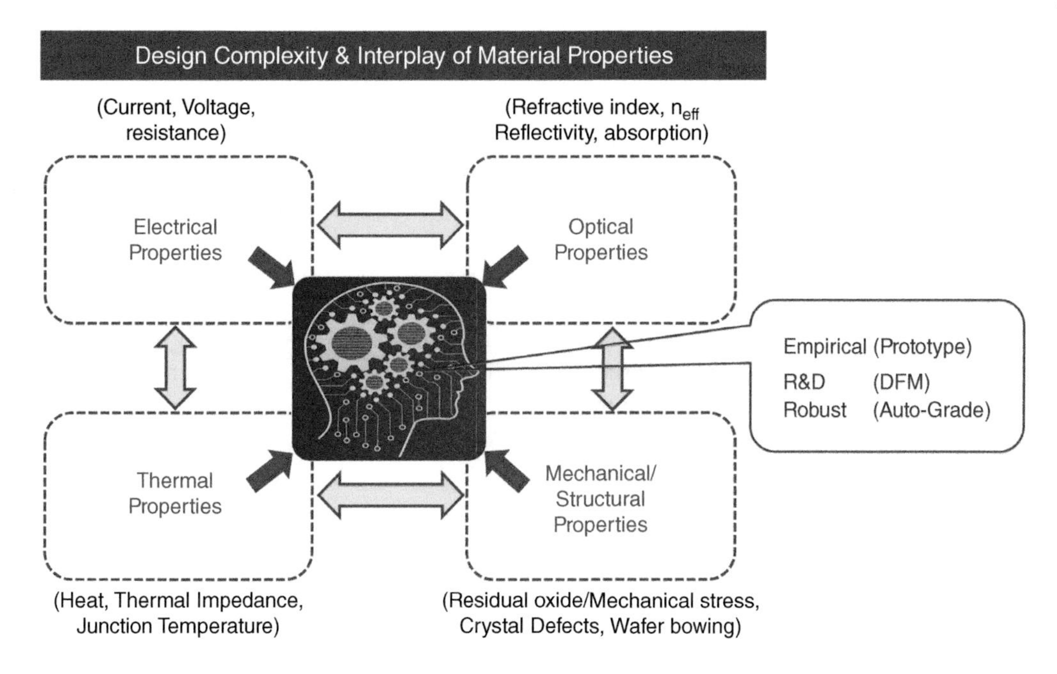

Figure A.8 Interplay of material properties with strength of simulations for various products. *Source:* Babu Dayal Padullaparthi © Components DFM Ltd.

fundamentals of optics, and electronics and its associated physics is necessary to simulate an optimized and robust VCSEL design. The levels of VCSEL design (empirical, R&D, or robust) for various applications indicates its suitability for product portfolios (protype, DFM, or auto-grade), respectively, as shown in callout of Figure A.8. The following sections briefly discuss the key design aspects of high-speed and high-power VCSELs for commercial product applications.

A.1.4.1 High-Speed Designs for 850 nm Datacom

The modulation speed of the VCSEL structure is governed by frequency response through its small-signal (S21) bandwidth, as presented in Section 2.9. A typical VCSEL bandwidth requirement is at least 60% of the modulation speed. For example, a 25 Gb/s VCSEL must have at least 15 GHz bandwidth and ideally more than 20 GHz bandwidth. Lower-bandwidth VCSELs may require electronic equalization, as described in Section 4.6.3. S21 response is dependent on several factors including the differential gain (dg/dN), optical confinement factor (Γ), photon lifetime (τ_p), parasitics in electrical equivalent circuits, and thermal limitations. All these parameters can be optimized to increase the bandwidth and hence speed. However, there are a few significant speed and thermal trade-offs that seriously affect VCSELs' performance. Let's see all these cases one by one with the relaxation oscillation function (f_{RO}) with its dependent parameters.

$$f_{Ro} \propto \sqrt{\frac{\eta_i . c . \frac{dg}{dN}\left(I - I_{th}\right)}{q . \frac{V_{ac}}{\Gamma}}} \quad f_{Ro} \propto \sqrt{\frac{1}{\tau_p}} \quad \begin{array}{l} \tau_p \propto \dfrac{1}{\alpha_m} \\ \alpha_m \propto \dfrac{1}{2d} \ln\left[\dfrac{1}{R_1 R_2}\right] \end{array} \tag{A.1}$$

A.1.4.1.1 Differential Gain (dg/dN)

The QW differential gain can be significantly improved over GaAs by using lattice strained $In_xGa_{1-x}As$ QWs ($x = 0.1$) [15]. The introduction of In (strain) in GaAs QWs creates reduction of density of states near the valance band and increases the separation of the light and heavy holes to nearly 200 mV. The result is a nearly 2x increase in the differential gain for five $In_xGa_{1-x}As$ QWs against three GaAs QWs. Systematic calculations with In (0, 10, 12, 15%) and experimental results are reported in [16].

A.1.4.1.2 Optical Confinement Factor (Γ)

In a typical 1.0λ cavity structure, the optical confinement factor (Γ) is ~2.9%, and it is further increased to ~4.4% when a 0.5λ cavity is used. The short cavity concept and its impact on speed can be seen in [17] and has a direct impact on the laser speed. The increase in the confinement factor for 1.0λ and 0.5λ optical cavities are shown in Figure A.9.

A.1.4.1.3 Photon Lifetime (τ_p)

From Eq. (A.1), reducing the reflectivity of DBRs, one can reduce the photon lifetime (τ_p) and thereby increase f_{RO}. By introducing a concept of surface relief (SR) in semiconductors [18], attempts were made to increase the bandwidth of VCSELs through the reduction of τ_p from 6.4 to 1.2 ps. The SR technology offers increased bandwidth and narrowed spectral linewidth at the expense of increased threshold currents [19]. The surface of VCSELs is grown at either in-phase or anti-phase thickness, and a shallow SR is etched to reduce or enhance the reflectivity of the top DBR. An example of SR and its impact on reducing photon lifetime are shown in Figure A.10a and b.

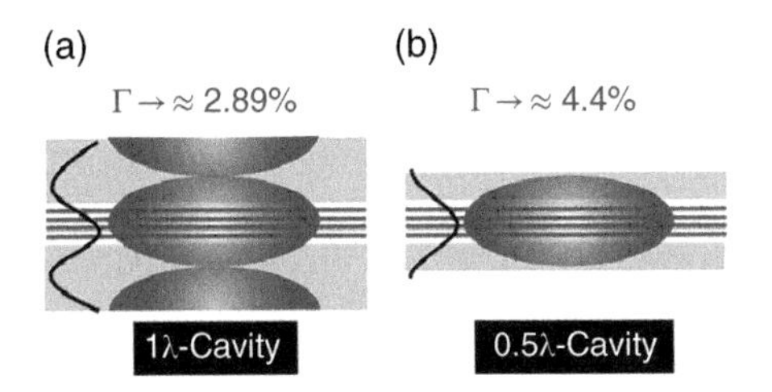

Figure A.9 Schematics of active regions with (a) 1.0λ and (b) 0.5λ cavity lengths. *Source:* Babu Dayal Padullaparthi © Photonic Components DFM Ltd.

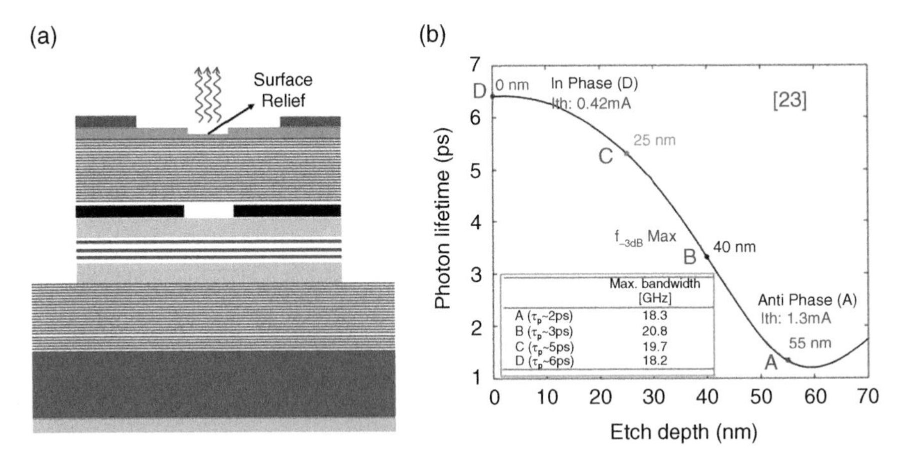

Figure A.10 (a) A schematic of surface relief integrated VCSEL. *Source:* Babu Dayal Padullaparthi © Photonic Components DFM Ltd. (b) A schematic of reduction of photon lifetime on shallow etch in emission window from in-phase to anti-phase direction. *Source:* Adapted from IEEE [23].

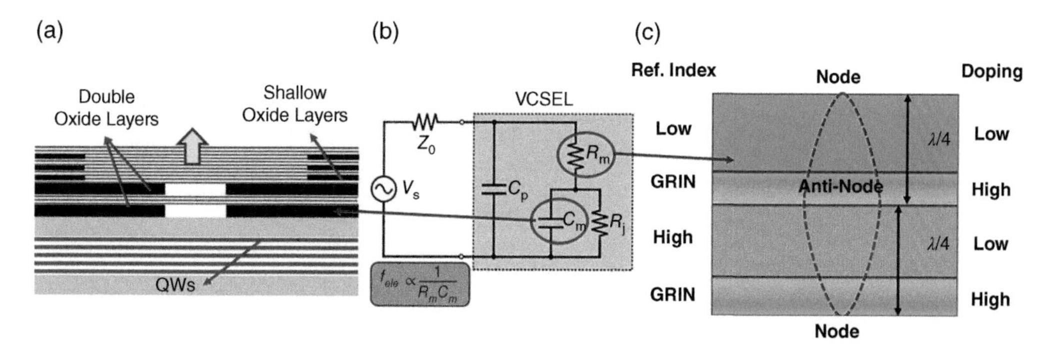

Figure A.11 Schematics of (b) electrical equivalent, (a) dual and shallow oxide layers, and (c) modulation doping scheme for a VCSEL structure. *Source:* Babu Dayal Padullaparthi © Photonic Components DFM Ltd.

One of the interesting aspects is that the etch depth of SR controls τ_p while its diameter controls the mode patterns and acts as a mode or tunable optical phase filter that improves the SNR of the laser [20]. In another approach, the systematic τ_p reduction from 7.1 to 2.4 ps through planar dielectric (SiN) layers resulted in devices with higher bandwidth and reduced damping [21, 22].

A.1.4.1.4 Electrical Parasitics (C_m and R_m)

In VCSELs multiple elements act as source of parasitics (capacitance, resistance, and inductance) and must be reduced to enhance frequency response through their equivalent circuits, as shown in Figure A.11b. The first one is mesa (oxide) capacitance, Cm, which is reduced by decreasing of the mesa size and is even further reduced by introduction of double and/or shallow oxide layers, as shown in Figure A.11a. Experimental studies show a significant increase in f_{RO} by decreasing C_m [24]. The capacitance of the device is further reduced by placing p-pad metal under the thick, low dielectric polymers such as BCB [25].

Similarly, the mesa resistance (Rm) can also be significantly reduced by judicious doping of the DBR. In particular, the doping can be increased at the antinodes of the standing wave of the electric filed without significant impact on the optical losses [3]. The doping concentrations are different for p- and n-DBRs and are carefully designed with relevant thickness of compositional graded layers as shown in Figure A.11c. This scheme is called modulation doping, and it is effective in reducing the mirror resistance and hence reduces the operating voltage. Modulation doping offers lower power consumption and increases the overall power conversion. Lastly, inductance can be reduced by using short wires for bonding, or possibly using flip-chip-bonded VCSELs. Details were also discussed in Section 4.6.1.

A.1.4.1.5 Narrow Linewidth and Low Noise Designs

For datacom applications, narrow linewidth and low beam divergence VCSELs are needed for long-distance (from 100 to 300 m or beyond) transmission of light in optical fibers with reduced MPN at the receiver. Similarly, for high-power applications, controlled beam quality (far-field) characteristics are needed for projecting the scene with a range of FOV, especially for consumer/automotive applications in a wide temperature range (−40 to 90/125°C). All these unique and key characteristics can be simultaneously obtained with use of the concept of Δn_{eff} in the oxide window of the VCSEL structure, as shown in Figure A.4a. The Δn_{eff} concept, PSC, and change of linewidth and beam divergence are shown in Figure A.12a–d. In the calculation of Δn_{eff}, a fully oxidized layer (non-crystalline Al_2O_3 with refractive index n = 1.6) is considered, as mentioned in Section 3.2.2.4. Here the phase-matching layers (PML) represent the layers above and below AR

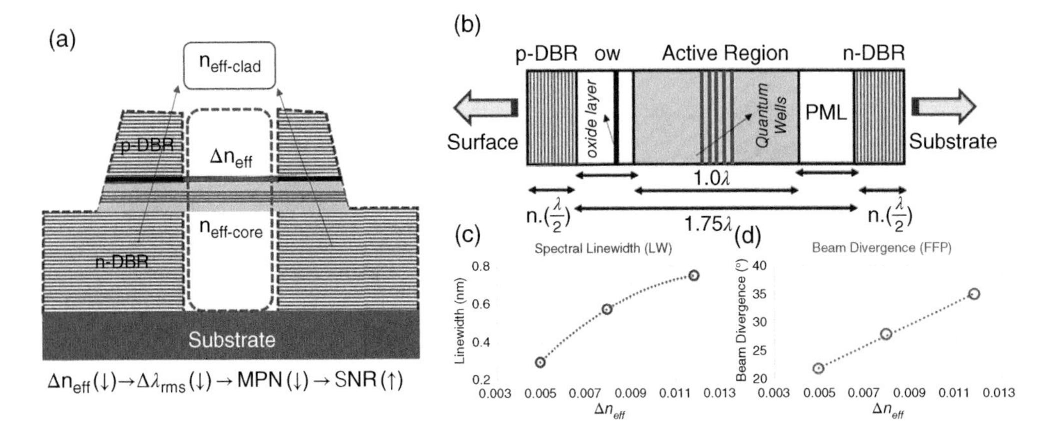

Figure A.12 (a) Concept of n_{eff}. (b) Phase conditions and experimental variation of (c) linewidth and (d) beam divergence in an optimized VCSEL structure. *Source:* Babu Dayal Padullaparthi © Photonic Components DFM Ltd.

with adjusted optical thicknesses to their nearest node/anti-nodes of SWE, as described in Section A.1.2.3. By using the optimized phase-shifted cavity (PSC) among OW, AR and phase-matched layers (PML) with reduced Δn_{eff}, narrow linewidth and low beam divergence can simultaneously be realized [26]. The reduced linewidth with shortened intermodal separations can create low MPN for long distance transmission [27]. The increase of Δn_{eff} also results in wide beam divergence and hence allows to project scenes with wide FOV through diffusers [28].

A.1.4.1.6 Junction Temperature

As described in Section A.1.2.4, thermal management plays a critical role in overall VCSEL performance and reliability. Proper heat sinking designs must be implemented in the epitaxial layer structure of the VCSEL to reduce the junction temperature (T_{jn}). T_{jn} and its measurement are further discussed in Section E.2. All the schemes mentioned in Section A.1.2.4 work well for reduction of T_{jn}. For example, increasing the top GaAs layer of a 940 nm VCSEL from λ/4 to 5λ4 decreases the junction temperature by more than 10°C at a nominal reliability stress condition of 12 mA. The challenge is the growth of thick (120–180 nm) heat sink layers (most likely GaAs) as surface termination layers at 850 nm with high absorption coefficient [inset of Figure A.5]. The result of these thick top heat sink structures is that they can significantly push power saturation (rollover) currents to much higher values providing a better margin for device operation and less power consumption (Figure A.13a). In 2D arrays of VCSELs, the device pitch and driving conditions can also influence the junction temperature. For active areas separated by more than 50 μm, the thermal cross-talk is minimal. Introduction of external high thermal conductivity heat sinks made from AlN, SiC, and diamond can help reduce the junction temperature too.

A.1.4.2 High-Power 940 nm VCSEL Design

The design of a VCSEL for high-power emission requires a different set of optimizations. Typically, the most important factor is the power conversion efficiency (PCE), which simultaneously maximizes the power extraction and minimizes the electrical power consumption. Optimization of the PCE also depends on the choice of driving conditions. In some cases, the VCSEL can be operated at the maximum PCE as is the case for structured light devices in 3D sensing. In LiDAR applications, the VCSEL is often driven with short electrical pulses and operated far from peak

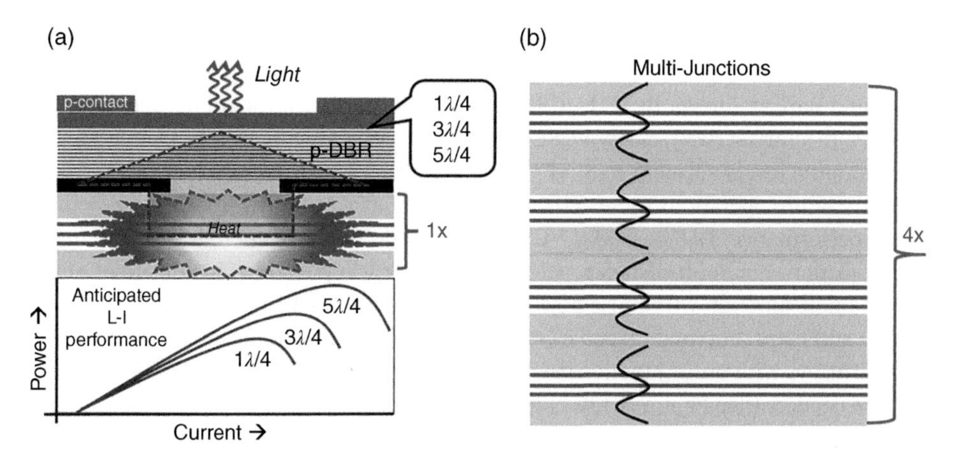

Figure A.13 (a) LIV characteristics of VCSELs with different top heat sink structures. (b) Schematic of multi-junction AR in VCSELs. *Source:* Babu Dayal Padullaparthi © Photonic Components DFM Ltd.

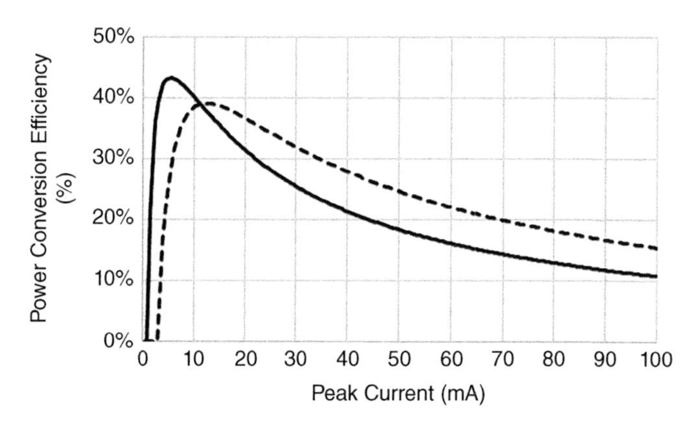

Figure A.14 Design (dotted line) and operation (continuous line) PCE characteristics of VCSELs as a function of peak current. *Source:* Courtesy of Jim A Tatum, Texas, USA (Copyright reserved).

PCE. Consider the two VCSEL design characteristics shown in Figure A.14 where the PCE is plotted as a function of the peak drive current.

The VCSEL design for maximum PCE is best operated at lower currents, but when designed for high peak drive currents, the VCSEL design shown in the dotted line is superior. In this case, the maximum PCE was reduced, but the PCE is improved at high peak power. The roll-off in PCE is inversely proportional to the resistance, and devices with larger diameter may be more appropriate for these applications. A detailed analysis of the PCE for various diameter VCSELs is needed to optimize the utility in the application [5]. The general principles for high-power VCSEL design are to maximize the slope efficiency and minimize the resistance. Since the slope efficiency is relatively insensitive to the diameter but the threshold and resistance scale with the area, the diameter of the emission region is an important consideration. This is a different optimization compared to a VCSEL for high-speed operation.

Thus, in-order to have laser operation at a particular bias current, the VCSEL layer structure must meet the following: (i) the QW gain must be positive to emit sufficient power output during laser operation, (ii) for top-emission, top-DBR reflectivity must be >95% and bottom DBR must be >99.99%, (iii) the F-P and PL offset (also called cavity detuning) must be chosen such that the threshold current must show lowest value at ~40°C (as shown in inset Figure A1.2.1(b), (iv) the position of the oxide layer must be fixed to have minimum leakage current (to avoid excess spontaneous emission), and (v) the p- and n-doping levels in respective p- and n-DBRs must be designed to have low series resistance (operating voltage) and to avoid high thermal resistance. High bandwidth and narrow linewidth MM VCSELs with reduced parasitics are needed for data center applications. In shortest (0.5λ) cavity VCSEL structures, one of the biggest challenges is the careful design of doping levels adjacent to SCH GRIN layers that are very close (~60 nm) to the QWs. At high-temperature operations these activated dopants may very well diffuse into QWs that degrade the laser performance! Datacom MM VCSELs simply do not have enough power density for long-range projection. For short-distance (<10 m) 3D sensing applications in consumer electronic products, a single junction with pulse power density of a few 10s W/mm^2, SE 0.7–0.9 W/A, and PCE ~45% are sufficient. For long-distance (~250 m) LiDAR systems in the automotive industry, high-power 2D arrays with multi-junction cavities with PCE 60% or higher and power densities of ~1 kW/mm^2 are needed. (See Figure A.13b.) [29, 30, 31].

A.1.5 Summary

This appendix has provided key stages of VCSEL mass production with emphasis on VCSEL design engineering. Essential elements along with major design optimizations have explained in high-speed and high-power VCSELs respectively for communication and sensing applications. Optimization schemes include electrical, optical and thermal managements that results in high performance margin in VCSELs. A case for numerical modeling of single VCSEL device is also presented.

References

1 A. Larsson, "Advances in VCSELs for communication and sensing," *IEEE J. Sel. Top. Quantum Electron.* **17,** 1552–1567 (2011).

2 A. Liu, P. Wolf, J. A. Lott, and D. Bimberg., "Vertical-cavity surface-emitting lasers for data communication and sensing," *Photonics Res.* **7**, 121 (2019).

3 A. Mutig., "High Speed VCSELs for Optical Interconnects," Ph. D. Thesis, TU. Berlin, July 2010.

4 S. Okur, M. Scheller, J. F. Seurin et al., "High-power VCSEL arrays with customized beam divergence for 3D sensing applications," *Proc. SPIE* **10938**, 109380F 1 17 (2019).

5 J. F. Seurin, C. L. Ghosh, V. Khalfin et al., "High-power, high-efficiency 2D VCSEL arrays," *Proc. SPIE* **6908**, 690808 1–14 (2008).

6 B. D. Padullaparthi., "Vertical cavity surface emitting laser device," US 9,014, 225 B2.

7 H. Motomura, N. Jikutani, and S. Sato, "Surface emitting laser element, surface emitting laser array, optical scanning apparatus, image forming apparatus, and optical communication system," US 2008/0056321 A1.

8 E. D. Palik., "*Handbook of Optical Constants of Solids*," Academic Press, 1998.

9 J. Piprek., "*Semiconductor Optical Devices: Introduction to Physics and Simulation*," Academic Press, Elsevier Science 2003, USA.

10 X. Wu., "Simulation Study of Epitaxially Regrown Vertical-Cavity Surface-Emitting Lasers," Master Thesis, Royal Institute of Technology, KTH Sweden, 2011.

11 E. K. Lau., "LaserMOD Tutorial," EE232 UC Berkeley 2006 https://people.eecs.berkeley.edu/~wu/ee232/RSoft/lasermod.pdf.

12 Synopsys, Active Devices Tool Part-III VCSELs, Video, 2016 https://players.brightcove.net/5748441669001/rka4xWwYG_default/index.html?videoId=6101589623001.

13 J. Wang, I. Savidis and E. G. Friedman, "Thermal analysis of oxide-confined VCSEL arrays," *Microelectron. J.* **42,** 820–825 (2011).

14 https://www.laserfocusworld.com/software-accessories/software/article/16548218/photonics-modeling-software-tcad-software-facilitates-comprehensive-vcsel-modeling.

15 S. B. Healy, E. P. O. Reily, J. S. Gustavsson et al., "Active region Design for High-Speed 850-nm VCSELs," *IEEE J. Quantum Electron.*, **46**, 506–512 (2010).

16 P. Westberg, J. S. Gustavsson, A. Haglund et al., "High-speed, low-current-density 850 nm VCSELs," *IEEE J. Sel. Top. Quantum Electron.*, **15**, 694–703 (2009).

17 E. Haglund, P. Westberg, J. S. Gustavsson et al., "High speed VCSELs with strong confinement of optical fields and carriers," *IEEE J. Lightwave Technol.* **34**, 269–277 (2016).

18 A. Haglund et al., "Single fundamental-mode output power exceeding 6 mW from VCSELs with a shallow surface relief," *IEEE Photon. Technol. Lett.* **16**, 368 (2004).

19 E. Haglund, A. Haglund, J. S. Gustavsson et al., "Reducing the spectral width of high speed oxide confined VCSELs using an integrated mode filter," *Proc. SPIE* **8276**, 8276-0L (2012).

20 B. D. Padullaparthi., "Tunable Optical Phase Filter," US 9716368 (B2), July 2017.

21 P. Moser, "Energy-efficient oxide-confined VCSELs for optical interconnects in data centers and supercomputers," Ph. D. Thesis, TU. Berlin, April 2015.

22 G. Larisch, P. Moser, J. A. Lott, and D. Bimberg, "Correlation of photon lifetime and maximum bit rate for 55 Gbit/s energy-efficient 980 nm VCSELs," MD3 p-16 978-1-5090-1874-1/16 (2016).

23 P. Westberg, J. S. Gustavsson, B. Kogel et al., "Impact of photon lifetime on high-speed VCSEL," *IEEE J. Sel. Top. Quantum Electron.* **17,** 1603–1613 (2011).

24 Y. Ou,J. S. Gustavsson, P. Westberg et al., "Impedance characteristics and parasitic speed limitations of high-speed 850-nm VCSELs," *IEEE Photon. Technol. Lett.* **21** 15 1840–1842 (2009).

25 B. D. Padullaparthi., "Semiconductor Light emitting element and method for manufacturing the same," US 9 484 495 (B2), November 2016.

26 B. D. Padullaparthi., "Surface emitting Semiconductor Laser," US 10008826 (B1), June 2018.

27 B. D. Padullaparthi., "Impact of Δn_{eff} of 850nm VCSEL cavity on low noise for 100G eSR4 transmission and its potential for ≥400G datacenter optical interconnects," *SPIE Proc.* **11704**, 11704-0R (2021).

28 B. D. Padullaparthi, T. Tamanuki, and D. Bimberg, "Field-of-view control of segmented NIR VCSEL arrays for next- generation flash LiDAR's," *Proc. SPIE* **11262**, 11262-0Z (2020).

29 J. M. Maillard et al., "Lasertel VCSEL development Progress for automotive Lidar," *Proc. SPIE* **11300**, 1130005 (2020).

30 https://www.lumentum.com/en/products/multi-junction-vcsel-array.

31 https://www.lumentum.com/en/media-room/videos/autosens-2020-high-power-vcsels-automotive-applications.

Appendix B

Epitaxial Growth Engineering

Babu Dayal Padullaparthi

B.1 Technologies and Materials for VCSELs

This section describes the considerations for selection of semiconductor materials for VCSEL growth based on existing technologies and market drivers for VCSELs; a detailed example is provided.

B.1.1 Existing Technologies

Monolithic epitaxial growth of laser materials on a given substate is the most critical part of VCSEL device manufacturing as the quality of growth dictates the lifetime and many of the device performance characteristics. There are four substrates used in making VCSELs, namely GaN, GaAs, InP, and GaSb. The substrate choice is determined by the desired emission wavelength, as shown in Figure B.1; this can be understood as active materials grown on respective substrates. For example, InGaAs/GaAs represents InGaAs quantum wells (QWs) grown on a GaAs substrate. In general, QW-based VCSELs are more common in product applications than quantum dots (QDs).

B.1.2 Selection of Materials

The selection of suitable materials for VCSELs mainly depends on the direct and indirect bandgap nature of respective materials, and this in turn depends on the composition of constituent atoms in respective binary, ternary, and quaternary alloy systems or compounds. The relationship

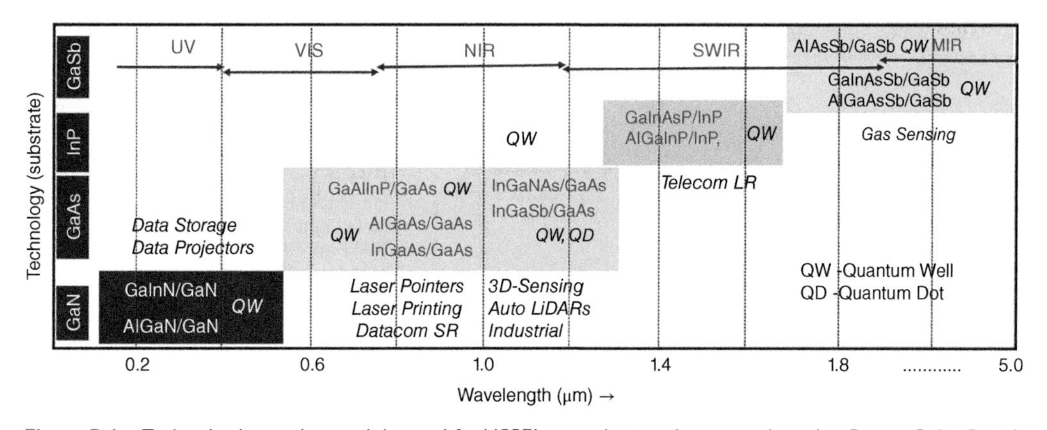

Figure B.1 Technologies and materials used for VCSEL growth at various wavelengths. *Source:* Babu Dayal Padullaparthi © Photonic Components DFM Ltd.

VCSEL Industry: Communication and Sensing, First Edition. Babu Dayal Padullaparthi, Jim A. Tatum and Kenichi Iga.

between material bandgap (emission wavelength) and lattice constant for III-V and II-VI groups of materials is shown in Figure B.2 [1].

B.1.3 Market Drivers

The vast majority of commercial VCSEL devices are grown on GaAs substrates and operate in the 650–980 nm (NIR) range. The forecasted growth in the 3D sensing market, LiDAR, and data communications will continue this trend, and the subsequent sections will focus on epitaxial growth of GaAs substrate VCSELs. Figure B.3 shows the forecasted growth in consumption of GaAs substrates and associated VCSELs [2]. GaAs-based EELs and VCSELs represent the largest growth rate (CAGR) in photonics.

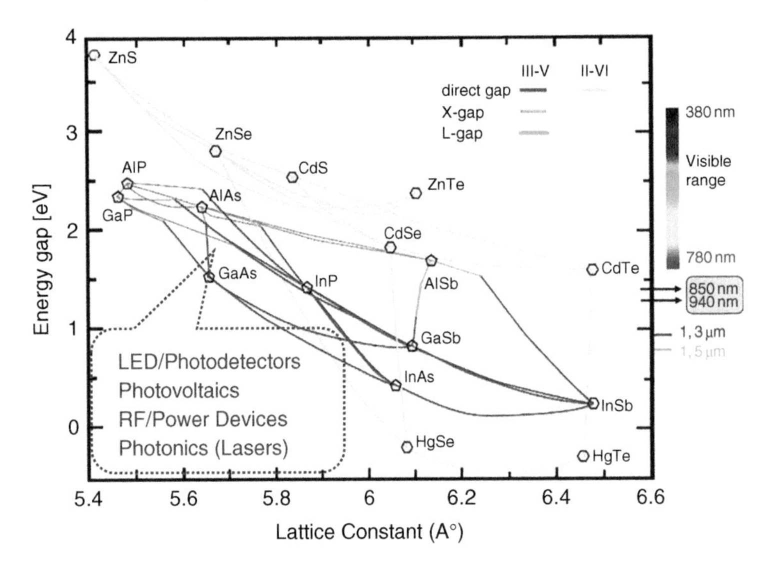

Figure B.2 Materials lattice constant vs bandgap at various wavelengths. Application fields are marked in callout. *Source:* Reprinted and modified with permission from Prof. Dr. Helmut Foll, University of Kiel, Germany.

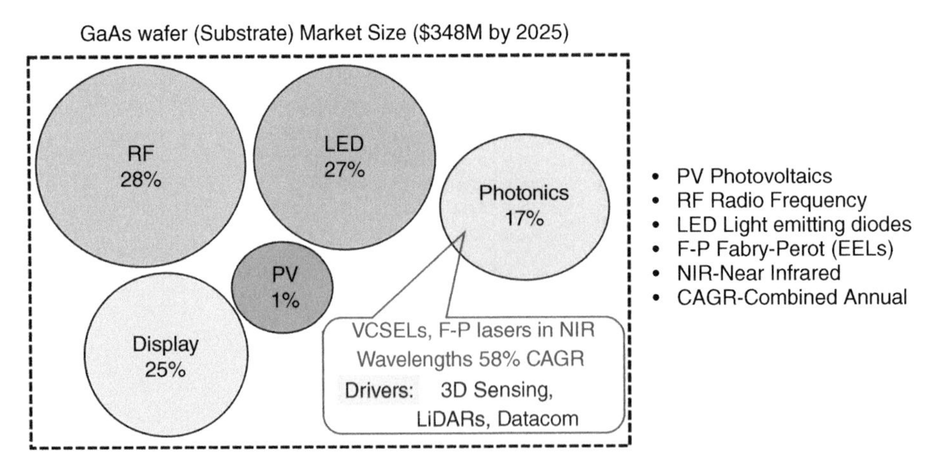

Figure B.3 GaAs epi-wafer share in photonics. *Source:* Babu Dayal Padullaparthi © Photonic Components DFM Ltd.

B.2 Epitaxial Growth

Epitaxial growth is the process of depositing materials on a crystalline substrate, often with single atomic layer resolution. The deposited thin film is called an epitaxial film, epitaxial layer, or epilayer and has a well-defined crystal orientation relative to the substrate. As the deposited films and substrate materials may have different lattice constants, the general condition for crystalline epitaxy is that the lattice mismatch must be <1% to facilitate a dislocation-free single crystal film. The thickness of the films must be below the critical thickness limit set by Matthews and Blakeslee (MB) theory [3]. Epitaxial growth is used in several semiconductor industries for commercial applications including LEDs, photovoltaics (PV), photonics, and RF components, as shown in Figure B.3. Epitaxial growth of semiconductor materials is of particular importance in III-V compounds to make semiconductor lasers. As shown in Figure B.1, for GaAs-based VCSELs, the epitaxial layers consist of ternary or quaternary compounds containing the group III elements of In, Ga, and Al and group V elements As and P. The primary p-type dopants are C and Zn, and the primary n-type dopants are Si and Te.

B.2.1 Growth Techniques: MBE and MOCVD

There are various methods for growing the epitaxial layers on crystalline substrates including liquid phase epitaxy (LPE), vapor phase epitaxy (VPE), chemical vapor deposition (CVD), metal organic chemical vapor deposition (MOCVD), molecular beam epitaxy (MBE), and atomic layer deposition (ALD). MBE and MOCVD are commonly used techniques for growing VCSEL structures [4]. VPE and LPE methods lack precision and growth stability, and ALD is primarily used for depositing high-quality and Si-based dielectric passivation materials [5, 6]. In MBE, the source materials are placed in an effusion cell and are heated to produce an evaporated flux of particles. The effusion rate is on the order of 1 Å/s, and deposition is controlled by shutters placed over the cell. The atoms travel through large mean free path, tens of centimeters, through an ultra-high vacuum (10^{-9}–10^{-11} torr) onto the substrate where they condense and form thin solid films [7]. Even though MBE is widely used for growing III-V group semiconductor materials, it is not widely used for high-volume manufacturing of VCSELs. On the other hand, MOCVD uses a combination of liquid organometallic compounds for the group III components (TMGa, TMAl, TMIn) and gas sources for the group V components (AsH_3, PH_3). The group III components are delivered to the reactor trough a carrier gas, typically H_2. MOCVD systems require complex gas handling systems to function. When the organometallic components and gases arrive at the substrate, they undergo pyrolysis, and the excess hydrogen is released to form atomically flat single crystalline films at moderate pressures (10–760 Torr). MOCVD systems are most commonly used to manufacture VCSELs [8]. The schematics of both of these growth processes are shown in Figure B.4, and some comparisons between them are also listed in Table B.1. MOCVD and MBE systems are capable of growing VCSELs on substrate diameters from 50 to 300 mm, with 100 and 150 mm being the most common today.

B.2.2 Key Reactor Parameters

Due to several advances in crystal growth techniques and epitaxial growth reactors, MOCVD has become a standard tool for VCSEL structure growth. Today there is wide availability of MOCVD-based VCSEL epi sources [9, 10]. With the expansion in the VCSEL market and its forecast for continued expansion, it is not surprising to see the epi-growth foundries such as IQE making huge investments to install multiple MOCVD systems to meet the burgeoning demand [11, 12]. Besides

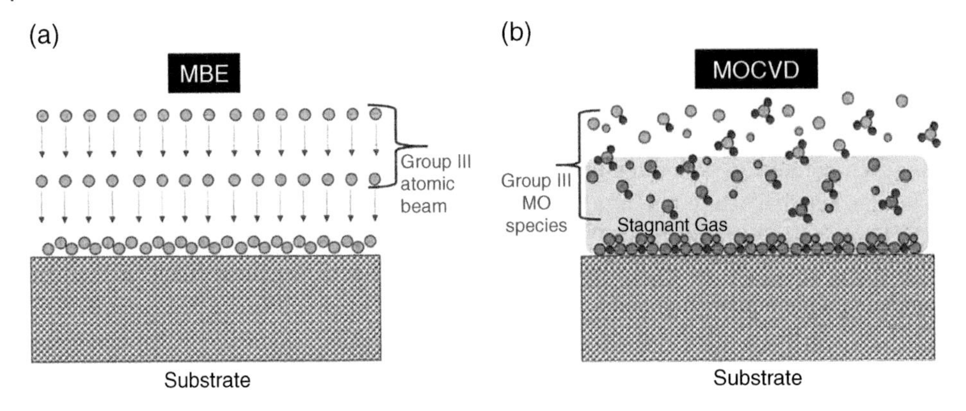

Figure B.4 Schematics of epitaxial growth in (a) MBE and (b) MOCVD. *Source:* Babu Dayal Padullaparthi © Photonic Components DFM Ltd.

Table B.1 Comparisons between MBE and MOCVD processes.

Feature	MBE	MOCVD
Source/flow	solid/atomic	liquid /molecular
Vacuum (torr)	UHV (10^{-9}–10^{-12})	H_2 atmosphere (25–760)
Deposition/thickness control	evaporation/excellent	pyrolysis/moderate
Reaction/growth temperature	kinetic/low	thermodynamic/high
Growth rate / lead time	low/long	high/short
In-situ monitoring	RHEED	RAS[a]
Mass production/cost	not compatible/expensive	Compatible/Moderate

Source: Babu Dayal Padullaparthi ©Photonic Components DFM Ltd.
[a] RAS: reflection absorption spectroscopy.

Black box: Know-hows

i) Growth Temperature
ii) Growth Rate
iii) V/III Ratio
iv) Doping levels
v) Growth interruptions
vi) Substrate parameters
vii) Layer design

Figure B.5 Key growth parameters as black-box items. *Source:* Babu Dayal Padullaparthi © Photonic Components DFM Ltd.

large-sized wafer level manufacturing challenges described in Section 3.2.3, there remain many epi-growth challenges at all stages of product development. The principal challenges are growth temperature control, growth speed (rate), control of the composition and film (thickness) uniformity, dopant control, resistivity, quality of epi-growth (photoluminescence linewidth), the cleanliness and purity of the surface (morphology and particles), contamination in reactor chambers, and so forth. Some of these key items are discussed in the following sections. However, most of these epitaxial growth parameters are closely guarded trade secrets to the epi source parties. A generic description of some of these critical items, not touching fine details, is shown in Figure B.5.

B.2.2.1 Growth Temperature

Stoichiometry of III-V epitaxial films depends on the surface chemistry and adsorption through the sticking coefficients of constituent elements. In the pyrolysis process, the required temperature increases with increasing chemical bond strength of the precursor OM compounds The several adsorbed elements (OM molecules) have different sticking coefficients, and growth difficulties

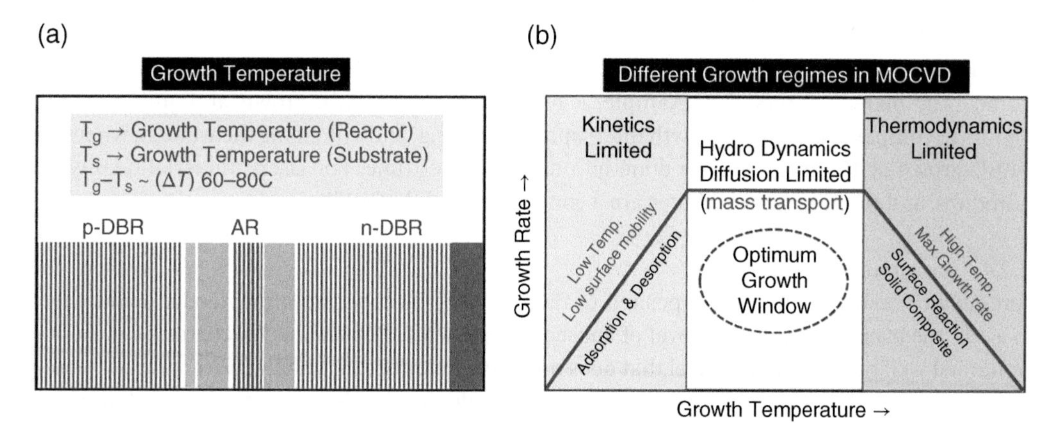

Figure B.6 (a) Multiple sections and growth temperature relationship in a VCSEL. (b) Model of growth temperature and its optimum growth window. *Source:* Babu Dayal Padullaparthi © Photonic Components DFM Ltd.

arise when the temperature changes. For stable growth of a VCSEL that contains hundreds of different layers and compounds with different sticking coefficients, multi-stage growths with different temperatures for different sections of the VCSELs is often implemented. A simple example in Figure B.6a shows the sections of VCSELs with different temperatures, where growth interruptions are implemented to avoid interdiffusion of elements across the interfaces. If the temperature is too low, the growth is kinetically limited, and incorporation of impurities with low surface mobilities creates structural defects. If the temperature is too high, the growth is thermodynamically limited and desorption of molecules can result in vacancies in the crystal. Thus, as shown in Figure B.6b, there exists a diffusion-limited growth temperature window driven by supersaturation of chemical species in the vapor phase that is the best regime for MOCVD reactors to operate [13]. As all of the materials used to form the VCSEL layers have different diffusion coefficients at different temperatures, maintaining optimum growth temperature is one of the most critical aspects of VCSEL layer growth. Growth temperatures for n-, p-, and AR are different and depend on layer compositions and dopant atoms. In a generic VCSEL growth with Si and C atoms dopants incorporated in n- and p-DBRs, one can say $Tg_{n\text{-DBR}} < Tg_{p\text{-DBR}}$.

B.2.2.2 V-III Ratio

The ratio of OM molecules from group III and group V (V/III ratio) during epitaxial growth strongly affects the morphology, crystal structure, and optical properties of the film deposited. If the V/III ratio is too low, it creates decomposition of OM molecules, nitrogen vacancies, and may raise background doping levels. If the V/III ratio is too high, with low surface mobilities, group III atoms create structural defects. For a diffusion-limited growth, the V/III ratios are generally well above 40 [14].

B.2.2.3 Growth Rate

The thickness of material grown per unit time is called growth rate and is represented in units of Å/s or μm/hour. The growth rate represents a figure of merit for quality of epi-wafer and varies for different products and different applications. A faster growth rate (typically 4–20 Å/s) may introduce defects and dislocations, which are common and not that critical for LEDs that emits spontaneous emission only. However, this becomes one of the most critical parameters for laser structure growth. For laser-grade quality, slow growth rate (typically 10–20x slower than LEDs) are needed to get the highest quality of grown single crystal without defects [15]. The tension between the desire for high growth throughput and material quality has been a challenge for many high-volume LED manufacturers. For example, a typical 850 nm VCSEL grown at 1 μm/hr takes ~8 hours to complete (including growth interruptions, loading, and unloading the wafers), whereas an LED grown at 10 μm/hr may be done in a much shorter time. For lasers, material quality is paramount, and the lower growth rates are a consequence of that reality.

B.2.2.4 Background Dopant Levels

There may be residual materials deposited in the MOCVD chamber from previous growths that can affect the background doping level of subsequent growths. The residual background must be maintained well below a certain level that does not affect the properties and performance of to-be-grown structures. Any excess amounts of background doping levels may seriously contaminate layers grown in the reactor and degrade the quality and performance of laser structures. In addition to H_2 and O_2 (byproducts released from pyrolysis), the C, Si, p-, and n-dopants for VCSELs must be maintained below 1e16, 1e16, 2e16, and 1e16 cm^{-3}, respectively [16]. For VCSELs, O_2 concentration should be strictly maintained below 1e16 cm^{-3} levels by using high-purity TMAl sources; otherwise, it can cause defects in AR and affect the optical, electrical, and thermal properties of the DBR.

B.3 Reactor Readiness and Calibrations

In anticipation of growing an epi-layer structure, the epi-growers must calibrate the MOCVD system and design elements to make sure that the system is ready for full VCSEL structure growth. This warrants the following sequence of steps: (i) composition of Al(x), (ii) multi-quantum-well PL linewidth, (iii) DBRs reflectivities and offset, (iv) doping level control in DBRs, (v) mini VCSEL growth, (vi) full VCSEL growth, and (vii) post-growth characterizations and epi-wafer acceptance to proceed to the next manufacturing stage. A flowchart of VCSEL epi-structure calibrations and MOCVD system readiness is shown in Figure B.7, and the procedures are described step-by-step in Sections B.3.1.1 to B.4.2. Before proceeding to calibrations, the design input of DBR reflectivity and stopband FP-dip are required.

B.3.1 Theoretical Estimations and Calibration Flow

B.3.1.1 Al(x) Mole Fraction and Doping Concentration

This step involves precise control of the Al(x) mole fraction or composition in DBR stacks, SCH layers, and oxide layer. In this example, an 850 nm VCSEL structure with p-DBR and n-DBR may have Al(x) from x = 0.15 to 0.9 and x = 0.15 to 1.0, respectively. The calibration structures are defined as shown in Figure B.8, and growth parameters including temperature, partial pressure, and V/III ratio are determined from previous attempts. After the growth, the wafers are ex-situ characterized by high-resolution crystal x-ray diffraction (HRXRD) spectra to identify the composition of Al(x) [17, 18, 19] and polaron depth profiling for doping concentration [20]. For Al(x) compositions in the active region, a similar procedure is followed, and the PL intensity and HRXRD techniques are simultaneously used. For the oxide layers with Al(x), with x in the range 0.96–1.0, the top-layer Al(x) must be adjusted as shown in Figure B.8b. If the composition and doping (C&D) levels are within 10% of their expected tolerance levels, parameter adjustments will be made at the final stage of full structure growth. For laser structures, an Al(x) mole fraction tolerance of ±0.02 or less is a reasonable specification for full VCSEL structure growth, as presented in Table 3.2.

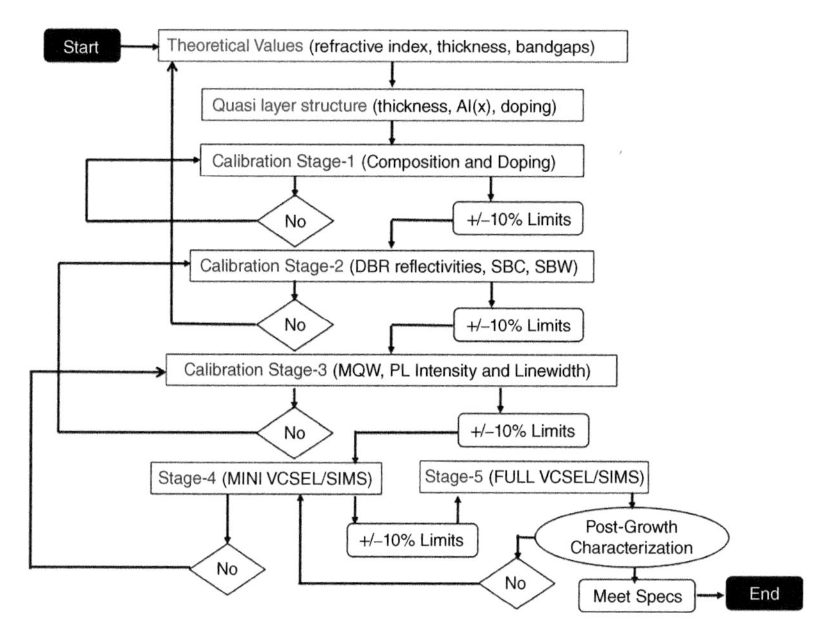

Figure B.7 Flowchart of MOCVD calibrations for a VCSEL structure. *Source:* Babu Dayal Padullaparthi © Photonic Components DFM Ltd.

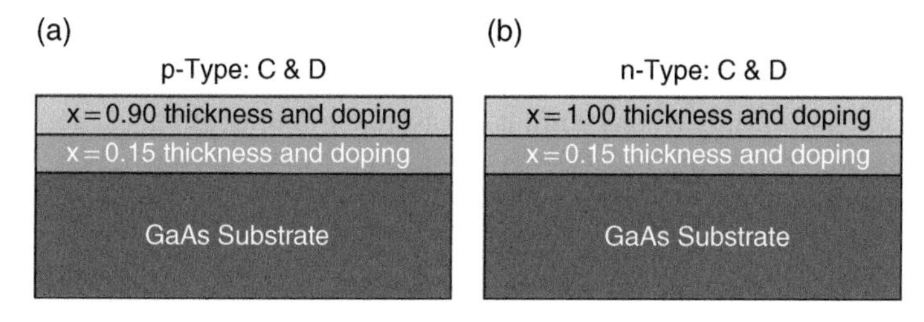

Figure B.8 Quasi layer structures for (a) Al(x) composition and (b) p- and n-doping calibrations. *Source:* © Photonic Components DFM Ltd.

An shown in Figure B.8, the epitaxial layer can be doped during MOCVD deposition by adding impurities to the source gases, and the concentration of impurity in the gas phase is to be determined from concentration of the deposited film. Often, high temperature growth can cause diffusion of dopants into adjacent layers in the wafer (the so-called out-diffusion) and affects the properties of the films. Strict doping level variation of $\pm 20\%$ or less (best for 10% level) should be maintained as a specification for full VCSEL structure growth, as presented in Table 3.2. An example of doping concentration profile from SIMS for a VCSEL structure is shown in B.3.2.

B.3.1.2 p- and n- Mirror Reflectivity

The control of thickness of low- and high-index semiconductor layers determines the reflectivity of DBRs, as shown in Section 2.5. Further, the introduction of thin-graded-index interface layers in between low- and high-index semiconductor layers helps to reduce resistance of the devices, as described in Section A.11 (c). Unlike the Al(x) composition case shown in Section B.3.1.1, for the quasi 850 nm VCSEL structure, the 10 pair p-DBR and 15 pair n-DBR with Al(x) from x = 0.15 to 0.9 and x = 0.15 to 1.0, respectively, are used with incorporation of GRIN layers in between each pair of DBRs, as shown in Figure B.9. After the growth, the wafers are ex-situ characterized by

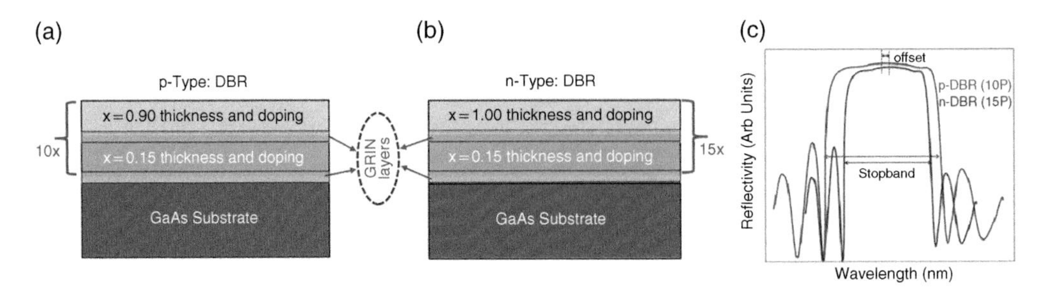

Figure B.9 Quasi layer structures for (a) p-DBR and (b) n-DBR reflectivities and (c) reflectivity spectra of grown mirrors. *Source:* Babu Dayal Padullaparthi © Photonic Components DFM Ltd.

mapping the PL and reflectivity spectra to identify the stopband center (SBC) wavelength, the width of the stopband, and reflectivity maximum. A schematic of the reflectivity of the calibration of two DBR stacks is shown in Figure B.9c. The PL spectrum is used to determine the Al composition of the high-index layer. If the center wavelength offsets between two DBRs is >2 nm, a recalibration growth of DBR stacks is required with fine-tuning the thicknesses or compositions of low- and high-index semiconductor layers until the offset meets the required specification.

B.3.1.3 QW Photoluminescence

High-quality growth of the active region is a critical step in achieving high laser gain and long lifetime of devices, and PL measurement of test structures is used to determine appropriate growth parameters. The undoped active region described in Section A.1.2.1 is grown on a GaAs substrate. After growth the room-temperature PL is mapped at the wafer level. For VCSEL manufacturing a PL wavelength tolerance of ±1.0 nm and variation <=3.0 nm is a reasonable specification, as presented in Table 3.2. An example of good control of PL wavelength for a 6″ (150 mm) 940 nm VCSEL is shown in Figure B.10a [15]. Large variation in the PL wavelengths is a result of either QW thickness (including barrier layers) or strain in the QWs [In(x) in case of $In_xGa_{1-x}As$ QWs] and should be adjusted to meet the required specifications. Further, the PL intensity linewidth (FWHM) is a direct measure of quality of epitaxial growth. For an 850 nm VCSEL with InGaAs MQW-based ARs, the PL linewidth is typically between 20 and 30 meV, with lower values being generally

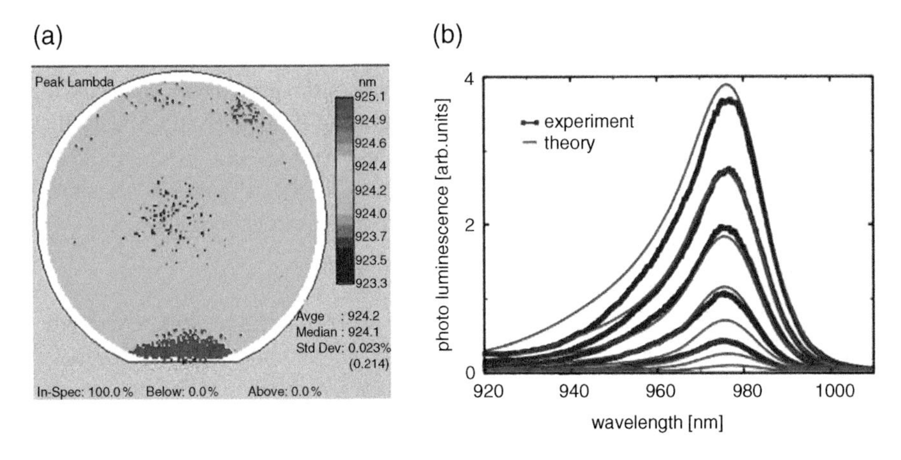

Figure B.10 (a) PL wavelength map of 6″ (150 mm) wafer for 940 nm VCSSEL. *Source:* Reprinted with permission from [15]. (b) PL spectra of InGaAs QWs. *Source:* Adapted from WILEY-VCH Verlag GmbH & Co. KGaA, Weinheim [21].

Table B.2 Substrate parameters to be considered for VCSEL layer growth.

Parameter	Unit	Specification	Remarks
Material		GaAs (100 mm or 4″)	
Conductivity type		n	SI for datacom
Dopant		Silicon	Undoped for SI
Carrier concentration	cm^{-3}	1–4e18	
Etch pit density (EPD)	cm^{-2}	<1000	0 for highest quality 100–500 product <10 000 DOE and calibrations
Diameter	mm	100 ± 0.3	100 for datacom 150 for 3D sensing
Defect Density	cm^{-2}	<40	defect size <2–3 μm^2
Thickness	μm	625 ± 20	675 ± 25 for 150 mm

Source: Babu Dayal Padullaparthi © Photonic Components DFM Ltd.

better. An example of PL intensity spectra for various excitation pump levels of three $In_{0.2}Ga_{0.8}As$ QWs surrounded by GaAs barriers demonstrating FWHM of 16 meV are shown in Figure B.10b [21].

B.3.1.4 Epitaxy Substrates

For GaAs-based VCSELs, 2° off-cut substrates with crystal orientation (001) is the industry standard, as presented in Table 3.2. Other crystal orientations are used for polarization control of VCSELs, as described in Section 8.2.2. The most common substrate is n-doped, and 100 and 150 mm are commonly used in both datacom and 3D sensing applications as shown in Table B.2. Semi-insulating and p-doped substrates are less commonly used in current GaAs VCSEL high-volume manufacturing. Details of 150 mm GaAs substrates are discussed in references [22, 23]. Items related to the wafer process such as substrate bow, major or minor flat orientation, and surface polish on both sides will be discussed in Appendix C.

B.3.1.5 Precise Dopant Source Analysis

Background and excessive dopants can have a serious impact on the epi-film properties of adjacent or underlying layers. Impurities in the MOCVD source gas, liberated by evaporation at the surface, may diffuse into the epitaxial layers (known as auto-doping) and affect the dopant profiles of both proceeding and subsequent layers. It is critical to evaluate the dopant source analysis before beginning calibrations to maintain high-quality epi-growth. This can be done by checking the chemical purity analysis certificates from the supplier and SIMS analysis of calibration wafers to precisely determine the background dopant levels. In particular, the oxygen concentration should be well below <1e16 cm^{-3}. As SIMS is expensive, it is often done only after mini or full VCSEL structure growth [24].

B.3.2 Mini and Full VCSEL Structure Growth

Once all calibration steps are completed, an intermediate step of mini VCSEL growth helps to reconfirm all growth parameters and proper functioning of the reactor before full VCSEL structure growth. The mini VCSEL structure consists of all elements of full VCSEL but with reduced number of both DBR pairs. The key purposes of this mini VCSEL structure are to check that cavity resonance (F-P dip) wavelength and stopband width are at designed values, as well as a detailed

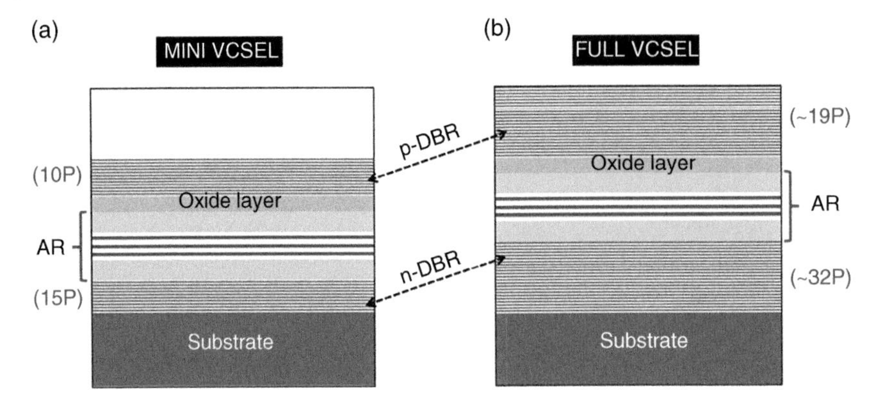

Figure B.11 (a) Mini and (b) full VCSEL structures to grow in a MOCVD system. *Source:* Babu Dayal Padullaparthi © Photonic Components DFM Ltd.

compositional analysis through SIMS. If the mini VCSEL growth results are according to the specification requirements, a full VCSEL growth is conducted without delay to avoid change of reactor chamber conditions. A schematic of both mini and full structures are shown in Figure B.11.

An example of SIMS depth profile of a full VCSEL structure with InGaAs QWs and carbon- and silicon-doped AlGaAs/AlGaAs DBRs is shown in Figure B.12. The high resolution of the system shows excellent and uniform variation of elemental concentrations of DBRs and AR of VCSEL layer structure.

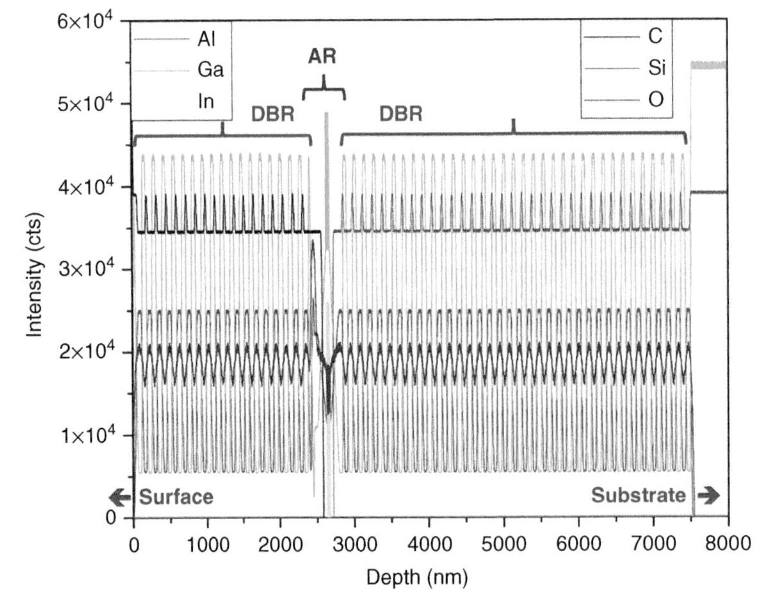

Figure B.12 SIMS doping profile of p-DBR, AR and n-DBR in a VCSEL structure. *Source:* Reprinted and modified with permission from Dr. Paweł Piotr Michałowski, Łukasiewicz Research Network—Institute of Microelectronics and Photonics, Warsaw, Poland.

B.4 Post-Growth Characterizations and Acceptance

After full wafer growth, optical scans are conducted on all wafers to identify residual particles or defects left over on the surface of wafers. The wafers are further subjected to detailed PL mapping and reflectivity measurements to identify variations of F-P dip stopband center and width, across multiple points on the wafers. To maximize final product yield, it is common to specify the F-P dip in terms of uniformity and percentage of the wafer that is within specification. An example of F-P dip wavelength for 850 nm VCSEL on 2″ (50 mm) GaAs substrate is given in references [25, 26]. It is important to consider within-wafer (WIW) and wafer-to-wafer (WTW) variations of all parameters of the epitaxial growth to meet the specifications listed in Table 3.2. Wafers that meet epitaxial specifications are moved to wafer processing, and the epi-growth is said to be accepted or to meet growth acceptance criteria (GAC), described in Section 3.2.2.1.

B.5 Epitaxial Growth on Large-Sized Wafers

Due to the volume of epi-wafer demands for photonics, there is an increasing interest to grow epi-layer structures on even larger wafers. So far, due to several mass production challenges in meeting growth tolerances and uniformities, VCSEL structure growths are largely limited to 6″ (150 mm) diameter wafers (Figure B.10a). Two interesting results have been reported recently. In March 2018, Allos Semi from Germany reported exceptional PL uniformity of GaN QWs grown on 8″ (200 mm) Si substrate for LED applications at the Compound Semiconductor International Conference [27], shown Figure B.13. This award-winning MOCVD growth reports a PL uniformity >99% (standard deviation of 0.56 nm) across the wafer and claims MOCVD is able to grow mini-LED structures on 12″ (300 mm) wafers [27]. Further, in October 2020, IQE announced its MOCVD VCSEL structure grown on 6″ (150 mm) Ge substrates and proposed to extend this VCSEL structure growth on 200 mm Ge wafers [28].

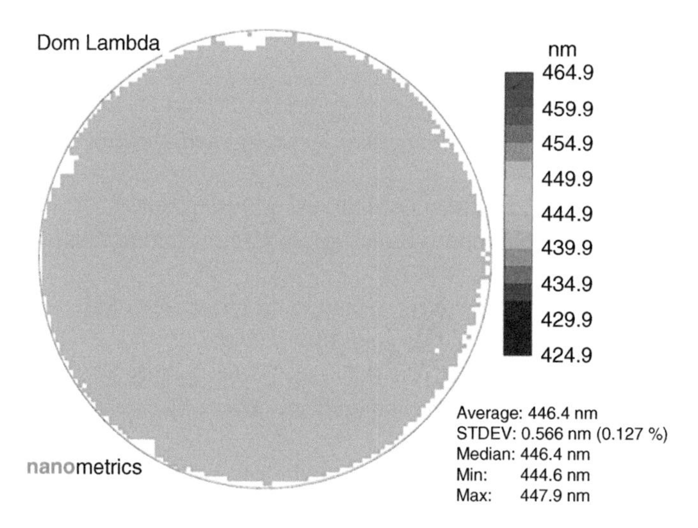

Figure B.13 PL map of GaN QWs grown on Si 8″ (200 mm) substrate. *Source:* Reprinted with permission from Allos Semiconductors, Germany.

B.6 Summary

This appendix has provided key technologies and materials for epitaxial growth engineering of VCSEL layer structures with emphasis on GaAs substrates. Essential features of VCSEL epitaxial growth using MOCVD such as reactor parameters, growth temperature, V-III ratios, doping levels, photoluminescence, reactor readiness through MINI VCSEL calibrations, full VCSEL growth and post growth acceptance criteria are briefly explained. Examples of highly uniform photoluminescence on InGaAs QW on- 6" GaAs (150mm) and GaN QW-on-8" Si (200mm) substrates are provided. A case for doping concentrations of a VCSEL structure with InGaAs QWs and AlGaAs/AlGaAs DBRs through SIMS depth profiles is also given.

References

1 H. Foll https://www.tf.uni-kiel.de/matwis/amat/semi_en and https://www.tf.uni-kiel.de/matwis/amat/semi_en/kap_5/backbone/r5_1_4.html.

2 https://s3.i-micronews.com/uploads/2020/05/YDR20083_GaAs_Wafer_and_Epiwafer_Market_RF_Photonics_LED_Display_and_PV_Applications_June2020_yole_Sample.pdf.

3 J. W. Matthews and A. Blakeslee, "Defects in epitaxial multilayers: I. misfit dislocations," *J. Cryst. Growth* **27**, 118 (1974).

4 C. Wilmsen, H. Temkin and L. A. Coldren, "*Vertical Cavity Surface Emitting Lasers: Design, Fabrication, Characterization and Applications*," Cambridge University Press, 1999.

5 S. Traut and S. Saintenoy, "A. semiconductor Laser device and a method for manufacturing a semiconductor laser device," GB. 249 4008 A, Feb 2013.

6 http://www.semiconductor-today.com/news_items/2020/may/acs-010520.shtml.

7 WS Knodle and R Chow, https://booksite.elsevier.com/9781437778731/past_edition_chapters/Molecular_Beam_Epitaxy.pdf.

8 J. L. Zilko, "Metal Organic Chemical Vapor Deposition: Technology and Equipment," Handbook of Thin Film Deposition Processes and Techniques (Second Edition) Chp-4, Pages 151–203, 2001.

9 M. Lapedus., https://semiengineering.com/mocvd-vendors-eye-new-apps.

10 https://www.aixtron.com/innovation/technologien/How_MOCVD_works.pdf.

11 https://compoundsemiconductor.net/article/107078/IQE_Mega_Foundry_Gets_First_VCSEL_Order%7BfeatureExtra%7D.

12 Photonics Overview September 2014, IQE, www.iqep.com.

13 H. O. Pierson, "*Handbook of Chemical Vapour Deposition: Principles, Technology and Applications*," 2nd, Noyes Publications, 1999.

14 I. L. Chen , W. C. Hsu, T. D. Lee et al., "Growth of highly strained InGaAs quantum Wells by metalorganic chemical vapor deposition with application to vertical-cavity surface-emitting laser," *Jpn. J. Appl. Phy.* **45**, L54–L56 (2006).

15 Y. Gou, J. Wang, Y. Cheng et al., "A. Modeling and experimental study on the growth of VCSEL materials using an 8 × 6 inch planetary MOCVD reactor ," *Coatings* **10**, 797 (2020).

16 A. Mutig., "High Speed VCSELs for Optical Interconnects," Ph. D. Thesis, TU. Berlin, July 2010.

17 M. A. Hayashi and R. Marcon, "High resolution X-ray diffraction to characterize semiconductor materials," *Revista Physicae* **1,** 21 (2000).

18 M. Zorn, F. Bugge, T. SChenk et al., "Feedback controlled growth of strain-balanced InGaAs multiple quantum wells in metal- organic vapour phase epitaxy using an in situ curvature sensor," *Semicond. Sci. Technol.* **21** L45–L48 (2006).

19 T. Veerland and B. M. Paine, "X-ray diffraction characterization of multilayer structure ," *J. Vac. Sci. Technol. A* **4**, 3153 1986.

20 R. E. Leoni III, "An Electrochemical Capacitance-Voltage Technique for the determination of pseudomorphic high electron mobility transistor material parameters," Masters Theis, Lehigh University, Aug. 1995.

21 J. V. Moloney, J. Hader and S. W. Koch., "Quantum design of semiconductor active materials: laser and amplifier applications," *Laser & Photon. Rev.* **1**, 24–43 (2007).

22 A. Kleinwechter, T. Bunger, T. Flade et al., "Mass Production of Large-Size GaAs Wafers at FREIBERGER," GaAs Mantech (2001).

23 T. Kawase, H. Yoshida, T. Sakurada et al., "Properties of 6-inch Semi-insulating GaAs Substrates Manufactured by Vertical Boat Method," GaAs Mantech (1999).

24 Y. K. Kim and K. Choquette, "Secondary ion mass spectrometry analysis of vertical cavity surface-emitting lasers ," *J. Vac. Sci. Technol. B* **22**, 949 2004.

25 Y. M. Song, B. K. Jeong, B H Na et al., "High-speed characteristics of vertical cavity surface emitting lasers and resonant-cavity-enhanced photodetectors based on intracavity-contacted structure," *Appl. Opt.* **48,** F11 (2009).

26 K. L. Chi, J. L. Yen, J. M. Wun et al., "Strong wavelength detuning of 850 nm vertical-cavity surface-emitting lasers for high-speed (>40 Gbit/s) and low-energy consumption operation," *IEEE J Sel Topics Quantum Electron*, **21**, 1701510 (2015).

27 https://www.allos-semiconductors.com.

28 http://www.semiconductor-today.com/news_items/2020/nov/iqe-021120.shtml.

Bibliography

[ALLOS-2] http://www.allos-semiconductors.com/wp-content/uploads/2020/03/200330-Press-release-ALLOS-shows-200-mm-reproducibility-and-300-mm-EN-1.pdf

Appendix C

Wafer Process Engineering

Babu Dayal Padullaparthi

C.1 Background of Wafer Processing

Fabrication of VCSELs is in many ways similar to CMOS processing of Si and GaN structures, but there are a few key differences. One key difference is that VCSELs are inherently vertical devices, and it is not common to connect one device sequentially to another on the wafer to form a circuit. Conduction is generally from the top of the wafer to the bottom. Another fundamental difference is that many of the device properties are determined by the epitaxial layers, as described in Appendix B, and fabrication is essentially only making electrical contacts to the device. The commonalities are more in the lines of process steps such as photolithography, deposition and etching of both metals and dielectrics, and thermal processing. The two unique steps in VCSEL fabrication are (i) the deep etch to access the oxidation layer and (ii) the oxidation process. Figure C.1 shows a comparison of cross-sectional images of (a) a HEMT and (b) a VCSEL. The deep trench, ~4 μm, often requires precise control, ±5%, and the subsequent oxidation process makes VCSEL fabrication a unique and complex matter. The thick epitaxial structure also presents the unique challenge of highly bowed wafers that creates stress in the material and adds to the already fragile nature of GaAs. The wafer bow can be several hundred microns and presents challenges for automated handling equipment and vacuum attachment. Fortunately, the wafer bow can be reduced to under 50 μm with the deposition of a stress-compensating dielectric on the back side of the wafer for most of the process steps. Unfortunately, this layer must be removed to make electrical contact to the backside of the wafer and presents difficulty during wafer testing and back-end process handling. VCSELs are also susceptible to ESD damage (see Appendix E), and

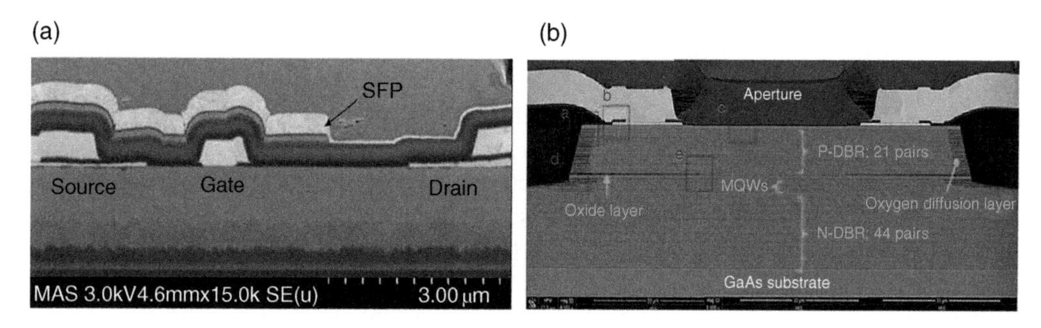

Figure C.1 (a) Cross-sectional image of HEMT. *Source:* Adapted from IEEE [2]. (b) Cross-sectional image of VCSEL. *Source:* Reprinted with permission from MSS Corp., Taiwan.

VCSEL Industry: Communication and Sensing, First Edition. Babu Dayal Padullaparthi, Jim A. Tatum and Kenichi Iga.

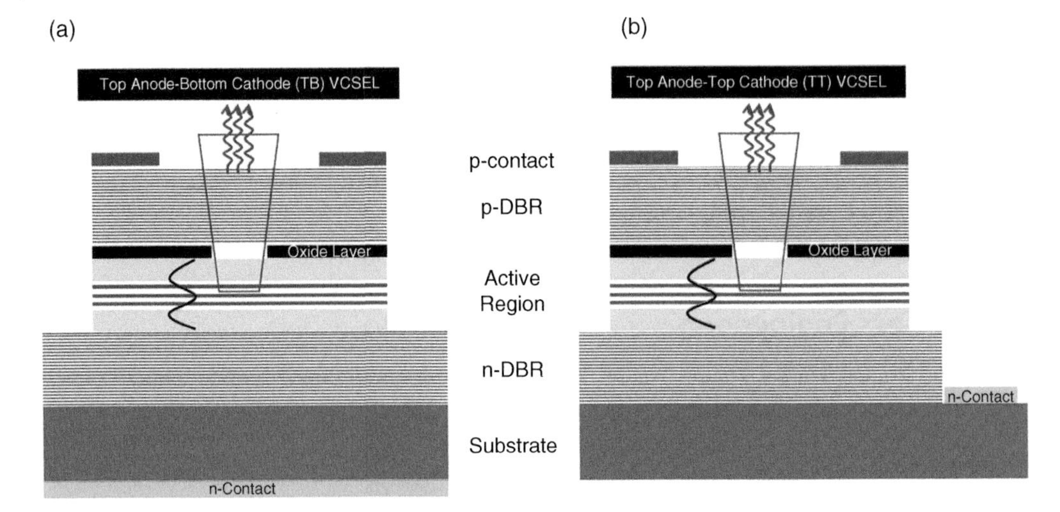

Figure C.2 (a) Top anode–bottom cathode (TB), and (b) top anode–top cathode (TT) configurations of VCSELs. *Source:* Babu Dayal Padullaparthi © Photonic Components DFM Ltd.

strict ESD protocols need to be maintained [1]. VCSEL wafer processing starts with considering the challenging items of etch depth, wafer bow, and ESD controls!

C.1.1 Process Configurations

There are commonly two types of top-emitting (TE) VCSEL configurations that are used for high-volume VCSEL processing, namely the top anode–bottom cathode (TB), and top anode–top cathode (TT), as shown in Figure C.2. Generic process flows for a TE datacom VCSEL and a TE high-power VCSEL are presented in Section C.2.3. High-speed datacom VCSELs prefer TT, while 3D sensor-based VCSELs use TB configurations. As a special case for LiDAR applications, a bottom-emitting (BE) VCSEL with TB configuration may be used but will not be shown here [3].

C.1.2 Photomask Design

To transfer the device design patterns onto the VCSEL wafers, standard photomask plates with desired elements are made by direct laser writing. Standard (drawing, writing, and mask) design rules apply to all mask plates based on critical dimension (CD) of patterns and process parameters such as depth, angle, and temperature. For VCSEL processing, mask design defines the chip dimensions (singlet, 1D, or 2D arrays), verniers and alignment marks/keys for alignment verification, test patterns for contact resistance measurement, individual devices with varying diameters for oxide aperture measurement and device characterization, and other test structures defined in process control monitors (PCM). In a stepper-based lithography mask set, the PCMs are often placed on the corners of each of the shot/block to facilitate process parameter monitoring across the wafers, as shown in Figure 3.10a. There are many different metrology points on the wafer depending on the process steps to be monitored. Analysis techniques include optical/NIR/laser microscopy, SEM, FIB, nano profilers, ellipsometry, and I-V measurements for TLM patterns. The number of total mask layers for a given product defines the cycle time for processing and varies from product to product, as explained in Figure 3.2.7 and Table C.1.

Table C.1 Process summary for various products.

	Process summary (# of repetitions)		
Process step	**Datacom**	**3D Sensing**	**LiDARs**
Photolithography	11	5	5 + 1*
ICP dry/wet etch	4	2	2
PECVD deposition	6	4	4
Metallization	3	3	3
Specialty	N/A	N/A	*microlens

Source: © Photonic Components DFM Ltd.

C.2 Wafer Processing Methods: DOE and Lot

Similar to epi-wafer growth, there are initial process verification steps defined through design of experiments (DOE) with mechanical wafers to confirm capability of individual (unit) process steps and process integration steps. After achieving the required specifications, all the individual steps will be carried out in full wafer processing flows, as described in Section C.2.3. VCSEL processing is grouped into three types of processing steps, front-end (FE), back-end (BE), and wafer level probing (WLP), which are conducted on a lot basis. (Note: Wafer testing is described in Appendix D.) A wafer lot is defined as a group of wafers (say, 6 or 10 pieces) that will be grouped and sequentially processed and called lot processing. Some of the process steps will involve multiple wafers and some may be individual wafers within the lot. Because of the extreme dependence of the oxidation process on the Al composition, a dummy wafer may be used to test the oxidation process. This dummy wafer is generally from the same epitaxial run. Some steps are process oriented while some are product or yield oriented, to be taken care of by technology and operation departments. All these steps are shown in Figure C.3 and individually explained in Sections C.2.1 through C.2.3.

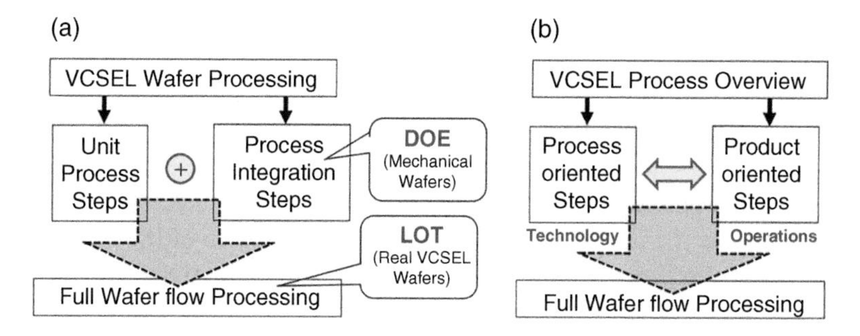

Figure C.3 (a) DOE and lot steps and (b) process and product oriented steps in VCSEL processing. *Source:* Babu Dayal Padullaparthi ©Photonic Components DFM Ltd.

C.2.1 Front-End Processing

Front-end processing (FEP) involves the major process steps of metallization, mesa etching, passivation, and oxidation; the details of these processes are described in the following sections.

C.2.1.1 Metallization

Similar to all electronic devices, VCSELs have p-contact (anode) and n-contact (cathode) ohmic metal contacts in either the TB or TT configuration. Industry standard metals such as Ti-Pt-Au and Au/AuGe are used for p- and n- contacts, respectively. To achieve good device performance, the contact resistance should be less than 1e−5 Ω/cm^2. These metals are deposited using either magnetron sputtering (for hard coatings), e-beam evaporation (most common for p- and n-contacts), or electro/electroless plating for thick Au layers (1–2 μm for datacom and up to 4 μm for 3D sensing for pad metal contacts). Patterns in the metal are defined using a lift-off process. The process flowchart in Figure C.5 shows p-, n-, and pad metal contacts.

C.2.1.2 Mesa Etching

The mesa-etching step defines the size of each individual emitter and is generally accomplished with dry plasma etching using either reactive ion etching (RIE) or inductively coupled plasma (ICP). This process step is critical to VCSEL performance and reliability. Smooth sidewall profiles are needed to promote dielectric coverage with large mesa shoe (Figure 3.2.2.3.1a and Section 3.2.2.3) and help to reduce stress in oxide layers. This step is only for angled mesas and not applicable for vertical mesas where side walls are not fully exposed (Figure 3.2.2.3.1b and Section 3.2.2.3). Wet etching of the mesa is not that common for large-scale VCSEL wafer processing, but wet etching of thin dielectric films may be warranted with properly optimized selectivity (etch rate of photoresists and dielectric films). Dry and wet etch selectivities are optimized for precise control of depth tolerances. The mesa etching is indeed a challenging process step for 4″ (100 mm) and 6″ (150 mm) wafers, as explained in Section 3.2.3.2, and its etch depth uniformity is crucial to maximizing yield and achieving highly reliable devices. An etch depth of 4–5 μm is needed for the first mesa for datacom VCSELs and 3D sensor/LiDARs VCSEL arrays to access the oxidation layer. A second mesa (etch depth ~4 μm) is often needed for TT configuration datacom VCSELs to make the cathode contact and further reduce mesa capacitance for high-speed operation. The appearance of mesa structures are shown in process schematics in Figures C.5 and C.7.

C.2.1.3 Passivation

Identical to other semiconductor electronic devices, mesa passivation (PASS) and/or top-surface protection (PROT) using dielectric materials is common for VCSEL devices. Standard SiO_2, SiOxN or SiN are used as passivation materials and are deposited by PECVD technique. For VCSELs on angled mesas the sidewall coverage with PASS layers is important for reducing residual stress in the oxide layer and may strongly affect device reliability. Similarly, for surface protection from environmental factors and avoiding mechanical scratches during pickup and testing stages, dielectric films are deposited on the top-most GaAs (emitting window) layer. However, a trade-off occurs between thickness control margins in dry etch patterns due to differences of refractive indices between SiO2 and SiN. Any dielectric deposited over the light-emitting region of the VCSEL becomes part of the DBR and must be accounted for in the device design. Unlike SiOxN as a standard passivation material for CMOS processing, it is often tricky to pick a suitable passivation material for VCSELs due to strict stress requirements for both PASS and PROT materials for highly reliable devices.

C.2.1.4 Wet Thermal Oxidation

Wet thermal oxidation is the unique step exclusively used in VCSEL processing. Oxidation creates the current conduction path to the AR and provides an optical guide for the laser modes. Obtaining a circular-shaped oxidation pattern (called aperture) and high uniformity across large scale wafers is indeed a challenge, as explained in Section 3.2.3.3. The circular shape and uniformity of oxidation aperture are very important to maintain fixed mode patterns and stable far-field beam profiles needed in datacom transmission and to reduce speckle to enhance image contrast in sensing applications. Datacom VCSELs use smaller aperture diameters to achieve higher bandwidth, while sensor/LiDARs VCSEL arrays may use larger aperture diameters to obtain higher peak powers. The appearance of aperture diameters are shown in Figures 3.2.2.4.2b and D.3.1

C.2.2 Back-End Processing

Back-end processing (BEP) involves the major steps of back-side metal contact, visual inspection, wafer thinning, dicing, and wafer packaging on tape. These steps are described in more detail in the following sections.

C.2.2.1 Back Side Contact and Thinning

In most cases wafer thinning is required, and the target thicknesses range from 100 to 200 μm with a tolerance of less than 10 μm. The exceptions to this are when the chip size is very large or in BE VCSELs where the backside surface may include micro-optical components such as a lens or diffractive elements [2]. In this case, the VCSEL will likely be the TT contact configuration, and no metallization is present on the back side of the wafer. In some cases, wafer testing may be conducted after a sacrificial contact is made to the VCSEL. Testing at this stage facilitates easier handling of the wafers. For the TB contact configuration, after thinning, the backside contact metal is deposited, and the wafer is subjected to a full wafer probe, as described in Section C.2.4 and more fully in Appendix D.

C.2.2.2 Dicing

After final wafer level probe tests, the fully processed wafer is diced into an individual VCSEL die. Dicing is typically done either mechanically using a saw and/or optically using specialized lasers. Cleaving typically done in EELs is not generally used in VCSELs. Mechanical blade dicing consumes more wafer area for the street (40–60 μm) and kerf (~20 μm) and may have a large amount of chipping and particles on the edges of dies, while laser dicing consumes less wafer area with smaller street width (~20 μm) and lower kerf (~10 μm) and may provide debris-less smooth edges that can increase the number of die per wafer. Another dicing technique called stealth dicing, where laser only interacts inside the substrate materials, is not that common for VCSELs, and laser dicing or mechanical sawing remain the preferred choice in the VCSEL industry.

C.2.2.3 Visual (Final) Inspection

Visual inspection of individual die is carried out to identify any defects and is generally done on 100% of the die. This step involves defining acceptable and non-acceptable visual criteria of the die according to custom inspection criteria adopted by chip manufacturing parties. This includes inspecting contamination (particles and residues), damage (scratches or cracks), color contrasts (discolorations), peeling-off areas, misalignments, and clear identification of die ID characters. Visual inspection is conducted in different locations of the die, at scribe lines at the edges of dies, the main chip area, bond pads, pad contacts, mesa/passivation (PASS) edges, and emission/surface protection (PROT) windows. Definition of the detailed inspection criteria is out of the scope of this appendix. A schematic of visual inspection areas by automatic pattern recognition through software for a datacom VCSEL and 3D sensing VCSEL arrays are presented in Figure C.4.

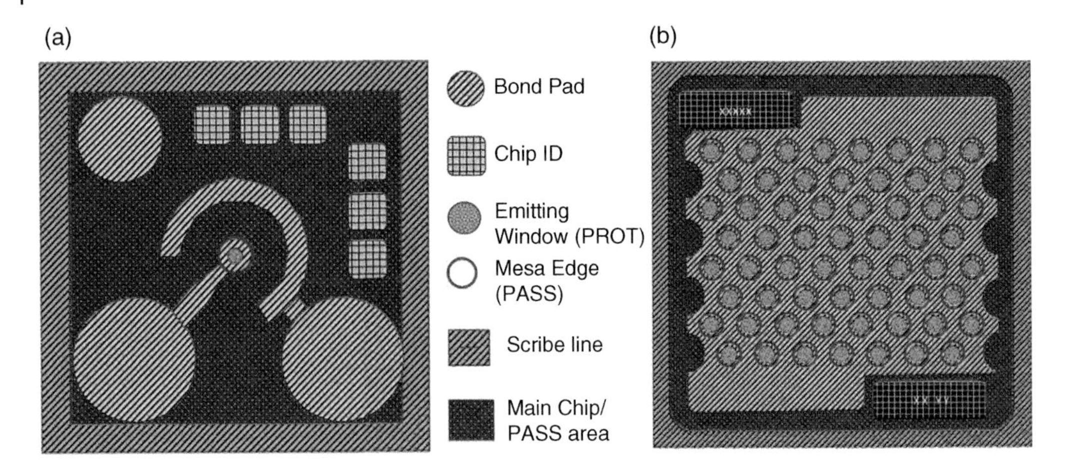

Figure C.4 Schematic of visual inspection patterns/contours for (a) datacom VCSEL and (b) VCSEL array for 3D sensing. *Source:* Babu Dayal Padullaparthi © Photonic Components DFM Ltd.

C.2.2.4 ESD Issues

Electrostatic discharge (ESD) is the leading cause of VCSEL failures and is discussed in detail in Appendix E. It can lead to both patent and latent failures [1, 4]. ESD can happen anywhere during process (especially after oxidation, when the current path is restricted), and is more prevalent in the testing, assembly, and packaging stages. Care must also be taken with ESD-sensitive tape and shipment bags during packaging with expanding/grip rings for UV exposure and assembly pickup. A special requirement of VCSEL fabrication is that all fab floor, machines, humans, and electrical cables must be grounded to achieve low ESD levels. Small aperture VCSELs can be damaged with less than 50 V.

C.2.2.5 WTS and WPT

Automatic wafer tracking systems (WTS) are implemented to monitor every process step including cleaning, process verification, and wait times. Dedicated software systems are used to record and instantaneously analyze process data in real time and may be a prerequisite to move from one stage to the next in the process. All process non-conformances and anomalies should be red-flagged and addressed (solved) before moving to the next step. A wafer process traveler (WPT) is used to record all parameters and serves as a history of all events during lot processing. WTS and WPS are essential parts of a complete quality management system implemented by manufacturers that employ dedicated engineering staff.

C.2.3 Generic Process Flows

In this section, the authors present two generic process flows of single TE VCSEL (emitter) for a high-speed (850 nm for datacom) and high-power (940 nm for 3D sensing imaging) applications. These flows do not represent any standards and are merely indicative of some of the key process steps in VCSEL manufacturing as real commercial products will have many customized and specialized trade secret steps. Both VCSEL flows start with VCSEL epi-wafers (4″ or 6″) grown on GaAs substrates by MOCVD after incoming wafer inspection (particle density, bow, adhesion

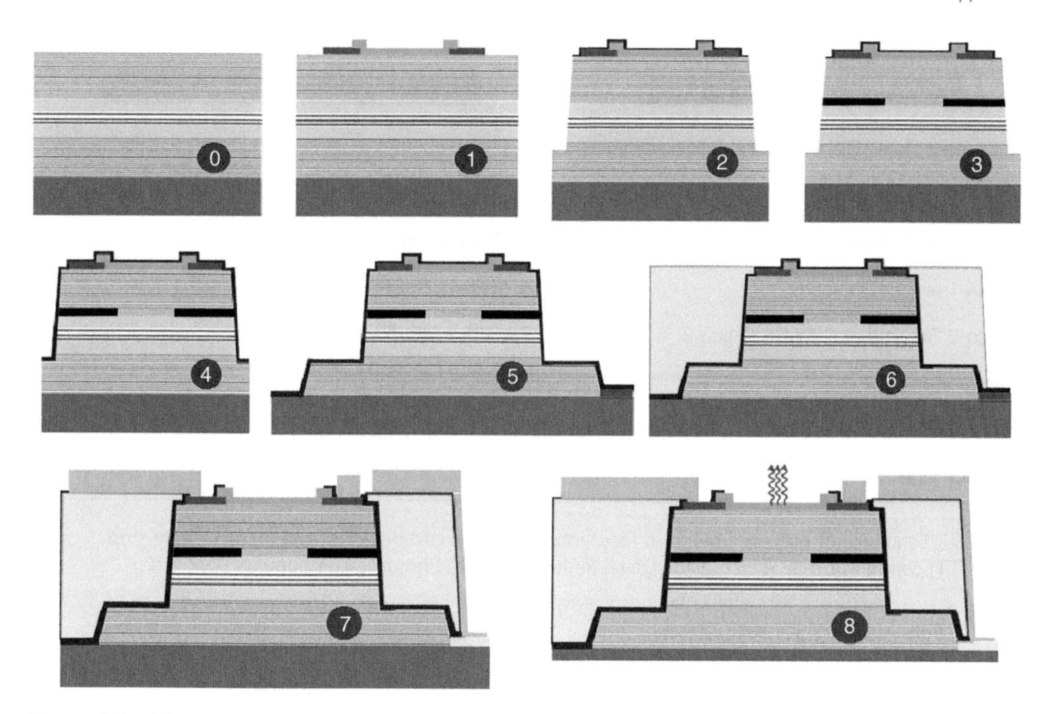

Figure C.5 Schematic generic process flow of high-speed (datacom) top-emitting VCSEL and top-top electrodes. *Source:* Babu Dayal Padullaparthi © Photonic Components DFM Ltd.

properties, etc.) as step-0 in Figure C.5. For all process steps shown in Figure 6, strict control of process specifications (tolerances and uniformities) applies as listed in Table 3.3.

C.2.3.1 High-Speed VCSEL Flow

A typical high-speed TE VCSEL flow starts with p-contact (Ti/Pt/Au) metal and PROT (SiN or SiOxN) window, as shown in step-1 in Figure C.5. This is followed by a mask-insulator (SiN) step to transfer the mask pattern to p-DBR stack for mesa-1 etching, as shown under step-2. After the mesa-1 etching step, the wafer is subjected to a wet thermal oxidation process to form the aperture for current confinement, as shown in step 3. The next step is mesa-1 passivation-1 (SiN), as shown in step-4.

As mesa capacitance plays a critical role in reduction of capacitance, a mesa-2 etching step is carried out (same as mesa-1 step) and subsequent passivation-2 (SiN). An n-contact (AuGe/Au) metal is deposited prior to passivation-2, as shown in step-5. Then for creating TT electrodes, both mesas are planarized with low dielectric constant photosensitive polymers, as shown in step-6. Next, wafers are subjected to a VIA-RIE step to expose the buried p- and n-contact metals, and then pad metal (evaporated Pt/Au or plated Au) deposition is carried with a lift-off process, as shown in step-7. After making all three contact metal depositions, an initial probe (sampling) test is carried out to functional characteristics fabricated VCSEL emitters. Next, the wafers are thinned to the required thickness for final wafer probe tests. All these steps are shown in Figure C.5 and with flow order in Figure C.6. An example of a fully processed 4″(100 mm) GaAs-based 850 nm VCSEL wafers for datacom applications is shown in Figure C.10.

(a)

High-Speed Datacom VCSEL

1. p-Contact metal, Device ID and PROTection layer
2. Mas RIE and Mesa ICP-RIE
3. Wet thermal Oxidation
4. Mesa-1 and PASSivation-1
5. Mesa-2 RIE, n-metal, PASSivation-2
6. Polymer and VIA-RIE
7. Pad-metal and thinning
8. Wafer Probe

(b)

High-Power VCSEL array for Sensing

1. p-Contact metal, Device ID and PROTection layer
2. Mas RIE and Mesa ICP-RIE
3. Wet thermal Oxidation
4. Mesa-1 and PASSivation-1

7A. VIA –RIE, Pad-metal

8A. Thinning and wafer Probe

Figure C.6 Sequence of process flow for (a) top anode–bottom cathode (TB), and (b) top anode–top cathode (TT) configurations. *Source:* Babu Dayal Padullaparthi © Photonic Components DFM Ltd.

C.2.3.2 High-Power VCSEL Array Flow

For the case of high-power TE VCSELs, steps from 1–4 are common in the datacom VCSEL process described in C.5. Immediately after step-4, wafers will be subjected to VIA RIE to expose buried p-contact and then to thick pad-metal (usually Au-plating) for top contact, shown in step-7A, as there is no need of mesa-2 and polymer planarization processes for high-power VCSELs. Then, for the TB configuration substate, thinning is carried out, and n-contact metal is deposited on the bottom side of the substrate, as shown in step-8A. All these steps are shown in Figure C.7 and with flow order in schematic C.6. An example of a fully processed 6″ (150 mm) GaAs-based 940 nm VCSEL wafers for 3D sensing (short distance) applications is shown in Figure 3.1.4.1 and Figure 1.12.

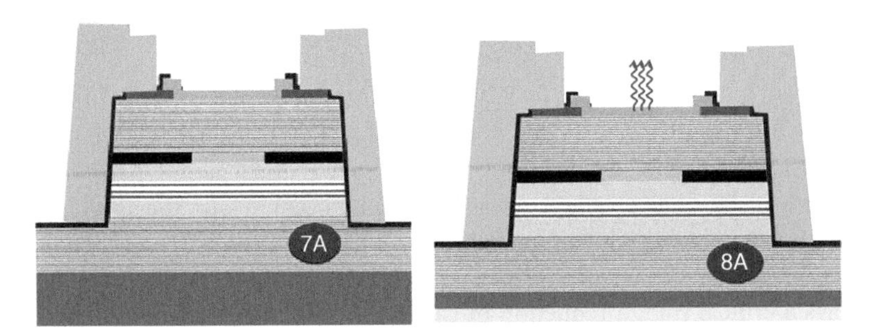

Figure C.7 Schematic generic process flow of high-power (3D sensing and imaging) top-emitting VCSEL and top-bottom electrodes. (Note: Steps 1–4 in Figure C.5 are common to high-power VCSEL flow.) *Source:* Babu Dayal Padullaparthi © Photonic Components DFM Ltd.

C.2.4 Wafer Process: Summary and Conditions

In summary, the VCSEL processes are tabulated in Table C.1 in terms of mask number, dielectric coatings, dry/wet etching, and metallization steps for datacom and 3D sensing arrays. It is obvious from Table C.1 that datacom chips are more complicated and incur longer cycle times to manufacture than high-power VCSELs. There are many manufacturing challenges in the control of various toxic and hazardous gases (such as SF_6, SiH_4, N_2O, NH_3, BCl_3, and Cl_2) that are used at different temperatures (room temperature to 370°C) as well as handling some carrier gases such as O_2 and N_2. However, similar to Figure B.2.2 for epi-layer growth, most of these process conditions are also know-how and proprietary in nature. The overall process summary is shown in Figure C.8.

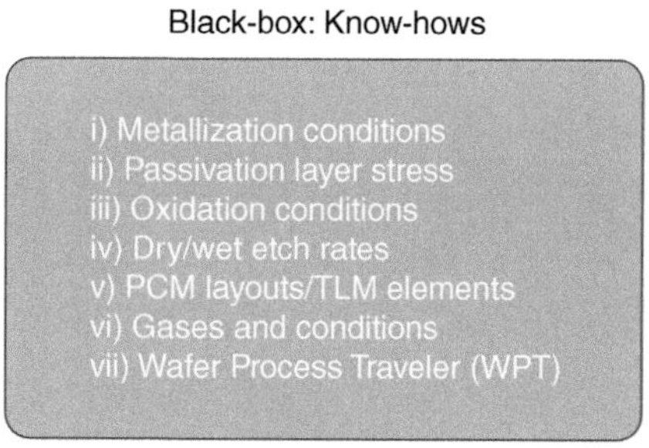

Figure C.8 Key process conditions for VCSEL manufacturing. *Source:* Babu Dayal Padullaparthi © Photonic Components DFM Ltd.

Black-box: Know-hows

i) Metallization conditions
ii) Passivation layer stress
iii) Oxidation conditions
iv) Dry/wet etch rates
v) PCM layouts/TLM elements
vi) Gases and conditions
vii) Wafer Process Traveler (WPT)

Table C.2 Critical points for various products.

	Critical Points (Tolerances)		
Product	**Significant Impact (Cpk)**	**Substantial Impact (pk)**	**Moderate/Slight Impact (p)**
Datacom	• mesa-1 etching • oxidation aperture size and uniformity • passivation stress	• mask insulator process • polymer (BCB) planarization • isolation etching	• p- and n-contacts • dry/wet etch selectivity • pad-metal thickness • wafer thinning
3D Sensing LiDARs	• PROT stress	• microlens (only for LiDARs)	

Source: Babu Dayal Padullaparthi © Photonic Components DFM Ltd.

C.2.5 Quality Gates: Critical Points and Data Handling

Similar to semiconductor electronics manufacturing, new product or technology introductions (NPI or NTI) stages also apply to VCSEL chip manufacturing. The quality gates include a mix of process demonstrations and product readiness addressed by technology and operating teams together, as outlined in Figure C.3b. During these quality test gates, all parameters that can significantly, substantially, modestly, and slightly affect the reliability and product performances are identified through process capability indices (Cpk and/or pk). Some Cpk and pk items for VCSEL-based datacom, sensing, and LiDARs products are given in Table C.2. As a part of quality management systems such as Six Sigma (6σ), standard methodologies are used together with statistical process control (SPC) to define the process capability and quality, as shown in Figure C.9.

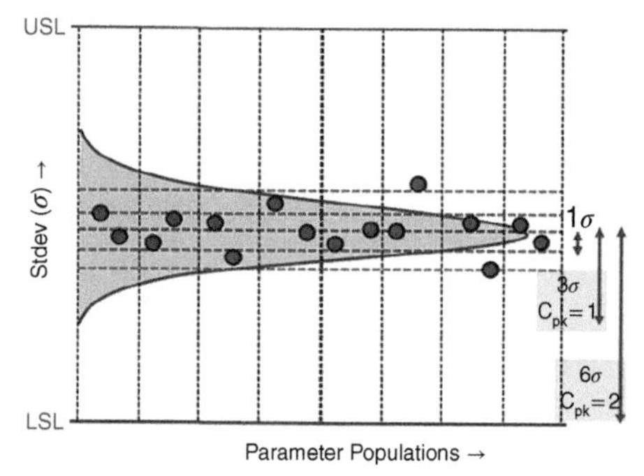

Figure C.9 Example of Six Sigma methodology used in parameter analysis. *Source:* Babu Dayal Padullaparthi © Photonic Components DFM Ltd.

C.3 Wafer Probe Measurements

As described in earlier sections, wafer level probing may be carried out at two stages: (i) before thinning and (ii) after thinning, called initial probe and final probe, respectively. Initial probe is generally a sampling of a select few devices on the wafer. LIV characteristics and peak wavelength measurements are conducted to ensure that proper electrical contacts were formed and expected device characteristics are achieved. Typically less than 1% of devices are tested

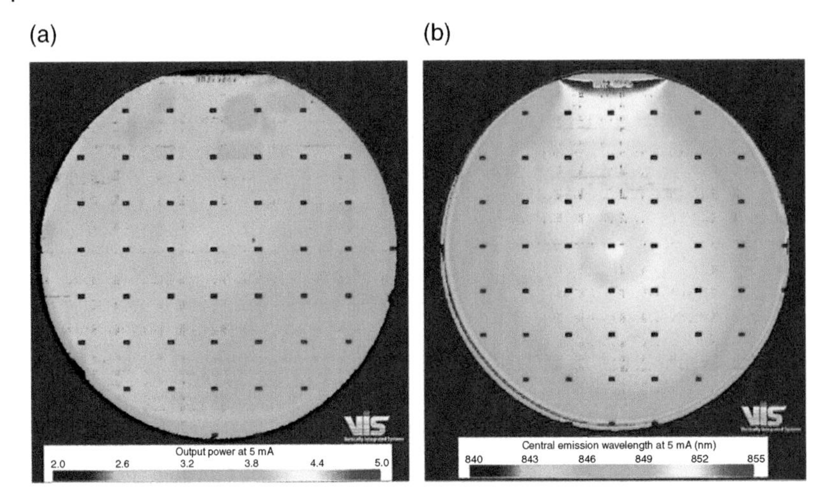

Figure C.10 Wafer probe (a) power and (b) wavelength maps for a fully processed 4″ (100 mm) GaAs-based 850 nm VCSEL wafer for datacom applications. *Source:* Reprinted with permission from VI Systems, Germany.

at this stage. At final probe stage, 100% of devices are tested. Statistical data of device parameters in the form of histograms are analyzed, and 6σ specification limits are determined, as shown in Figure C.9. If the overall probe yields cross certain threshold defined by individual manufacturers, the wafer is declared as meeting process acceptance criteria (discussed in Section 3.2.2.2) and then released for more component level validation, as described in Appendix D.

An example of final wafer probe (power and wavelength) mapping data for a fully processed 100 mm GaAs-based 850 nm VCSEL wafer used in high-speed datacom applications is shown in Figure C.10. This wafer shows excellent uniformities of power and wavelength across full wafer size with high yields (> 95%).

References

1 J. J. Krueger, R. Sabharwal, S. A. McHugo, et al., "Studies of ESD-related failure patterns of Agilent oxide VCSELs'," *Proc. SPIE* **4994** 162 (2003).

2 R. Therrien, S. Singhal, J. W. Johnson et al., "A. 36nm GaN-on-Si HFET Producing 368W at 60V with 70% Drain Efficiency," IEEE International Electron Devices Meeting (IEDM), p.p. 568–571, (2005) ISBN: 078039268.

3 M. E. Warren, D. Podva, P. Dacha et al., "Low-divergence high-power VCSEL arrays for Lidar application *Proc. of SPIE* **10552**," 105520E-1 (2018).

4 H. C. Neitzert, A. Piccirillo, and B. Gobbi., "Sensitivity of proton implanted VCSELs to electrostatic discharge pulses," *IEEE J. Sel. Top. Quantum Electron.*, **7,** 231 (2001).

Bibliography

[2008WEI] J. R. M. Weiss., "Nanometer-scale CMOS circuits and packaging for electro- optical high density interconnects up to 40 Gb/s," PhD. Thesis, 2008, doi:10.3929/ethz-a-005565086.

Appendix D

Wafer Level Testing

Jim Tatum

D.1 Introduction

As has been mentioned in this book many times, one of the principal advantages to VCSELs over other laser structures is the ability to test complete device performance at the wafer level. This includes measurement of the complete L-I and V-I characteristics, spectral properties, optical beam properties, high-speed characteristics, and RIN as well as the parametrics extracted from these measurements. It is routine to do the wafer level measurements at multiple temperatures. This massive amount of data collection enables the VCSEL maker to ensure final product compliance without incurring any packaging cost and transforms manufacturing into operations more reminiscent of LEDs and ICs than traditional lasers. A further advantage of the wafer level testing is the capability to detect shifts in the manufacturing process, rapidly develop new designs, and to troubleshoot material- or performance-related issues. Figure D.1 depicts a VCSEL manufacturing cycle with lumped data collection points at epitaxial growth, wafer fabrication, performance testing, and reliability. The data collected at these test points is fed back to previous process steps and fed forward to predict yield. The data is heavily utilized during product design to select the final configuration and to establish manufacturing controls and limits. Control is then passed to manufacturing yield to manage the product through its lifecycle. In the following sections, a more detailed review of the data collection in the major process steps is presented. The test level, either wafer level or package level, is shown by the colored boxes. In many cases, the final product for the VCSEL manufacturer is at the die level, so performance testing is typically done at the wafer and package level to develop test limits and correlations to final customer assembly. A schematic with the several different test levels of performance and reliability is shown in Figure D.2.

D.2 Post-Epitaxial Characterization

Immediately after epitaxial growth, a VCSEL wafer can be non-destructively characterized and mapped to estimate the manufacturing yield. This is an amazingly powerful tool to manage high-volume production. The most common characterization test at the epi level is to measure the white light reflectance spectrum of the wafer. The spectrum is acquired at several hundred points across the wafer and is analyzed to extract physical properties such as the laser resonance wavelength and then mapped onto the wafer area. Figure D.3a shows a typical measured reflectance spectrum of a high-power 940 nm VCSEL wafer, and Figure D.3b is a map of the measured FP-dip

VCSEL Industry: Communication and Sensing, First Edition. Babu Dayal Padullaparthi, Jim A. Tatum and Kenichi Iga.

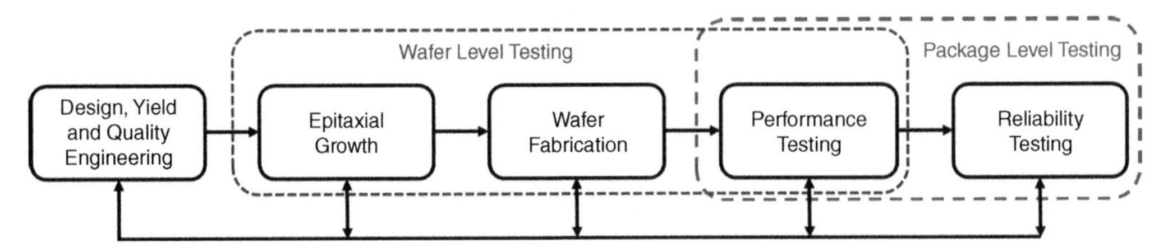

Figure D.1 Schematic of a VCSEL manufacturing process showing test aggregation points with design and yield feedback to enhance manufacturability.

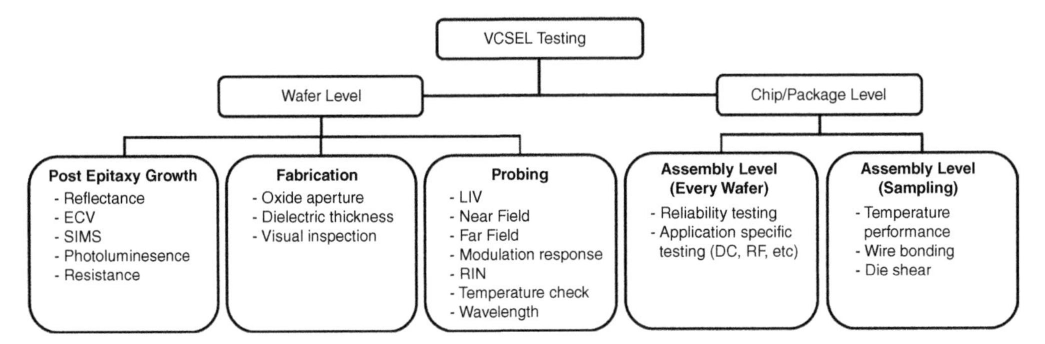

Figure D.2 Schematic of VCSEL testing stages.

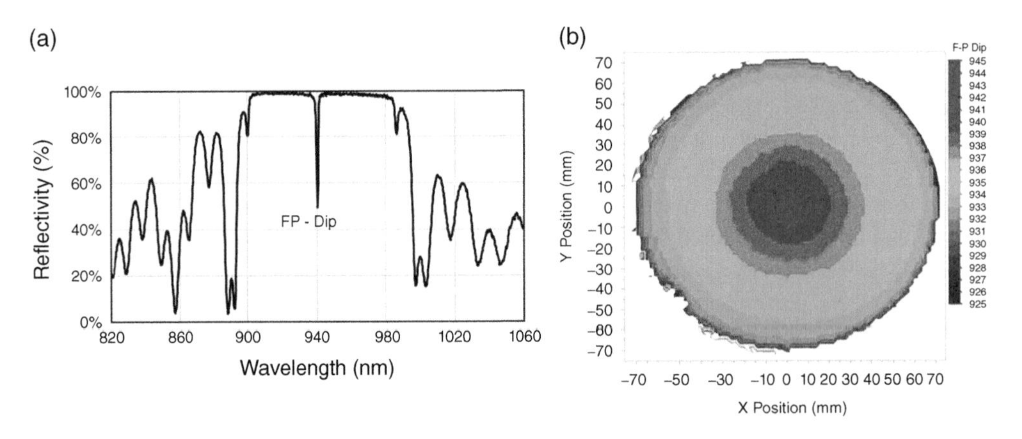

Figure D.3 (a) Measured VCSEL reflectance spectrum on wafer immediately after epitaxial growth. (b) Wafer map of the extracted FP-dip (laser resonance) across the 150 mm wafer. The standard deviation of the FP-dip is < 2 nm with a 2 mm edge exclusion zone.

(laser resonance) across a 150 mm wafer. The mean and standard deviation of the FP-dip are 936 nm and 1.8 nm, respectively. Other figures of merit such as the stopband width, the deviation of the FP-dip from the center of the stopband, locations of various satellite peaks, and many others can be mapped across the wafer from the measured reflectance spectrum.

The measured white light reflectance spectrum is valuable for the laser designer and the epitaxial grower because it can be fit to the laser design model (Appendix A). Any deviation in growth compared to design can be quickly analyzed and corrected. Extracted parameters from the measured spectrum, such as the reflectance at the FP-dip wavelength, can be correlated to subsequent measurements of the VCSEL threshold and slope efficiency. The wafer uniformity map is used by the epitaxial engineers to optimize the reactor growth uniformity; fitting of the measured spectrum can be a precise measurement of growth rate of the many VCSEL layers. Other wafer level maps of the VCSEL structure such as sheet resistance, surface defects, and photoluminescence are often gathered and offer further tools to guide the design of and control the manufacturing process.

D.3 Wafer Fabrication Data

During the VCSEL fabrication process there are many process areas where data can be collected and used by process engineers to control individual steps. VCSEL fabrication process details can be found in Appendix C. Control of the many processes is critical to maintain overall VCSEL performance and yield. The most critical steps for device performance are the oxidation and dielectric deposition steps, as described in Appendix C. The final oxidation aperture has a profound effect on all of the VCSEL performance characteristics but most critically on threshold current, series resistance, and optical mode structure. Because oxidation aperture measurements can be tricky and time-consuming to make accurately, they are generally made on a small sample of points across the wafer and then interpolated across the entire wafer area. The measurement of the oxide aperture utilizes a microscope with narrowband light emission that is tuned to find a spectral window that penetrates the DBR mirror but has reflectance contrast with the oxidation region. The spectral width of the window is fairly small, ~5 to 10 nm, and the character of the image can change dramatically over a very narrow wavelength. Figure D.4 shows the image of an oxide aperture in a VCSEL taken after complete device fabrication using an illumination source with 5 nm of spectral width and then tuned across 15 nm in central wavelength. Notice that the oxidation aperture appears black at shorter wavelength, then turns nearly invisible, and then finally appears bright over this narrow tuning range. The spectral window to view the oxide aperture can be predicted from the reflectance spectrum shown in Figure D.3a; note the minimum in reflected light near 860 nm that can be used to see the contrast of the oxidation layer. This behavior and non-circular appearance can make aperture measurements a challenge to manufacturing. The oxide aperture shape is controlled by the photolithography method of accessing the oxidation layer, the VCSEL epi design, and the oxidation conditions. This critical process is further described in Appendix C.

There has been considerable effort in this area to automate the wavelength tuning and use advanced image recognition to determine the oxidation aperture. (Note in this image, the presence of metal makes the contrast adjustment more difficult.) The measured oxidation data can be used to predict areas of the wafer where the aperture is appropriate for single-mode operation, for example, or determine areas that may perform differently in high-speed applications, reliability, or simply correlated to customer acceptance criteria. Figure D.5 is an example of an interpolated aperture map on a 150 mm diameter wafer using 9 measured points.

The second most important fabrication process that directly affects the VCSEL performance characteristic is the total amount of dielectric deposited over the emission aperture. The amount of dielectric deposition depends on the underlying DBR design and is typically an integer multiple of $\lambda/4$ to maintain proper phase of the optical field in the VCSEL. The thickness of the dielectric can have a profound effect on the VCSEL threshold and slope efficiency if it is either too thick or too thin. The total reflectance of the mirror, and thus the

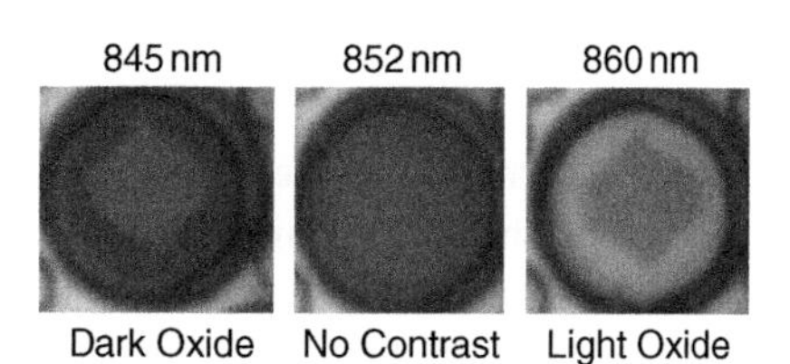

Figure D.4 VCSEL oxide aperture imaged with different center wavelengths showing the variation in appearance of the oxidation and semiconductor material. The spectral width of the illumination source is approximately 5 nm in all cases.

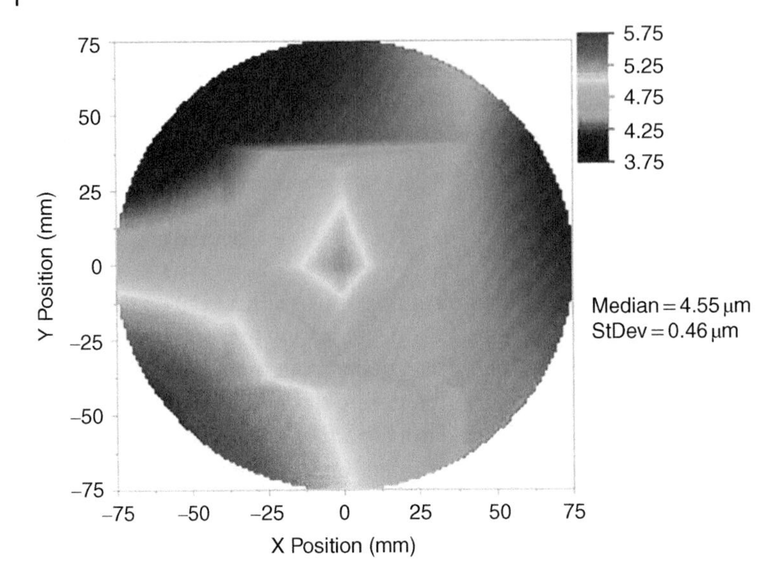

Figure D.5 VCSEL oxide aperture measurement interpolated onto a 150 mm diameter wafer from 9 measured data points. The quality of the interpolation improves with more measurements and more-broadly spaced points.

threshold and slope efficiency, will vary through the optical periodicity of the dielectric thickness. If multiple dielectric types and thickness are used in the fabrication process, it is best to treat them as thin films to calculate the overall mirror reflectance in the DBR design. The other consideration for dielectric thickness is to achieve low stress impermeable films to better facilitate non-hermetic applications. A typical final film thickness is 1λ, which becomes an optically disappearing layer to the VCSEL. Figure D.6 shows the dependence of the laser threshold current, slope efficiency, output power, and wall plug efficiency of a VCSEL as a function of the dielectric thickness variation. The total thickness needs to be well centered and variation across the wafer minimized. The dielectric thickness can be used to tailor the VCSEL power and efficiency depending on the use condition. Similar to the oxidation interpolation, the measured dielectric thickness can be mapped across the wafer and correlated to VCSEL performance characteristics. The total film thickness is measured using an ellipsometer, and variations of +/− a few percent can be tolerated if the total thickness is well centered.

D.4 Wafer Level Performance Test

Complete turnkey wafer-level test systems are now available from multiple vendors (MPI, Cascade Microtech, Chroma, among many others) or can be configured from off-the-shelf electronics and wafer probers. The basic configuration is a traditional XY probe station configured with a temperature-controlled chuck and vacuum to hold the wafer in place. One significant challenge in VCSEL manufacturing is handling and probing wafers after they have been thinned to final thickness, typically 100–200 μm. The inherent built-in stress from the many DBR layers, the oxidation, and the various dielectric coatings can make the wafers extremely fragile when compared to GaAs ICs and LEDs. There have been recent developments using carriers to mount the VCSEL wafer by several test system makers, but these carriers may limit the testing to pulsed application because of the reduction in thermal capacity.

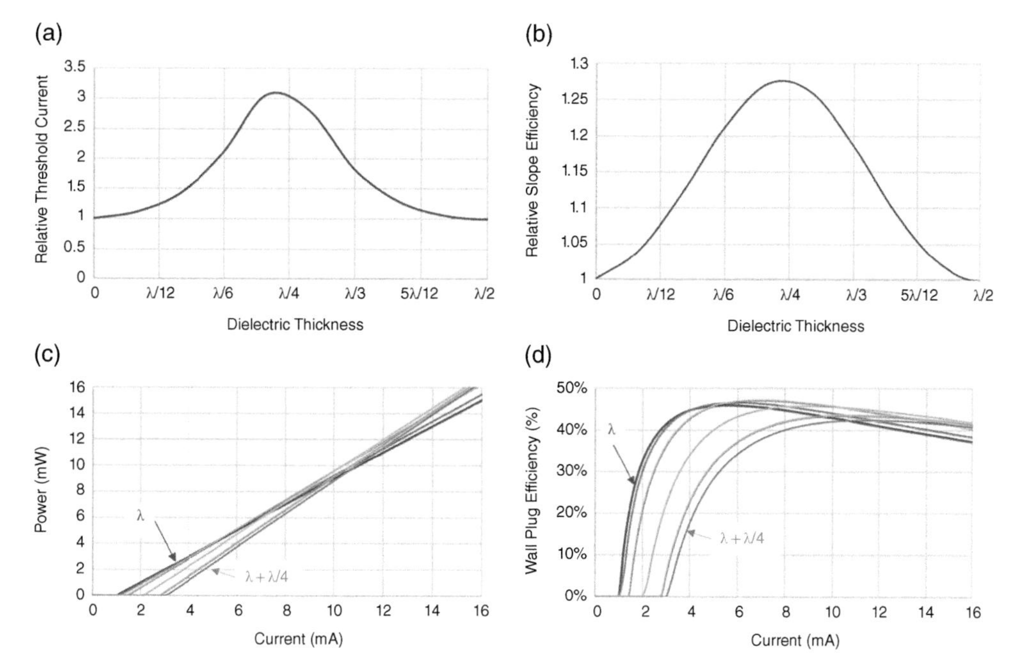

Figure D.6 (a) Variation of VCSEL threshold current and (b) slope efficiency as a function of the total dielectric thickness. In this example, a Si_3N_4 film with index of refraction n = 2 is assumed. (c) Power output as a function of current for different dielectric thickness and (d) wall plug efficiency as a function of current for different dielectric thickness.

Nearly all the tests described in Chapter 2 of this book can be conducted at the wafer level. This is an extremely powerful tool in both the design and production lifecycle and is not readily available in EEL laser manufacturing. A complete data set of DC performance parameters can be collected in a few hundred milliseconds for each VCSEL. As an example, Figure D.7 shows the wafer-level variation of threshold current, slope efficiency, power output, resistance, wall plug efficiency, and wavelength measured on a 150 mm wafer that contains a multitude of VCSEL designs.

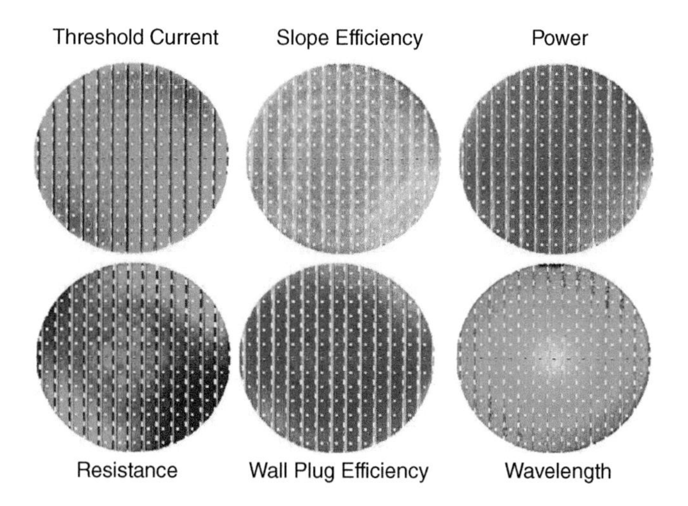

Figure D.7 Wafer maps showing the variation in DC-measured parameters on a 150 mm wafer with multiple designs included.

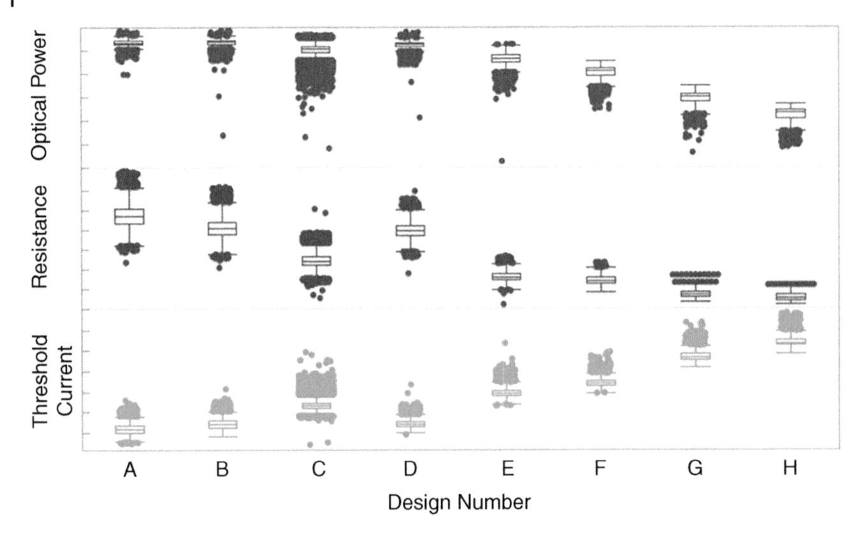

Figure D.8 Box plots showing the variation in DC measured parameters on a 150 mm wafer with multiple designs included.

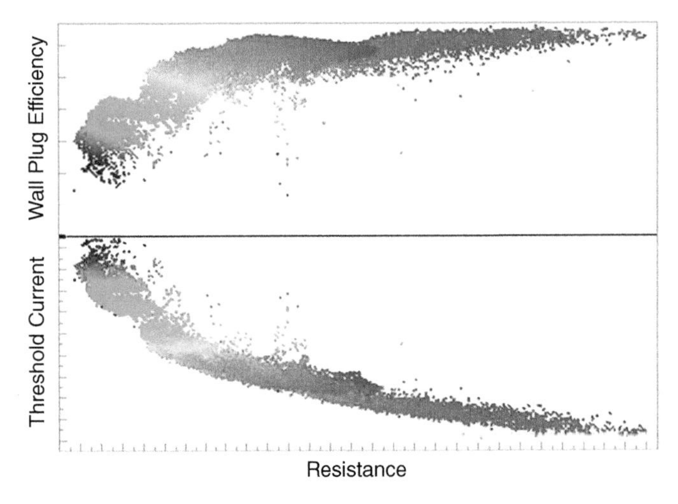

Figure D.9 Correlation of threshold current and wall plug efficiency as a function of the series resistance. The color axis is the measured optical power.

When multiple designs are included on the wafer, the relative performance between them can be compared in a statistically meaningful way. This is often done when optimizing 2D VCSEL array layout, for example. Figure D.8 shows the dependence of the laser threshold current, resistance, and optical power on each of the design variants.

In the design phase, the wafer level data is also used to determine functional dependence of the fabrication process, and the dependence of performance parameters on the design. Figure D.9 is an example of the dependence of the laser threshold current and wall plug efficiency on series resistance. In this case, the resistance is a proxy for the total emission area. This data is collected on every VCSEL on every wafer and creates a rich database for optimizing the VCSEL design and associated performance parameters. The data can also be used to set practical limits for data sheet parameters.

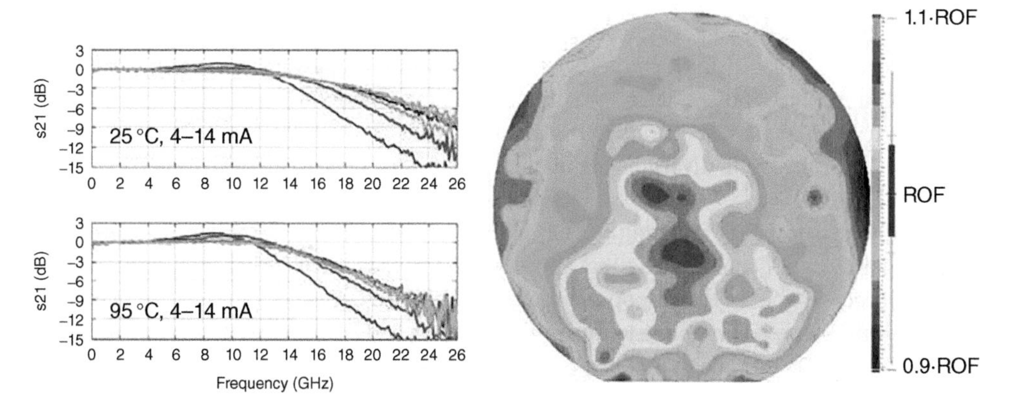

Figure D.10 Wafer map of relaxation oscillation frequency sampled and mapped across a 100 mm wafer [1].

Wafer level measurements are not limited to DC parameters. On-wafer measurement of parameters critical to high-speed VCSEL performance such as the relaxation oscillation frequency (ROF) and relative intensity noise (RIN) are also routinely made. Figure D.10 shows a wafer map of the ROF on a 100 mm production wafer [1]. The test time for a complete small-signal measurement is under 1 second and is similar for RIN. It is often not necessary to collect this data on every single VCSEL. The AC data can be correlated to the DC measurements, and thus only a sample of the VCSEL wafer needs to be tested. The measurement can be spatially and parametrically mapped to the remaining devices. Like the DC measurements, the data is used to refine the design and control the manufacturing process.

In high-power VCSEL array products, the beam divergence, emission power, and uniformity are critical to performance. The beam divergence can be measured for the overall die (far field) or even at the individual emitter level (near field and far field). The near-field images can be analyzed to determine the power in each of the individual emitters. Figure D.11 shows the uniformity of emission from a 2D array of VCSEL emitters. Uniformity of emission power is an important figure of merit in structured light sensors. Figure D.11 shows the emission pattern with a 1 ms pulse and a 10 ms pulse with the data collected at the beginning and the end of the pulse. The uniformity of the emission changes as the die heats. This nonuniform emission translates to a variation in the temperature of the individual emitters, which has impact on both the uniformity and potentially the reliability of the array, as further discussed in Appendix E.

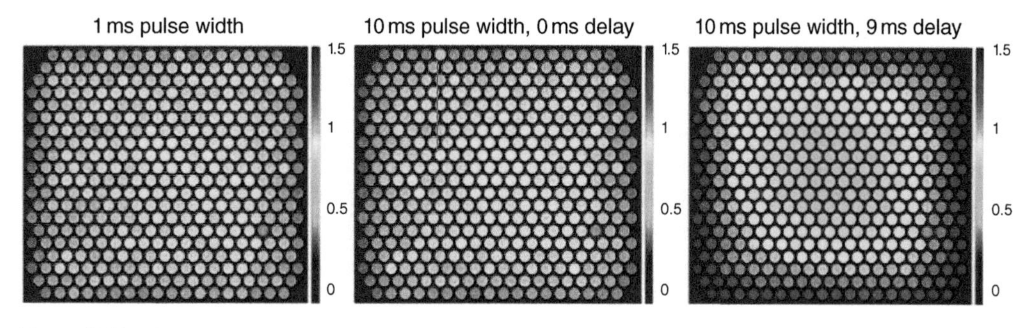

Figure D.11 Power emission form a single die on a wafer under different pulse drive conditions and time delay before image acquisition [2].

To mitigate heating during wafer level testing, pulse durations of 100 μs or with less than 1% duty cycle can be employed. This is particularly important for LiDAR applications where the pulse widths are more typically in the nanosecond range. Wafer test systems with high peak currents and nanosecond-level pulses are not readily available, so often it is necessary to correlate performance at the actual use conditions to the measurements at the wafer level to ensure compliance. Test times for the beam divergence and die level uniformity are typically under 1 second per measurement.

When the wafer level test data from epitaxy, fabrication, and device testing is taken in aggregate, correlations of performance can be used to set appropriate process control limits at all stages of the manufacturing process. The feedback loop as described in Figure D.1 is used for control and continuous process improvement programs to maximize product yields. Using this data has driven die level yields well above 90% in many applications, a critical metric to achieving cost objectives in high-volume consumer applications.

D.5 Package Level Validation

For many VCSEL makers, the final product may be at the die level, and packaging will be performed by the next level customer. It is possible for VCSEL characteristics to change after wafer dicing and packaging due to mechanical stress or thermal processing. Therefore, it is common for the VCSEL manufacturers to package devices in a similar fashion to the final customer and collect performance data to ensure suitable performance is maintained. For example, in a data communications application, VCSELs are often packaged in TO-can assemblies and tested for compliance of parameters such as relative intensity noise, modulation response in the form of rise/fall times or optical eye diagrams, spectral width, and fiber coupling efficiency, as described in Chapter 4. For large area 2D array die, the thermal properties of the die attach medium and the package thermal characteristics are critical to performance. The 2D arrays may be packaged similarly to the final customer assembly and tested for uniformity of emission and beam divergence (both near and far field), and the data is used to develop appropriate wafer level test limits. Reliability testing requires the VCSEL supplier to package and test devices at multiple increments in time and under a variety of stress conditions, as described in Appendix E. The changes in operating characteristics during these tests are used to establish the beginning of life wafer test limits to guarantee compliance to data sheet parameters at the end of life. Alternatively, the aging characteristics may be derated from the beginning of life values by the end user. In either case, it is important to consider how the VCSELs will change during operation, and this must be incorporated in the full system design.

D.6 Summary

One of the differentiating features of VCSELs is the ability to do complete DC, AC, and beam parameter testing at the wafer level. The amount of data collected is limited only by the desired data and/or the required test time. The data is used by all facets of the VCSEL design and manufacturing organizations to define products, control the manufacturing process, and to troubleshoot the inevitable drift in performance over time. Deep data collection is not readily available in EEL manufacturing at the wafer level and is typically only collected at higher levels of assembly. This drives manufacturing cost, where it is essential to take yield losses as early in the process as possible.

References

1 P. Westbergh, D. Gazula, G. Landry, T. Gray, E. Shaw and J. Tatum, "Turbocharging the Datalink," Compound Semiconductor, 2018.

2 Graham, H. Chen, J. Cruel, J. Guenter, B. Hawkins, B. Hawthorne, D. Kelly, A. Melgar, M. Martinez, E. Shaw and J. Tatum, "High-power VCSEL arrays for consumer electronics," *Proc. SPIE* vol. **9381** (2015).

Appendix E

Reliability and Product Qualification

Jim Tatum

E.1 Introduction

The development of reliability models and the resulting lifetime predictions across a wide variety of operating conditions was critical to the adoption of VCSELs in data communications systems. The control of the many complex design and manufacturing processes is a necessity in maintaining the quality and fidelity of the VCSEL over manufacturing life. This is reflected in the various qualification requirements dictated by several standards organizations and associated applications. Reliability may be expressed in many forms, and one common method is the so-called bathtub curve, as shown in Figure E.1. The total observed failure rate may be broken into time frames of failures. Early failures are generally a result of defects created during the manufacturing or assembly processes. Examples of these failure modes might be defects in the semiconductor crystal, electrostatic discharge, or improper connections. These failures are generally eliminated through various tests during manufacturing or through initial burn-in of products and is characterized by a decreasing failure rate with time. Random failure rates happen constantly over time, and it is often difficult to ascertain specific causes. In VCSELs this might be a result of a defect propagation in the semiconductor crystal. Failures that increase over time or length of device operation are known as wear-out failures. Reliability metrics are often expressed in terms of failures in time (FIT), which is defined as the number of failures per billion device-hours. FIT can be also defined in terms of MTTF as 10^9/MTTF, where MTTF is the mean time to failure. For single-channel VCSELs at normal operating conditions, typical FIT values are typically less than 1000, or MTTF on the order of 10^6 hours. In this appendix, an approach to reliability modeling is presented, and the qualification requirements in several application segments are presented.

E.2 Reliability Model Development

The purpose of reliability model development is to determine how the operating characteristics of VCSELs will change over device lifetime and different environmental conditions. The most commonly used metric is the change in the operating power expressed as a function of the VCSEL current and junction operating temperature. The relative acceleration factor (AF) between the operating condition and a reference condition is modeled using the modified Arrhenius relationship

$$AF = \left(\frac{I_{OP}}{I_{REF}}\right)^n \exp\left(\frac{E_A}{k_b}\left(\frac{1}{T_{REF}} - \frac{1}{T_{OP}}\right)\right),$$

Figure E.1 Reliability bathtub curve showing the early failure rate (red), random failure rate (blue), wear-out failure rate (green), and the total failure rate (black).

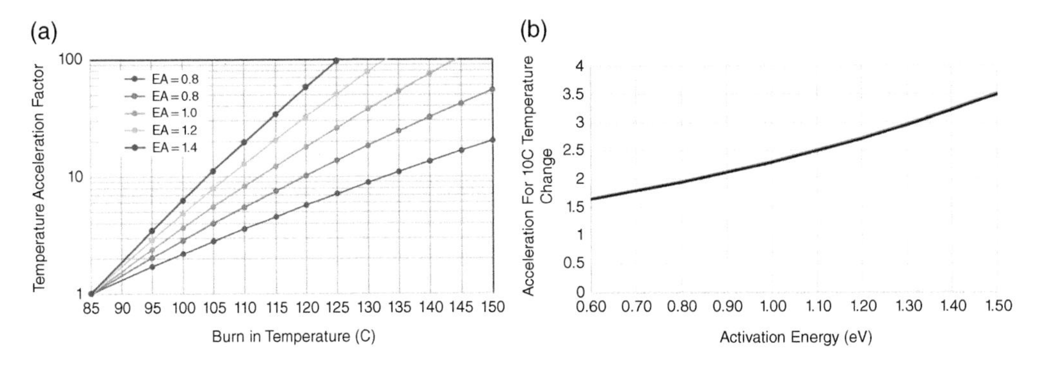

Figure E.2 (a) Plot of acceleration factor as a function of temperature for several values of activation energy assuming T_{REF} is 85°C. (b) The acceleration factor for a 10°C rise in the temperature as a function of activation energy.

where I_{OP} is the operating current, I_{REF} is the reference current, n is the current acceleration fact, E_A is the activation energy, k_b is Boltzmann's constant, T_{OP} is the operating temperature, and T_{REF} is the reference temperature. I_{REF} and T_{REF} are generally the nominal operating point. (Note: The temperatures in this equation are in kelvins.) Reliability test studies are designed to determine the values of n and E_A for a given VCSEL design. The AF for several values of E_A are shown in Figure E.2a, and the AF for a change in operating temperature of 10°C is shown in Figure E.2b.

The currents are simple to define and measure, but the temperature is more complicated, as seen in Figure E.2, and can lead to large changes in the estimation of the activation energy and lifetime predictions. For simplicity to the user, the temperature specified for reliability is generally defined as the VCSEL substrate temperature. However, the modeling must be done using the actual temperature of the VCSEL pn semiconductor junction, $T_{JUNCTION}$, which is a function of both the ambient temperature and the injection current. Determination of $T_{JUNCTION}$ is critical to accurate modeling of the reliability. Because of the temperature dependence of the thermal impedance of semiconductor materials such as GaAs, the junction temperature should be measured at the actual reliability test conditions. The junction temperature can be estimated by measuring the emission wavelength as a function of the ambient temperature and the total power dissipation, P_{DISS} (Figure E.3a). The dissipated power is related to the operating current, I_{OP}, the operating voltage, V_{OP}, and the emitted optical power, P_{OPT}, by $P_{DISS} = I_{OP} \times V_{OP} - P_{OPT}$. An example data set is shown in Figure E.3a for two ambient temperatures. From the plot of wavelength as a function of P_{DISS},

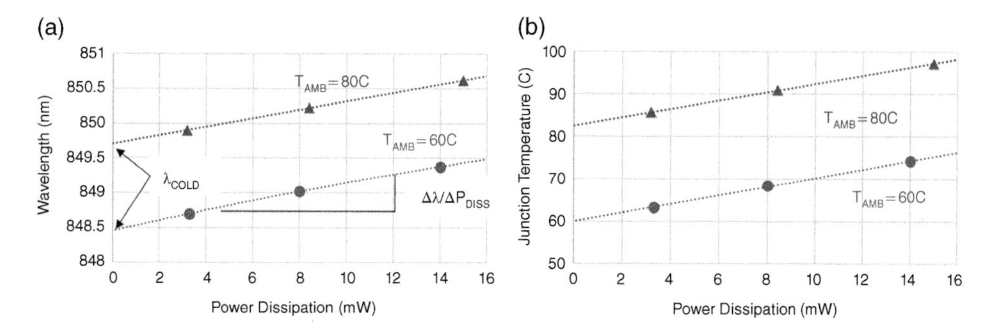

Figure E.3 (a) Plot of wavelength as a function of P_{DISS} and (b) the calculated junction temperature as a function of P_{DISS} for ambient temperatures of 60 and 80°C.

the cold cavity wavelength λ_{COLD} can be determined for each temperature, and the slope of the fit line determines the (temperature dependent) change in wavelength as a function of power dissipation, $\Delta\lambda/\Delta P_{DISS}$. Finally, the junction temperature can be determined using $T_{JUNCTION} = T_{AMB} + (\lambda_{OP} - \lambda_{COLD})/(\Delta\lambda/\Delta P_{DISS})$, where λ_{OP} is the wavelength measured at I_{OP} and T_{OP}. Figure E.3b shows the junction temperature as a function of the laser current for ambient temperatures of 60 and 80°C. (Note: The analysis here is correct strictly for single-mode VCSELs. When multi-mode VCSELs are analyzed, a single mode, typically the longest-wavelength mode, should be isolated and tracked over temperature and current.)

The analysis presented here is for a single emitter under continuous wave operating conditions. When applied to two-dimensional VCSEL arrays, the junction temperature of each of the elements may be different, and the difference depends on the electrical drive conditions. As an example, a thermal imaging camera was used to measure the temperature of individual VCSEL elements in a two-dimensional array and is reproduced in Figure E.4 [1]. The analysis shows more than 15°C variation in the emitter temperature from the center to the edge of the array. The linear analysis of wavelength as a function of temperature presented above may need to be modified to include non-linear terms to accurately model the reliability. When modeling reliability of two-dimensional arrays, the hottest emitter is generally used in the lifetime predictions. Under pulsed operation, the variation in temperature across the array can be a strong function of the pulse width. The thermal time constant of an individual active region is on the order of a few microseconds, but the entire two-dimensional array may have significantly longer settling times, up to a few milliseconds, depending on the packaging. For low-duty cycle operation, and short pulse widths ($t_{PULSE} < 1$ μs) the thermal gradient may be negligible. For this reason, two-dimensional VCSEL arrays are often

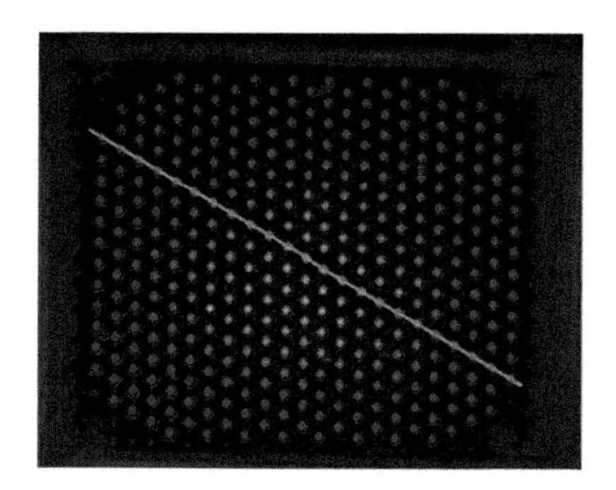

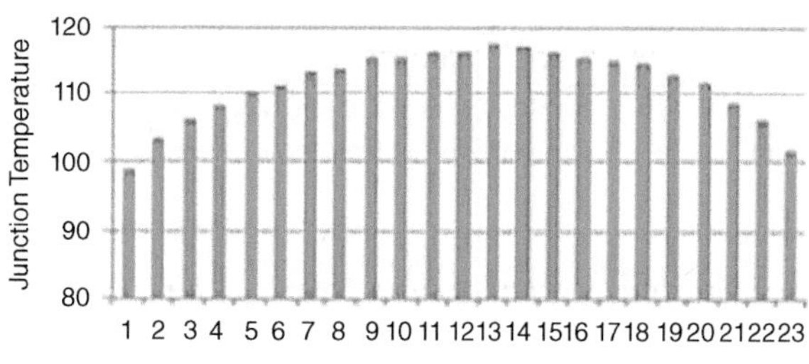

Figure E.4 Thermal image of a two-dimensional VCSEL array and the measured temperature profile across the diagonal line [1].

Table E.1 Summary of published reliability models.

Wavelength	VCSEL type	QW type	EA	n	References
670	oxide	AlInGaP	0.6	2	[5]
750–850	oxide	AlGaAs	0.6	2	unpublished
850	proton	GaAs	0.7	2	[3]
850	oxide	GaAs	0.7	2	[4, 6, 7, 8]
850	oxide	InGaAs	1.3	3	[9, 10]
850 2D array	oxide	GaAs	0.5	5	[11]
940	oxide	InGaAs	1.3	3	[1]
1060	oxide	InGaAs	0.7	1	[12]
1310	oxide	InGaAsN	1.17	4	[13]
1550	tunnel	InGaAsP	0.79	1	[14]

tested using pulsed drive currents to manage the thermal profile of the emitters and to better mimic actual operating conditions.

To create a reliability data set to fit the Arrhenius relationship often requires highly accelerated test temperatures and currents [2]. It is important to not overstress the VCSELs because different failure modes can occur, particularly at highly accelerated currents. A test matrix of multiple currents and temperatures is required, and often more than six conditions are needed [2, 3, 4]. It is good practice to have at least two of the expected junction temperatures to be the same to aid in determining the current acceleration factor. This can be estimated using the $T_{JUNCTION}$ model presented earlier. Currents should span the expected operating range of the VCSEL in normal operating regimes. Operation at very large currents outside of the normal operating regime can induce new failure modes that may not be present in ordinary operation conditions. Once the test conditions are established, the burn-in process begins with readings of the optical and electrical parameters taken over time. Burn-in testing proceeds until enough parts have degraded to failure, generally defined as 1 or 2 dB decrease in the optical power. Note that the VCSELs are generally tested at room temperature and the nominal operating current. Once sufficient degradation has occurred to the point of failure in a significant number of parts, the lifetime at the use condition can be estimated. The most common failure distribution used in VCSEL reliability analysis is lognormal, though Weibull distributions are sometimes used. In a lognormal distribution, the natural logarithm of the individual failure times is fit to a normal distribution with a mean value of μ, giving a mean lifetime of e^{μ}, and a standard deviation of the lifetime, σ, for each test condition. The mean lifetime (or other metric such as TTX%F) is then used to fit the data to the Arrhenius relationship, and the standard deviation is used to calculate the confidence in the model. The reported values for E_A and n for several published reliability studies are summarized in Table E.1. Note that the discussion here has centered on high-temperature operating lifetime (HTOL), but other metrics such as low-temperature operating lifetime (LTOL), temperature cycling, and other test methods can also be explored.

E.3 Failure Modes in VCSELs

To make meaningful reliability model predictions, it is necessary to understand the degradation mechanism of the VCSELs in accelerated life testing, as described in E.2. Care should be taken when defining the accelerated stress conditions because different failure modes can be observed in high-stress burn-in compared to normal operating conditions, and catastrophic events can lead to rapid failures. The failure modes in VCSELs can be broken into two categories, (i) long-term wear out characterized by a gradual decrease in the optical power over aging and (ii) rapid degradation generally caused by a catastrophic event and subsequent defect propagation. The susceptibility to the several failure modes can be influenced by the VCSEL design, operating conditions, and material handling during the manufacturing process and device operation. In this section, a brief description of the failure mode types and impact on device operating lifetime is presented. It is important to note that when VCSEL operating lifetime model is reported as in Table E.1, it is for the wear-out lifetime and does not include rapid degradation mechanisms. Rapid degradation due to catastrophic events is outside the laser manufacturer's control, and in any case rare damaging events of multiple types cannot be encompassed in any single reliability model. Rapid degradation due to built-in defects likewise resists modeling unless such defects are in a large fraction of the population, but that is unlikely in any actual product. In the following sections, the failure modes of VCSELs in both rapid and long-term reliability studies are discussed. It is useful to consider the failure modes in terms of being thermally or electrically generated. Events that happen on time scales larger than the thermal time constant of the VCSEL active region ($t_{THERMAL,ACTIVE} < 10\ \mu s$) or the entire VCSEL chip ($t_{THERMAL,CHIP} < 1$ ms) tend to create damage that is thermally induced, whereas events much faster than $t_{THERMAL,ACTIVE}$ tend be caused by material breakdown caused by extremely high electric fields.

E.3.1 Electrical Overstress (EOS)

EOS is most often associated with high-current conditions for long periods of time ($>> t_{THERMAL,ACTIVE}$) and result in melting or intermixing of the semiconductor layers that make up the VCSEL mirrors and active region. The VCSEL IV generally fails to a short circuit, and damage to the emission region is often visible on the surface of the VCSEL near the emission aperture. Current densities >100 kA/cm^2 for times longer than $t_{THERMAL,CHIP}$ can produce EOS events. An example top view from a transmission electron microscope (TEM) and cross-section scanning electron microscope (SEM) images are shown in Figure E.5.

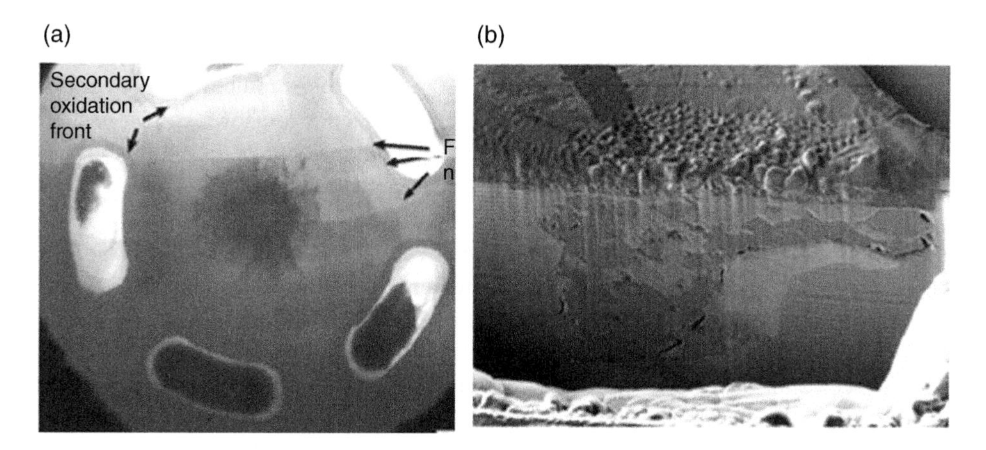

Figure E.5 (a) top view from a TEM cross-section of a VCSEL active region that has failed due to electrical overstress (b) Cross-section SEM view of the VCSEL layers showing intermixing and melting of the individual layers [15].

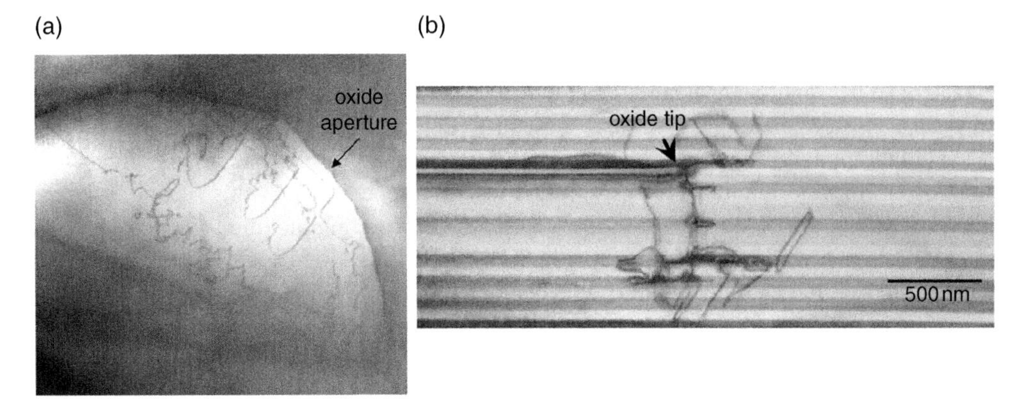

Figure E.6 (a) Plan-view TEM image of an oxide isolated VCSEL showing the propagation of DLDs across the quantum well active region. (b) Cross-section view of the same device in (a) showing the DLD propagation in the quantum well active region [17].

E.3.2 Electrostatic Discharge (ESD)-Induced Damage

Like all semiconductor devices, VCSELs are susceptible to damage form ESD events. The damage threshold depends on the type of ESD (charge device model [CDM], machine model [MM] or human body model [HBM]), the size of the VCSEL emission region, the type of quantum wells, and the placement and size of the oxide aperture. The general failure mode as a result of ESD-induced dislocation damage is the propagation of dark line defects (DLDs) from the damage location to the quantum well active region. To visualize ESD events in a VCSEL, a novel plan-view transmission electron microscope (TEM) sample preparation technique was developed and described in [16]. Figure E.6 shows a plan-view image of an oxide isolated VCSEL and a corresponding cross section showing the active region. In this example, an ESD event has created a damage point, and the DLDs have propagated to the active region causing a reduction in the optical output power.

The ESD models represent different amounts of stored charge and rate of discharge modeled by a capacitor holding the charge. Both HBM and MM ESD are modeled by the charge stored in a 100 pF capacitor. In HBM, the capacitor is discharged through a 10 μH inductor in series with a 1500 ohm resistance connected in series with the device under test. The MM ESD source is modeled with a 500 nH inductor in series with the device under test. CDM is modeled as a voltage pulse with rise and fall times of 400 ps and a pulse duration of 600 ps. To understand how an ESD pulse flows through a VCSEL, it can be modeled using lump circuit elements, as shown in Figure E.7. When the ESD event happens, current will flow in the different lump elements depending on the type of ESD event and thereby generate damage in different regions of the device, as indicated in Figure E.7.

The signature of damage caused by ESD events can be directly observed with both destructive and nondestructive methods. The most common measurement of ESD-related damage (or, more generally, damage to the quantum well active region) is to measure the leakage current as a function of reverse bias voltage. Figure E.8 is a plot of the reverse I-V characteristic for a normal VCSEL and one that has been subjected to ESD damage. Note the increase in leakage current prior to avalanche breakdown of the device. The leakage occurs where defects or dark lines have damaged the active area and created a shunt current path in parallel with the junction. The amount of leakage current will be somewhat proportional to the amount of damage caused, or at least the areal extent of the damage. HBM, MM, and CDM events can lead to slightly different character of the

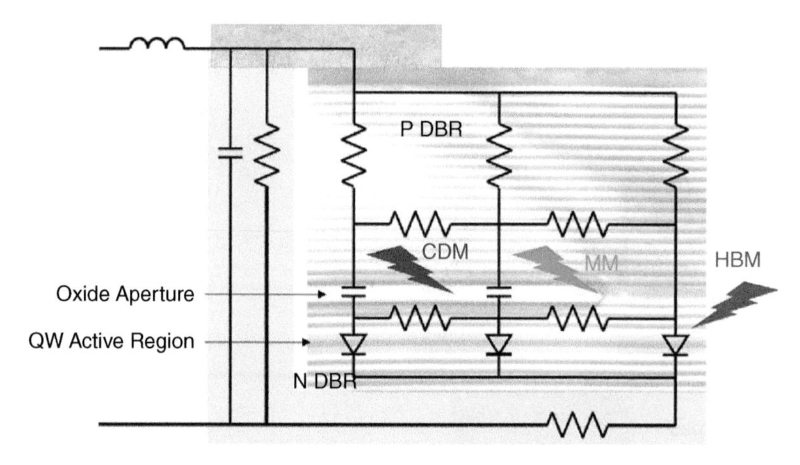

Figure E.7 Cross-section view of a typical oxide isolated VCSEL with a lumped equivalent circuit model superimposed. The location of damage for CDM, MM, and HBM ESD events are noted on the figure. Analysis follows from [15].

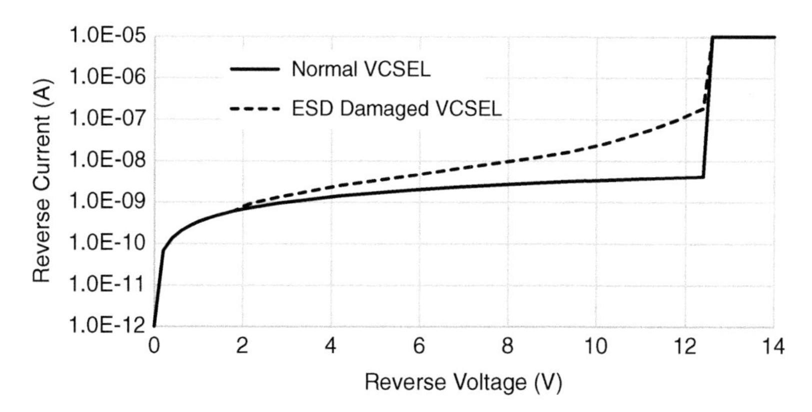

Figure E.8 Reverse I-V characteristic of a normal VCSEL and one that has been exposed to an ESD damage event.

leakage current, but the exact differences are not consistent enough to be diagnostic. To determine the ESD event type precisely, destructive analysis is likely required.

Another nondestructive technique of visualizing ESD damage in a VCSEL is forward and reverse bias electroluminescence. Here the VCSEL is operated in reverse bias in the voltage region where the leakage current has increased relative to a normal part. In forward bias the VCSEL is operated at extremely low currents, on the order of 1% of threshold. The emission aperture is imaged with a high-resolution and high-sensitivity infrared camera. Light will be emitted from areas in the VCSEL that have not been damaged, but the areas with ESD damage will appear dark in forward bias, as indicated in Figure E.9. In reverse bias, light is emitted from the damage areas but is generally very broad and difficult to image. The ESD damage sites appear like small holes in the image, as previously discussed (see also Figure E.6).

TEM plan and cross-section views are the definitive way to view ESD- and EOS-related damage but are destructive and generally used as a last resort. With a plan-view TEM, the source location of the damage can sometimes be pinpointed and the defect propagation mapped. More often, the general location of the ESD damage can be identified, and as discussed earlier, the nature of the ESD event can be ascertained. This can be especially useful when trying to troubleshoot failures in

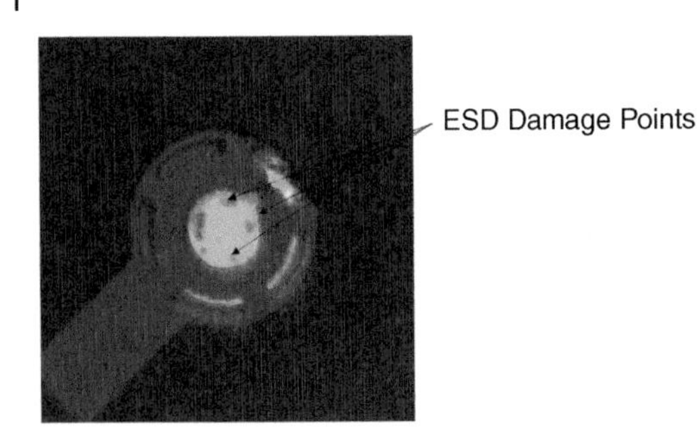

Figure E.9 Forward bias electroluminescence image showing ESD damage locations. [18].

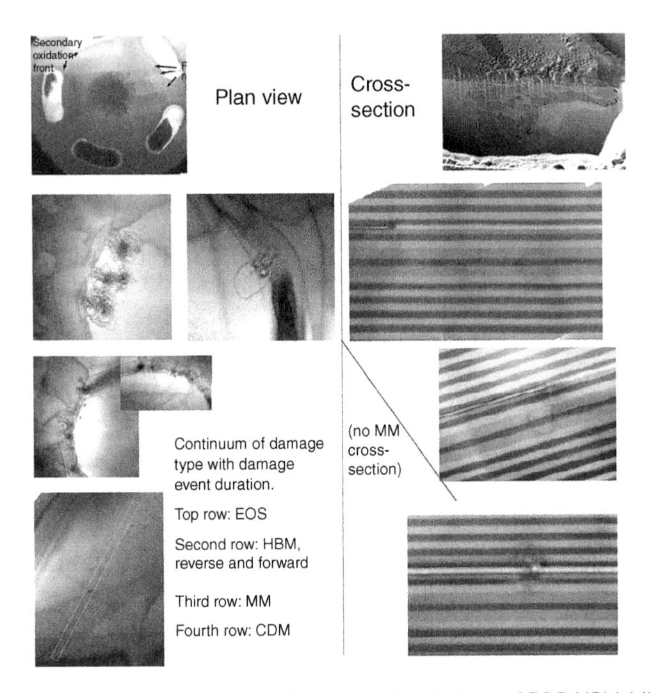

Figure E.10 Plan-view and cross-sectional views of EOS, HBM, MM, and ESD damage. [15].

a manufacturing environment or any place where VCSEL die or components are handled. Generally, once the VCSEL is packaged and inserted into a system, the device is protected by the surrounding elements. As an example of how TEM analysis can be used to determine ESD character, Figure E.10 is a summary of different ESD events and the corresponding images. Note that as the ESD pulse width decreases, damage moves from the center of the VCSEL (EOS) to near the oxide edge (HBM), the oxide edge (MM), and finally to the oxide layer itself (CDM), as indicated in Figure E.7.

The ESD damage level depends on several design aspects of the VCSEL including the emission diameter, the thermal and electrical impedance, the active region type, reverse breakdown voltage, and the thickness and placement of the oxidation layer. The junction damage to VCSELs is generally dominated by the reverse bias direction. This is due to the large voltage build up in reverse bias and the breakdown of points in the quantum wells versus current flowing in the entire active diameter in forward bias. To a reasonable approximation, the HBM ESD damage level scales with the diameter of the emission area. Figure E.11 shows the HBM ESD damage level for oxide VCSELs with a range of aperture diameters [4]. Also shown in the figure is a proton isolated VCSEL

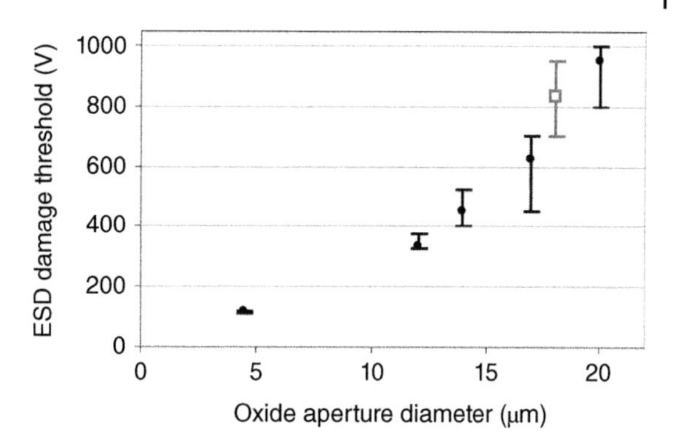

Figure E.11 Damage threshold for HBM ESD events as a function of the emission area. A proton isolated VCSEL is shown with the gray bar for reference. [4].

structure showing a similar damage level threshold. For CDM events, the damage is caused by breakdown of the oxidation layer, and the damage threshold tends to scale with the thickness of the oxidation layer. MM ESD events are somewhere in between on the damage continuum.

The previous discussion on ESD events has focused on damage in a single-emitter active region, but many of the 3D sensing applications will have VCSEL die with hundreds of individual active areas connected in parallel. In this case, the ESD damage scales with the number of apertures, or effectively, total emission area. In an ESD event, there is a limited amount of charge that is available to flow through the device. As the emission area increases, the total charge is dissipated over a wider region, lowering the amount of power dissipated in any single location. This increases the robustness of the overall 2D array of VCSEL emitters. For example, VCSEL arrays with tens of elements can be safe to several kilovolts of ESD, whereas the individual apertures may only withstand a few hundred volts.

E.3.3 Humidity-Accelerated Failures

The many diverse application areas and desire for low-cost packaging options have resulted in VCSELs being operated in non-hermetic packages and other uncontrolled environments. Testing of VCSELs in high temperature and high humidity has produced a range of failure mechanisms that must be considered during the design and operation of the system. The manufacturing process for oxide-confined VCSELs requires exposure of an $Al_xGa_{1-x}As$ ($x > 0.96$) layer within the mirror structure to a high temperature (~400°C) and steam environment to convert the oxidation layer to aluminum oxide, as described in Appendix C. Inevitably other parts of the mirror are also partially oxidized during this process. The final material quality of the oxidation layer depends on the material design and oxidation process. It is nearly impossible to make an oxide layer that is completely impervious to environmental conditions and to prevent the inherent built-in stress from the reduction in material volume going from $Al_xGa_{1-x}As$ to aluminum oxide. The sidewalls of the mesa or trench used to access the oxidation layer are generally covered in a conformal dielectric such as Si_3N_4, but there can still be points at which moisture can find its way into the VCSEL. Humidity-related failures can show many of the same nondestructive signatures of EOS and ESD: an increase in leakage current, reduction of optical power, and an increase in forward voltage. However, unlike EOS or ESD, when closely examined using TEM techniques, the source of the dislocation networks isthe edge of the VCSEL mesa or trench or directly from the oxidation layer with no other local damage. These failure signatures can also be observed as maverick early random failures in reliability testing, as shown in Figure E.12a and b.

Another failure mode associated with damp heat exposure is semiconductor cracking. In this case, the material stress induced by the oxidation process and subsequent humidity exposure can

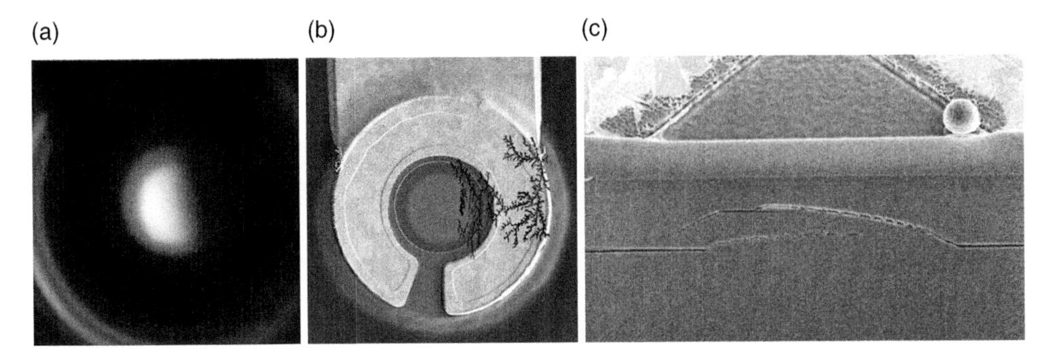

Figure E.12 (a) shows low-current luminescence image of a VCSEL with a darkened area. (b) Image is a VCSEL with the dark line defect network superimposed on the image. The source of the DLD networks is at the edge of the mesa and likely due to excessive mechanical stress or humidity exposure [19]. (c) Cross-section SEM of a VCSEL that failed in damp heat testing demonstrating the semiconductor cracking failure mechanism [20].

cause the oxidation layer to delaminate from the surrounding material and to crack the DBR mirror structure. This failure is generally observable in a microscope image and will produce a very elevated voltage. A cross-section SEM image of this failure mode is shown in Figure E.12c [20]. A few other failure modes for VCSELs exposed to damp heat environments have been reported [20] such as surface contamination and erosion, or the migration and collection of ionic contaminants during testing [21]. These failures are infrequent and thought to be a result of the interaction of the environment and the assembly materials, which can be preferentially deposited or grown in the presence of electrical bias or optical energy. Humidity-related failures can be eliminated by careful consideration of the material design, the oxidation process, and passivation coatings applied to exposed surfaces during device fabrication.

E.3.4 Radiation Tolerance

One unique application requirement for VCSEL reliability has been the use in high-radiation environments including inside CERN's Large Hadron Collider (LHC), space exploration, and even in some aircraft. With a relatively small active area volume, VCSELs have demonstrated robustness to exposure to high radiation levels and are generally more reliable than other components in the system. VCSELs have been tested under proton and gamma ray radiation [22], and the primary failure mode was found to be a result of dislocation of an atom in the semiconductor lattice. In radiation exposure the optical power decreases at a fixed bias current and is a result of both an increase in the threshold current and a decrease in the slope efficiency. The radiation-induced point defects in the crystal have also been shown to anneal with continued operation or storage in high temperature after the radiation exposure [22, 23]. A similar effect on the optical power and subsequent annealing was also found under neutron irradiation [24]. Figure E.13 shows the power output as a function of current of a VCSEL before irradiation, after exposure to 30 MeV protons, and after exposure and operation for 1 week at 10 mA. The recovery of optical power with continued operation is unique to radiation exposure and an indication that the point defects do not lead to the formation of permanent DLDs but to a temporary loss of carrier injection efficiency. VCSELs have been in use in the ATLAS detector (in the LHC) and on space qualified missions without significant loss of utility for more than 10 years.

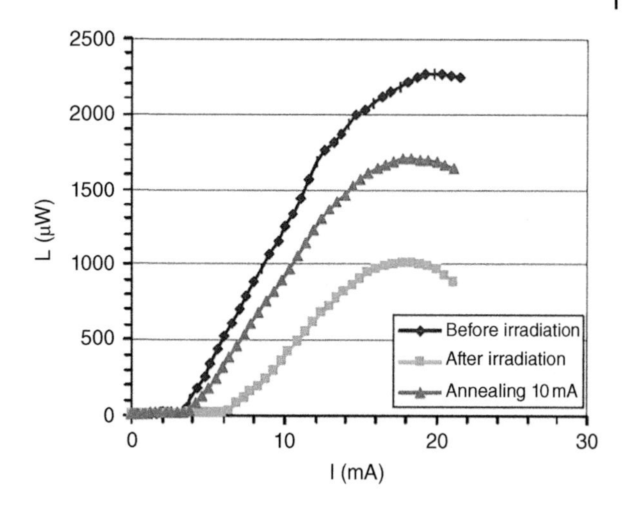

Figure E.13 Measured power output as a function of current for a VCSEL prior to irradiation, after irradiation from a 30 MeV source, and after irradiation and operating at 10 mA for 168 hours [22].

E.3.5 Long-Term Wear-Out Failures

The failure mode resulting in gradual reduction in optical power during accelerated aging tests is not well understood and does not provide a signature in modern failure analysis techniques. The prevailing theory is that over long periods, vacancies, dopants, or residual elements such as hydrogen ions and compounds from the epitaxial growth process can drift in the crystal. The migration under forward bias moves the point defects toward the VCSEL active region. The defects lead to an increase in carrier loss or reduce carrier injection efficiency into the active region. The signature of the degradation is primarily an increase in the threshold current [25] and has been observed from 780 to 980 nm. The use of power monitoring photodiode and control circuits can somewhat extend the lifetime by compensating the drive current [25].

E.3.6 Difference between GaAs and InGaAs

Table E.1 shows a wide range of E_A and n values obtained from several different wavelength VCSELs and multiple active region designs. VCSELs with InGaAs quantum wells tend to show higher E_A and n values while those containing AlGaAs quantum wells tend to have lower E_A and similar n values to devices with GaAs quantum wells. The addition of In to the quantum well creates compressive strain in the crystal lattice structure. The strain reduces the mobility of certain dislocations and other crystal defects and results in a higher activation energy in reliability testing. In some cases, the presence of the In-induced strain can even cause dislocation loops to circle back and pin to themselves thereby preventing further spread and degradation. Al-containing quantum wells do not have much impact on strain and do not prevent DLD or other dislocation propagation and thus have similar E_A to GaAs.

E.4 Array Reliability and Sparing

When considering the design of VCSEL-based systems, particularly those where the lasers are deeply embedded such as on an electronics board inside a computing or storage network, employing a spare laser may be attractive. Most failure modes in VCSELs are device specific and do not affect any neighboring devices. This is true for ESD- and EOS-induced dark line defects and long-term wear-out modes. In some cases, defects that are propagated from the substrate (threading

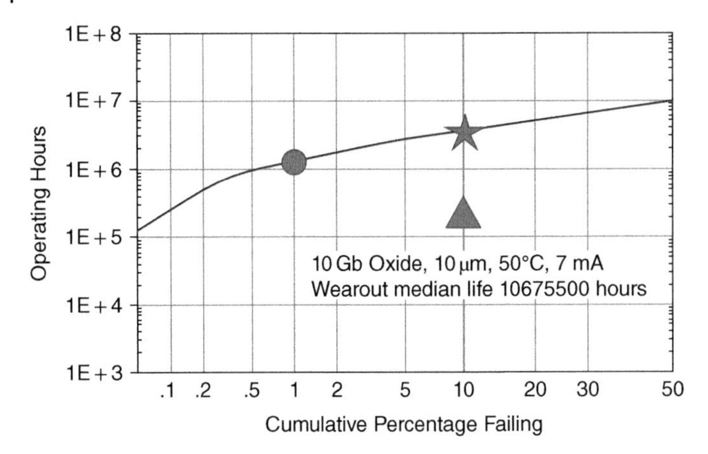

Figure E.14 Damage threshold for HBM ESD events as a function of the emission area. A proton-isolated VCSEL is shown with the gray bar for reference. [26].

dislocations) or particulates during the epitaxial growth process may span multiple VCSEL emission areas. Fortunately, these defects are rare and in one case study were found to be less than 1% of VCSEL field failures [15]. Since the lasers operate essentially independently, the failures will also be independent in time. Consider the case of a 10-element VCSEL array used in a data communication system. The reliability of these lasers is critical as the transceivers may be soldered onto motherboards deeply embedded in hyper-scale computers. The simple approach to analyzing failure would be to take the TT1%F for a single element and divide it by the number of elements in the array. In this case it would be TT11%F/10. However, this is not correct because of the independent nature of the failures. The correct extrapolation is to look at the time to 0.10% failures. This difference can be orders of magnitude in expected lifetime and is shown in Figure E.14 [26]. In this example, the predicted lifetime is plotted as a function of the cumulative percentage of parts failing. The star represents a reliability prediction of 3.8 M hours for 10% of single elements to reach failure at 55°C and 7 mA. If exponential statistics are used to estimate the reliability of the 10-element array, the predicted life time would be 0.38 M hours, represented by the triangle on the chart. This is an overly pessimistic view of reliability. In reality, with lognormally distributed failures and given that each device operates independently, the reliability of the 10-element array is scaled by moving down the failure curve to 1% fails, represented by the circle, which gives an operating life prediction of 1.6 M hours, more than 4x that predicted with exponential statistics.

One way to utilize the independent failure of VCSEL elements is to add extra elements to the array, known as sparing. The spare VCSELs may or may not operate concurrently with the other elements. When the concept of sparing is included, the overall system reliability can be dramatically improved. To demonstrate the value of hot sparing (where the spare VCSEL is powered along with the other elements) in the VCSEL system, the following model can be used [26]. Let P_S be the probability of failure for a single element; then the probability of an N-element array failing when any single element in the array fails, P_N, can be written as $P_N = 1 - (1 - P_S)^N \sim NP_S$. Now if there are R redundant (spare) elements in the array, the probability of a single element failure can be written as $P_R = P_S^R$; then the probability of failure of an N-element array with R redundant (spare) elements, P_{NR}, can be written as

$$P_{NR} = \prod_{a=0}^{R-1} (N - a) P_S$$

Table E.2 Impact on wear-out reliability as spare channels are added to VCSEL arrays. In this table, the array count is the total number of VCSEL emitters operated together on a single die, though the analysis is consistent with hot sparing done at a system level as well.

	Array count			
Spares	**1**	**12**	**24**	**168**
0	100%	43%	36%	24%
1		78%	64%	40%
2		102%	78%	48%
3		117%	91%	52%

The results of the sparing calculation for several array sizes and sparing quantities is summarized in Table E.2. One of the unique advantages of VCSELs in 3D sensing and LiDAR applications is that the emitter consists of many individual elements. Thus, the system is not dependent on a single point of failure such as the emission facet in an EEL. The 2D distribution of emission elements is also very beneficial in helping to meet the eye-safety requirements described in Appendix F.

E.5 Considerations for High-Power 2D VCSEL Arrays

Much of the previous discussion was applicable to both single VCSEL emitters and large array 2D arrays used in 3D sensors and LiDARs, and there are some aspects and challenges that are specific to 2D arrays to consider [11, 27]. The reliability and performance of large array 2D VCSEL arrays depends critically on the packaging and specifically on the quality of the thermal contact between the VCSEL die and the thermal sink. Eutectic die attach is a preferred option, and care must be taken to eliminate any voids in the solder. Voids can lead to hot spots on the die and significantly degrade reliability. In large area arrays, there can be some allowance for weak or dead emitters. These defects may be caused by particles from the epitaxial growth or threading dislocations from the substrate. In general, these are limited to one or a few localized emitters and they generally do not propagate to neighbor devices. Near-field testing of all die used in reliability testing and tracking of the power in each individual emitter is a best practice to ensure good thermal conduction and that weak or failed emitters do not propagate defects to surrounding emitters. Excellent process uniformity of dielectric passivation layers is also critical to ensure robust damp heat performance as final packaging in consumer electronics will seldom be hermetic. Automotive applications will require even more strict guarantees of quality as they move from infotainment systems to LiDAR applications. Reliability and product qualification testing for LiDAR and 3D sensors may need to be conducted under high-current pulse conditions to ensure no new failure modes are introduced as a result of the application use case. When driven pulsed, the peak current can be 10 to 100 times higher than in continuous operation and can lead to failures not revealed in lower-current testing. It can be an engineering challenge to design production burn-in systems to operate with nanosecond pulses.

E.6 Qualification Standards

With the emergence of VCSELs into so many disparate applications, several different qualification standards and requirements have been employed over the years. In the data communications environment, the focus is on the Telcordia GR468 process [28]. This standard was developed in the

1990s, and its primary objective was to insure the longevity of fiber optic components in undersea and terrestrial communications networks. Due to the high cost of implementation, many of these installed components must operate continuously for more than 10 years (87 600 hours) in potentially harsh environmental conditions. Telcordia qualification reflects the need for a long operating lifetime. With the emergence of VCSELs into several different consumer electronic devices, such as the Apple iPhone, the qualifications requirements shifted to the JEDEC 77 standard, which has a primary focus on shorter-term operating lifetime and random failure modes [29]. To understand the difference in operating lifetime, consider the on-time for a VCSEL array in a facial recognition system. Let's assume a five-year operating lifetime, that the user unlocks the phone 100 times per day, each of the unlock cycles takes 2 seconds, the VCSEL array is driven with a 50% duty cycle, the total operating time for the VCSEL is approximately 50 hours, and typically use temperatures are not too extreme. (Humans don't typically spend extended times in temperatures above 40°C or below −10°C.) Thus a quality system for consumer electronics should focus on random failures by testing a larger number of parts over a shorter time. As VCSELs have found applications in the automotive industry with driver awareness and emerging LiDAR markets, the automotive standard AEC-Q102 has become a new requirement for manufacturers [30]. The AEC-Q102 can be thought of as essentially a combination of all the components of Telcordia and JEDEC, with the addition of even a few more tests. The operating use condition of the VCSEL system requires testing and qualification at the operating extremes of −40°C to +95°C and over a relatively long operating lifetime compared to consumer electronics. A typical requirement for automotive is 100 000 miles, which translates to typically 3000 hours of engine-on time, for conservativeness usually rated to 5000 hours. This operating time is in between the Telcordia and JEDEC focus. In addition, the VCSEL could be in a safety-critical system in an automobile and could be in relatively extreme environmental conditions, so the AECQ102 requirements reflect this reality. A typical automotive qualification also comes with significant increase in the attention to quality systems requirements and may require system changes that can take more than a year to fully implement. For VCSEL suppliers, it may be sufficient to maintain level 3 AECQ requirements. Table E.3 is a superset of the various qualification tests for Telcordia, JEDEC, and AECQ102. Note that the standards are evolving, and Table E.3 should be considered as guidance only. As an example of a recent qualification set, Figure E.15 shows published results reproduced from [27].

E.7 Ongoing Quality Verification

In addition to testing the performance characteristics at the wafer level as described in Appendix D, VCSEL manufacturers must also implement ongoing reliability and application-specific performance testing to ensure continued quality and reliability. In some cases, a sample of every wafer used in an application may be tested for long-term reliability, damp heat exposure, and other reliability metrics. This is often the case in data communications applications where long-term reliability is critical to performance. In other cases, such as in consumer electronics where the wafer volume can be quite high, a regular sampling of the production wafers may be implemented to asses any drift in reliability. Application-specific testing may also be required, such as measurement of the small-signal modulation bandwidth and relative intensity noise in VCSELs for data communications. Periodic verification of operation across the rated temperature range may also be required in both consumer electronics and automotive applications. Specially, AEC-Q102 contains detailed requirements on qualification and notification for any changes in the manufacturing process.

Table E.3 Superset of tests from Telcordia, JEDEC, and AECQ qualification tests.

Mechanical Tests	**Storage Tests**	**Operational Tests**
chip level	high temperature storage	high temperature operating life
solderability	low temperature storage	
die shear	high temperature and humidity storage	low temperature operating life
wire bond pull	cyclic moisture resistance	high temperature and humidity operating life
chip and/or package level	temperature cycling	
solderability	hermeticity	cyclic moisture resistance
mechanical shock	flammability	temperature cycling
vibration	internal moisture	
thermal shock		
reflow preconditioning		
fiber pull and retention		
connector durability		

Stress	Pre-Conditioning	Sample Size (*)	Hours/Cycles (**)	Total Cumulative Hours/Cycles (****)	Failures (***)
Power Temperature Cycling (PTC)	YES	69	1,000 cycles	1,000 cycles	0
Unpowered Temperature Cycling (TC)	YES	89	1,016 cycles	1,016 cycles	0
Low Temperature Opening Life (LTOL)	NO	96	1,075 hours	1,075 hours	0
High Temperature Opening Life #1 (HTOL1)	NO	96	1,034 hours	4,093 hours	0
High Temperature Opening Life #2 (HTOL2)	NO	96	1,034 hours	5,346 hours	0
Pulsed Operating Life (PL)	NO	95	1,605 hours	4,856 hours	0
Wet High Temperature Operating Life #1 (WHTOL1)	YES	96	1,093 hours	3,500 hours	0
Wet High Temperature Operating Life #2 (WHTOL2)	YES	96	1,065 hours	1,065 hours	0
Dew (high humidity temperature cycling) (DEW)	NO	96	1,011 hours	4,734 hours	0

Stress	Sample Size (*)	Hours/Cycles (**)	Failures (***)
ESD (CDM/HBM)	63	Completed	0
H2S Exposure	92	336 h	0
Flow Mixed Gas (FMG) Exposure	96	500 h	0
Mechanical Tests (CA -> VVF -> MS)	30 + 10	Completed	0

Figure E.15 Results from an automotive AECQ-102 qualification [27].

E.8 Summary

VCSELs are highly reliable lasers that are used in a wide variety of demanding applications. It is common for lifetimes to exceed decades at nominal operating conditions while simultaneously producing extremely low early failure rates. Extensive work has gone into understanding and characterizing the failure modes of VCSELs. Manufacturing operations now have comprehensive quality control procedures in place to assure product reliability and performance.

References

1 Graham, H. Chen, J. Cruel, J. Guenter, B. Hawkins, B. Hawthorne, D. Kelly, A. Melgar, M. Martinez, E. Shaw and J. Tatum, "High-power VCSEL arrays for consumer electronics," *Proc. SPIE* vol. **9381** (2015).

2 R. Herrick, "Reliability of vertical-cavity surface-emitting lasers," *Jpn. J. Appl. Phys.* **51** (2012).

3 J. Guenter, B. Hawthorne, D. Granville, M. Hibbs-Brenner, R. Morgan, "Reliability of proton-implanted VCSELs for data communications," *Proc. SPIE* vol. **2683**, pp. 102–113 (1996).

4 B.M. Hawkins, R.A. Hawthorne III, J.K. Guenter, J.A. Tatum, J.R. Biard, "Reliability of Various Size Oxide Aperture VCSELs," Proceedings of the 52nd Electronic Components and Technology Conference, pp. 540–550, IEEE, Piscataway, NJ, (2002).

5 K. Johnson, M. Hibbs-Brenner, W. Hogan, and M. Dummer, "Advances in red VCSEL technology," *Adv. Opt. Technol.*, Vol. 2012 (**2012**).

6 R. Herrick, "Oxide VCSEL reliability qualification at Agilent Technologies," *Proc. SPIE* **4649**, pp. 130–141, (2002).

7 I. Aeby, D. Collins, B. Gibson, C. Helms, H. Hou, W. Luo, D. Bossert, C. X. Wang, "Highly reliable oxide VCSELs for datacom applications," *Proc. SPIE* vol. **4994**, pp. 152–161,(2003).

8 L. A. Graham, M. Schnoes, K. Maranowski, T. Fanning, M. Crom, S. A. Feld, M. A. Gray, K. Bowers, S. L. Silva, and K. Cook, "New developments in 850 and 1300nm VCSELs at JDSU," *Proc. SPIE* **7229**, pp. 11–20 (2009).

9 L. Graham, H. Chen, D. Gazula, T. Gray, J. Guenter, B. Hawkins, R. Johnson, C. Kocot, A. MacInnes, G. Landry, and J. Tatum, "The next generation of high speed VCSELs at Finisar," *Proc. SPIE* **8276** (2012).

10 T. Fanning, J. Wang, Z. Feng, M. Keever, C. Chu, A. Sridhara, C. Rigo, H. Yaun, T. Sale, G. Koh, R. Murty, S. Aboulhouda, and L. Giovane, "28-Gbps 850-nm oxide VCSEL development and manufacturing progress at Avago," *Proc. SPIE* **9001**, (2014).

11 J. Seurin, S. Wilton, A. Miglo, "Reliability Report of Princeton Optronics' High-Power 860nm VCSEL Array Chip-on-Submount," (Initial release 05 December 2013 Princeton Optronics).

12 S. Kamiya, K. Takai, S. Imai, J. Yoshida, M. Funabashi, Y. Kawakita, K. Hiraiwa, T. Suzuki, H. Shimizu, N. Tsukiji, T. Ishikawa, and A. Kasukawa, "Highly Reliable 1060nm Vertical Cavity Surface Emitting Lasers (VCSELs) for Optical Interconnects," Proceedings of MRS, vol. Reliability and Materials Issues of III-V and II-VI Semiconductor Optical and Electron Devices and Materials II (2012).

13 L. Graham, J. Jewell, K. Maranowski, M. Crom, S. Feld, J. Smith, J. Beltran, T. Fanning, M. Schnoes, M. Gray, D. Droege, V. Koleva, M. Dudek, J. Fiers and R. Patterson "LW VCSELs for SFP+ applications," *Proc. SPIE* vol **6908** (2008).

14 K. Rhew, S. Chang Jeon, O. Kwon, D. Lee, B. Yoo, and I. Yu, "Reliability assessment of 1.55mm Vertical cavity surface emitting lasers for optical communications systems," 45th Annual International Reliability Physics Symposium (2007).

15 J. Guenter, J. Tatum, R. Hawthorne, R. Johnson, D. Mathes and B. Hawkins, "A plot twist: the continuing story of VCSELs at AOC," *Proc. SPIE* vol **5737**, pp 20–34, (2005).

16 D. Mathes, "Materials issues for VCSEL operation and reliability," Ph.D. Disseration, University of Virginia (2002).

17 D. Mathes, R. Hull, K. Choquette, K. Geib, A. Allerman, J. Guenter, B. Hawkins, and Bobby Hawthorne, "Nanoscale materials characterization of degradation in VCSELs," *Proc. SPIE* vol. **4994**, pp. 67–82, (2003).

18 J. Guenter, J. Tatum, R. Hawthorne, B. Hawkins and D. Mathes, "VCSELs at Honeywell: the story continues," *Proc. SPIE* vol **5364**, pp. 34–46, (2004).

19 C. Helms, I. Aeby, W. Luoa, R. Herrick, and A. Yuen, "Reliability of oxide VCSELs at Emcore," *Proc. SPIE* **5364** (2004).

20 S. Xie, R. Herrick, D. Chamberlin, S. Rosner, S. McHugo, G. Girolami, M. Mayonte, S. Kim, and W. Widjaja, "Failure mode analysis of oxide VCSELs in high humidity and high temperature," *J. Lightwave Technol.* **21**, 1013–(2003).

21 J. Guenter, B. Hawkins, R. Hawthorne, R. Johnson, G. Landry, K. Wade, "More VCSELs at Finisar," *Proc. SPIE* **7299** (2009).

22 P.K. Teng, T. Weidberg, M.L. Chu, T.S. Duh, I.M. Gregor, L.S. Hou, S.-C. Lee, P.S. Song, D.S. Su, "Radiation hardness and lifetime studies of the VCSELs for the ATLAS semiconductor tracker," *Nucl. Inst. Methods Phys. Res. A* **497** (2003).

23 C. Barnes, J. Schwank, G. Swift, M. Armendariz, S. Guertin, G. Hash and K. Choquette, "Proton Irradiation Effects in Oxide-Confined Vertical Cavity Surface Emitting Laser VCSEL) Diodes," RADECS 2000, Session L-10 (2000).

24 F. Berghmans, M. Van Uffelen, and M. Decreton, "High-total-dose gamma and neutron radiation tolerance of VCSEL assemblies," Proc. SPIE 4823, Photonics for Space Environments VIII, (2002).

25 J. Guenter, B. Hawkins, B. Hawthorne, "Phenomenological study of VCSEL wearout reliability," *Proc. SPIE* vol **79522**, pp. 58–65, (2011).

26 J. Tatum, R. Johnson, J. Guenter, D. Gazula and A. MacInnes, "High data throughput VCSELs," *Proc. SPIE* **7720** (2010).

27 Thomas R. Fanning, et al, "Performance, manufacturability, and qualification advances of high-power VCSEL arrays at TriLumina corporation ," *Proc. SPIE* vol. **11300** (2020).

28 Telcordia GR-468-CORE, Issue 2 September 2004.

29 JESD47G "Stress-test-driven qualification of integrated circuits," and associated clauses in JEDEC-22 (2009).

30 AEC-Q102 "Failure mechanism based stress test qualification for discrete optoelectronic semiconductors in automotive applications," (2017).

Appendix F

Eye Safety Considerations

Jim Tatum

A major concern for any commercial product is the need to control the laser output power to maintain a safe operating condition over all time, environmental conditions, and reasonably foreseeable fault conditions. In this appendix a review of the laser classifications, testing methodology, and design considerations are presented. Many VCSELs operate in the infrared and are essentially invisible to the human eye and require strict adherence to the eye-safety requirements defined in IEC 60825-1 laser safety standard [1]. (NOTE: In the US, the standard is ANSI Z136.1 [2].) The standards may be revised from time to time; this appendix is based on edition 3 of IEC 60825-1. This appendix is not intended to be a complete review of all necessary requirements to meet eye-safety requirements. Review of product data and requirements must be completed and certified by a laser safety officer. The material presented here is only intended as guidance to the standard and is in no way considered a substitute for certification. For a more complete treatment of the laser classifications, the reader is referred to [3, 4] and the standards noted above.

Laser are classified in four major categories according to the total power emitted, the wavelength of the laser, and the laser beam divergence. The classes of lasers are summarized in the table below (Table F.1).

VCSELs will generally fall in classes 1, 2, or 3 but can in some cases be considered class 4 lasers in some industrial applications. For classification purposes, the optical power is measured at a distance of 100 mm and into a 7 mm diameter aperture from the emission point accessible to the user and must include any optical elements in the system. Some examples of emission points are the end of an optical fiber, the front face of a mobile phone, and the end of a laser pointer. A schematic of the test configuration is shown in Figure F.1.

Lasers with low divergence, or collimating optical components, that have divergence less than 2° (35 mrad) will completely pass through the aperture. In any case, the aperture must be moved across the emission area to maximize the power falling on the detector. The test is intended to mimic a typical viewing condition and the iris opening of a human eye. The maximum power level must also be considered under all operating environments and with the possibility of at least one fault in the control system. For example, what happens to the laser power if the current to the laser is increased in the event of a short circuit in the control system, or if moisture is present on the emission point, and so forth. For class 1 certification, the maximum permissible exposure (MPE) limit is a function of both the laser wavelength and the exposure (laser pulse) duration. For most VCSEL applications, the wavelength is between 700 and 1050 nm; the MPE values for both skin and eye damage are summarized in Table F.2. The power correction factor, C_A, for this wavelength range is given by $C_A = 10^{2(\lambda - 0.700)}$, where λ is the wavelength in microns, and the units of C_A are W/cm^2.

VCSEL Industry: Communication and Sensing, First Edition. Babu Dayal Padullaparthi, Jim A. Tatum and Kenichi Iga.

Table F.1 IEC classification of lasers.

Classification	Description
1	Lasers in this classification are considered safe under al operating conditions.
1M	Lasers in this classification are considered safe when viewed with the naked eye but may be hazardous when viewed with an optical instrument. These include magnifying glasses, microscopes, binoculars, telescopes, etc.
2	This class is reserved for visible lasers where the person has a natural aversion (blink response) to looking at the light and is generally safe under all operating conditions. However, in continued viewing for longer than 250 ms, or by overcoming the eye aversion response, these lasers may become dangerous.
2M	This class is reserved for visible lasers as described in class 2 but may become dangerous when viewed with an optical instrument as described in class 1M.
3R	Lasers in this classification are generally considered low risk, with the power limits set at five times the level for equivalent class 1 lasers for invisible lasers and class 2 levels for visible lasers.
3B	Lasers in this classification are considered eye safety hazards. The maximum power of a continuous wave laser is 500 mW. The laser radiation may be hazardous to both the eye and skin. Viewing laser radiation reflected from a diffuse surface is generally considered safe.
4	Lasers in this classification are considered hazardous to both the skin and eye. Viewing of diffuse reflections may also be dangerous. Class 4 lasers generally require special enclosures and operating procedures.

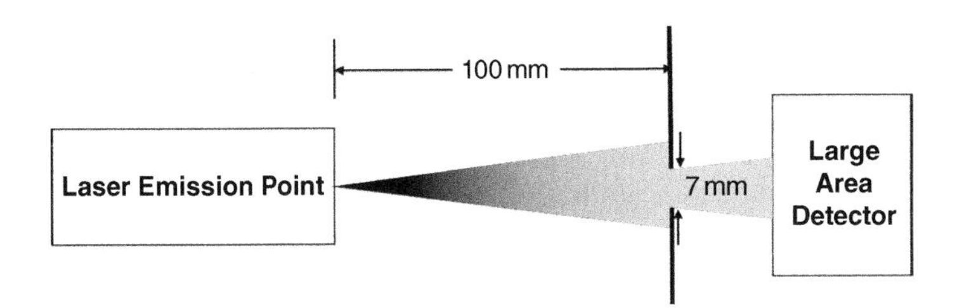

Figure F.1 Measurement configuration for eye-safety classification.

Table F.2 MPE formulas for lasers between 700 and 1050 nm.

Damage Area	t = Exposure Time (s)	MPE	Units
	$10^{-13} < t < 18 \times 10^{-11}$	3.8×10^{-8}	J/cm^2
Eye	$10^{-11} < t < 5 \times 10^{-6}$	$7.7 \times C_A \times 10^{-8}$	J/cm^2
	$5 \times 10^{-6} < t < 10$	$7 \times C_A \times t^{0.75} \times 10^{-4}$	J/cm^2
	$10 < t < 1000$	$3.9 \times C_A \times 10^{-4}$	W/cm^2
Skin	$10^{-9} < t < 10^{-7}$	$2 \times C_A \times 10^{-2}$	J/cm^2
	$10^{-7} < t < 10$	$1.1 \times C_A \times t^{0.25}$	J/cm^2
	$10 < t < 3000$	$0.2 \times C_A$	W/cm^2

As an example, a typical VCSEL used in data communications operates at 850 nm (0.85 μm) and has $C_A = 2$ W/cm^2. When measured in the system shown in Figure F.1, this gives a maximum power level of 767 μW on the detector. Note in this example we have used the average power since the modulation of the laser is expected to be very fast. Further examples of eye-safety limits for data communications systems including multi-wavelength sources and parallel optical fibers can be found in [5]. As another example, a typical 3D TOF sensor might operate with a pulse width of 10 ns, and at 940 nm (0.94 μm), $C_A = 3.01$, which would allow 150 W/cm^2 of peak power (or 1500 nJ, Joules = Watts × pulse width). When measured with the system shown in Figure F.1, this results in a maximum peak power of 58 mW. This dramatic increase in peak power enables lasers to be operated with the very high pulse peak powers required in the 3D sensors and LIDAR applications. Figure F.2 is a plot of the MPE levels for 850 and 940 nm as a function of exposure time and pulse width. When considering a pulsed laser application such as LiDAR, it is important to consider the details of the cumulative pulse train, frequency, pulse width, and total number of pulses. A detailed analysis of the changes introduced by the IEC, and its effect on the eye-safety requirements on a LiDAR application has been summarized in [6].

The preceding analysis has assumed that the power into the aperture emanates from a single point source of radiation, a correct assessment for a single laser source that would focus to a single point in the eye. In many 3D sensors, illuminators, and night vision systems, the laser light is coming from an array of lasers, and these form a so-called extended source. In other words, the light is not coming from a single point but many points, which will focus to a larger area on the human eye. A more complex analysis must be done when considering extended sources, and the MPE levels are generally higher. Consider the optical diagram shown in Figure F.3. For the purpose of extended source calculations, the human eye is modeled with a 17 mm distance from the lens to the image point and creates an image size d_{IMAGE} that subtends a full angle α. When extended to

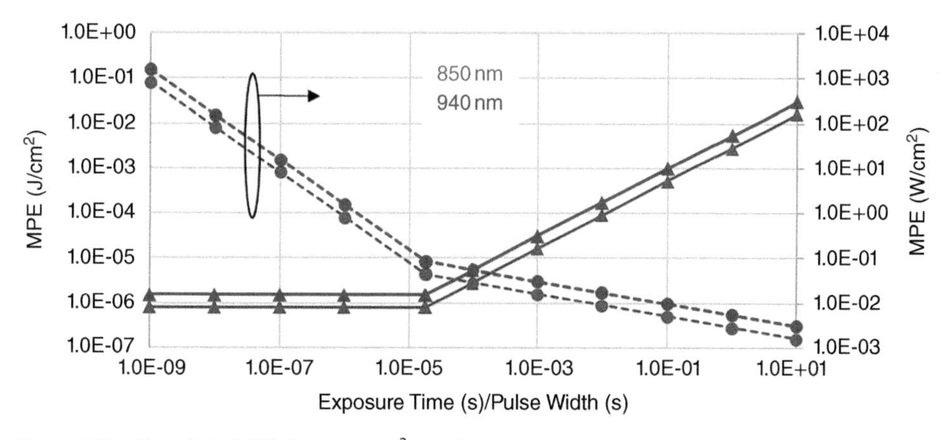

Figure F.2 Plot of the MPE limit in J/cm^2 as a function of exposure time (left axis) and the MPE in W/cm^2 as a function of pulse width (right axis) for 850 nm (blue lines) and 940 nm (red lines).

Figure F.3 Schematic of the optical system used to determine extended source emission properties.

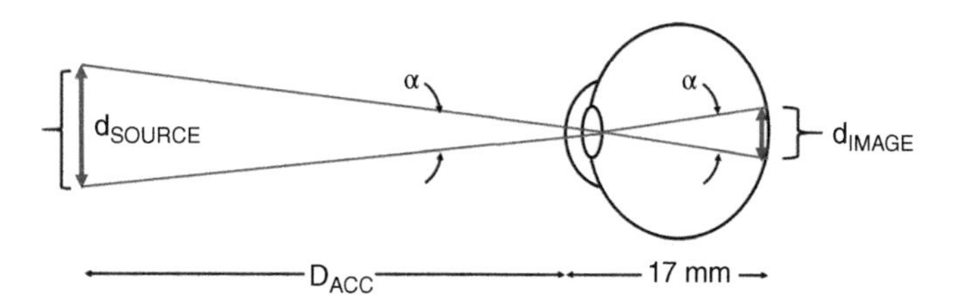

the object side of the eye, the angular subtense and the distance to the access point of the radiation, D_{ACC}, determine the size of an emitter that can be considered as an extended source. Currently, sources with 1.5 mrad $< \alpha <$ 100 mrad may be considered extended sources. For example, if D_{ACC} is 100 mm, and the source image > 75 μm, then the emission may be considered to come from an extended source. For VCSEL emitters in a data communications array spaced at 250 μm centers, the extended source rules can apply [5, 7].

However, in a 3D sensor array, where the emitter spacing can be less than 30 μm, the extended source will not apply to the entire array but to those emitters within ~70 μm radius. For a hexagonal array, this may include many emitters. The number of emitters will also depend on any optics prior to the laser beam and the user access point. If a source can be classified as extended with an angular subtense $\alpha_{EXTENDED}$, the eye-safety correction factor C_A is increased by the ratio of angular subtense, or $C_{A,EXTENDED} = C_A \times \alpha_{EXTENDED}/\alpha_{MIN}$, where $\alpha_{MIN} = 1.5$ mrad. The determination of proper eye-safety limits can be very complicated depending on the source, and it is highly recommended that a laser safety professional be consulted. The methodology presented here may be too simplistic for real optical systems, and the authors make no warranty on its accuracy or suitability for laser-safety designations.

When considering laser eye safety, it is a requirement that the laser and system be tolerant to any reasonable fault condition that might occur. For example, if the power supply is incorrect, if the laser pulse driver stops working and the laser is running CW, a fiber is unplugged, and so forth. To this end, safety features such as a power monitoring photodiode are often included in the laser package and used to shut the laser off in the event of an overpower condition. Other features are often included in the laser driver circuitry such as current limitation, overvoltage protection, and so forth, to further tolerance to reasonably foreseeable fault conditions. It is imperative that laser safety be a part of any application design, and it is up to the laser and system supplier to properly certify any components and systems.

Over the last several years, laser eye-safety limits have been relaxed significantly as they have become a part of everyday life. Eye safety regulations will continue to change periodically and can be different across the many countries a product may be used. It is the responsibility of the laser and system manufacturer to ensure that the latest requirements have been met wherever the laser product or system is sold.

References

1 IEC 60825-1 Ed. 3.0 b:2014 Safety of laser products - Part 1: Equipment classification and requirements.
2 ANSI Z136.1 Guidelines for Implementing a Safe Laser Program
3 *Lasers and Optoelectronics: Fundamentals, Devices and Applications*, 1. Anil K. Maini, © 2013 John Wiley & Sons Ltd. Published 2013 by John Wiley & Sons Ltd.
4 K. Schulmeister, The new edition of the international laser product safety standard IEC 60825-1," White paper available at https://www.scribd.com/document/381800392/whitepaper-iec-60825-1-v1d-pdf.
5 J. Castro, R. Pimpinella, B. Kose, P. Huang, A. Novick and B. Lane, "Preliminary Evaluation of OFCS Hazards for VCSEL MMF Channels," Presented at IEEE 802.3cm meeting May 2018.
6 S. Keller, F. Matteini, B. Penlae, and L. Carrara, "Novel illumination strategy for lidar enabled by update in the laser product standards," *J. Laser Appl.* **30**, 012011 (2018); doi:10.2351/1.5024300.
7 J. Castro, R. Pimpinella, B. Kose, P. Huang, A. Novick and B. Lane, "Optical Power Level Limits for Eye Safety: Spreadsheet calculator," Presented at IEEE 802.3cm meeting November 2018.

Appendix G

Laser Displays and TV

Kenichi Iga

G.1 What is a Laser Display?

In this appendix, laser displays are briefly summarized. Displays are used to project images onto a surface [1]. There are two basic types of displays [2, 3]. The first uses direct projection of the laser on a surface, and the beam is raster-scanned to project the image. In this system, the several different wavelength laser beams are mixed on the surface to create the color image. The second projects a white light laser (or several different wavelength lasers) onto a liquid crystal that controls the color of the image.

The primary advantages of lasers in displays are brightness and chromaticity. Several light sources are used for the red, green, and blue (RGB) including LEDs, diode lasers, second harmonic generation (SHG) from solid-state lasers, and gas lasers.

VCSELs are generally low-power devices and are not directly used in current displays. The red-emitting VCSEL described in Appendix H and GaN-based blue and green VCSELs introduced in Appendix I may enable future displays. For a small head-mounted display [4], a microwatt-class low-power surface-emitting laser is a key issue for laser eye glasses for virtual reality, as introduced in Chapter 6.

G.1.1 Laser Display Format

The readers may have a question: What is a laser display? Most television and projector displays are made using liquid crystal panels with LED backlight and organic electroluminescence (OEL) panels.

Then, the laser display is simply defined as display that uses a laser as a light source. Then the natural follow-up question is: What are the features? So let's consider the features of the laser display [5].

a) Good color reproducibility: By mixing RGB lasers, one can widely reproduce the various colors in the chromaticity diagram.
b) Laser light can be made into a beam. Therefore, it can be projected in large-scale formats such as a movie theater screen. In addition, it is possible to project onto a building of arbitrary shape in the distance, such as projection mapping.
c) Power efficiency. This was achieved by a high-power semiconductor laser, which is superior to conventional gas laser systems.

VCSEL Industry: Communication and Sensing, First Edition. Babu Dayal Padullaparthi, Jim A. Tatum and Kenichi Iga.

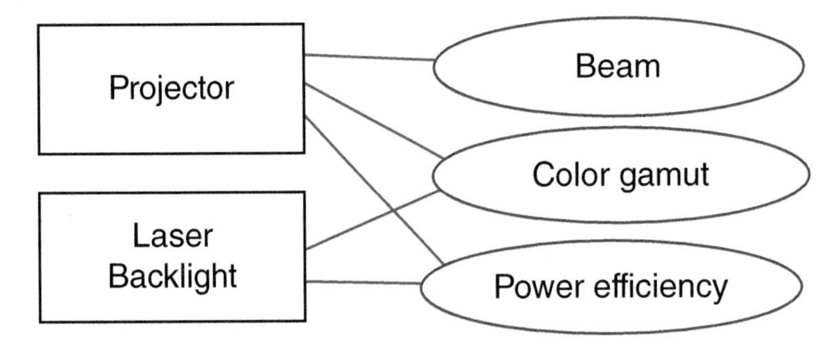

Figure G.1 Features and format of laser display [6].

As shown in Figure G.1, the laser projector makes full use of these three features with particular emphasis on color reproducibility and power efficiency. A relatively simple system is a TV or panel that uses a laser as a backlight for a liquid crystal panel. Lasers produce coherent light, and speckle noise is always a concern when they are used in displays. (Speckle was introduced in Chapter 7.) Speckle can be reduced by using broadened linewidth sources or by combining many emitters of a similar color.

G.1.2 Laser Projector Principle

Figure G.2 shows the principle of the laser projector [5]. In Figure G.2a, three color lasers are prepared, and in the case of a semiconductor laser, the image signal is directly modulated. The beams of each of the RGB lasers are combined into one beam by the optical system. The raster scanning is done with a micro-electromechanical mirror or other means. The intensity and color gamut of the display is controlled by directly modulating each of the RGB lasers. For example,

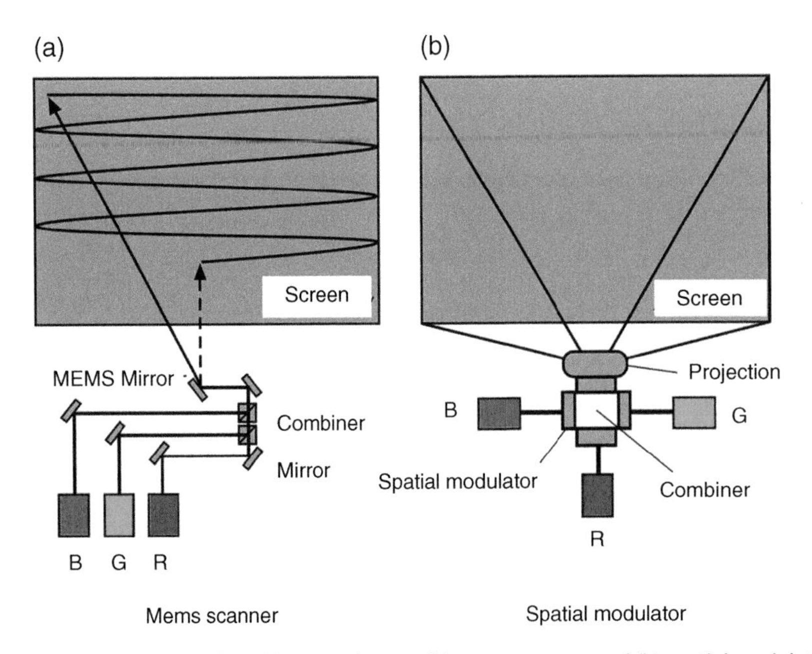

Figure G.2 Principles of laser projectors: (a) mems scanner and (b) spatial modulator. [5, 6].

Table G.1 Laser display development history and events.

Decade	Event
1970s	gas lasers
1980s	red semiconductor lasers
2000s	blue Semiconductor lasers, green from SHG of diode pumped solid-state lasers
2010s	InGaN-based green lasers
2020s	large-format displays such as in the Tokyo Olympics

when red and blue are both on, the resulting color is pink in appearance. This composed full-color image is scanned on the screen, and when viewed from a distance, the full color image is produced.

Next, in the case of Figure G.2b, the RGB lasers are combined to produce white light, and a spatial light modulator (SLM) divides the entire screen into pixels. The SLM modulates the transmittance and reflectance for each pixel, and it can be projected onto the screen as it is without scanning. The key is how much the number of pixels can be increased and whether the SLM can handle moving images with high modulation speed. The key to making practical laser-based televisions is the availability of RGB sources with direct semiconductor lasers. In the 1970s laser displays were made with ion gas lasers, and because of the very large size of the lasers, they were only sparsely used. In the 2000s, red semiconductor lasers were widely available, and with the advent of InGaN-based blue and green lasers, the color gamut became available. It is anticipated that VCSELs in the RGB spectrum may become practical in the future and be used widely in displays. Table G.1 shows the history of major social events and the development of displays; we refer the interested reader to Professor Kazuhisa Yamamoto's chronology [5].

G.2 Displays and Color Gamut

Table G.2 shows the categorization of various displays and application areas. Among them, there may be a chance for VCSELs to participate in laser displays. The most promising system may be a laser backlight LCD used in a head-mounted displays [3, 7, 8]. In any case, VCSELs emitting RGB visible spectra are indispensable.

The range of color that can be displayed on a display or print is called the color gamut, and there are various standards and specifications. Figure G.3 shows the color gamut using the xy chromaticity diagram for each of these standards and specifications. Among them, ACES and ProPhoto RGB include colors that do not actually exist in the color gamut, as can be seen from the figure. ACES has R, G, and B chromaticity coordinates set so that it can express all visible light.

As can be seen from Figures G.3 and G.4, the BT.2020[a] standard (the area filled in in Figure G.3), which is expected to be introduced in 8 K/SHV broadcasting, has a significantly wider color gamut than the conventional standard. The vertices of the color gamut triangle are almost on the spectral

a ITU-R recommendation SHV (super-high vision, pixel count: 7680 × 4320) specifications in BT.2020. The first edition was published by the ITU in 2012.

Table G.2 Categorization of displays [6].

Type of Device	Panel Display	Projection Display	HUD (Heads-Up Display) Half mirror	Retina projection
Cathode-ray tube	TV, measurement, PPI scope for RADAR	none	none	none
LCD (liquid crystal display)	TV, PC smartphone	TV, projector, theater	eyeglasses	none
Organic EL	TV, smartphone	none	eyeglasses	none
LED	digital signage micro-LED	LED-lamp for LCD, projector	none none	none none
Laser	TV (Laser-Backlight LCD)	theater, TV dashboard projection mapping	laser HMD	laser HMD
MEMS digital mirror (DLP)	none	theater mobile projector	HMD	HMD

Note: PPI: plan position indicator, EL: electroluminescence, HUD: heads-up display, DLP: digital light processing.

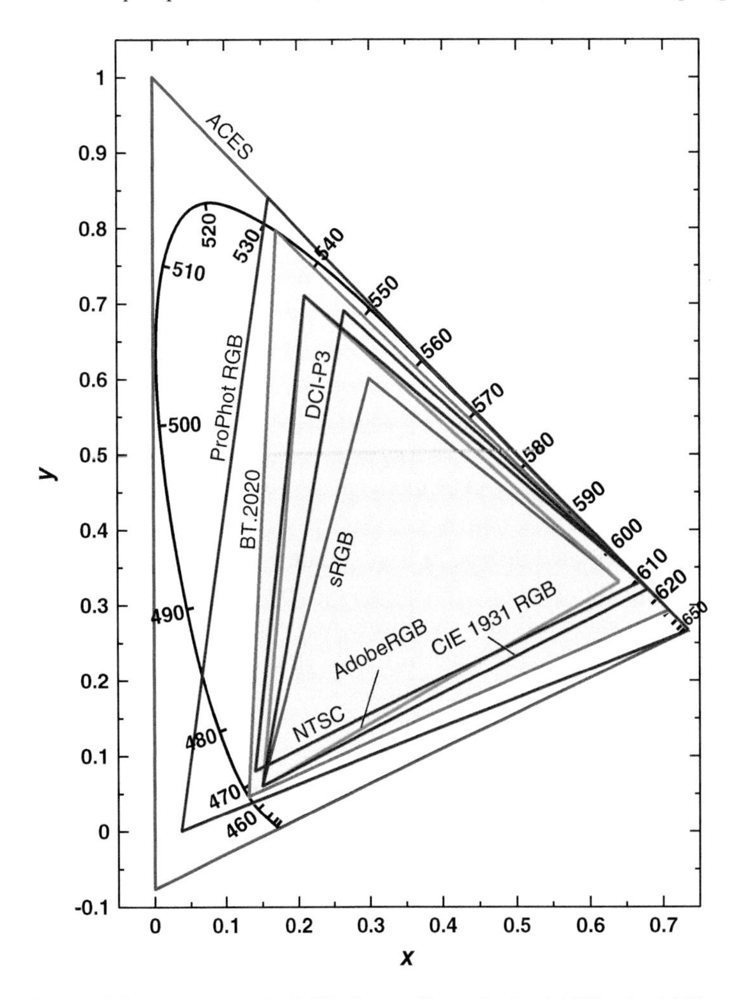

Figure G.3 Gamut standards [9]. *Source:* Figure by Genichi Hatakoshi (Copyright reserved).

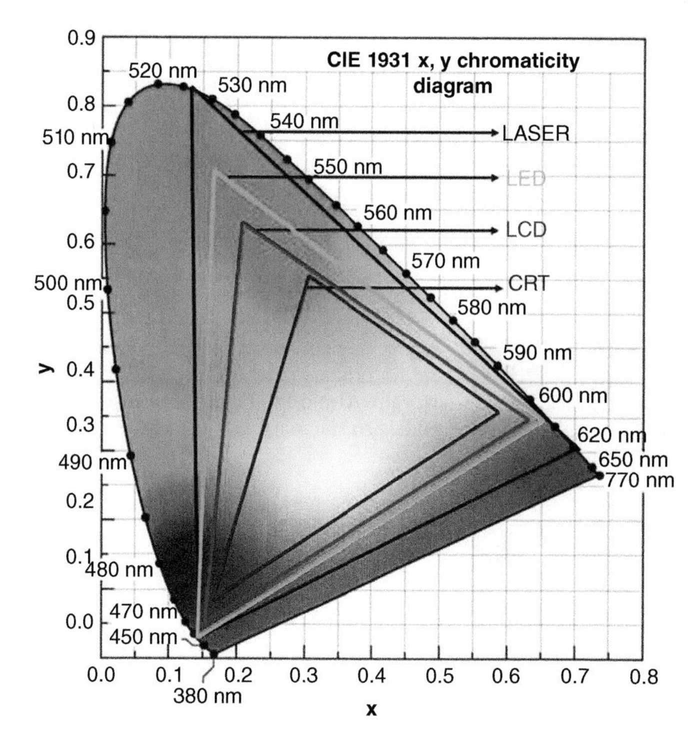

Figure G.4 Gamut examples. (The color of each system depends on devices used.) *Source:* https://doi.org/10.1364/AO.49.000F79

trajectory, and therefore the stimulation purity is almost 1 and light is nearly monochromatic. The corresponding monochromatic light wavelengths are as follows:

R(0.708,0.292): 630 nm
G(0.170,0.797): 532 nm
B(0.131,0.046): 467 nm

This color gamut can be achieved only with lasers.

G.3 Laser Light Sources for Display

G.3.1 Displays and Lasers

To make a color display, lasers at all three RGB wavelengths are needed. The development of both red and blue lasers was driven by the optical storage market (CD-ROM an DVD technologies). The difficulty in displays was the blue wavelength and the industry had to wait for the InGaN-based LED to be commercialized around 1995; the semiconductor laser followed in the 2000s. However, the increase in direct green output from semiconductor lasers is still lagging due to the difficulty of crystal growth technology. The green laser light at 532 nm can be made from SHG of diode-pumped Nd:YAG 1064 nm solid-state lasers. Table G.3 shows the wavelength band and the material of the semiconductor laser. A combination of direct lasing and SHG from a VECSEL have been used to create an RGB source, as shown in Figure G.5.

Table G.3 Laser display and semiconductor laser.

	Red	Green	Blue
Wavelength (nm)	650	530	450
Material	InGaAlP	InGaN	InGaN
Substrate	GaAs	sapphire, GaN	sapphire, GaN

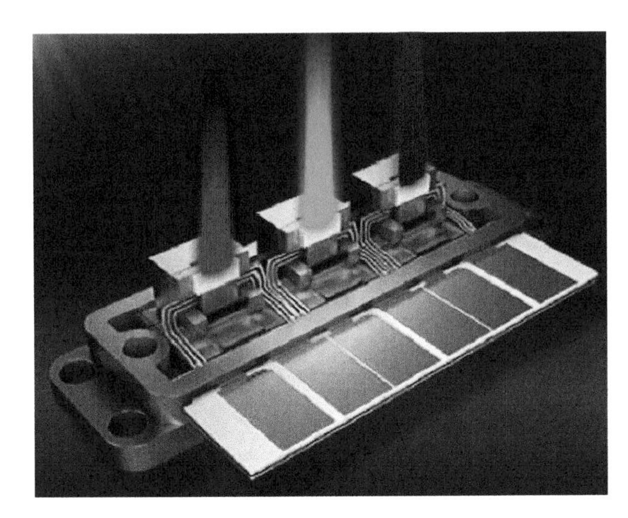

Figure G.5 Image of semiconductor laser set for primary color display. *Source:* By Novalux/Ushio, Inc. Figure: Courtesy by Hidekazu Hatanaka, Ushio, Inc.

G.3.2 RGB Light Sources

The red in BT.2020 corresponds to 630 nm, and this wavelength is readily available in a direct diode laser using InGaAlP quantum well lasers. The development of this wavelength was driven by the DVD industry and has grown to be used in many other applications including printers, pointers, automobile brake lamps, and red traffic lights. Red VCSELs are becoming available, as described in Appendix H. Development of direct diode emission in the blue wavelength range was again driven by the optical storage industry (Blu-ray, for example). Blue in BT.2020 corresponds to 467 nm, and the lasers developed for optical storage are between 445 and 460 nm, which is a slightly shorter wavelength than BT.2020. The semiconductor laser is made by using InGaN/GaN quantum wells and is further described in Appendix I. Availability of direct diode laser emission at green wavelength has lagged in development for displays. To achieve green light, SHG from a diode-pumped Nd:YAG laser has been used; green light has the highest luminosity factor in the human eye. The green wavelength in BT.2020 corresponds to 532 nm. Direct green emission from InGaN quantum wells with a large proportion of In is used, but if the proportion of In is large, crystal growth tends to be difficult. Direct diode green lasers are currently limited in power and led to the development of VECSELs. A schematic of a VECSEL is shown in Figure G.5. The method shown here was proposed by Aram Mooradian, who started the former Novalux company [10, 11, 12, 13]. The company's green laser has been taken over by Ushio, Inc. The method introduces a nonlinear crystal into the resonator of a surface-emitting laser (VECSEL) and extracting the second harmonic of green. Figures G.6 and G.7 show the image and principle of this semiconductor laser display. The internal cavity SHG uses the external resonator VECSEL. Figure G.7 shows the green beam of the output light.

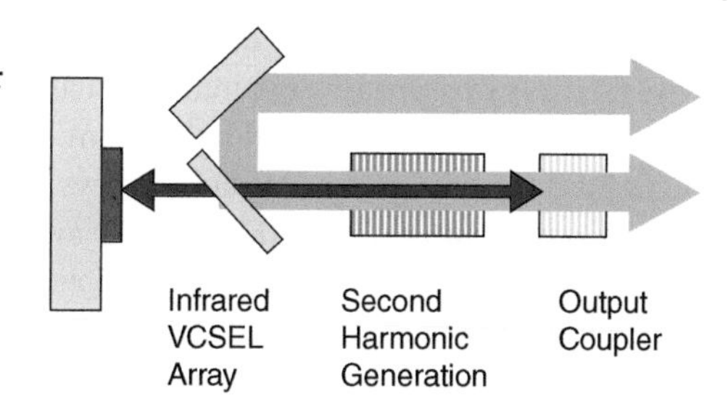

Figure G.6 Method of extracting green color from nonlinear crystals placed in the VCSEL resonator. *Source:* By Novalux/Ushio, Inc. Courtesy of Hidekazu Hatanaka.

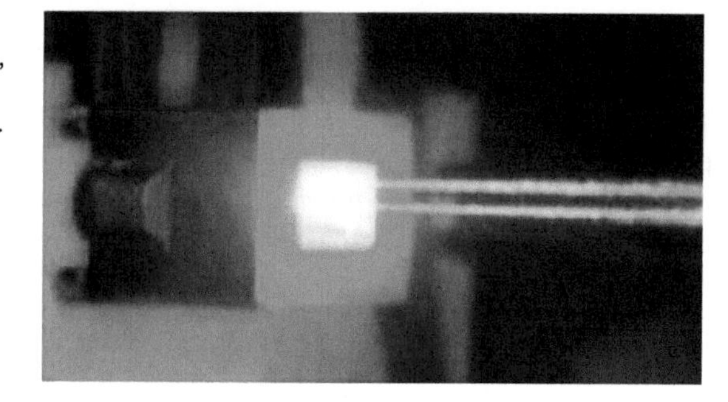

Figure G.7 Second harmonic generation VCSEL generating green in the resonator [6, 12, 13,]. *Source:* Photo: Courtesy of Novalux/Ushio, Inc., courtesy of Ushio, Inc., Hidekazu Hatanaka. In the projector for premium cinema in the Osaka Expo Memorial Park, 24 green emitters are used in 2 stages and 48 beams, 10 sets of which are used.

G.3.3 Laser Projector Components

a) Speckle reductions:
Several methods are considered for this, including widening of the laser spectral width by modulation of the laser, using a vibrating diffuser plate, vibrating a screen in lateral direction, etc.
b) Spatial light modulator (SLM):
In this component, an image is created by turning on/off a two-dimensional array-shaped modulator placed in space and projecting the reflection, which can only be done with a spatial light modulator. One example of an SLM is a liquid crystal on silicon (LCOS).
c) Combiner (optical combiner):
The optical combiner combines multiple light beams including the three primary colors into one beam.
d) Scanner:
A beam scanner is a device that scans a laser beam on a screen. There are mechanical polygon mirrors and micro-electromechanical systems (MEMS).
e) Projection lens:
Literally, it is the lens that projects the light beam toward the screen at the final stage. A large-diameter lens with little aberration is used. High-power projector lens need to withstand significant ambient heat from the laser sources.

G.3.4 Laser Projector and its Applications

For details on the application of laser displays, refer to the description in the reference [5].

A primary use of laser displays are in large-scale projectors and projection mapping used outdoors. Various images are projected on buildings such as large-scale theme parks and events. In addition, 4K high-definition images have emerged in high-definition 3D movie theaters. For medium-sized TVs, lasers are expected to develop in the future as a TV display for business use and home use, taking advantage of its low power consumption and wide color gamut. Organic EL and laser backlight methods compete for this. As the size of the display area decreases heads-up displays (HUDs) are attractive in aircraft and automobiles. Although liquid-crystal displays are often used for this purpose, future development is expected to favor lasers. Higher output and lower prices for semiconductor lasers are the keys. In addition, microwatt-class low-power semiconductor lasers come into play for HMDs for VR and AR, and the development of surface-emitting lasers hold the key to future innovation [14]. Applications include home TVs, medical and medical monitors, entertainment, pico-projectors for PC presentations, visual aids, and human interfaces in HUDs, in VR and AR environments [15].

G.4 Laser Backlight Method

The laser backlight liquid-crystal display replaces the LEDs with RGB semiconductor lasers; the principle is shown in Figure G.8. The beams of the three colors are combined and guided to the light guide plate. This was previously mentioned, in the LCD section. As an application system, there are medium to large projectors. In addition, an example of a laser display for TV application is given in reference [16].

G.5 Summary of Laser Displays

The main benefits of lasers in displays are summarized below.

a) Laser displays are considered to be an important method in the future from the viewpoint of color reproducibility. A standard called BT.2020 has been established, and laser displays are compatible.

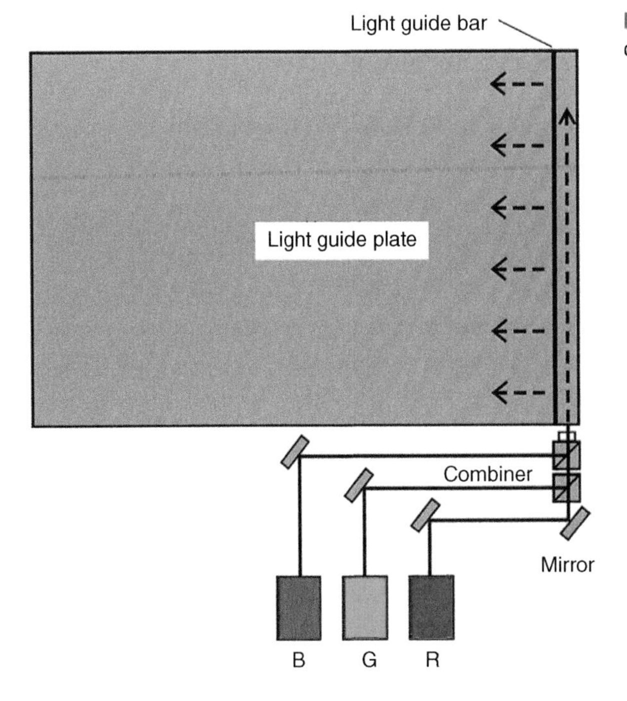

Figure G.8 Laser-backlight liquid-crystal display [5, 6, 16].

b) There are many application systems such as large projectors for movie theatres, TVs, and even smaller HUDs and HMDs.
c) The method that combines LCD and laser backlight is also interesting. LCD technology is inherited as it is, and the color tone of the laser beam and its good power efficiency can be utilized.
d) At present, HUDs are not limited to the laser method, but it will be an indispensable technology for high-performance displays in future vehicles.
e) HMDs will continue to develop as VR and AR components in computer vision.
f) The applications that are attracting attention are medical and visual aids, and the range will expand with higher definition.

References

1. I. E. Sutherland, "A head mounted three dimensional display," *Proc. AFIPS'68 (Fall Part I)*, pp. 757–764 (1968).
2. Y. Suematsu, K. Kobayashi, "Photonics-Optoelectronics and its development," *Ohmsha* (2007).
3. K. V. Chellappan, E. Erden, and H. Urey, "Laser-based displays: a review," *Appl. Opt.* **49** F79–98 (2010).
4. T. Kenno: "Head Mounted Display," edited by Chizuka Tani, Highly Realistic Display, 2.3, Kyoritsu Shuppan (2001).
5. K. Kuroda, and K. Yamamoto Edited, "Laser Lighting / Display," Optronics (2016).
6. K. Iga and G. Hatakoshi, "The Principle and Application Systems of Vertical Cavity Surface Emitting Laser," Adcom-Media Co. Ltd. Tokyo, Sept. 25, 2020. (PDF Japanese language version).
7. T. Kagami, "Technical History of Head-mounted Display (HMD) optical systems-from the viewpoint of technical information collectors," *O plus E*, Vol. **137**, No. 12, pp. 990–998 (2015).
8. M. Sugawara, "Wearable Visual Information Terminal-Retinal Projection Laser Eyewear-," Section 5.3, Kazuo Kuroda, Kazuhisa Yamamoto ed.: Laser Lighting / Display, Optronics (2016).
9. K. Iga and G. Hatakoshi, "Treasure Microbox of Optoelectronics," Adcom-Media Co. Ltd. Tokyo, April 25, 2020. (PDF Japanese language version).
10. A. Mooradian, S. Antikochev, B. Cantos, G. Carey, M. Jansen, S. Hallstein, D. Lee, J. -M. Pelaprat, R. Nabiev, G. Niven, A. Shchegrov, A. Umbrasas and J. Watson, "High power extended vertical cavity surface emitting diode lasers and arrays and their applications," 11th Microoptics Conference (MOC'05), Tokyo, M1 (2005).
11. S. Blohm, "Laser TV: Laser beams could revolutionise television," Inside Display Technologies (2008); http://www.prad.de/en/tv/specials/laser-tv-technology.html.
12. H. Hatanaka, G.T. Niven, "Laser cinema application of high brightness laser light source," *O plus E*, Vol. 36, No. **6** (2014).
13. H. Hatanaka, G.T. Niven, "Visible/external resonator/current injection high brightness surface emitting laser and lighting application by Second Harmonic Generation," *Laser Res.*, Vol. **42**, No. 7, pp. 534–538 (2014).
14. https://www.sekisui.co.jp/news/2015/1274415_23166.html.
15. T. Nakamura, and F. Kawamura, "Head-mounted display-a bright and clean"usable"device," Hitachi, Ltd. Corporate Information, http://www.hitachi.co.jp/rd/portal/contents/story/hmd/ (2015).
16. K. Kojima, "Development of RGB laser-backlit liquid crystal display," *MICROOPTICS NEWS*, pp. 25–30 (2016).

Appendix H

Red VCSELs

Jim Tatum

H.1 Introduction

VCSELs operating in the visible wavelengths have been commercialized in several applications such as blood oximetry, photodynamic therapy, and communications using plastic optical fiber. The required material system presents a multitude of challenges that limit the achievable performance. First the active region is generally made from InGaP quantum wells and AlInGaP barriers (see Chapter 2). The conduction band offset is about half that of VCSELs at 850 nm. This reduces the carrier confinement and reduces the maximum temperature operation. Second, the high-index material for the mirrors must be transparent at the operating wavelength requiring $Al_xGa_{1-x}As$ content above $x = 0.4$. The low index contrast of the DBR requires a large number of pairs to obtain the reflectivity needed for lasing. The situation is further exacerbated by the reduction in thermal conductivity of the high-index layers. This combination of low conduction band offset and low thermal conductivity of the mirrors conspires to limit red VCSELs to lower-temperature applications. For similar reasons, the wavelength tends to be above 650 nm and typically in the 670–680 nm range.

H.2 Data Communications

Plastic optical fiber (POF) has long been touted as an attractive economical alternative to glass optical fiber. The main advantage is the simple termination of the fiber and large core region (~1 mm diameter) that enables very low-cost assemblies. From a consumer perspective, it is also convenient to be able to visually verify the presence of light. The main detraction to POF is the relatively high attenuation of the light in the fiber. Figure H.1 shows the attenuation of a typical POF [1]. Note that glass fiber has attenuation <5 dB/km over this entire wavelength range.

The attenuation has a local minimum near 650 nm, and this is the target wavelength for the VCSEL. Data links are limited in distance by the attenuation, and even a short 10 m link will have nearly 3 dB of loss. The link distance is further limited by the modal dispersion (see Chapter 4) of the large core. More-advanced versions of POF have focused on non-polymethyl methacrylate (PMMA)–based constructions to reduce the absorption at longer wavelengths and then reduce the core size and grade the index profile to increase the modal bandwidth. All of these features add cost in either the fiber or the packaging constraints on the VCSEL. A few standards have emerged for POF and red emitter technology. Generally, the emitter definition is left to the implementor and can be a VCSEL, EEL, or LED, as appropriate. VCSELs tend to be more attractive in low-power and

VCSEL Industry: Communication and Sensing, First Edition. Babu Dayal Padullaparthi, Jim A. Tatum and Kenichi Iga.

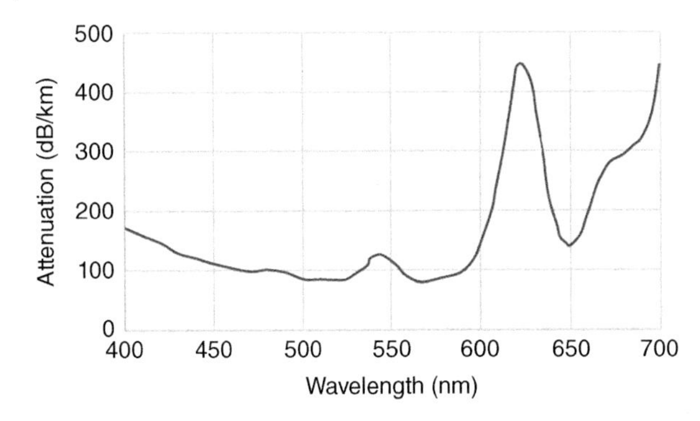

Figure H.1 Attenuation of typical POF as a function of wavelength.

Table H.1 Summary of current POF-based fiber networking standards.

Application area	Standard	Speed
Industrial control	Profibus	12 Mbps
	Profinet	10/100 Mbps
	Sercos	100 Mbps
Consumer electronics	IEEE 1394b	400 Mbps
Automotive	MOST	150 Mbps

higher-speed links. Table H.1 is a summary of several relevant standards. Note that there are many others and still more in development. The focus of POF is at the consumer level and embedded in connectivity that is either fixed or not likely to need increased bandwidth in the future. Red VCSELs continue to have limited drive into this market because of the limited range of temperature operation, relatively low reliability, and lack of bandwidth scalability of POF. Red VCSELs can certainly have the base bandwidth to support 10 Gbps operation [2], but the link distances and speeds are severely limited by the POF.

H.3 Blood Oximetry

Single- (or few-times-) use measurement of the oxygen content in blood is an area of considerable interest in the medical community, particularly in the oxygen measurement of venous or capillary vessels. The measurement relies on the difference in absorption of the various hemoglobin compounds at several different wavelengths. The normalized absorption of several hemoglobin compounds is shown in Figure H.2. The absorption is a variable function of wavelength and when measured with broad sources such as an LED can lead to errors. Multiple VCSELs emitting in the range of 650–1000 nm have been used to make an oximeter and packaged into a single element [3, 4]. The packaging capability, vertical emission, and small die size enable this application.

It is interesting to note the most recent version of the Apple Watch incorporates a blood oxygen sensor, as depicted in Figure H.3. The current implementation utilizes four red LEDs [5]. Traditionally, blood oximetry measurements are done on a fingertip or earlobe where the light can

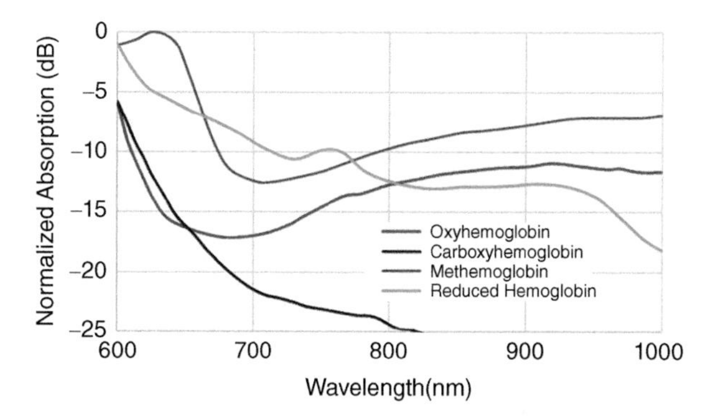

Figure H.2 Normalized absorption spectra of several blood hemoglobins as a function of wavelength.

Figure H.3 Implementation of a blood oximeter in the Apple Watch using four red LEDs [5].

be measured in transmission. With the placement of the watch on the wrist this is not possible, and the sensor uses a combination of reflection and algorithms to determine oxygen saturation. This limits the accuracy of the sensor and is perhaps a new application area for red VCSELs.

H.4 Hair Regrowth

The use of lasers in medicine has been growing dramatically in the last decade, and in particular the use of low-level photodynamic therapy (LLPT) for home use [6]. One particular example for red VCSELs is in regrowth of hair [7]. Several companies have now commercialized helmet- or hat-type devices with many tens of VCSELs in the device. The requirements on the laser are modest, a few milliwatts of power at essentially room temperature. The emission can be multi-mode, and the lifetime requirements are a few thousand hours of operation. In the future, more LLPT devices are likely to be commercialized. Already there is research into teeth whitening and arthritic pain relief among many others.

References

1 R. Ribeiro, V. Silva, and A. Barbero, "Material dispersion interplay with spectral filtering on plastic optical fiber (pof) links," *Microw. Opt. Technol. Lett.*, vol. **51**, no. 7, (2009).

2 K. Johnson, W. Hogan, M. Dummer, C. Steidl, and M. Hibbs-Brenner, "Advances in high-speed red vcsel performance," The 21st International Conference on Plastic Optical Fibers, (2012).

3 K. Johnson, M. Hibbs-Brenner, W. Hogan, and M. Dummer, "Advances in red VCSEL technology," *Adv. Opt. Technol.*, vol. 2012 (**2012**).

4 D. Kollmann, W. Hogan, C. Steidl, M. Hibbs-Brenner, D. Hedin, and P. Lichter, "VCSEL based, wearable, continuously monitoring pulse oximeter," Annual Int Conf IEEE Eng Med Biol Soc. (2013).

5 "Should You Trust Apple's New Blood Oxygen Sensor?" IEEE Spectrum (2020).

6 M. Hibbs-Brenner, K. Johnson, and M. Bendett "VCSEL technology for medical diagnostics and therapeutics," *Proc. SPIE* **7180**, (2009).

7 R. Lanzafame, R. Blanche, A. Bodian, R. Chiacchierini, A. Fernandez-Obregon, and E. Kazmirek, "The growth of human scalp hair mediated by visible red light laser and LED sources in males." *Lasers Surg. Med.*, **45**(8), (2013).

Appendix I

GaN-Based VCSELs

Kenichi Iga

I.1 AlGaInN Laser Materials

GaN-based edge-emitting lasers (EELs) have been developed for use mostly in optical disks. Now they are available from ~300 nm to longer than 600 nm of wavelength and are considered to be the sources for many other applications. The material challenges in the nitride-based lasers took several decades to resolve and make commercially viable lasers. Figure I.1 shows the lattice constant and bandgap energy of compound semiconductors including the AlInGaN material variants [1]. The bandgap energy ranges from the ultraviolet region at 6.2 eV (AlN) through the visible region at 3.4 eV (GaN) all the way to the infrared region of 0.7 eV (InN). The crystalline structure of AlGaInN is with wurtzite or sphalerite, and both are shown in the figure. The wurtzite lattice configuration is preferred for optoelectronic devices.

The ability to make such a wide range of wavelengths is unique to this material system and makes it very attractive for semiconductor lasers. For comparison purposes, the range of AlInGaAs materials is also shown in Figure I.1. The materials based on Zn and Mg are also shown in Figure I.1, but they are not generally used in practical commercial semiconductor lasers due to unstable material properties.

I.2 AlGaInN Laser Development

Development of AlInGaN optoelectronic devices began in the 1990s. There were many technical challenges with growing the semiconductor materials in the quality needed to make practical LEDs and semiconductor lasers. Breakthroughs during this period included the improvement in crystal growth technology using the MOCVD (metal organic chemical vapor deposition) method by Isamu Akasaki and Hiroshi Amano [2]. Crystal growth of high-quality GaN is achieved by introducing an AlN buffer layer [2].

Practical p-n junctions in GaN based materials was realized with the development of Mg-doping and its electron beam irradiation activation technology. With this, InGaN compound crystal by Shuji Nakamura and co-workers in Nichia Chemicals was obtained, and high-intensity quantum well light emission and thermal annealing technology was developed [3]. These advancements drove the development of a blue light-emitting diode using an InGaN/GaAlN quantum well as a light-emitting layer [4, 5]. As a blue semiconductor laser, continuous operation at room temperature was achieved; it became possible to operate it reliably and was installed in Blu-ray disks [6].

VCSEL Industry: Communication and Sensing, First Edition. Babu Dayal Padullaparthi, Jim A. Tatum and Kenichi Iga.

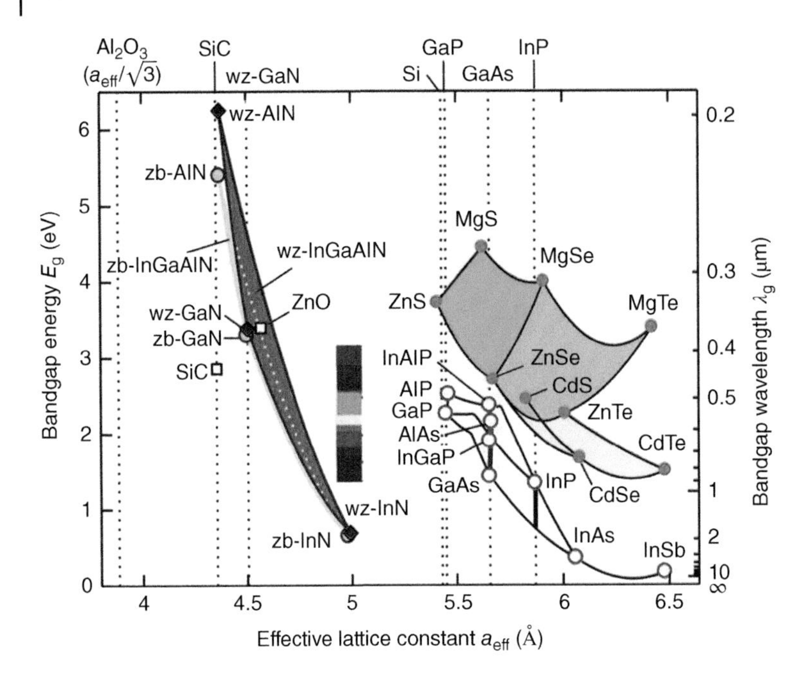

Figure I.1 Bandgap energy vs. lattice constants. *Source:* Genichi Hatakosi [1].

In late 2010, large-diameter, high-quality GaN substrates were commercialized, and it has become possible to grow wafers for semiconductor lasers based on them.

There are many technical drawbacks to using the AlGaInN material system in VCSELs compared to AlGaAs on lattice-matched GaAs substrates [7]:

i) Difficulty of growing GaN/InGaN on sapphire substrates prior to GaN wafers becoming available in 2010;
ii) Relatively low refractive index contrast in AlInN and GaN for DBR materials; large numbers of DBR pairs are needed to get enough reflectivity for laser oscillations in VCSELs;
iii) There is no efficient current confinement structure like AlGaAs oxidation;
iv) Electrical conductivity of AlGaN/GaN-based DBRs is relatively low;
v) The high effective mass of the carriers requires relatively high carrier density for gain; and
vi) The GaN wafer processing technique is not mature.

Even with these challenges, research on AlGaInN-based VCSELs has progressed. Figure I.2 shows a summary of reported laser wavelength and threshold current in visible wavelength VCSELs [1]. Room-temperature continuous oscillation has been achieved in the VCSELs based on InGaAlP (red), GaAlAs (~ 700 nm), and InGaN (blue) materials, respectively. The "green gap" in the green region also existed in the surface-emitting laser. As will be described later, in 2016, continuous room-temperature oscillation at 490 to 560 nm was achieved using a quantum dot active layer.

I.3 AlGaInN Blue VCSELs

Current injection lasing was first achieved in blue-emitting VCSELs in 2008 [2]. This laser operated continuously at 77 K, with an oscillation wavelength of 462 nm. Also in 2008, continuous oscillation of a current injection VCSEL at room temperature was achieved [8]. This

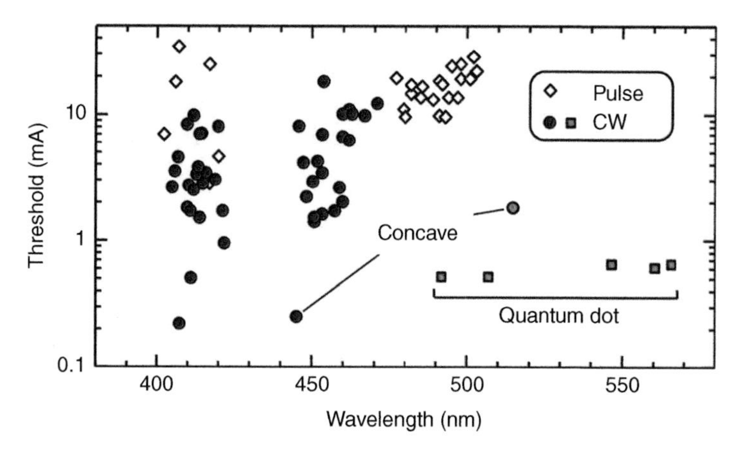

Figure I.2 InGaN-VCSEL reports. *Source:* Figure by Genichi Hatakoshi [1].

surface-emitting laser uses SiO_2/Nb_2O_5 dielectric multilayer film DBR on both sides. Subsequent developments have produced lower threshold currents of 1 to several milliamperes at wavelengths of 410–460 nm. Watt-class output powers have been obtained [9]. Figure I.3 shows emission from a blue VCSEL [10].

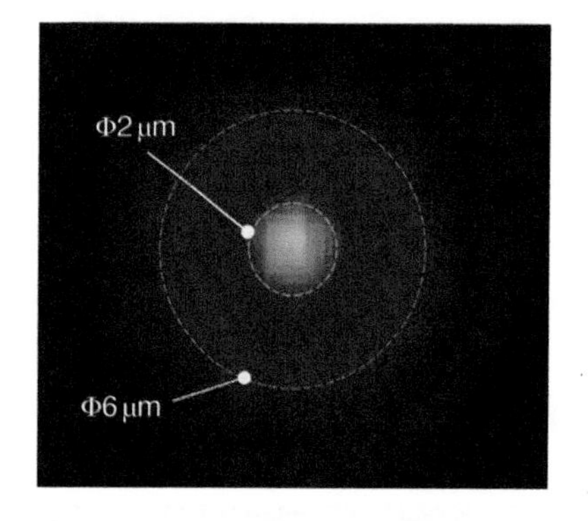

Figure I.3 Lasing light from blue VCSEL. *Source:* Sony Corporation, Japan.[10].

I.4 AlGaInN Green VCSELs

The challenges in developing green emission from VCSELs (and EELs) are even more acute than for blue emission. The second-harmonic generation has been employed to obtain green laser emission and commercially applied to displays [11, 12, 13, 14, 15, 16, 17, 18, 19, 20, 21, 22, 23].

The first green VCSEL operating at room temperature with current injection was reported after blue lasing was achieved [24]. This paper demonstrated room-temperature continuous oscillation characteristics of a blue VCSEL and the room-temperature pulse lasing of a green VCSEL [24]. Lowering the threshold value by increasing the reflectance while lowering the electrical resistance of the semiconductor DBR and performing current injection in a small area are issues for practical use.

In 2016, room-temperature continuous laser oscillation from 491.8 to 565.7 nm was reported in a VCSEL using quantum dots as the active layer [25, 26]. Quantum dots are formed when the In composition becomes large causing distortion in the crystal growth due to lattice mismatch. The threshold current density is as low as 0.6–0.8 kA/cm^2, and the low threshold characteristics are realized even when compared with the quantum well VCSEL [25, 26, 27, 28].

For the above-mentioned DBR material, a method of arranging a dielectric DBR with high reflectance on the bottom surface of a GaN substrate was studied. However, there is a problem that the diffraction loss becomes large because the resonator length becomes large. To solve this, an approach to make the DBR a concave mirror was reported. This method provides CW oscillation at a low threshold of 0.25 mA at 445 nm [10, 29]. Furthermore, in 2020, a device integrated with a concave mirror DBR was reported to operate continuously at room temperature with a wavelength of 515 nm and a threshold of 1.8 mA [30].

Figure I.4 shows a reported example of an InGaN-based VCSEL including quantum wells VCSEL and a concave mirror DBR integrated VCSEL. The data of the literature [27, 29, 30] has been added to the data of the InGaN in Figure I.2. The oscillation wavelength covers a large part of the "green gap." If a green VCSEL can emit the power high enough to operate for a flash LiDAR, one possibility would open for sensing in water or through water.

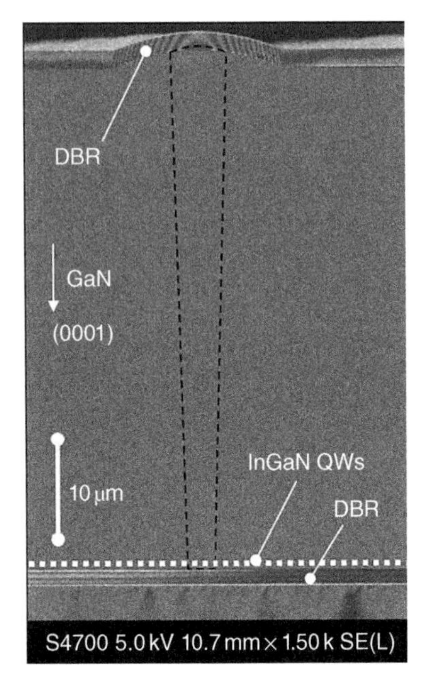

Figure I.4 VCSEL cross-section with curved dielectric mirrors [29]. *Source:* Sony Corporation, Japan.

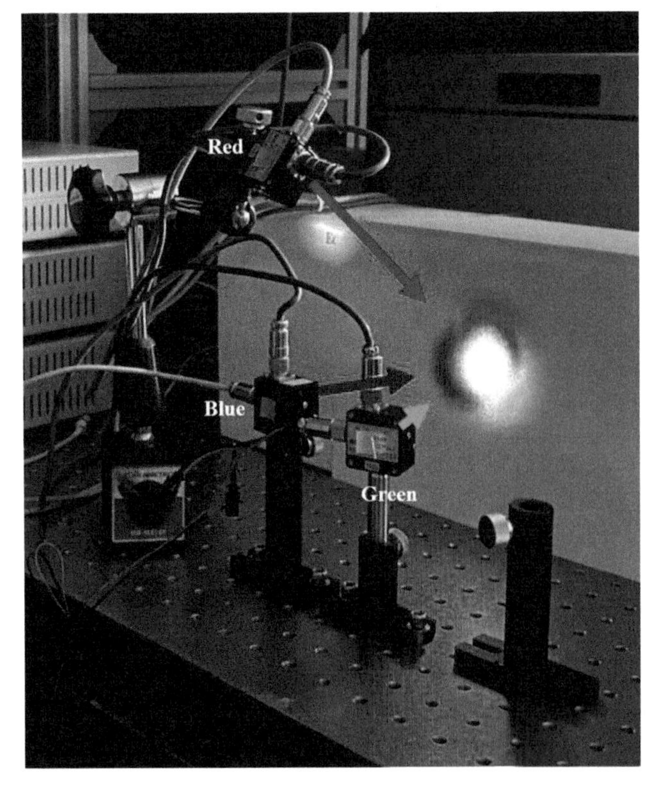

Figure I.5 Full (white) color generation from red/green/blue VCSELs. *Source:* Sony Corporation, Japan. [32].

I.5 White Light Generation by VCSELs

In order to obtain multi-color and white light [31], some novel attempts have been made. An example of white light generation using red, green, and blue VCSELs is shown in Figure I.5 achieved by Sony [32]. The VCSELs in this demonstration used curved dielectric mirrors on the top and bottom sides of the VCSEL active region. This is an impressive feat of VCSEL engineering and shows the promise of AlGaInN-based VCSELs in the display applications described in Appendix G.

According to Yole Dévelopement interview on October 30, 2020, SONY achieved 1% of power conversion efficiency in 2016 and 10% in 2018 using GaN-based VCSELs by using curved dielectric mirrors, as shown in Figure I.4. By taking advantage of BT.2020 color standard compatibility, VCSEL-based displays may offer an attractive alternative to micro-LEDs and OLEDs in large format displays.

References

1 K. Iga and G. Hatakoshi, "The VCSEL: Its principle and applied systems," Adcom-Media Co. Lts. Edition, Design-egg Publishing, Tokyo. https://contendo.jp/store/contendo/Product/Detail/Code/J0010425BK0101116001/, Oct 2020.

2 H. Amano, N. Sawaki, and I. Akasaki, "Metalorganic vapor phase epitaxial growth of a high quality GaN film using an AlN buffer layer," *Appl. Phys. Lett.*, vol. **48**, no. 5, pp. 353–355 (1986).

3 S. Nakamura, "GaN light-emitting devices, Review," *AJpn J. Appl. Phys.*, vol. **65**, no. 7, pp. 676–686 (1996).

4 S. Nakamura, "InGaN multiquantum-well-structure laser diodes with GaN–AlGaN modulation-doped strained-layer superlattices," *IEEE J. Sel. Top. Quantum Electron.*, vol. **4**, no. 3, pp. 483–489 (1998).

5 S. Nakamura, "InGaN/GaN/AlGaN-based laser diodes grown on GaN substrates with a fundamental transverse mode," *Jpn. J. Appl. Phys.*, vol. **37**, no. 9A/B, pp. L1020–L1022 (1998).

6 S. Nakamura and G. Fasol, *The Blue Laser Diode*, Springer (1997).

7 K. Iga, "Possibility of green/blue/UV surface emitting lasers," Int. Symp. Blue Laser and Light Emitting Diodes, Chiba Univ., Th11, pp. 263–266 (1996).

8 Y. Higuchi, K. Omae, H. Matsumura, and T. Mukai, "Room-temperature CW lasing of a GaN-based vertical-cavity surface-emitting laser by current injection," *Appl. Phys. Express*, vol. **1**, 121102 (2008).

9 M. Kuramoto, S. Kobayashi, T. Akagi, K. Tazawa, K. Tanaka, K. Nakata, and T. Saito, Watt-class blue vertical-cavity surface-emitting laser arrays, *Appl. Phys. Express* **12**, 091004 (2019)

10 T. Hamaguchi, M. Tanaka, and H. Nakajima, 'A. review on the latest progress of visible GaN-based VCSELs with lateral confinement by curved dielectric DBR reflector and boron ion implantation' *Jpn. J. Appl. Phys.* **58**, SC0806 (2019)

11 T. D. Raymond, W. J. Alford, M. H. Crawford, and A. A. Allerman, "Intracavity frequency doubling of a diode-pumped external-cavity surface-emitting semiconductor laser," *Opt. Lett.*, vol. **24**, no. 16, pp. 1127–1129 (1999).

12 D. Lee, A. V. Shchegrov, E. M. Strzelecka, J. P. Watson, M. K. Liebman, C. A. Amsden, A. Umbrasas, B. D. Moran, V. V. Doan, J. G. McInerney, and A. Mooradian, "Second harmonic generation at 488 nm by intracavity doubling of extended-cavity surface-emitting lasers," *Proc. SPIE*, vol. **4972**, pp. 102–111 (2003).

13 J. P. Watson, A. V. Shchegrov, A. Umbrasas, D. Lee, C. A. Amsden, W. Ha, G. P. Carey, V. V. Doan, A. Lewis, and A. Mooradian, "Laser sources at 460nm based on intracavity doubling of extended cavity surface emitting lasers," *Proc. SPIE*, vol. **5364**, pp. 116–121 (2004).

14 A. Mooradian, S. Antikochev, B. Cantos, G. Carey, M. Jansen, S. Hallstein, D. Lee, J.-M. Pelaprat, R. Nabiev, G. Niven, A. Shchegrov, A. Umbrasas, and J. Watson, "High power extended vertical cavity surface emitting diode lasers and arrays and their applications," 11th Microoptics Conference (MOC'05), Tokyo, M1 (2005).

15 L. Fan, T. Hsu, M. Fallahi, J. T. Murray, R. Bedford, Y. Kaneda, J. Hader, A. R. Zakharian, J. V. Moloney, S. W. Koch, and W. Stolz, "Tunable watt-level blue-green vertical-external-cavity surface-emitting lasers by intracavity frequency doubling," *Appl. Phys. Lett.*, vol. **88**, 251117 (2006).

16 J. E. Hastie, L. G. Morton, A. J. Kemp, M. D. Dawson, A. B. Krysa, and J. S. Roberts, "Tunable ultraviolet output from an intracavity frequency-doubled red vertical-external-cavity surface-emitting laser," *Appl. Phys. Lett.*, vol. **89**, 061114 (2006).

17 J. Lee, S. Lee, T. Kim, and Y. Park, "7 W high-efficiency continuous-wave green light generation by intracavity frequency doubling of an end-pumped vertical external-cavity surface emitting semiconductor laser," *Appl. Phys. Lett.*, vol. **89**, 241107 (2006).

18 J. Y. Kim, S. Cho, S. J. Lim, J. Yoo, G. B. Kim, K. S. Kim, J. Lee, S. M. Lee, T. Kim, and Y. Park, "Efficient blue lasers based on gain structure optimizing of vertical-external-cavity surface-emitting laser with second harmonic generation," *J. Appl. Phys.*, vol. **101**, 033103 (2007).

19 I. Kardosh, F. Demaria, F. Rinaldi, M. C. Riedl, and R. Michalzik, "Electrically pumped frequency-doubled surface emitting lasers operating at 485 nm emission wavelength," *Electron. Lett.*, vol. **44**, no. 8, pp. 524–525 (2008).

20 H. Matsubara, S. Yoshimoto, H. Saito, Y. Jianglin, Y. Tanaka, and S. Noda, "GaN photonic-crystal surface-emitting laser at blue-violet wavelengths," *Science* vol. **319**, pp. 445–447 (2008).

21 S. Ishizawa, K. Kishino, R. Araki, A. Kikuchi, and S. Sugimoto: "Optically pumped green (530–560 nm) stimulated emissions from InGaN/GaN multiple-quantum-well triangular-lattice nanocolumn arrays," *Appl. Phys. Express*, vol. **4**, no. 5, 055001 (2011).

22 J. D. Berger, D. W. Anthon, A. Caprara, J. L. Chilla, S. V. Govorkov, A. Y. Lepert, W. Mefferd, Q.-Ze Shu, and L. Spinelli: "20 watt CW TEM_{00} intracavity doubled optically pumped semiconductor laser at 532 nm," *Proc. SPIE*, vol. **8242**, 824206 (2012).

23 H. Kahle, T. Schwarzbäck, M. Eichfelder, R. Roßbach, M. Jetter, and P. Michler: "UV laser emission around 330 nm via intracavity frequency doubling of a tunable red AlGaInP-VECSEL," *Proc. SPIE*, vol. **8242**, 82420M (2012).

24 D. Kasahara, D. Morita, T. Kosugi, K. Nakagawa, J. Kawamata, Y. Higuchi, H. Matsumura, and T. Mukai: "Demonstration of blue and green GaN-based vertical-cavity surface-emitting lasers by current injection at room temperature," *Appl. Phys. Express*, vol. **4**, 072103 (2011).

25 G. Weng, Y. Mei, J. Liu, W. Hofmann, L. Ying, J. Zhang, Y. Bu, Z. Li, H. Yang, and B. Zhang: "Low threshold continuous-wave lasing of yellow-green InGaN-QD vertical-cavity surface-emitting lasers," *Opt. Express*, vol. **24**, no. 14, pp. 15546–15553 (2016).

26 Y Mei, G-E Weng, B-P Zhang, J-P Liu, W Hofman, L-Y Ying, J-Y Zhang, Z-C Li, H Yang and H-C Kuo, 'Quantum dot vertical-cavity surface-emitting lasers covering the 'green gap', *Light: Sci. Appl.* **6** 1 (2017) doi:10.1038/lsa.2016.199

27 H. Yu, Z. Zheng, Y. Mei, R. Xu, J. Liu, H. Yang, B. Zhang, T. Lu, and H. Kuo: "Progress and prospects of GaN-based VCSEL from near UV to green emission," *Prog. Quantum Electron.*, vol. **57**, Jan. pp. 1–19 (2018).

28 B. P. Zhang: "Green VCSELs based on nitride semiconductor," 24th Microoptics Conf., Nov., Toyama, Paper F-6 (2019).

29 T. Hamaguchi, H. Nakajima, M. Tanaka, M. Ito, M. Ohara, T. Jyoukawa, N. Kobayashi, T. Matou, K. Hayashi, and H. Watanabe: "Sub-milliampere-threshold continuous wave operation of GaN-based vertical-cavity surface-emitting laser with lateral optical confinement by curved mirror," *Appl. Phys. Express*, vol. **12**, No. 4, 044004, Apr. (2019).

30 T. Hamaguchi, Y. Hoshina, K. Hayashi, M. Tanaka, M. Ito, M. Ohara, T. Jyoukawa, N. Kobayashi, H. Watanabe, M. Yokozeki, R. Koda, and K. Yanashima: "Room-temperature continuous-wave operation of green vertical-cavity surface- emitting lasers with a curved mirror fabricated on {20−21} semi-polar GaN," *Appl. Phys. Express*, vol., **13**, no. 4, 041002, pp. 1-5, Mar. (2020).

31 J. B. Wright, S. Liu, G. T. Wang1, Q. Li, A. Benz, D.1 D. Koleske1, P. Lu, H. Xu, L. Lester, T. S. Luk, I. Brener, and G. Subramania: "Multi-colour nanowire photonic crystal laser pixels," *Sci. Rep.* vol. **3**, 2982 (2013).

32 T. Hamaguchi, R. Koda, and P. Boulay: "AR & VR displays: a target for GaN-based VCSELs- An interview with Sony Corporation." https://www.i-micronews.com/ar-vr-displays-a-target-for-gan-based-vcsels-an-interview-with-sony-corporation, Oct 2020.

Appendix J

Photodetectors

Babu Dayal Padullaparthi

J.1 Introduction

J.1.1 Classification of Photodetectors

All optical systems require the ability to generate and detect light. The majority of this book has focused on the generation of light, but modern photodetectors (PDs) are an indispensable component in all of the applications described in this book. In data communications systems, the PD is a single device associated with a single optical channel. Advances in data rate have driven the need for higher bandwidth devices, and PDs are now reaching some physical limits where compromises in design and performance are being made as the speed increases [1]. 3D sensing and LiDAR applications have been enabled by the convergence of advanced camera and VCSEL technologies. The capture of photons with the shortest possible pulses and high frame rates and the ability to detect near-single photon reflections are key advances in camera technology for both 3D sensor and LiDAR systems. The responsivity of a PD is determined by its constituent materials and wavelength of the collected light. PDs operate wavelengths from UV to MIR and can be

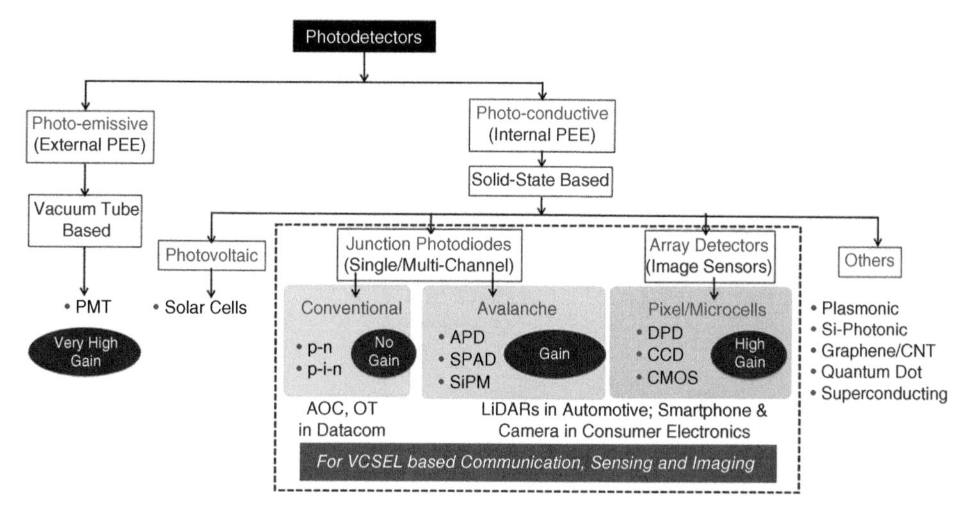

Schematic J.1.1 Classification of photodetectors for light-based communication and sensing [Babu Dayal Padullaparthi ©Photonic Components DFM Ltd.]. For abbreviations, please see paragraph below.

VCSEL Industry: Communication and Sensing, First Edition. Babu Dayal Padullaparthi, Jim A. Tatum and Kenichi Iga.

classified as shown in Figure J.1.1, which also shows the different types of PDs that are available for photodetection.

PDs can be classified into two major categories, vacuum based (photomultiplier tubes, or PMT) and solid-state based. PMTs operate on the external photoelectric effect (PEE) that leads to emission of an avalanche of electrons when a photon strikes the surface. PMTs are bulky and rarely used in compact-sized sensing modules in consumer electronics. The vast majority of PDs are semiconductor based and are further classified into either photovoltaic (used in solar cells) or junction photodiodes. The photodiode type is of primary use with VCSELs and is further subclassified into conventional PDs that do not have gain (more than 1 electron per photon) and those with gain, as shown in Figure J.1.1. 3D sensors and some LiDAR applications use arrays of photodetectors in a camera module as the sensor element. VCSEL-based light detection and sensing applications in the NIR wavelengths primarily use p-i-n PDs, APD, DPD, SPAD, and Si-PM because of their high sensitivities. More details can be found in the references [2, 3, 4].

J.1.2 Photodetectors Materials and Operation Principles

The primary semiconductor materials used to manufacturer PDs are Si, Ge, GaAs, and InGaAs, which cover the wavelength range from 0.3 to 1.7 μm. These PDs are fabricated as either p-n or p-i-n junction devices and operate in reverse bias. (Note: DPDs are an exception and operate from reverse to forward bias.) In this configuration the photo-generated carriers (electrons and holes) drift away from the junction creating current at the terminals. If the bias voltage is too high, the p-n junction will break down and current will flow without photo-generation, which is known as avalanche breakdown. A special class of photodiodes, APDs, operate right on the edge of avalanche breakdown and by doing so can generate more than one electron per photon, resulting in improved light sensitivity. Figure J.1.2.1 shows the schematic of reverse-biased p-n junction as a function of voltage (V).

The amount of current generation, or responsivity, in a PD is material dependent. Different materials have different absorption rates, as shown in Figure J.1.2.2 (a). The proportion of incident light absorbed by the PD material, A, depends on the absorption coefficient (α) and the thickness of the material (d) as $A = 1 - e^{-\alpha d}$. The responsivity, R, measured in amperes per watt, is related to the absorption and the incident photon energy by $R = A\lambda/1242$, where λ is measured in nanometers. The quantum efficiency, QE, of a photodiode is defined as the number of carriers generated for a given number of incident photons, $QE = R^{*}1242/\lambda$. Both R and QE are common figures of

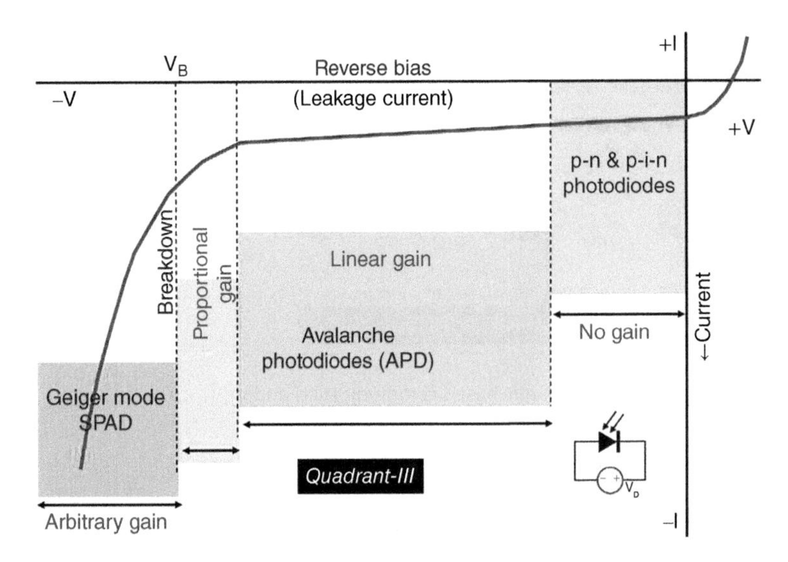

Figure J.1.2.1 Schematic of reverse-biased p-n junction (light characteristics) showing operating regimes of PDs [Babu Dayal Padullaparthi ©Photonic Components DFM Ltd.].

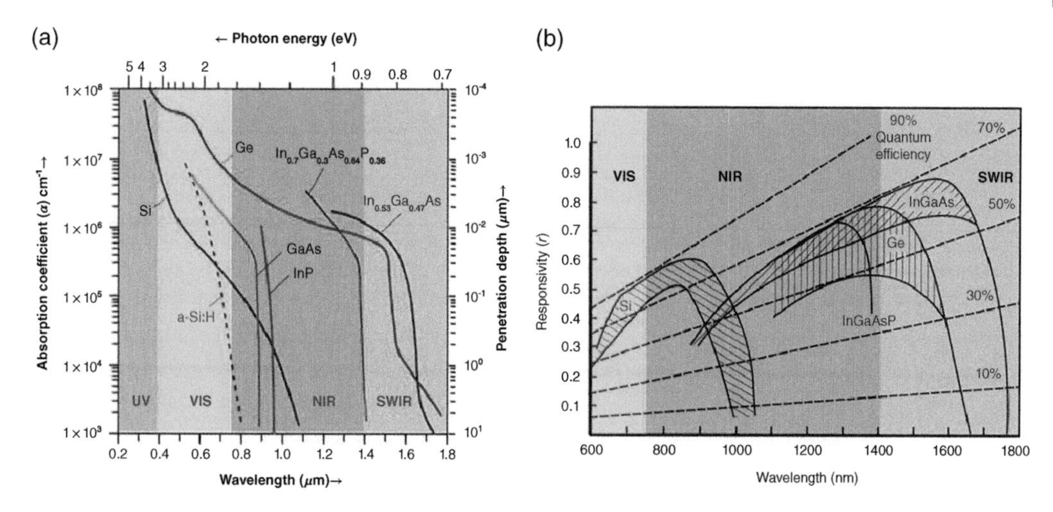

Figure J.1.2.2 (a) Absorption coefficient as a function of wavelength for different PD materials (modified from [3]), (b) Responsivity as a function of wavelength for different PD materials (modified from [5]).

merit in photodiodes and are shown for some common materials in Figure J.1.2.2(b). In standard p-n and p-i-n PDs, there is a single electron-hole (e-h) pair generated for each incident photon. Some photodiodes, such as APDs, have the ability to generate more than one e-h pair per incident photon and thereby increase the base responsivity of the material. This is known as the multiplication factor, or more generally as gain. The operation principles of APDs are further discussed in section J.3.2. All semiconductor devices and circuits have noise, which will be described in more detail in section J.2. One other figure of merit at the system level is the signal-to-noise ratio (SNR), which determines the sensitivity of the optical system. SNR is defined as the amount of photocurrent generated divided by the total noise current with no incident signal. Note that SNR can also include any unwanted light entering the system and generating photocurrent. When the SNR is greater than 1, a real signal can be measured. The SNR can be greatly enhanced when gain is present in the photodiode. This is a critical capability that enables LiDAR systems, particularly for longer-distance applications, adverse weather conditions, and low object reflectivity.

J.2 Noise in Photodetectors

Photodetectors have noise, which can be created either internally (σ_{INT}) or externally (σ_{EXT}) and taken together are the total noise, σ_{TOT}. External noise comes from ambient incident light and other components in the detection system electronics that may create radiatively coupled currents in the PD. Optical radiation noise (σ_{RAD}, sometimes called σ_{BACK}) includes solar and other ambient light that may be in the detector response wavelength range. Often, to minimize ambient noise, σ_{EXT}, wavelength filters that are closely matched to the source laser are placed in the optical path. This concept was further described in Section 7.2 where the advantages of a VCSEL light source were clear. Another source of σ_{EXT} is readout, or speckle, noise ($\sigma_{READOUT}$, $\sigma_{SPECKLE}$), which comes from fluctuations in the generation of photoelectrons or speckle fluctuations among received photons. There are several internal sources of noise, including thermal noise and shot noise (collectively called fixed system noise), low-frequency noise, photon noise, and thermal noise (σ_{TH}) (also called Johnson noise, Nyquist noise, or receiver circuit noise), which comes from random collisions among electrons and atoms inside the semiconductor, exists in all resistive elements, and depends on temperature. Shot noise (σ_{SHOT}) (also called photocurrent noise, dark current noise,

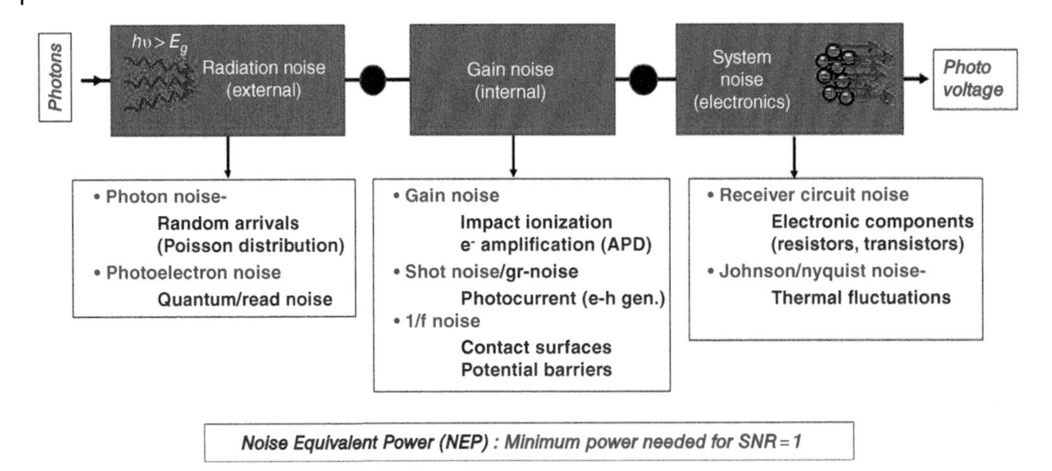

Figure J.2.1 Schematic of sources of noise in photodetectors [Babu Dayal Padullaparthi ©Photonic Components DFM Ltd.].

gr-noise) is related to the incident optical signal and is a result of random variations in the e-h generation and recombination within the detector in low light levels. Other internal noises such as low frequency (1/f) noise ($\sigma_{1/f}$) are likely to come from contacts, potential barriers, and surfaces. Photon noise (σ_{PHOTON}) comes from random arrival of photons and is modeled with Poisson statistics. All of these noise sources contribute to the total PD noise and effect the SNR [5, 6, 7].

Figure J.2.1 shows the sources of noises in the photodetection system and their origins. The minimum power needed for SNR = 1 is called noise equivalent power (NEP). If $(\sigma_{SHOT})^2 > (\sigma_{TH})^2$, the PD is called functional in quantum regime, and when $(\sigma_{TH})^2 > (\sigma_{SHOT})^2$, it is called functional in thermal regime.

J.3 Photodetectors for Communication and Sensing

In an optical sensor, the light source and photodetector need to be matched in wavelength. This puts limitations on what materials can be used in certain systems. For example, a Si photodiode does not respond to wavelengths longer than about 1.1 μm. VCSEL-based communication and sensor systems in the SWIR wavelength typically use either Si or GaAs materials and are either p-n, p-i-n, or APD photodiodes. For example, p-n and p-i-n PDs operating in the linear regime are used in data communications, while APD, DPD, SPAD, and PMs are used in low-level light and may be capable of single-photon counting. A special class of diamagnetic superconducting photodiodes (SNSPD) can provide exemplary single-photon detection rates (1 billion counts per second with sub-20 ps response times) and are used in the SWIR and MIR bands for LiDAR and quantum (communication, key distribution, and computing) applications.

J.3.1 p-n and p-i-n PDs

Both p-n and p-i-n are junction PDs used as detectors in communication systems that operate in the linear photoconductive regime of I-V spectra (as shown in Figure J.1.2.1) and provide no gain from their materials. In p-n PD, photons are absorbed in the depletion region formed under reverse bias of the heavily doped p- and n- semiconductors. Typical depletion widths are less than 5 μm. As the reverse bias is increased, the depletion region width increases and so the junction capacitance

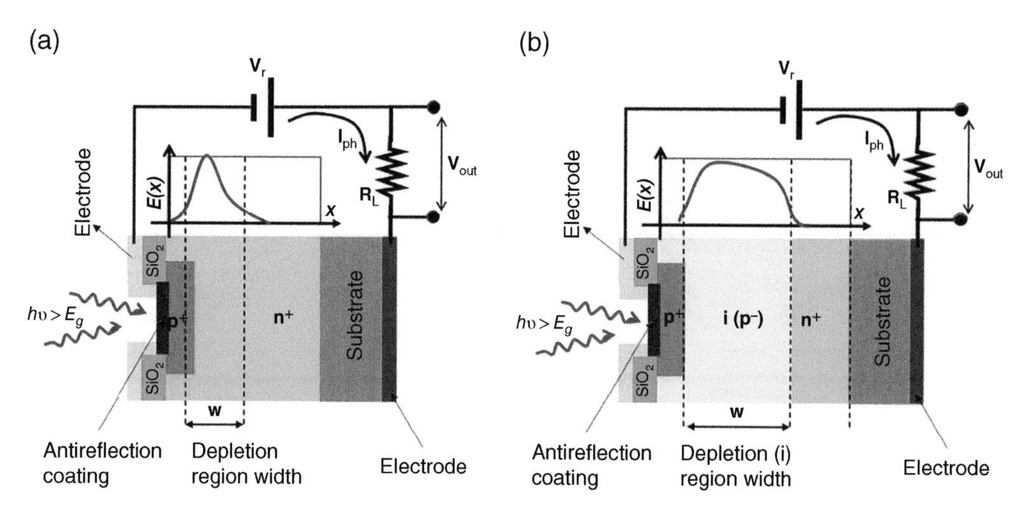

Figure J.3.1.1 Schematic x-section of (a) p-n and (b) p-i-n PDs with associated field regions [©Photonic Components DFM Ltd.].

drops. Figure J.3.1.1 (a) shows the schematic of p-n PD with associated field regions. p-n PDs are generally made from either Si or Ge and together can cover the entire UV–SWIR band.

In contrast to p-n PDs, p-i-n PDs have an un-doped or intrinsic (*i*) region sandwiched between p- and n- heavily doped regions, and the depletion region is almost completely defined by the intrinsic region. The intrinsic region need not have to be truly intrinsic, but it has to be sufficiently resistive, and all the light absorption must take place in this region. The depletion region stays fully within the intrinsic area and can be much larger than that of a p-n PD. The depletion width does not vary significantly with bias, making the capacitance essentially independent of the bias voltage, which paves the way to a stable operation. Increasing the depletion width increases the absorption probability of incoming photons and reduces junction capacitance and thereby the RC time constant. However, as described in Section 4.7.3, this can reduce the bandwidth when the frequency response becomes limited by the time it takes for the carrier to traverse the *i* region. Figure J.3.1.1 (b) shows the schematic of a p-i-n PD with associated field regions. Most of the p-i-n PDs have double hetero-structures; examples include (i) p+AlGaAs/GaAs/n+AlGaAs for light detection between 0.7 and 0.87 μm and (ii) p+-InGaAs/InGaAs/n+InP for light detection between 1 and 1.6 μm. The p-i-n hetero-junction PDs with low bandgap *i*-region and high bandgap n+/p+ regions offer flexibility in optimizing performance of PDs. Some design parameters are the long wavelength cut-off, λ_H, and short wavelength cut-off, λ_S, so that the quantum efficiency and responsivity in between λ_H and λ_S can be optimized. The p-i-n diodes do not provide material gain but can offer very high bandwidths (up to 100 GHz). One of the drawbacks of p-i-n PDs is their large response time and so the sensitivity is not enough for low-light applications with high background light (solar ambient) situations.

J.3.2 Avalanche Photodiode (APD)

The issue of detection of photons at low-light applications can be overcome by use of avalanche photodiodes. APDs work based on an internal gain mechanism with large reverse bias conditions through the process of impact ionization. At large reverse bias, the strong electric field accelerates the carriers and generate secondary electrons through impact ionization leading to an avalanche process (gain) that is limited to the intrinsic region. This ionization process repeats as more carriers

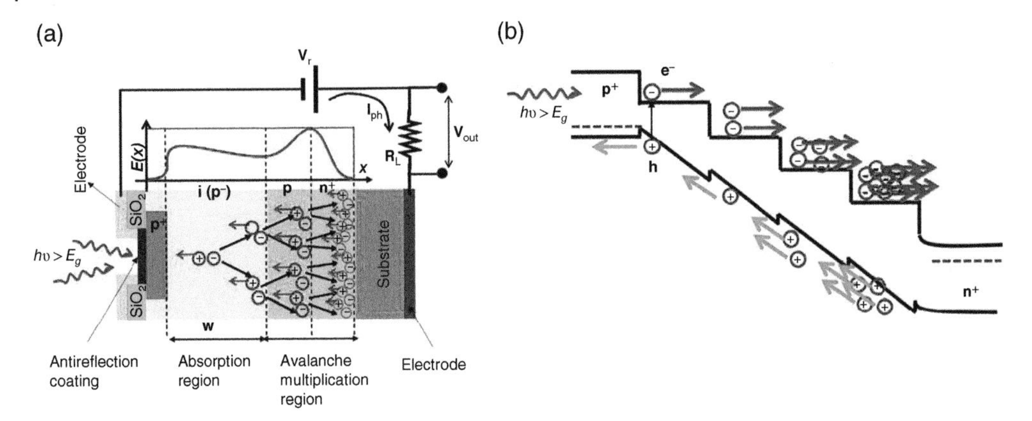

Figure J.3.2.1 (a) Schematic structure of APD with impact ionization and (b) multilayer staircase APD under reverse bias condition [©Photonic Components DFM Ltd.].

are generated in the i-region, amplifying secondary electrons. The gain of an APD can be up to a few 100s and is directly proportional to incoming incident photons; the reverse bias is shown in Figure J.3.2.1 (a).

APDs are sensitive and have high responsivity. They can detect each single photon incident on them. Beyond a particular reverse bias and maximum gain condition, the junction breaks down, the current flows through load resistance, and APD operates in non-linear Geiger mode, as shown in Figure J.1.2.1. As the avalanche process creates many fluctuations of generated carriers, the SNR can degrade, which often hinders the PD's performance. The noise in APDs from its avalanche multiplication can be reduced by the use of multi-layer staircase or superlattice APDs, where the band gap of p+ side to n+ side is continuously graded (typically 10–20 nm), allowing band discontinues and reducing hole-induced ionization [7]. The superlattice structure offers lower operating voltage due to reduced potential barriers, faster response time, and reduced avalanche build-up. A multi-stage staircase APD is illustrated in Figure J.3.2.1 (b) with reverse biased condition. APDs are better devices with comparable bandwidths to those of conventional p-n and p-i-n PDs. Their high detection sensitivity in a low-light ambient makes them a good choice in low light signal systems such as TOF LiDARs. The gain of an APD is limited by avalanche breakdown and the width of the depletion region. A figure of merit for an APD is the gain-bandwidth product, or the speed frequency where the gain is unity. For thin active regions (higher speed) there is not enough intrinsic material to create secondary carriers, whereas for thick intrinsic regions there is more opportunities for collision, but the bandwidth becomes transit-time limited. The performance of APD is characterized by the ionization ratio, $\beta = \alpha_h/\alpha_e$, where α_h and α_e are the ionization coefficients (ionization probabilities per unit length, measured in cm^{-1}) of carriers holes and electrons, respectively. β is zero for the ideal case of single carrier multiplication process. The details of impact ionization process are well explained in [7]. Single element, 1D and 2D arrays of Si- and InGaAs-based APDs are commercially available but are relatively expensive and may not be suitable for single-photon detection applications.

J.3.3 Single-Photon APD (SPAD)

Linear APDs that are operated in the non-linear Geiger-mode region are called single-photon avalanche diode (SPAD). SPADs operate well beyond junction breakdown voltage with extraordinary levels of electric fields, $> 5 \times 10^5$ V/cm, to generate single electron-hole pairs that are injected into

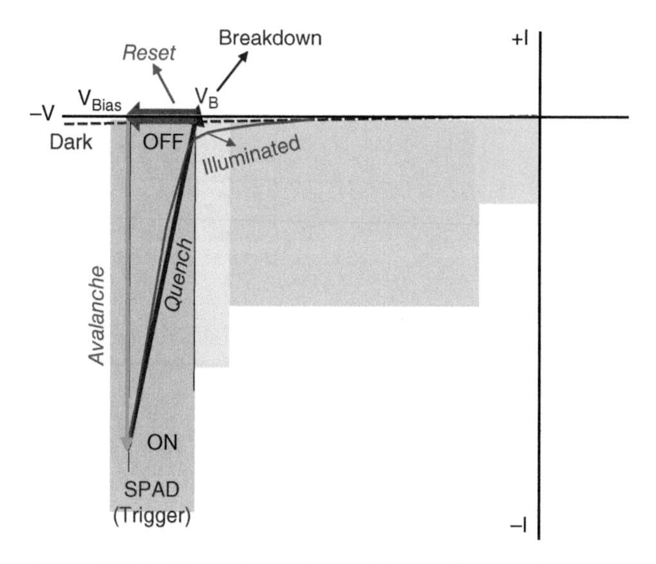

Figure J.3.3.1 Operating regimes with non-linear avalanche processes in SPAD [Babu Dayal Padullaparthi ©Photonic Components DFM Ltd.] (Loadline not shown intensionally).

the depletion region to trigger a self-sustained avalanche, called a Geiger discharge, throughout the device volume. In the Geiger-mode region, as the current swiftly reaches macroscopic level, the avalanche process must be quenched in the non-linear amplification process to measure the current value through an ON-OFF circuit, as shown in Figure J.3.3.1. Efficient and fast-quenching mechanisms must be used to withstand the repeated avalanche process. Passive quenching with large resistors and active quenching with current feedback loops with quenching times on the order of a few hundred nanoseconds are generally used. During the quench, the SPAD is not detecting incoming photons and represents a dead operating time. After quenching the charge, the SPAD is returned to Geiger mode operation to detect incoming light. Even though SPADs are nearly noiseless due to their digital detection threshold from quenching circuits and extremely efficient in low-noise detection such as LiDARs [8, 9, 10] and biological imaging [11, 12], thermal fluctuations among carrier generation-recombination processes and thermal delays in releasing carriers that induce the avalanche process are troublesome. SPADs can be integrated with CMOS circuits and made into both 1D and 2D arrays. For example, Fraunhofer IMS Germany uses a CMOS integrated SPAD array for low-light detection for long-range LiDAR applications. [13] (See Chapter 6 for details.)

J.3.4 Silicon Photomultiplier (Si-PM)

SPADs are extremely sensitive to back reflections that can saturate the detector and render it non-functional for short periods, a major drawback when large-scale reflective objects such as traffic signs are present. Besides, the photon-triggered ON-OFF switch results in a binary output, and a proportional output is not available. Multi-pixel photon counters (MPPC), also called silicon photomultipliers (SiPMs) may offer a better solution in these instances. SiPMs are essentially independent and dense SPAD arrays each with its own quench resistor, the basic unit called a *microcell* or *pixel*. SiPMs have microcell densities from a few hundred to several thousand per square millimeter. The microcells are connected vis CMOS circuitry that combines the outputs of individual SPAD microcells into an analog signal that is proportional to the number of microcell SPADs

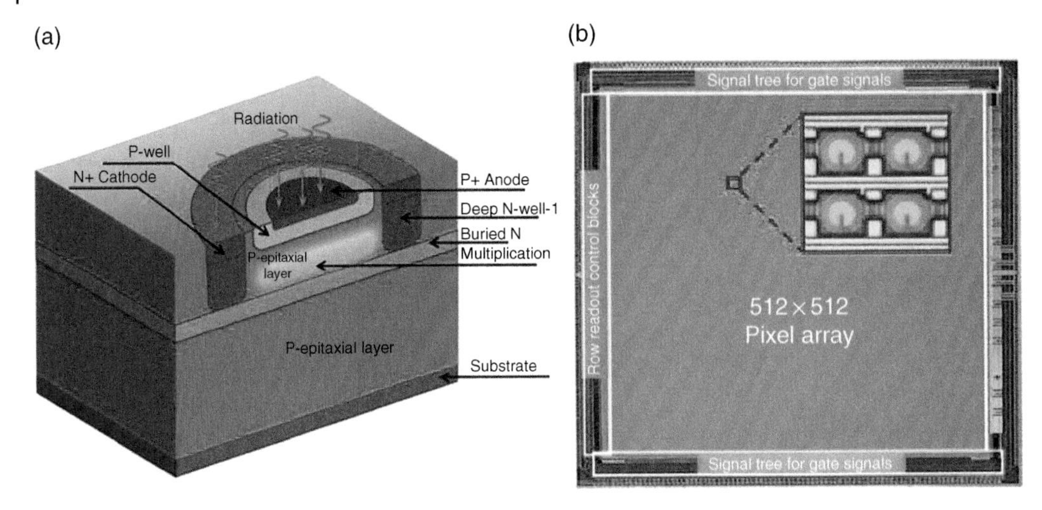

Figure J.3.4.1 (a) Cross section of single SAPD as microcell SiPM, and (b) its processed die with superimposed building blocks. (Adapted from IEEE [14].)

triggered. This enables photon counting/detection capability through quenched ON-OFF switching. Each microcell has a capacitively coupled output with a fast internal switching mechanism and detects the photons identically and independently. When a microcell in the SiPM absorbs a photon, a Geiger avalanche is immediately initiated and causes photocurrent to flow through the microcell. The avalanche is confined to the single microcell where it was initiated while other microcells remain fully charged and ready to detect photons. The sum of the photocurrents from each of the individual microcells combine as a quasi-analog output and is thereby capable of producing a proportional response to the incident photon flux. A schematic of SiPM array, its microcell, and cross-sectional structure are shown in Figure J.3.4.1.

SiPMs have many practical features including compact size, high gain ($> 10^5$), large spectral responses (100 to 1000 nm), low operation voltage, low power consumption, analog photon counting capability, and availability in many commercial (module) sizes when compared to conventional PMTs. This makes SiPMs particularly suitable for single-photon counting in LiDAR applications. However, as each of the SiPM microcell outputs is binary and all of the microcells read in parallel and then are superimposed to obtain a final analog photocurrent distribution, SiPM sensitivity, defined by photon detection efficiency (PDE), is a function of the wavelength of incident light, applied over-voltage, and the microcell fill factor. PDE for array detectors is slightly different than QE for p-i-n PDs and APDs. PDE is a measure of probability of avalanche in a given active volume of microcells, while QE is a measure of likelihood of e-h creation in the full volume of sensor. There are few sources of noise in SiPMs to consider. The main contributor is dark current rate, which is primarily a result of thermal electrons generated in the active volume; it is a function of active area, over-voltage, and temperature. Additional noise contribution comes from optical crosstalk between microcells and secondary avalanche induced by emitted photons from primary avalanche. Both are dependent on the over-voltage and fill factor of the SiPM.

J.3.5 Dynamic Photodiode (DPD)

Photodiodes with gain, such as APDs and SPADs, require high voltage and complex integration. In 2020, an innovative dynamic PD (DPD) was reported that operates on switched (negative to positive) bias mode at around 1V and generates high forward current with low noise [15, 16]. This new switched mode operation induces a large positive bias after time lag dependent on the amount of light and is controlled by applied voltage (independent of light intensity). It could be a game-changing technology in the light sensing market, especially for 3D sensing and LiDARs where the high SNR (~90 dB) and low power consumption offer compelling advantages. Its low voltage operation also favors power scaling among pixels. DPDs are fully compatible with advanced Si-CMOS processes, offer pixel sizes as small 3 μm, are sensitive to NIR (850/940 nm) wavelengths, and can directly integrate with digital circuits without analog amplification. They can also be used for single-photon counting events and can easily replace SPADs without altering the measurement principles and major changes in the circuits. The high dynamic range is limited by dark current (weak signal) and resolution of time-to-digital converter (strong signal). DPDs are promising for applications in mainstream 3D TOF sensing (smartphones, LiDAR, FR, autofocus), wearables in health monitoring (low power and weak photoplethysmography [PPG] signal sensing for heart rate, blood oxygen saturation [SpO2], and blood pressure), and hearing devices. An example of a 1V DPD in an i-TOF system operating at up to 5 m has been reported by ActLight Inc [17]. It will be interesting to see whether the sensing functionality of DPDs with a VCSEL source can replace LEDs for human blood oxygen and hemoglobin sensing [18].

J.3.6 PDs Integrated with Image Sensors

Photodetectors integrated on both analog and digital image sensor platforms offer high resolution and fast response time and are widely used in electronic imaging such as consumer cameras and smartphones, medical imaging, robotics, 3D sensors, and LiDAR. There are two types of image sensor technologies, namely charge-coupled device (CCD) and complementary metal-oxide-semiconductor (CMOS). The image sensors consist of a number of pixels and convert the light into voltage through electronics such as A/D converters, shift registers, or amplifiers; the stronger the light, the more electrons are generated for image processing. Image sensors register light from bright to dark and are color-blind. Small filters in front of the sensors allow color to be assigned to each pixel and can be mapped to human visibility. Multiple color models such as primary RGB (widely used with Bayer array configuration), and complementary CMYG (less common) are available.

A CCD pixel shown in Figure J.3.6.1 (a) has a series of MOS capacitors, and by controlling gate voltages, the charge can be transferred from pixel to pixel, as shown in Figure J.3.6.1 (b–e) [Wiki-Charge-Coupled Device]. A CMOS pixel (the structure, and its charge readout shown in Figure J.3.6.1 (f)] consists of a buried photosensitive element pinned PD (PPD), a transfer gate (TG), and a floating diffusion (FD) with double p+np junction. The p+ surface (pinning) implant significantly reduces dark current and isolates the PPD from e-h recombination under the pre-metal dielectric-PMD (or Si-SiO_2) interface. It is also surrounded by a p-well that isolates the SCR (space charge region) from p+ shallow trench isolation (STI). TG acts as a switch to transfer read from PPT to FD, as shown in Figure J.3.6.1 (g). The PPD has been the primary technology for CCD and CMOS image sensors; detail operations of PPD pixel (4T) with CMOS can be found in comprehensive reviews [19, 20, 21, 22, 23].

J.3.7 CMOS-Based SPAD Arrays

In the recent past, solid-state single-photon CMOS sensors have emerged as one of the most versatile and easy to use image sensors. The introduction of PPDs in CMOS image sensors (CIS) offers low noise and high sensitivity characteristics that have made CIS a mainstream detector in the commercial camera industry, finding high-end applications in TOF and fluorescence lifetime

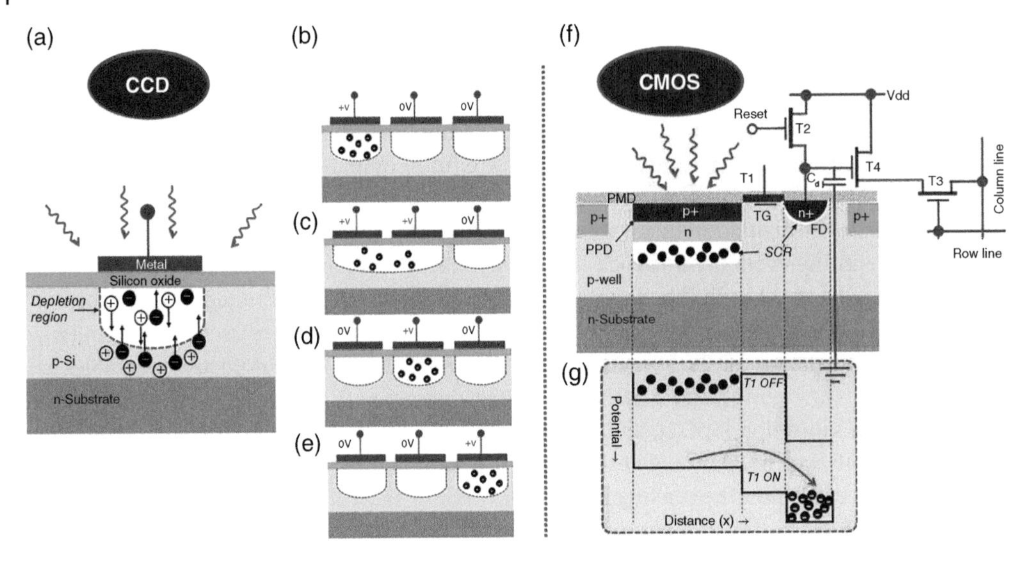

Figure J.3.6.1 Structure and charge transfer in (a–e) CCD and (f–g) 4T CMOS pixels [Modified from Wiki Charge-Coupled Device, [19, 20]

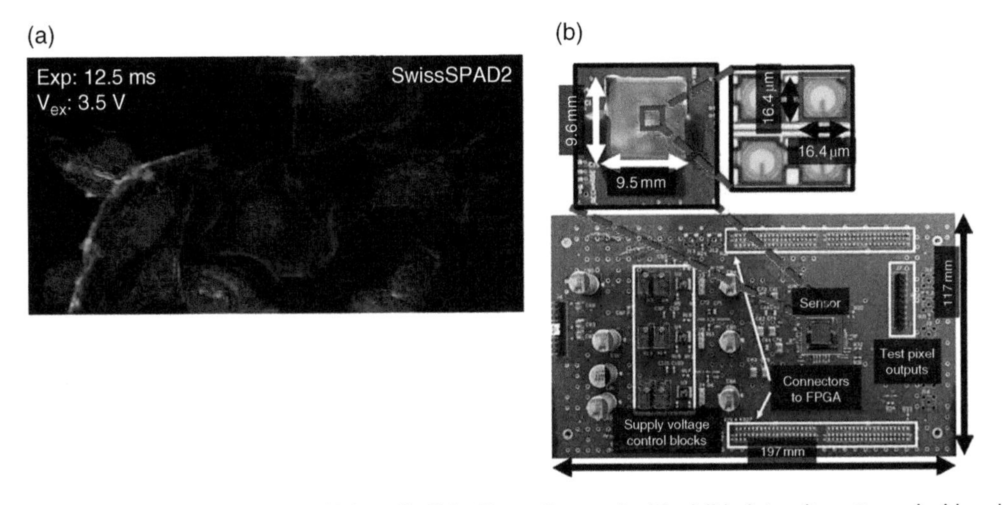

Figure J.3.7.1 (a) FLIM images HeLa cells (20–40 μm diameter) with visible intensity patterns inside cells due to low SNR, and (b) image sensor camera module with FPGA. (Adapted from IEEE, [14].)

imaging microscopy (FLIM), as shown in Figure J.3.7.1 (a). APD sensors integrated with CMOS can meet demanding applications with time resolution and extremely low photon flux applications with time periods of ~100 ps. One example is time-resolved TOF image sensing, where each pixel in a camera module can acquire the ambient surrounding and reconstruct a real-world 3D image while simultaneously measuring distance using a pulsed light source, as shown in Figure J.3.7.1 (b).

SPADs integrated with CMOS circuitry have attracted a lot of attention as image sensors. The CMOS SPAD technology was introduced in 2003 [24], and since then massive, fully integrated CMOS SPAD sensors were made for linear and image-sensing applications [11]. Key performance metrics such as optical crosstalk, dark current, after-pulsing, and deadtime were optimized, and the technology is pushing deep into smaller feature (pixel) size CMOS technologies 25, 26]. Engineering efforts on integration and increasing the number of pixels are ongoing, and with strong demand for photonic components, new materials such as Ge-on-Si, InGaAs, and InP are becoming mainstream for products and industry [27]. Recent reports on backside illumination and block illumination of laser transmitters integrated with 64 × 68 and 32 × 128 SPAD arrays, respectively, show their potential to use in TOF LiDARs [28, 29].

J.4 Summary of Photodetectors

A summary of all PDs (except emerging plasmonic, Si-photonic, CNT, and quantum dot–based PDs) is given in Table J.4.1. The user needs to critically review the sensitivity, efficiency, and response time for relevant product applications. The choice of a PD for a given application is based on its operating wavelength, responsivity, noise sources, and ultimately the required SNR.

Table J.4.1 Summary of key performance parameters/characteristics of PDs [Babu Dayal Padullaparthi ©Photonic Components DFM Ltd.].

Character/ Parameter	p-i-n, SBD*	APD	SPAD	SiPM	DPD	PMT	SNSPD*
Type			Semiconductor			Vacuum Based	Super conductor
Gain	1 or less	~50-200 (Linear Impact Ionization)	10^4-10^5 (Gieger Avalanche)	10^6 (Gieger Avalanche)	Switched bias	10^5-10^7 (Avalanche)	10^9 per second (Detection rate)
QE/PDE/ SDE (%)	40	>65		>50		~40	~100
Min detectable Power		Good		Better	Good	Best	
Temperature Effects		Strong		Strong	Strong	Weak	Strong
Advantages	Fast High photon flux ambient High Bandwidth	Linear gain control from bias, Compact Rugged Insensitive to B	SPD	SPD Compact Rugged Insensitive to B Low noise	Low Noise Low voltage operation CMOS compatible SPD	High Gain UV Detection High Bandwidth	SPD Noiseless low response time Zero dark current
Drawbacks		Excess Noise	Response time control	Saturable, overvoltage dependent		Bulky, Delicate Low QE Interact with B	Cryogenic cooling

SBD-Schottky Barrier Diode
SPD - Single Photon Detection, B - Magnetic field, * Not discussed in this appendix.

References

1 J. Tatum, D. Gajula, L. A. Graham et al., "VCSEL-Based Interconnects for Current and Future Data Centers," in *J. of Lightwave Technol.*, **33**, 727–732 (2015).
2 D. Vasileska, https://nanohub.org/resources/9143/download/PHOTODETECTORS.pdf.
3 S. O. Kasap, "*Optoelectronics and Photonics: Principles and Practices*," 2nd Edition, Pearson Education Inc. NJ, USA, 2013.
4 https://www.sonoma.edu/users/r/rahimi/courses/es485. https://www.sonoma.edu/ssu-search?query=kasap&op=Search&form_build_id=form-z1cO94BBUtnUT6LxoDW3e58j2gowTz8SOu022yQqNUA&form_id=ssusearch_search_block_form#gsc.tab=0&gsc.q=kasap&gsc.page=1 (KasapPP5_01 and KasapPP5_02.pptx).
5 W. C. Wang http://courses.washington.edu/me557/sensors/detector.pdf.
6 http://catalogue.pearsoned.ca/assets/hip/us/hip_us_pearsonhighered/samplechapter/0130203378.pdf.
7 B. E. A. Saleh and M. C. Teich, "*Fundamentals of Photonics*", Chp-17-Semiconciductor Photon Detectors, Page 644–695, John Wiley Sons, Inc. USA, 1991.
8 https://www.epfl.ch/labs/aqua/page-26733-en-html/page-157534-en-html/.
9 M-J. Lee, A. R. Xiemens, P. Padmanabhan et. al., "High-Performance Back-Illuminated Three-Dimensional Stacked Single-Photon Avalanche Diode Implemented in 45-nm CMOS Technology," *IEEE J Sel. Topics Quantum Electron.* **24**, 3801809 (2018).
10 M. Sanzaro, P. Gattari, F. Villa et. al., "Single-Photon Avalanche Diodes in a 0.16 μm BCD Technology With Sharp Timing Response and Red-Enhanced Sensitivity," *IEEE J Sel. Topics Quantum Electron.* **24**, 3801209 (2018).
11 C. Bruschini, H. Homulle, I. M. Antolovic et. al., "Single-photon avalanche diode imagers in biophotonics: review and outlook, Light: Science & Applications," (2019) https://doi.org/10.1038/s41377-019-0191-5.
12 https://www.laserfocusworld.com/detectors-imaging/article/14074048/spads-offer-possible-photodetection-solution-for-tof-lidar-applications.
13 W. Brockherde https://www.ait.ac.at/fileadmin/mc/digital_safety_security/downloads/1040_BROCKHERDE_CMOS-SPADs-for-LiDAR-Applications.pdf.
14 A. C. Ulku, C. Bruschini, I. M. Antolovic et al., "A 512 x 512 SPAD image sensor with integrated grating for widefield film," *IEEE J. Sel. Topics in Quantum Electron.* **25**, 6801212 (2019).
15 https://www.laserfocusworld.com/detectors-imaging/article/16556344/photodiodes-dynamic-photodiodes-reach-singlephoton-sensitivity-at-low-voltages-with-minimal-noise.
16 https://www.i-micronews.com/digital-photodiodes-enable-a-whole-new-level-of-sensitivity-to-3d-sensing-an-interview-with-actlight/.
17 https://act-light.com.
18 https://spectrum.ieee.org/view-from-the-valley/biomedical/devices/should-you-trust-apples-new-blood-oxygen-sensor.
19 Y. Xu, "Fundamental Characteristics of a Pinned Photodiode CMOS Pixel," Ph.D. Thesis, Technische Universiteit Delft, 20 November 2015, ISBN: 978-94-6233-140-2. https://repository.tudelft.nl/islandora/object/uuid%3Ae0371a0e-7d0a-4ce2-9cdc-20ad49916c44
20 N. Teranishi https://indico.cern.ch/event/522485/contributions/2174996/attachments/1282603/1906227/2016_5FEE_teranishi_ver5.pdf.
21 N. Teranishi, "Effect and Limitation of Pinned Photodiode," *ITE Technical Report* **38**, 52 (2014).
22 E. R. Fossum, and D. B. Hondongwa, "A Review of the Pinned Photodiode for CCD and CMOS Image Sensors," *IEEE J. Ele. Dev. Soc.* **2**, 33 (2014).

23 A. Pelamatti, "Estimation and modeling of key design parameters of pinned photodiode CMOS image sensors for high temporal resolution applications," Ph.D. Thesis, University of Toulouse Midi-Pyrenees, 7 November 2015. https://core.ac.uk/display/33664540.

24 A. Rochas, M. Gani, B. Furrer et al., "Single photon detector fabricated in a complementary metal–oxide–semiconductor high-voltage technology," *Rev. Sci. Instrum.* **74**, 3263 (2003).

25 E. Charbon., "Single-photon imaging in complementary metal oxide semiconductor processes," *Phil. Trans. R. Soc. A* **372**, 1 (2014) http://dx.doi.org/10.1098/rsta.2013.0100.

26 J. Zhang, M. A. Itzler, H. Zbinden and J-W. Pan., "Advances in InGaAs/InP single-photon detector systems for quantum communication," *Light: Science & Applications* **4**, e286 (2015), https://doi.org/10.1038/lsa.2015.59.

27 M. Keller https://indico.cern.ch/event/837899/contributions/3570849/attachments/1916472/3168565/Bergen_LW_CMOSSPAD_Talk_MKeller.pdf.

28 J. Ruskowski, C. Thattil, J. H. Drewes and W. Brockherde., "64x68 pixel backside illuminated SPAD detector arrays for LiDAR applications," *Proc. SPIE* **11288**, 1128805 (2020) https://doi.org/10.1117/12.2550634.

29 S. Jahromi, J-P. Jansson, P. Keränen, and J. Kostamovaara., "A 32 × 128 SPAD-257 TDC Receiver IC for Pulsed TOF Solid-State 3-D Imaging," *IEEE J. Solid-State Circuits*, **55**, 1960 (2020).

Image Gallery

Datacom

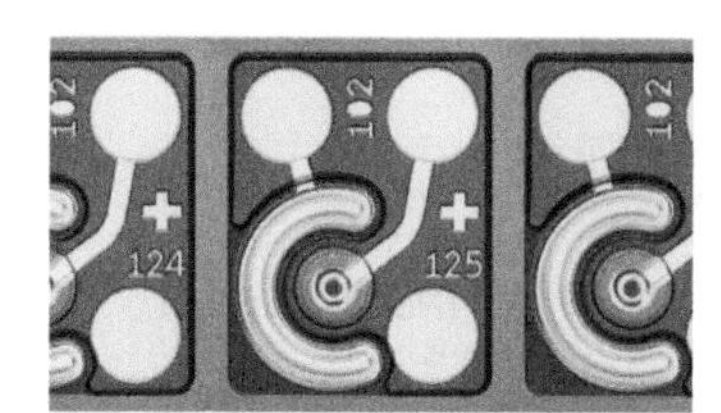

56G_PAM4 Datacom VCSEL © II-VI Inc.

A top view of VCSEL image (3d) from VI Systems © VI Systems GmbH

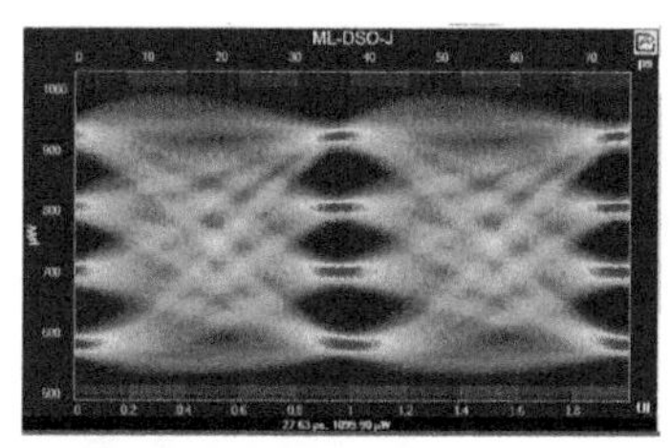

26 GBd PAM4 eye from commercial VCSEL products © Multi Lane Inc.

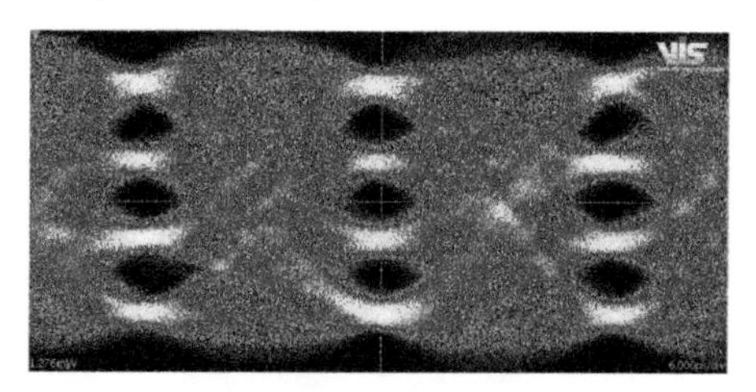

850nm 100G PAM4 eye from VI Systems © VI Systems GmbH

Datacom

850nm 25G 4x VCSEL Chip for Datacom
© Sino-semiconductor Photonics Integrated Circuit Co.,Ltd

Cosemi USB Type A/A Active Optical Cable, capable of 10G, USB 3.2 Gen 2x1 up to 100 meters in length.
© Cosemi Technologies Inc (Now Mobix Labs)

Cosemi USB Type A/C Active Optical Cable, capable of 10G, USB 3.2 Gen 2x1 up to 100 meters in length.
© Cosemi Technologies Inc (Now Mobix Labs)

Cosemi USB Type C/C Active Optical Cable, capable of 10G, USB 3.2 Gen 2x1 up to 100 meters in length.
© Cosemi Technologies Inc (Now Mobix Labs)

VCSEL Industry: Communication and Sensing, First Edition. Babu Dayal Padullaparthi, Jim A. Tatum and Kenichi Iga.

3D Sensing (Consumer/Mobile)

940nm single emitter for TWS & Proximity Sensing © Sino Semiconductor Photonics Integrated Circuit Co.,Ltd

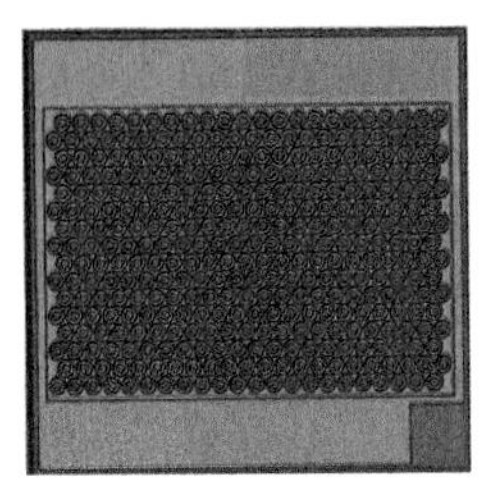

3W 940nm array for TOF (Time of Flight) © Sino Semiconductor Photonics Integrated Circuit Co.,Ltd

Automotive LiDAR

8W AEC-Q102 Qualified VCSEL Chip © TriLumina (Now Lumentum)

VCSEL on Board © TriLumina (Now Lumentum)

Automotive LiDAR/ Industrial

180124_3D LiDAR_64x256_LCA3_v02_02 © LeddarTech

LCA3 chip side-by-side © LeddarTech

Shuttle pixell ville © LeddarTech

940nm LiDAR Tx Chip Assembly © Sense Photonics (Now Oster Auto)

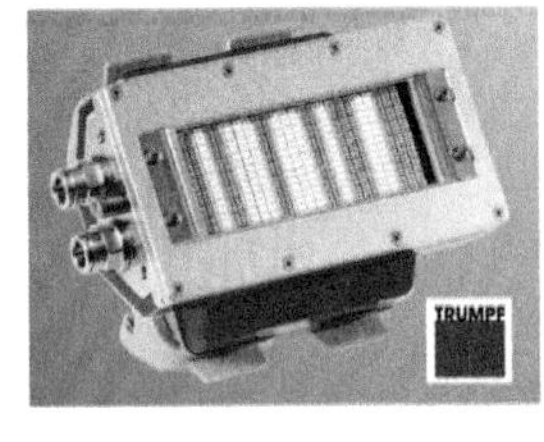

5kW heating module Small RGB-T_6304 © Trumpf Photonic Components GmbH

Index

VCSEL Industry: Communication and Sensing, First Edition. Babu Dayal Padullaparthi, Jim Tatum and Kenichi Iga.

THE COMSOC GUIDES TO COMMUNICATIONS TECHNOLOGIES

The ComSoc Guide to Next Generation Optical Transport: SDH/SONET/OTN
Huub van Helvoort

The ComSoc Guide to Managing Telecommunications Projects
Celia Desmond

WiMAX Technology and Network Evolution
Kamran Etemad and Ming-Yee Lai

An Introduction to Network Modeling and Simulation for the Practicing Engineer
Jack Burbank, William Kasch, and Jon Ward

The ComSoc Guide to Passive Optical Networks: Enhancing the Last Mile Access
Stephen Weinstein, Yuanqiu Luo, and Ting Wang

Digital Terrestrial Television Broadcasting: Technology and System
Jian Song, Zhixing Yang, and Jun Wang

TV White Space: The First Step Towards Better Utilization of Frequency Spectrum
Ser Wah Oh, Yugang Ma, Edward Peh, and Ming-Hung Tao

Digital Services in the 21st Century: A Strategic and Business Perspective
Antonio Sanchez and Belen Carro

Toward 6G: A New Era of Convergence
Amin Ebrahimzadeh and Martin Maier

VCSEL Industry: Communication and Sensing
Babu Dayal Padullaparthi, Jim A. Tatum and Kenichi Iga

Printed and bound by CPI Group (UK) Ltd, Croydon, CR0 4YY

16/04/2025

14658588-0001